Contents

Preface

This new edition is a companion to the book *Construction Technology*, and has been revised to reflect many of the changes and advances since the fourth edition was published. A logical sequence associated with the stages of construction has been adopted. The contractor's plant section has been completely revised, and many of the enabling works sections have been combined within an encompassing part, which covers the temporary and enabling works needed before work starts on a construction site.

The content covers many aspects of the commercial and or industrial applications of advanced construction technology, moving away from the domestic and low-rise construction methods covered in *Construction Technology*. Many of the existing parts have been refreshed in terms of health and safety legislation and the introduction of new technologies. The part covering substructure has been restructured and now includes more modern techniques of basement construction in non-domestic situations, as well as a new section on excavating vertical shafts and tunnelling methods. Roofing, cladding, stairs, rooflights and structural glazing have all been updated.

Part 11 Access and facilities for people who require additional support: buildings other than dwellings

Much of the content is covered in section 9 of Construction Technology 5th edition pages 450–468. However, a free downloadable PDF file of this Part can be found on the Pearson website: www.pearsonfe.co.uk

The uniqueness of Chudley and Greeno remains essential to this edition. These core texts are renowned for the style of their illustrations, so these have been retained as far as possible, with additional footnotes if needed to comply with the current legislation framework. Modern illustrations have replaced only those considered to be outdated or obsolete.

Simon Topliss QTLS MIFL
and
Mike Hurst ICIOB MIFL Cert ED

Acknowledgements

The book originated in 1973, conceived and authored by Roy Chudley with invaluable assistance from Colin Basset as General Editor. This book and its companion *Construction Technology* have been continually updated and reprinted and have gained a world-wide readership. Roger Greeno worked on updating the content from the second to the fourth editions maintaining the agreeable style and quality of illustration and clear and accessible text. Roger was not involved in compiling the 5th edition.

Roy Chudley, MCIOB.
Roy is a chartered builder and was formerly senior lecturer in construction at Guildford College. Roy's prominently illustrated building technology textbooks have become the established standard reference resource for students. His numerous books represent building practice and procedures in a unique style of comprehensive text supplemented with simple illustrations. This interpretation of the subject combined with an understanding of student needs evolved from many years on site and in the design office.

Roger Greeno, BA(Hons.), FCIOB, FCIPHE, FRSA.
Roger is a chartered builder and registered plumber. He has lectured in several further and higher education colleges and Portsmouth University. Roger is a published author, editor, writer and illustrator of numerous construction papers, including well-known and accredited course readers and textbooks as study guides and industry references. He has been an examiner and moderator for many examination boards, including City & Guilds, Edexcel and the Chartered Institute of Building with ongoing commitments to the College of Estate Management, University of Reading and the Open University.

5th Edition Advanced Construction Technology

As with the companion edition Construction Technology, the considerable updating to this book has been carried out by Mike Hurst and Simon Topliss.

Mike Hurst MSc. BSc.(Hons) ICIOB CertEd MIfL
Mike is currently a senior lecturer in Civil Engineering at the University of East London and has over 25 years teaching experience of a variety of construction and civil engineering qualifications at both FE and HE level. He is also an FE External Verifier for Edexcel and an HE External Examiner for a number of Universities. His working career spans both local authority and private practice firstly in building control then in various engineering consultancies. With his interest in historic buildings, he is currently involved in research into environmental and energy issues of upgrading current building stock, building pathology and the use of traditional construction materials and skills. He has contributed to a number of construction related publications, textbooks and study guides including the BTEC Level 3 Construction & the Built Environment standard course text.

Simon Topliss BSc(Hons) PGCE. IFL. CIEA.
Simon is the Team Leader for Construction at the Grimsby Institute for Further and Higher Education in North East Lincolnshire, where he has taught on the Edexcel BTECs at Levels 2, 3 and 4 for over 15 years. He has written extensively for Pearson Education, contributing to a suite of construction textbooks from the Foundation Learning Tier through to Level 3. He currently holds the posts of Internal Standards Verifier, Senior Standards Verifier for Construction, Chief Examiner and Moderator on the Level 2 Construction and Built Environment Diploma and External Examiner for the Higher Education Construction and Civil Engineering programmes. He has written a substantial number of specification units for Edexcel assisting with the rewriting from the NQF to QCF framework

Pearson Education would like to thank the following:

Roy Chudley as founding author and Roger Greeno for his work in updating and revising over the years.

Mike Hurst and Simon Topliss for their dedication and enthusiasm in working tirelessly on this new edition.

Damian McGeary from The College of Haringey, Enfield and North London and Peter Lakin from the University of East London for their perceptive and careful technical reviews.

Sarah Christopher for the work with the authors in preparing the manuscript for this edition.

Mike Hurst would like to thank his wife and daughters for their patience,understanding and encouragement.

Simon Topliss would like to thank his wife Linda for her unwavering support and Paul Monroe who introduced him to Edexcel. He would especially like to thank Paul Brown for his support and interest.

Pearson Education would like to thank the following for their kind permission to use their material in illustrations:

ACT Acknowledgements

We are grateful to the following for permission to reproduce copyright material:

p.23: Open Government Licence; taken from page 8 of pdf file gs6, http://www.hse.gov.uk/pubns/priced/gs6.pdf; p.28: Permission to reproduce extracts from British Standards is granted by the British Standards Institution (BSI). No other use of this material is permitted. British Standards can be obtained in PDF or hard copy formats from the BSI online shop: www.bsigroup.com/Shop or by contacting BSI Customer Services for hard copies only: Tel: +44 (0)20 8996 9001, Email: cservices@bsigroup.com; p.40:© Defender Power & Light; p.40: Screwfix; p.40: QVS Discount Electrical Wholesale; p.40: Essential Supplies (UK) Ltd; p.57: Sitebox Ltd; p.65: Illustration of façade retention shoring courtesy of RMD Kwikform www.rmdkwikform.com; p.74: Drawing provided by www.longreachhighreach.com; p.75: Soosan Heavy Industries Co Ltd; p.75: Genesis GmbH; p.75: NPK Europe; p.84: TENAX SpA - Geosynthetics Division; p.85: Illustration courtesy of Idaho National Laboratory; p.86: Naval Facilities Engineering Command NAVAC; p.98 DeWalt Power Tools; p.109: Selwood Group Limited; p.113 Komatsu UK Ltd; p.115: AB Volvo; p.118 123 129 132 136 137 JCB; p.123: Hymac; p.127: Reprinted courtesy Caterpillar Inc; p.130: Delden CSE Limited; p.136, 137: DAF Trucks Limited; p.139, 140 Wheelwash Limited, Reprinted courtesy Caterpillar Inc; p.146: Terex Corporation; p.155, 157: Clarke Chapman Group; p.160: Manitou; p.172, 173: Multi Marque Production Engineering Ltd; p.179: Schwing America Inc; p.187; Harsco Infrastructure; p.189, 190: Terex Corporation; p.237: Copyright©Giken Seisakusho Co. Ltd. All Rights Reserved; p.246, 247, 248: Buchan Concrete Solutions Limited; p.246,247,248: Macrete; p.251: http://www.hse.gov.uk/construction/pdf/pjaguidance.pdf, Open Government Licence; p.259: Sprayed Concrete Association (SCA); p.259: Adapted from HSE Safety of NATM Tunnels pdf pg 16 http://books.hse.gov.uk/hse/public/saleproduct.jsf;jsessionid=791137E8C809CB72A82487835A884F96.plukweb4?catalogueCode=9780717610686, Open Government Licence; p.261 262: Crossrail; p.359: British Gypsum; p.363: Source: Kingspan Limited – contact www.kingspan.com for latest details; p.363: Milbank; p.363 Concrete Flooring Systems Ltd; p.373: http://www.planningportal.gov.uk/uploads/br/BR_PDF_ADB1_2006.pdf, Part B Volume 2 Building Regulations page 18 Open Government Licence; p.382: http://www.planningportal.gov.uk/uploads/br/BR_App_Doc_B_v2.pdf Part B Volume 2 Building Regulations page 37 Open Government Licence; p.389,414,416: Source: Kingspan Limited – contact www.kingspan.com for latest details; p.441: Euroclad Facades Limited; p.448: Tremco; p.450: Courtesy Pilkington Glass Limited; p.451 ,452,455: Linox Technology Pty Ltd; p.453: Lisus Technology PTE Ltd; p.456 Metro GlassTech: p.460: Architen Landrell Associates; p.462: Tricel Sewage Treatment; p.473: Picture Courtesy of Polysteel UK; p.479.480: Harsco Infrastructure; p.526:

DVS Limited; p.527, 528: Brett Martin Daylight Systems Ltd; p.529: Reproduced by kind permission of Xtralite (Rooflights) Limited; p.543: Rodeca GmbH; p.558: By kind permission of The Wood Marketing Federation; p.570: Printed with permission from Delta Structures; p.588, 589: Hillaldam Coburn Ltd; p.593: British Gypsum; p.594; 622: Otis Elevator Company; p.623 Fraunhofer Institute for Systems and Innovation Research ISI; p.644: Taken from Government's DMRB Manual Vol 4 Section 2 their Figure 2.2 http://www.dft.gov.uk/ha/standards/dmrb/vol4/section2/ha10306.pdf, Open Government Licence.

Introduction

This new edition of Advanced Technology extends the knowledge and understanding obtained from the accompanying edition *Construction Technology* and lifts it into the more complex technology that is applied to the construction of commercial low rise and multi storey buildings and structures.

The Advanced Construction Technology edition is directed at the higher level 4/5 learners. These learners may be undertaking their Higher National Certificates and Diplomas in Construction or the initial years of a Degree in a Construction discipline where they are studying modules in construction techniques covering aspects of high rise and commercial buildings.

Elements of the content are also extremely useful for completion of some Level 3 Construction courses that contain construction, civil engineering and building technology.

The format closely follows that of the revised edition of *Construction Technology*, providing concise notes and generous illustrations to elaborate on the text content, and the use of chapter colour identification. Many of the previous drawings and sketches have been retained to continue the unique Chudley approach throughout the two new revised editions. The older and less common method illustrations have been removed or updated within this edition. The reader should appreciate that the illustrations are used to emphasise a point of theory and must not be accepted as the only solution. A study of construction documentation, drawings and details from building case studies given in the various construction journals will add to background knowledge and comprehension of this advanced construction technology edition.

This edition has a positive sustainability theme woven into many chapter aspects, to promote the sustainability of our environment which is now a major consideration in any construction project. The introduction of low voltage technology into electrical supplies, the use of contaminated brownfield sites, sustainable external envelope finishes, and the use of BREEAM for the sustainable assessment of commercial buildings are just a few of the new topics that have been covered. The setting up of a construction site along with all the enabling works, such as demolition that might have to be undertaken, have been condensed into a larger initial chapter within this new edition. Also with the increased mechanisation of construction work the section on contractor's plant has been completely revised to include a range of modern technological developments from portable hand tools through to wheel washers and the latest mini-excavators and material handling equipment; not forgetting the ubiquitous mobile elevating platforms and other health and safety inspired plant. Substructure works now include a revised section on underground engineering in soft to firm soils, including modern systems for both temporary and permanent shaft sinking and techniques for sprayed concrete linings in tunnel construction. The use of brownfield sites for re-development is now also covered in this volume describing the use of remediation techniques, health and safety control methods and ground improvement methods. Horizontal and vertical access has been extended with the inclusion of a section on lifts, escalators and travelling walkways. The chapter on external claddings to the external envelope of a commercial building now includes the use of modern facade systems that utilise an insulation core which is then covered with a range of aesthetical finishes from sustainable timber panels through to pressed aluminium tiles. Glass external finishes have been updated with the use of spider, and cable truss supported glass facades, and a large section has been introduced on insulated composite panelling.

The standards and legislative elements of the previous edition have been fully revised, many now adopting the EN European based standard. The new CDM Regulations 2007 and the Regulatory Reform (Fire Safety Order) 2005 have been included.

This edition obviously cannot cover every technological aspect that is utilised within the modern commercial building. Readers are encouraged to supplement their understanding by researching all sources of reference on any particular topic of study, to maximise information and to gain a thorough comprehension of the subject.

Simon Topliss
Mike Hurst

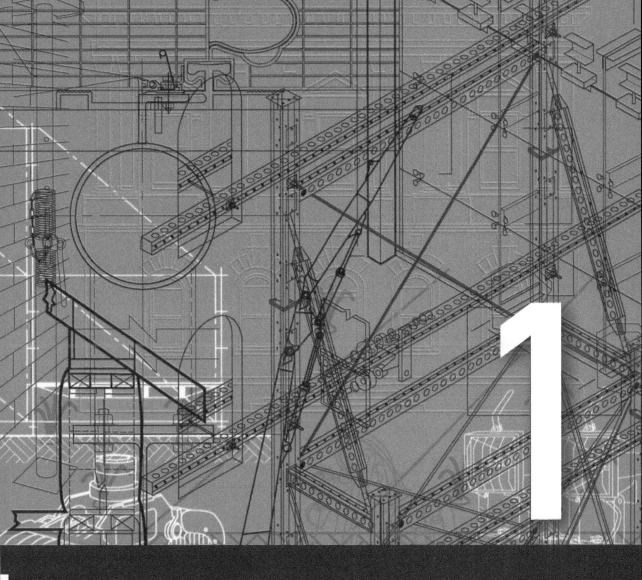

1

Site works

Site layout   1.1

The construction of a building can be considered as a temporary production site, the building site being the 'factory' in which the contractor will produce the product. To enable this activity to take place the contractor requires several resources to be coordinated – namely labour, materials, plant and subcontractors. All of these have to be carefully controlled so that:

- the operatives have the right plant and equipment in the most efficient position;

- the materials are stored so that they are readily available next to the location required and not interfering with the general site circulation, with adequate temporary storage space;

- sufficient site accommodation is available for the workforce during non-working periods.

There is no standard size ratio between the free site space required to construct a building and the total size of the site on which the building is to be erected. Rather a balance needs to be struck according to the constraints of the site size and the footprint of the design produced for the client. The space left is that which is available for the movement of the resources and temporary structures required during the construction phase. Therefore each site must be considered as a separate problem in terms of allocating space for construction operatives, materials, plant and equipment. To obtain maximum efficiency there is an optimum way of laying out the site and also a correct amount of expenditure to support the proposed site layout. This is the key process within a project's planning. Time and resources spent in planning the site layout will greatly aid the efficient construction of the building project, driving economies in minimising movements and transportation on site, thus reducing preliminary expenditure. Any planned layout should be reviewed periodically and adjusted to suit the changing needs of the site activities. Efficient site management of the planning of the site activities will greatly increase profitability for a construction organisation and will reduce the likelihood of accidents on site.

Before any initial planning of the site layout can take place, certain preliminary work must be carried out, preferably at the pre-tender stage.

- The decision to tender will need to be taken by the managing director or, for small works, by the senior estimator up to a contract value laid down by the managing director.

- A pre-tender construction programme will need to be prepared, usually by the estimator and the contract manager, to ascertain the most time-efficient methods of completing the contract using the company's resources, to give a competitive edge to the tender submission.

- A thorough study of the bill of quantities will give an indication of the amount and quality of the materials required, and of the various labour resources needed to carry out the contract.

A study of the tender issue drawings, together with the bill of quantities and/or the specification, will enable the contractor to make a preliminary assessment of the size and complexity of the contract. Once the full size and complexity of a potential contract has been ascertained, the senior management can assess the risk element associated with the potential award of the contract. Risk is the complexity and likelihood that the company will maintain the profit margin that it has placed on the submitted tender.

Before the estimator can make a start on calculating unit rates, a site visit must be carried out, preferably by the site manager. This is done to investigate any site constraints that will not be obvious within the tender documents or the drawn information. The site visit report should include a range of information.

- **Access to site** On- and off-site access, pavement crossings, road, water and rail facilities, travel distances involved, one-way systems, overhead power lines, rights of way restrictions, local authority or police restrictions and bridge weight or height limitations on approach routes.

- **Services** Existing power, drainage and water supplies, location of service connections, diversions required, and the lead-in periods involved in service authority diversions/connections, together with cost implications.

- **Layout** General site conditions such as nature of soil, surface obstructions, height of water table, flooding risks, tidal waters, neighbouring properties, site boundaries, preservation orders, trees, demolition problems and special insurance considerations.

- **Staff** Travel distances, availability of local skilled contractors, specialist contractors, local rates of pay, cost of temporary accommodation.

- **Security** Local vandalism and pilfering record, security contractors' facilities, need for night security, fencing and hoarding requirements.
- **Waste management** Skip hire suppliers, rates, distance to licensed disposal tips, recycling facilities.

With the knowledge and data gained from contract documents and site investigations, and any information gained from the police and local authority sources, pre-tender work can now be carried out.

- **Pre-tender programme** This is usually in the form of a bar chart showing the proposed time allowances and sequencing for all the major activities.
- **Pre-tender health and safety plan** This is prepared by the project coordinator to enable tendering contractors to consider the practical and cost implications and the adequacy of their resources with regard to the requirements of the CDM Regulations.
- **Cost implications** Several programmes for comparison should be made to establish possible break-even points, giving an indication of required bank loan, possible cash inflow and anticipated profit.
- **Plant schedule** This can be prepared in the form of a bar chart and method statements showing requirements and utilisation, which will help in deciding how much site maintenance, equipment and space for plant accommodation will be needed on site.
- **Materials schedule** Basic data can be obtained from the bill of quantities. The buyer's knowledge of the prevailing market conditions and future trends will enable estimates to be made of usage and delivery periods and the amount of site space and/or accommodation required.
- **Labour summary** Basic data obtained from the bill of quantities, site investigation report and pre-tender bar chart programme can be used to establish the number of subcontract trades required.
- **Site organisation structure** This is an organogram chart showing the relationships and interrelationships between the various members of the site team, and is normally required only on large sites where the areas of responsibility and accountability must be clearly defined.
- **Site layout** This includes site space allocation for materials storage, working areas, quantity and type of site accommodation, plant positions and general circulation areas, as well as access and egress for deliveries and emergency services.
- **Protection** Protection should be assured for adjacent buildings and structures (including trees) with preservation orders. Fencing or hoardings will be required to prevent trespass and to protect people in the vicinity. It is also necessary to check the current level of insurances required.

When planning site layouts the following must be taken into account:

- the activities that will be carried out on site;
- efficiency of traffic movements around the site;
- movement to avoid double handling of resources;
- control to prevent any wastage or loss of resources;
- facilities for health, safety and welfare provision;
- accommodation for staff and storage of materials.

SITE ACTIVITIES

The time required for undertaking each of the tasks within the pre-tender programme can now be estimated from the data obtained previously for preparing the material and labour requirements. Each of the site activities will have a duration that has to be accommodated within the pre-tender programme, in between the start and completion dates supplied by the client. Normally, where plant and equipment are involved in a construction activity, it is the plant element that the labour output will have to be aligned against. For example, the rate of placing concrete will be determined by the output of the mixer and the speed of transporting the mix to the appropriate position. Alternatives that could be considered are:

- more than one mixer;
- regulated supply of ready-mixed concrete;
- on large contracts, pumping the concrete to the placing position.

All alternative methods for any activity will give different requirements for staff numbers, material storage, access facilities and possibly plant types and capacities.

EFFICIENCY

Maximum efficiency has to be obtained on site by ensuring that the required outputs are achieved against each of the bar chart tasks. The factors that will influence efficiency on site include:

- avoidance of the double handling of materials;
- the structure of any bonus system in operation and the use of fixed price subcontracting;
- proper storekeeping arrangements to ensure that the materials are of the correct type, in the correct quantity, and available when required, reducing wastage losses;

- keeping walking distances to a minimum to reduce the non-productive time spent in covering the distances between working, rest and storage areas;

- providing adequate protection for unfixed materials on site, thereby preventing time loss and cost of replacing damaged materials;

- avoidance of loss by theft and vandalism by providing security arrangements in keeping with the value of the materials being protected and by making the task difficult for the would-be thief or vandal by having adequate hoardings and fences;

- minimising on-site traffic congestion by planning delivery arrivals, having adequate parking facilities for site staff cars and mobile machinery when not in use, and by having sufficient turning circle room for the types of delivery vehicle likely to enter the site;

- installation of temporary roads and hardstandings for ease of traffic and operative movement around the site during bad weather and excavation works.

MOVEMENT

Apart from the circulation problems mentioned above, the biggest problem is one of access on and off the construction site. Vehicles delivering materials to the site should be able to do so without difficulty or delay. Delays may incur additional charges from a supplier: for example, when discharging ready-mixed concrete you are allowed so many minutes for discharging concrete; if you go over this or delay the delivery, waiting time may be charged by the supplier. The weight of a fully laden supplier's vehicle can be considerable and must be adequately supported on the temporary road surfaces around the site. For example, a fully laden ready-mixed concrete lorry can weigh 20 tonnes, and lorries used for delivering structural steel can be 18.000 metres long, weigh up to 40 tonnes and require a large turning circle. If it is anticipated that heavy vehicles will be operating on site, it will be necessary to consider the road surface required. This is especially important for heavy cranes that require a level firm surface to operate upon. If the roads and paved areas are part of the contract and will have adequate strength for the weight of the anticipated vehicles, it may be advantageous to lay the roads at a very early stage in the contract; however, if the specification for the roads is for light traffic, it would be advisable to lay only the base hoggin or hardcore layer at the initial stages, because of the risk of damage to the completed roads by the heavy vehicles. As an alternative, it may be considered a better policy to provide only temporary roadways composed of railway sleepers, metal tracks or mats until a later stage in the contract, especially if such roads will only be required for a short period.

SUPERVISION

The efficiency of the site will be highly influenced by the control imposed by the site management appointed to supervise the project. A well-qualified and experienced site manager will increase the efficiency of the site and minimise the amount of movements required to construct the project. The site manager will use their experience and knowledge to coordinate all site activities and maintain the progress in accordance with the contract programme. Continual monitoring through site meetings with the contracts manager will ensure that the most efficient methods are used to complete the project to time, cost and quality.

FACILITIES

These must be planned for each individual site, but certain factors will be common to all sites – not least the implications of the Construction Design and Management Regulations 2007, the Work at Height Regulations 2005 and the Health and Safety at Work etc. Act 1974. The main contractor has to provide:

- a safe, healthy place of work;
- safe systems of work;
- plant and equipment that are not a risk to health;
- information, training and instruction on the safe use of equipment.

Both regulations are wide-ranging and set goals or objectives relating to risk assessment to ensure reasonably practicable steps are taken to ensure safety provision. Prescriptive requirements for such provisions as scaffold guard rail heights and platform widths are scheduled in the Work at Height Regulations.

The principal considerations under the **Construction Design and Management Regulations** are summarised below.

- **Regulation 26:** *Safe places of work.* This regulation covers the safe access and egress into a place of work, including any working space that is required while working. Each place of work shall be safe and be maintained so that there are no risks to persons working there as far as is reasonable practicable.

- **Regulation 28:** *Stability of structures.* This regulation covers the temporary nature of construction work where any part of a structure may become unstable. Measures must be taken to prevent any weakness or instability in a structure so that it does not collapse, or is overloaded.

- **Regulation 29:** *Demolition or dismantling.* This requires that the demolition or dismantling of a structure is planned in a safe and organised way such that the risk of such work is lowered to as reasonably practical a level as possible.

- **Regulation 31:** *Excavations.* Any work within an excavation must be supported to prevent any collapse, and any materials, persons or plant must not be able to fall into the excavation. There is also a requirement under this regulation for a competent person to undertake an inspection of the excavation.

- **Regulation 35:** *Prevention of drowning.* This is not applicable to all sites, but if there is a danger from water or other liquids in any quantity, then every practical means possible must be taken to prevent people falling into it. Personal protective and rescue equipment must be available, maintained in good order, and water transport to and from the place of work must be provided under the control of a competent person.

- **Regulation 36:** *Traffic routes.* This makes provision for the segregation of vehicles and pedestrians, with definition of routes. It requires adequate construction and maintenance of temporary traffic routes, use of suitable signage, warnings (audible or otherwise) of vehicle movements, prohibition of misuse of vehicles, and safeguards for people using powered doors and guards, such as those on hoist facilities.

- **Regulations 42 to 44:** *Fresh air and weather protection.* These are general requirements to ensure fresh air availability at each workplace, make sure reasonable temperatures are maintained at internal workplaces, provide protection against inclement weather and ensure adequate levels of lighting (including emergency lights).

The provision of welfare facilities on a construction site is covered by the requirements of Schedule 2 of the Regulations, summarised below.

- **Sanitary conveniences** Suitable conveniences shall be provided that are adequately ventilated and lit, and separate rooms shall be provided for male and female operatives. They must be cleaned regularly and kept in good order.

- **Washing facilities** These shall be provided adjacent to sanitary conveniences and any changing rooms, with an adequate supply of hot and cold water, soap and towels for drying. Separate facilities for male and female operatives shall be provided.

- **Drinking water** An adequate drinking water supply must be provided at suitable places, indicated with appropriate signage along with sufficient cups, unless the supply is a jet.

- **Changing rooms** These must be provided for operatives to change into work wear and for the drying of clothes. Suitable seating and lockers need to be provided to keep PPE, personal effects and work wear within.

- **Rest facilities** During rest breaks suitable facilities should be provided that have adequate seating, tables and arrangements for the preparation of meals, including the provision to boil water. The facilities must be maintained at an appropriate temperature. Separate facilities must be provided for smokers outside and away from the entrance.

The principal considerations under the **Work at Height Regulations** apply to any place at or above ground level, as well as below ground. They also include the means of gaining access to and egress from that place of work. Measures taken by these regulations are designed to protect a person from injury caused by falling any distance. This may be from plant and machinery or from equipment such as scaffolding, trestles and working platforms, mobile or static.

- **Regulation 4:** *Organisation and planning.* It is the employer's responsibility to ensure that work at height is planned, supervised and conducted in a safe manner. This includes provision for emergencies and rescues, and regard for assessing risk to persons working during inclement weather.

- **Regulation 5:** *Competence.* It is the employer's responsibility to ensure that persons engaged in any activity relating to work at height are competent. This would include checking that they do not suffer from vertigo.

- **Regulation 6:** *Risk avoidance.* This primarily deals with the evaluation of the work task relative to its situation. What this means is evaluating whether the work can be undertaken at ground level and not at height with the associated risks involved: for example, a winch arrangement for changing bulbs in a lighting rig. Provision needs to be in place for preventing persons sustaining injury from falling.

- **Regulation 7:** *Work at height equipment.* This covers further requirements for assessment of risk relative to the selection of plant and equipment suitable for collective rather than individual use. For example, a fall arrestor stops injury but cuts off blood supply to vital organs unless the operative is rescued within a certain time period.

- **Regulation 8:** *Specific work equipment.*

Scaffold and working platforms shall be constructed in accordance with the following requirements:

- Top guard rail, to be min. 950 mm high.
- Intermediate guard rail, positioned so that no gap between it and top rail or toe board exceeds 470 mm.
- Toe board must be sufficient to prevent persons or materials falling from the working platform.
- Platform should be stable and sufficiently rigid for the intended purpose.
- Dimensions must be adequate for a person to pass along the working platform, unimpeded by plant or materials storage.
- No gaps in the working platform and surface must be resistant to slipping or tripping.
- Platform must be designed to resist anticipated loading from personnel, plant and materials.
- Scaffold frame must be of sufficient strength and stability.

- If the scaffold is unconventional in any way, calculations are required to prove its structural integrity.
- During assembly, alteration, dismantling or non-use, suitable warning signs to be displayed as determined by the Health and Safety (Safety Signs and Signals) Regulations. Means to prevent physical access also required.
- Assembly, alteration and dismantling under the supervision of a competent person qualified by an approved training scheme.

Nets, airbags or other safeguards for arresting falls should be used where it is considered not reasonably practical to use other safer work equipment without it. A safeguard and its means for anchoring must be of adequate strength to arrest and contain persons without injury, where they are liable to fall. Persons suitably trained in the use of this equipment, including rescue procedures, must be available throughout its deployment. Where personal fall protection equipment is considered necessary, it should be correctly fitted to the user, adequately anchored and designed to prevent unplanned use by the user's normal movements.

For ladders, the following requirements apply.

- Used solely where a risk assessment indicates that it is inappropriate and unnecessary to install more substantial equipment. Generally, this applies to work of a short duration.
- The upper place of support is to be firm, stable and strong enough to retain the ladder without movement. Position to be secured by rope lashing or other mechanical fixing.
- Inclination is recommended at approximately 75° to the vertical, that is, in the vertical to horizontal ratio of 4:1.
- A suspended ladder is to be secured and attached to prevent displacement and swinging.
- The extension of a ladder beyond a place of landing should be sufficient for safe bodily transfer – normally taken as 1.050 m minimum measured vertically.
- Where a ladder ascends 9.000 m or more vertically, landing points are to be provided as rest platforms.

- **Regulation 9:** *Fragile surfaces*. It is the employer's responsibility to ensure that no person works on or near a fragile surface. Where it is impossible to avoid, then sufficient protection, such as platforms and guard rails, are to be provided. Location of fragile surfaces is to be indicated by positioning of prominent warning signs.
- **Regulation 10:** *Falling objects*. Suitable provisions, such as fan hoardings, are required to prevent persons suffering injury from falling objects or materials. Facilities are to be provided for safe collection and transfer of

materials between high and low levels, such as chutes. No objects are to be thrown. Materials are to be stacked with regard to their stability and potential for movement.

- **Regulation 11:** *Dangerous areas*. Areas of work of specific danger, such as around demolition work, are to be isolated to ensure that persons not engaged in that particular activity are excluded. Warning signs are to be displayed.

- **Regulations 12 and 13:** *Inspection*. These Regulations specifically apply to scaffolding, ladders and fall protection equipment. After installation or assembly, no equipment may be used until it has been inspected and documented as safe to use by a competent person. Further inspections are required where conditions may have caused deterioration of equipment, or alterations or changes have been made. Following an interval, every place of work at height should be inspected before work recommences.

- **Regulation 14:** *Personnel duties*. Persons working at height should notify their supervisor of any equipment defect. If required to use personal safety/protective equipment (PPE), individuals should be adequately trained and instructed in its use.

Under the **Health and Safety at Work etc. Act**, employers must have defined duties, which include providing:

- a safe place of work;
- safe access to and egress from places of work;
- safe systems of work;
- safe items of plant and equipment;
- suitable and adequate training, supervision and instruction in the use of equipment;
- suitable and appropriate PPE applicable to head, hands, feet, eyes and mouth;
- materials and substances that are safe to use (COSHH Regulations 2002);
- a statement of health and safety policy.

Employees and the self-employed have duties to ensure that they do not endanger others while at work. This includes members of the public and other operatives on site. They must:

- cooperate with the health and safety objectives of their employer (the main contractor);
- not interfere with any plant or equipment provided for their use, other than its intended use;
- report any defects to equipment and dangers relating to unsafe conditions of work.

The preceding section on provision of facilities under the Construction Design and Management Regulations, the Work at Height Regulations and the Health and Safety at Work etc. Act is intended as summary comment for guidance only. For a full appreciation, the reader is advised to consult each specific document. These can be viewed at www.legislation.gov.uk; see also www.hse.gov.uk/construction/cdm.htm

ACCOMMODATION

The suite of temporary offices that supports the construction of a project is an important component of the site set-up. Resources well spent on modern accommodation and welfare facilities will enable the operatives and management to work efficiently in any location or environment. Many different types of facilities are available for site accommodation. The basic requirements under the Construction Design and Management Regulations are examined in detail below.

CANTEEN FACILITIES

These are for the purposes of preparing, heating and consuming food during meal breaks, which may require the following service connections: drainage, light, power and cold water supply. To provide a reasonable degree of comfort a floor area of 2.0 to 2.5 m² per person should be allowed. This will provide sufficient circulation space, room for tables and seating, and space for the storage of any utensils.

Often on large construction sites a system of staggered meal breaks is introduced to reduce congestion during breaks. Canteens should be sited so that they do not interfere with the development of the site but are positioned so that travel time is kept to a minimum, as meal breaks are normally 30 minutes in duration. On sites that by their very nature are large, it is worthwhile considering a system whereby tea breaks can be taken in the vicinity of the work areas. It is the principal contractor's responsibility to ensure that reasonable welfare facilities are available on site, although they do not necessarily have to provide these: it may be part of subcontractors' conditions of engagement that they provide their own.

DRYING ROOMS

These are used for the purposes of depositing and drying wet clothes. Drying rooms generally require a lighting and power supply for background heating, and lockers or hanging rails for wet clothes. A floor area of 0.6 m² per person should provide sufficient space for equipment and circulation. Drying rooms should be sited near, or adjacent to, the canteen or mess room.

TOILETS

Contractors are required to provide at least adequate washing and sanitary facilities as set out in Schedule 2 of the Construction Design and Management Regulations. All these facilities will require light, ventilation, water and drainage connections. If it is not possible or practicable to make a permanent or temporary connection to a drainage system, the use of tanked storage units (which require emptying and disposal off site) should be considered. Sizing of toilet units is governed by the facilities being provided. If female staff are employed on site, separate toilet facilities must be provided for men and for women. Toilets should be located in a position that is convenient to both offices and mess rooms, which may mean providing more than one location on large sites.

FIRST AID ROOMS

These are only required on large sites as a specific facility, or may be provided by the client's own organisation; otherwise a reasonably equipped first aid facility will suffice. A first aid box correctly stocked and equipment for emergency first aid must be provided along with a trained first-aider on site who has responsibility for accounting for the contents of the first aid box and its use.

SITE LAYOUT

Before the proposed site layout is planned and drawn, the contracts manager and the proposed site manager should visit the site to familiarise themselves with the existing site conditions. During this visit the position and condition of any existing roads should be noted, and the siting of any temporary roads considered necessary should be planned. If necessary a dilapidation survey may be required. Information regarding the soil conditions, height of water table and local weather patterns should be obtained by observation, site investigation, soil investigation and local knowledge, or from the local authority. The amount of money that can be spent on this exercise will depend on the size of the proposed contract and possibly on how competitive the tenders are likely to be for the contract under consideration.

Figure 1.1.1 shows a typical small-scale general arrangement drawing, and needs to be read in conjunction with Fig. 1.1.2, which shows the proposed site layout.

The following is an exercise that illustrates how the site would be planned by the contracts and site managers to ensure the efficient running of the contract. The amount of time and effort put into this will equate to savings in the long run on preliminary costs associated with the construction period. The following data has been collected from a study of the contract documents and by carrying out a site investigation.

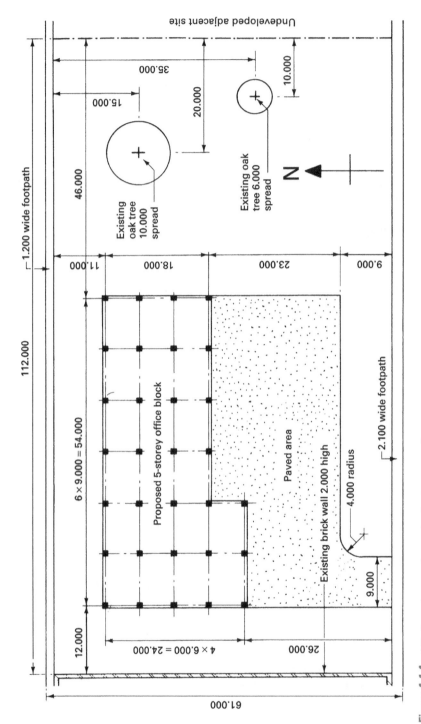

Figure 1.1.1 Site layout example: general arrangement

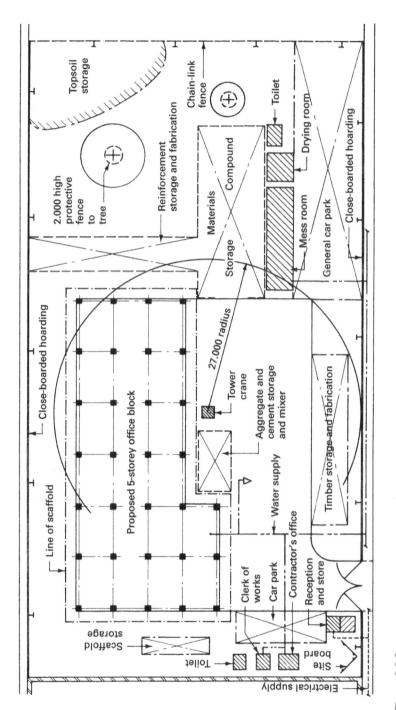

Figure 1.1.2 Site layout example: proposed layout of accommodation and storage

- Site is in a typical urban district within easy reach of the contractor's head office and therefore will present no transport or staffing problems.
- Subsoil is a firm sandy clay with a water table at a depth that should give no constructional problems.
- Possession of site is to be at the end of April, and the contract period is 18 months. The work can be programmed to enable the foundation and substructure work to be completed before adverse winter weather conditions prevail.
- Development consists of a single five-storey office block with an in-situ reinforced concrete structural frame, in-situ reinforced concrete floors and roof, precast concrete stairs, and infill brick panels to the structural frame with large hardwood timber frames fixed into openings formed by the bricklayers. Reduced-level dig is not excessive, but the topsoil is to be retained for landscaping on completion of the building contract by a separate contractor; however, the paved area in front of the office block forms part of the main contract. The existing oak trees in the north-east corner of the site are to be retained and are to be protected during the contract period.
- Estimated maximum number of staff on site at any one time is 40, in the ratio of 1 supervisory staff to 10 operatives, plus a resident clerk of works.
- Main site requirements are as follows:
  - 1 office for 3 supervisory staff;
  - 1 office for resident clerk of works;
  - 1 office for reception and materials checker/assistant site agent;
  - 1 hutment as lock-up store;
  - 1 mess room for 36 operatives;
  - 1 drying room for 36 operatives;
  - toilets;
  - storage compound for major materials;
  - timber store and formwork fabrication area;
  - reinforcement store and fabrication area;
  - scaffold store;
  - car parking areas;
  - 1 tower crane;
  - area for concrete deliveries, sand and cement storage, and site mixer.

Sizing and location of main site requirements can be considered in the following manner.

- **Offices for contractor's supervisory staff** Area required = 3×3.7 m^2 = 11.1 m^2. Using plastic-coated galvanised-steel prefabricated cabins based on a 2.400 m-wide module gives a length requirement of $11.1 \div 2.4 = 4.625$ m: therefore use a hutment 2.400 m wide $\times$ 4.800 m long, giving an area of 11.52 m^2. Other standard internal widths are 2.7 and 3.0 m, and standard internal lengths range from 2.4 to 10.8 m in 1.2 m increments.

- **Office for resident clerk of works** Allowing for one visitor area required = 2×3.7 m^2 = 7.4 m^2. Using same width module as for contractor's office length required = $7.4 \div 2.4 = 3.08$ m, therefore using a 2.400 m wide $\times$ 3.600 m long cabin will give an area of 8.64 m^2. The contractor's office and that for the clerk of works need to be sited in a position that is easily and quickly found by visitors to the site and yet at the same time will give a good view of the site operations. Two positions on the site in question seem to meet these requirements: one is immediately to the south of the paved area and the other is immediately to the west of it. The second position has been chosen for both offices because there is also room to accommodate visitors' cars in front of the offices without disturbing the circulation space given by the paved area.

- **Office for reception and materials checker** A hut based on the requirements set out above for the clerk of works would be satisfactory. The office needs to be positioned near to the site entrance so that materials being delivered can be checked, directed to the correct unloading point, and – most important – checked before leaving to see that the delivery has been completed. It also needs to be easily identifiable so visitors can sign into the site and be inducted.

- **Lock-up store** This needs to be fitted with racks and storage bins to house valuable items, and a small unit of plan size 2.400 m $\times$ 2.400 m has been allocated. Consideration must be given to security, and in this context it has been decided to combine the lock-up store and the site manager's/materials checker's office, giving a total floor plan of 2.400 m $\times$ 6.000 m. This will enable the issue of stores only against an authorised and signed requisition to be carefully controlled, the assistant site agent fulfilling the function of storekeeper.

- **Mess room** Area required = 36×2.5 m^2 = 90 m^2, using a width module of 3.000 m, length required = $90 \div 3 = 30$ m: therefore a number of combinations based on the standard lengths listed are possible. Perhaps three modules of 10.8 m = 32.4 m total length (97.2 m^2) or five modules of 6 m length = 30 m total length (90 m^2), the choice depending to some extent on the disposition of the site staff, and the number and size of subcontractors involved. The mess room needs to be sited in a fairly central position to all the areas of activity, and the east end of the paved area has been selected.

- **Drying room** Area required = 36×0.6 m^2 = 21.6 m^2, using a width module of 3.000 m, length required = 21.6 ÷ 3 = 7.2 m: therefore select a single length or two modules of 3.600 m. The drying room needs to be in close proximity to the mess room and has therefore been placed at the east end of the mess room. Consideration could be given to combining the mess room and drying room into one unit.

- **Toilets** On this site it is assumed that connection can be made to existing drains. If this is not convenient, temporary (or preferably permanent) drain branches can be connected to a main sewer. Two such units are considered to be adequate, one to be sited near to the mess room and the other to be sited near to the office complex. Adequate sanitary conveniences are required in the Construction Design and Management Regulations 1996. For the mess toilet unit catering for 36 operatives, two conveniences are considered minimum, but a three-convenience toilet unit will be used, having a plan size of 2.400 m × 3.600 m. Similarly, although only one convenience is required for the office toilet unit, a two-convenience unit will be used with a plan size of 2.400 m × 2.400 m.

- **Materials storage compound** An area to be defined by a temporary plywood hoarding 2.400 m high and sited at the east end of the paved area giving good access for deliveries and within reach of the crane. Plan size to be allocated 12.000 m wide × 30.000 m long.

- **Timber storage** Timber is to be stored in top-covered but open-sided racks made from framed standard scaffold tubulars. Maximum length of timber to be ordered is unlikely to exceed 6.000 m in length: therefore, allowing for removal, cutting and fabricating into formwork units, a total plan size of 6.000 m wide × 36.000 m long has been allocated. This area has been sited to the south of the paved area, giving good access for delivery and within the reach of the crane.

- **Reinforcement storage** The bars are to be delivered cut to length, bent and labelled, and will be stored in racks as described above for timber storage. Maximum bar length to be ordered assumed not to exceed 12.000 m: therefore a storage and fabrication plan size of 6.000 m wide × 30.000 m long has been allocated. This area has been sited to the north of the storage compound, giving reasonable delivery access and within reach of the crane.

- **Scaffold storage** Tube lengths to be stored in racks as described for timber storage, with bins provided for the various types of coupler. Assumes a maximum tube length of 6.000 m, a plan size of 3.000 m wide × 12.000 m long. This storage area has been positioned alongside the west face of the proposed structure, giving reasonable delivery access and within reach of the crane if needed. The scaffold to be erected will be of an independent type around the entire perimeter positioned 200 mm clear of the building face and of five-board width, giving a total minimum width of 200 + (5 × 225) = 1.325, say 1.400 m total width.

- **Tower crane** To be sited on the paved area in front of the proposed building alongside the mixer and aggregate storage position. A crane with a jib length of 27.000 m, having a lifting capacity of 1.25 tonnes at its extreme position, has been chosen so that the crane's maximum radius will cover all the storage areas, making maximum use of the crane.

- **Car parking** Assume 20 car parking spaces are required for operatives, needing a space per car of 2.300 m wide × 5.500 m long, giving a total length of $2.3 \times 20 = 46.000$ m and, allowing 6.000 m clearance for manoeuvring, a width of $5.5 + 6.0 = 11.500$ m will be required. This area can be provided to the south of the mess room and drying room complex. Staff car-parking space can be sited in front of the office hutments, giving space for the parking of seven cars, which will require a total width of $7 \times 2.3 = 16.100$ m.

- **Fencing** The north and south sides of the site both face onto public footpaths and highways. Therefore a close-boarded or sheet hoarding in accordance with the licence issued by the local authority will be provided. A lockable double gate is to be included in the south-side hoarding to give access to the site. The east side of the site faces an undeveloped site, and the contract calls for a 2.000 m-high concrete post and chain-link fence to this boundary. This fence will be erected at an early stage in the contract to act as a security fence during the construction period as well as providing the permanent fencing. The west side of the site has a 2.000 m-high brick wall, which is in a good structural condition, and therefore no action is needed on this boundary.

- **Services** It has been decided that permanent connections to the foul drains will be made for convenient site use, thus necessitating early planning of the drain-laying activities. The permanent water supply to the proposed office block is to be laid at an early stage, and this run is to be tapped to provide the supplies required to the mixer position and the office complex. A temporary connection is to be made to supply the water service to the mess room complex, because a temporary supply from the permanent service would mean running the temporary supply for an unacceptable distance. An electrical supply is to be taken onto site, with a supply incoming unit housed in the reception office along with the main distribution unit. The subject of electrical supplies to building sites is dealt with in Section 1.2. Telephones will be required to the contractor's and clerk of works' offices. It has been decided that a gas supply is not required.

- **Site identification** A V-shaped board bearing the contractor's name and company symbol is to be erected in the south-west corner of the site in such a manner that it can be clearly seen above the hoarding by traffic travelling in both directions, enabling the site to be clearly identified. The board will also advertise the company's name and possibly provide some revenue by including on it the names of participating subcontractors.

As a further public relations exercise it might be worthwhile considering the possibility of including public viewing panels in the hoarding on the north and south sides of the site.

■ **Health and safety** Attached to the hoarding at the site entrance is a board displaying the employer's policy for corporate site safety. Some examples of the standard images that could be used are shown in Fig. 1.1.3.

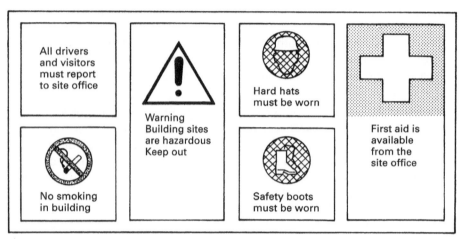

Figure 1.1.3 Site safety board

Table 1.1.1 Sign colours, shapes and indications

Sign colours	Geometric shape	Indication
Red on white background, black image	Circular with a diagonal line	Prohibition
Yellow with black border, black image	Triangular	Warning
Blue with white image	Circular	Mandatory
Green with white image	Oblong or square	Safe condition

References:
Health and Safety at Work etc. Act.
Health and Safety (Safety Signs and Signals) Regulations.
Management of Health and Safety at Work Regulations.
European Directive 92/58 EEC.

The extent to which the above exercise in planning a site layout would be carried out in practice will depend on various factors, such as the time and money that can reasonably be spent and the benefits that could accrue in terms of maximum efficiency compared with the amount of the capital outlay. The need for careful site layout and site organisation planning becomes more relevant as the size and complexity of the operation increase. This is particularly true for contracts where spare site space is very limited.

Electricity on building sites　1.2

A source of temporary electricity is often required in advance of the more permanent service supply during the temporary construction works. This will be needed to provide power and lighting to the site accommodation and to feed any electrical plant and equipment in use, such as a hoist. Two sources of electrical supply to the site are possible:

- portable self-powered generators;
- metered supply from the local section of the national grid distribution network.

As a supply of electricity will always be required in the final structure, the second source is usually adopted, because it is generally possible to connect a permanent supply cable to the proposed development for construction operations, thus saving the cost of laying a temporary supply cable to the site. Clients are normally happy with this approach as long as the supply to the contractor is metered.

To obtain a metered temporary supply of electricity, a contract must be signed between the main contractor and a local area electricity company. The company will require the following information:

- address of site;
- site location plan;
- maximum anticipated load demand in kW for the construction period (note any cost increase due to a bigger capacity cable as a result of this larger demand may be borne by the contractor);
- final load demand of the completed building to ensure that the correct rating of cable is laid for the permanent supply;
- date on which temporary supply will be required;

- name, address and telephone number of the building owner or their agent, and of the main contractor.

To ensure that the supply and installation are available when required by the contractor, it is essential that an application for a temporary supply of electricity is made at the earliest possible date, due to the long lead-in periods required.

On any construction site it is possible that there may be existing electricity cables, which have been isolated and protected, from the demolition of an existing structure; these can be advantageous or may constitute a hazard or nuisance. Any overhead cables will be visible, whereas the routes and depths of underground cables can be ascertained from the records and maps kept by local area supply companies. Overhead cable voltages should be checked with the local area suppliers, because these cables are usually uninsulated and are therefore classed as a hazard, due mainly to their ability to arc over a distance of several metres.

High-voltage cables of over 11 kV rating will need special care, and any of the following actions could be taken to reduce or eliminate the danger:

- apply to the local area supplier to have the cables re-routed at a safe distance or height (this may be a costly option);
- apply to have the cable taken out of service, which may again be a costly temporary decommission;
- erect warning barriers to keep site operatives and machines at a safe distance.

Figure 1.2.1 illustrates the requirements taken from the HSE Guide *Avoidance of danger from overhead electric power lines.*

The position and depth of underground cables given by electricity suppliers must be treated as being only approximate, because historical records show only the data regarding the condition as laid, and since then changes in site levels may have taken place. Before any excavation work commences the location of the cable must be ascertained through a CAT (cable avoidance tool) scan of the area, which will provide an accurate indication of the cable location. When excavating in the locality of an underground cable extreme caution must be taken, which may even involve careful hand excavation to expose the cable. A banksman must be present at all times during the excavation work to act as an extra set of eyes for the driver who may not be able to see hazards. Exposed cables should be adequately supported, and suitable barriers with warning notices should be erected. Any damage, however minor, must be reported to the electricity supplier for the necessary remedial action. It is worth noting that, if a contractor damages an underground electric cable that was known to be present, and possibly causes a loss of supply to surrounding properties, the contractor can be liable for negligence and uninsured losses.

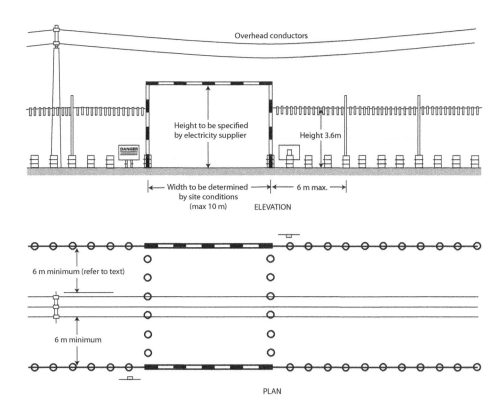

Overhead conductors

Height to be specified by electricity supplier

DANGER

Height 3.6m

|←— Width to be determined →|←— 6 m max. —→|
by site conditions
(max 10 m)   ELEVATION

6 m minimum (refer to text)

6 m minimum

PLAN

Figure 1.2.1 **Requirements of HSE Guide** *Avoidance of danger from overhead electric power lines*

SUPPLY AND INSTALLATION

In Great Britain electrical installations on construction sites are subject to the requirements of a series of regulations and standards as follows:

- Electricity Supply Regulations 1988;

- Electricity at Work Regulations 1989;

- Health and Safety at Work etc. Act 1974;

- Construction Design and Management Regulations 2007;

- BS 7671: *Requirements for electrical installations* (Institution of Electrical Engineers Wiring Regulations 17th Edition) – Section 704 details provision for temporary installations and installations on construction sites;

- BS 7375: *Code of practice for distribution of electricity on construction and building sites*;

- BS 4363: *Specification for distribution assemblies for reduced low voltage electricity supplies for construction and building sites* – this covers the equipment suitable for the control and distribution of electricity from a three-phase four-wire AC system up to a voltage of 400 V;

- BS EN 60309-2, which specifies plugs, socket outlets and cable couplers for the varying voltages recommended for use on construction sites.

The application of these regulations ensures that electricity can be used safely during the construction phase of a project. The appliances and wiring used in temporary installations on construction sites may be subjected to extreme abuse and adverse weather conditions: therefore correct circuit protection, earthing and frequent inspection are most important, and this work, including the initial installation, must be entrusted to a qualified electrician or to a specialist electrical contractor.

Electrical distribution cables contain three live wires and one neutral, which can give either a 400 V three-phase supply or a 230 V single-phase supply. Records of accidents involving electricity show that the highest risk is encountered when electrical power is used in wet or damp conditions, which are often present on construction sites. This is because water conducts electricity very efficiently. To use 240 V supplies on site is therefore considered to be dangerous and it is therefore recommended that the distribution voltage on building sites should be 110 V. It should be noted that 110 V electric shock can still be fatal to an operative.

The recommended voltages for use on construction sites are given below.

Mains voltage

400 V three-phase:

- supply to transformer unit, heavy plant such as cranes and movable plant fed via a trailing cable;

- hoists and plant powered by electric motors in excess of a 3.75 kW rating.

230 V single-phase:

- supply to transformer unit using short lengths of cable;

- supply to distribution unit where it can be lowered to another voltage using a transformer;

- installations in site accommodation buildings wired in accordance with electrical regulations through a consumer unit;

- fixed floodlighting where cables can be enclosed in conduits;

- small static machines protected by RCDs (residual current devices).

Reduced voltage

110 V single-phase:

- portable and hand-held tools;
- small items of plant;
- site floodlighting other than fixed floodlighting;
- portable hand-lamps;
- local lighting up to 2 kW.

It is worth considering the use of 18 V lithium battery-supplied hand-lamps if damp situations are present on site. All supply cables must be earthed, and in particular 110 V supplies should be centre-point earthed so that the nominal voltage to earth is not more than 65 V on a three-phase circuit and not more than 55 V on a single-phase circuit.

Protection to a circuit can be given by using MCBs (miniature circuit breakers) and RCDs. Adequate protection should be given to all main circuits and subcircuits against any short-circuit current, overload current and earth faults.

Protection through earthing may be attained in two distinct ways:

- provision of a path of low impedence to ensure over-current device will operate in a short space of time;
- insertion in the supply of a circuit breaker with an operating coil that trips the breaker when the current due to earth leakage exceeds a predetermined value.

BS 4363 and BS EN 60309-2 recommend that plug and socket outlets are identified by a colour-coding as an additional safety precaution to prevent incorrect connections being made. The recommended colours are:

25 V – violet
50 V – white
110 V – yellow
230 V – blue
400 V – red
500/650 V – black.

Here is a list of the equipment that can be used to distribute an electrical supply around a construction site.

- **Incoming site assembly (ISA)** Supply, control and distribution of mains supply on site – accommodates supply company's equipment and has one outgoing circuit.
- **Main distribution assembly (MDA)** Control and distribution of mains supply for circuits of 400 V three-phase and 230 V single-phase.

- **Incoming site distribution assembly (ISDA)** A combined ISA and MDA for use on sites where it is possible to locate these units together.
- **Transformer assembly** Transforms and distributes electricity at a reduced voltage; can be for single-phase, three-phase or both phases and is abbreviated TA/1, TA/3 or TA/1/3 accordingly.
- **Socket outlet assembly (SOA)** Connection, protection and distribution of final subcircuits at a voltage lower than the incoming supply.
- **Extension outlet assembly (EOA)** Similar to outlet assembly except that outlets are not protected.
- **Earth monitor unit (EMU)** Flexible cables supplying power at mains voltage from the MDA to movable plant incorporate a separate pilot conductor in addition to the main earth continuity conductor. A very low voltage current passes along these conductors between the portable plant and the fixed EMU. A failure of the earth continuity conductor will interrupt the current flow, which will be detected by the EMU, and this device will automatically isolate the main circuit.

The cubicles or units must be of robust construction, strong, durable, rain-resistant and rigid to resist any damage that could be caused by transportation, site handling or impact shocks likely to be encountered on a construction site. All access doors or panels must have adequate weather seals. Figure 1.2.2 shows a typical supply and distribution system for a construction site. Figure 1.2.3 shows this in a typical multi-storey construction detail where electrical supplies are required on each floor along with lighting.

The routeing of the supply and distribution cables around the construction site should be carefully planned. A trailing cable is a risk in that it can become a trip hazard. Cables should not be allowed to trail along the ground unless suitably encased in a tube or conduit; it is better to tie cables at a high level, out of harm's way, while the supply is being used. Overhead cables should be supported by hangers attached to a straining wire and suitably marked with flags or similar visual warning.

Cables that are likely to be in position for a long time, such as the supply to a crane, should preferably be sited underground at a minimum depth of 500 mm and protected by tiles, or alternatively housed in clayware or similar pipes.

In the interest of safety, and to enable first-aid treatment to be given in cases of accident, all contractors using a supply of electricity on a construction site for any purpose must display, in a prominent position, extracts from the Electricity at Work Regulations. Pictographic safety signs for caution of the risk of electric shock are applicable under the Health and Safety (Safety Signs and Signals) Regulations 1996. Suitable placards giving instructions for emergency first-aid treatment for persons suffering from electrical shock and/or burns are obtainable from RoSPA, the St John Ambulance Association and stockists of custom-made signs.

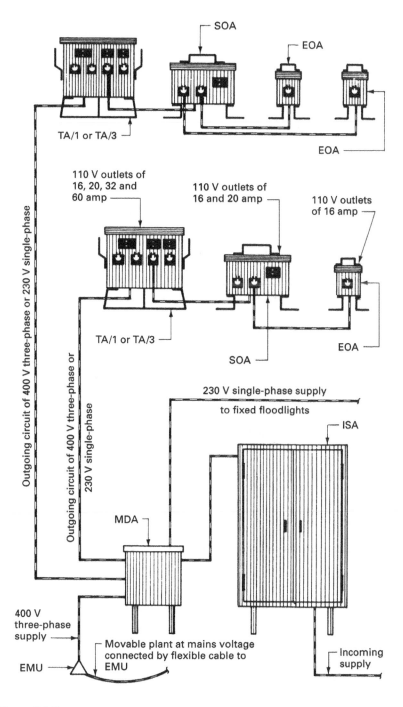

SOA

EOA

TA/1 or TA/3

EOA

110 V outlets of
16, 20, 32 and
60 amp

110 V outlets of
16 and 20 amp

110 V outlets
of 16 amp

Outgoing circuit of 400 V three-phase or 230 V single-phase

Outgoing circuit of 400 V three-phase or 230 V single-phase

TA/1 or TA/3

EOA

SOA

230 V single-phase supply
to fixed floodlights

ISA

MDA

400 V
three-phase
supply

Movable plant at mains voltage
connected by flexible cable to
EMU

EMU

Incoming
supply

Figure 1.2.2 Typical distribution sequence of site electricity

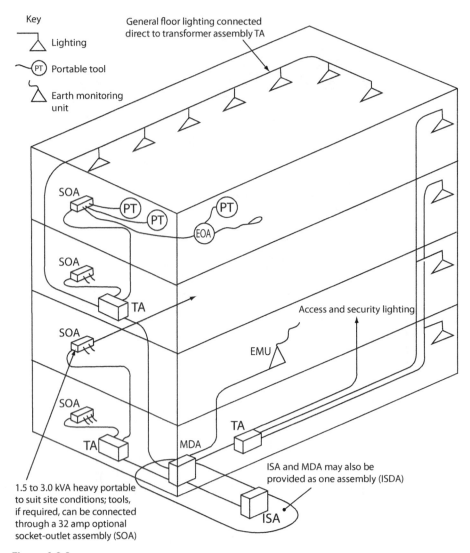

Key

Lighting

(PT) Portable tool

Earth monitoring unit

General floor lighting connected direct to transformer assembly TA

SOA

SOA

TA

SOA

SOA

TA

MDA

TA

Access and security lighting

EMU

ISA

(PT) (PT) (PT)

EOA

ISA and MDA may also be provided as one assembly (ISDA)

1.5 to 3.0 kVA heavy portable to suit site conditions; tools, if required, can be connected through a 32 amp optional socket-outlet assembly (SOA)

Figure 1.2.3 **Typical distribution sequence for a multi-storey construction**

Temporary construction lighting 1.3

The UK sits in a northern latitude, which dictates that during the winter months we experience reduced lighting levels and shorter days. During the early evening and at the start of a working day artificial lighting will need to be provided as a matter of health and safety. Inadequate light can have a marked effect on the level of production on UK construction sites between the months of November and February and when seasonal time adjustments are made.

Inadequate lighting is a hazard: it increases the risk of an accident to operatives and creates the threat of theft on the site. The initial costs of installing a system of artificial lighting for both internal and external activities can usually be offset by more efficient output, better-quality defect-free work and a more secure site. The initial costs can be apportioned over a number of contracts or priced accordingly within the preliminaries for the contract.

There are many reasons for installing a system of artificial lighting on a construction site, including:

- because the quality of the light in winter months makes working impractical;
- to prevent personal injury to operatives;
- to reduce losses in productivity;
- to reduce the wastage of resources that often results from working in poor light;
- to avoid short-time working and longer programme periods due to the inability to see clearly enough for accurate and safe working;
- to improve the general security of the site, especially at night-time.

There are benefits to installing and using a system of artificial lighting on a construction site.

- Site activities can be coordinated and conducted independent of the availability of natural daylight.
- Cost savings are made by not starting a working day early at 7.30 which would invoke overtime.
- Deliveries and collection of materials or plant can be made outside normal site working hours, thus helping to avoid delays and/or congestion.
- The amount of wastage and the consequent rectification caused by working under inadequate light can be reduced.
- It provides an effective deterrent to the would-be trespasser or thief.
- Employees will benefit both financially and in working relationships from working a full week.

Planning the lighting requirements depends on:

- site layout in terms of which areas require the most lighting;
- size and shape of site;
- geographical location, as the further north a site is, the shorter the period of natural lighting;
- availability of a constant electrical supply;
- the planned programmed activities for the winter period.

Figure 1.3.1 shows two charts covering various regions in the UK from north to south. These illustrate how many hours of daylight are available during the summer and winter months (for example, in Glasgow there are seven hours of daylight in the middle of winter). Any form of temporary artificial site lighting should be easy to install and modify as needs change, and should be easy to remove while works are still in progress.

ILLUMINATION

Illumination can be considered as the measure of light or illuminance falling on a surface. The standard of measurement is the **lux**, which can be measured with a small portable lightmeter, which consists of a light-sensitive cell generating a small current proportional to the light falling on it.

The level of illuminance or lux at which an operative can work in safety and carry out tasks to an acceptable standard, in terms of both speed and quality, is quite low, because the human eye is very adaptable and efficient at low levels of light. A suitable risk assessment should always be undertaken when operating in low light levels as there are many variables to account for, concerning not only the operative but also the task undertaken.

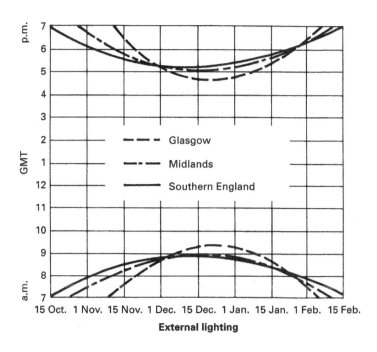

External lighting

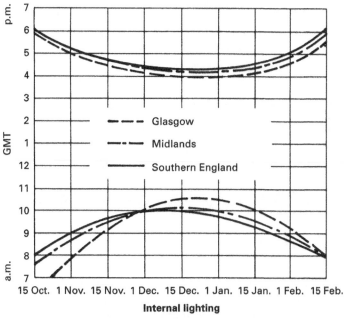

Internal lighting

Figure 1.3.1   Approximate times for site lighting

Table 1.3.1 presents typical service values of illuminance when undertaking various activities; the more detailed the work, the higher the lux level that is required.

The values shown in the table do not allow for deterioration, dirt, bad conditions or shadow effects. Therefore, in calculating the illuminance required for any particular situation, a target value of twice the service value should be used.

Table 1.3.1   Typical service values of illuminance

Activities	Illuminance (lux)
External lighting	
Materials handling	200
Open circulation areas	100
Internal lighting	
Circulation	100
Working areas	200
Reinforcing and concreting	300
Joinery, bricklaying and plastering	500
Painting and decorating	500
Fine craft work	1,000
Site offices	500
Drawing board positions	750

When deciding on the type of installation to be used, two factors need to be considered:

- type of lamp that is to be used and its energy efficiency;
- nature and type of the activity to be undertaken.

The following examines the various types of lamp that are available to provide temporary internal and external lighting to the construction of a project.

LAMPS

INCANDESCENT

This is a tungsten-filament lamp that is ideal for short periods. It consists of a traditional glass bulb with bayonet or screw-type end cap. The main recommended uses are for general interior lighting and low-level external movement. They are cheap to buy but are relatively expensive to run. This type of bulb is steadily being phased out and replaced by the more energy-efficient compact types.

HALOGEN

The tungsten halogen lamp has a compact fitting with high light output, and is suitable for all general area floodlighting. It is easy to mount, and has a more effective focused beam than the filament lamp. These lamps generally have twice the life of ordinary filament lamps, and quartz lamps have a higher degree of resistance to thermal shock than glass filament lamps. They cost more than filament lamps and are still relatively expensive to run, but should be considered if the running time is in the region of 1,500 hours annually.

FLUORESCENT

Tubular fluorescent lamps are uniformly bright in all directions, and are used when a great concentration of light is not required. They are efficient, with a range of colour values. These lamps have a long life and are economical to run. They are manufactured as tubes to various lengths and sizes, and must be encapsulated for construction use as they can be easily broken.

COMPACT FLUORESCENT LAMPS

The move towards energy efficiency to reduce the effects of CO_2 emissions has produced some technical revolutions in the design of lighting lamps. The technology of the large fluorescent tube has now been reduced and compacted into the standard shape of the small light bulb. These are now mass-produced in a variety of wattages and shapes, and can be effectively used in fittings for temporary lighting during construction. They are available in a screw or bayonet fitting, and as a small tube or spiral type.

HIGH-INTENSITY DISCHARGE LAMPS

High-pressure discharge lamps are compact and efficient, with a long life. Examples are metal halide, mercury vapour and high-pressure sodium. For the best coverage without glare they should be mounted above 13.500 m high. Costs for lamp and control gear are high, but running costs are low, which makes them suitable for area lighting.

LOW-PRESSURE SODIUM DISCHARGE LAMPS

This is the most energy-efficient lamp type available, providing more lux for each watt of power used but at a cost in the quality of light produced. It emits a yellow light that makes the distinguishing of colour difficult. They are the main source used for street lighting at night, as they do not cause glare or strobing. They can be used on construction sites for low-level external lighting of temporary roads and storage areas.

LIGHT-EMITTING DIODES (LEDs)

Technological developments in light-emitting diodes (LEDs) have enabled their use within the energy-efficient bulb market. They are very energy-efficient, and will last over three times longer than a traditional light bulb. The diodes can be shaped and collected into any format of light fitting: for example, an LED lamp can now be manufactured to replace a standard fluorescent tube.

Apart from the cost of the lamps and the running charges, consideration must be given to the cost of the electrical supply cables, controlling equipment, mounting poles, masts or towers. A single, high tower may well give an overall saving against using a number of individual poles or masts in spite of the high initial cost for the tower, and provides a better solution against vandalism of the lamp fitting. Consideration can also be given to using the scaffold, the incomplete structure or the mast of a tower crane for lamp-mounting purposes, each subject to earthing.

SITE LIGHTING INSTALLATIONS

When deciding on the type and installation layout for construction site lighting, consideration must be given to the nature of the area and work to be lit, and also to the type or types of lamp to be used. These aspects can be considered under the following headings:

- external and large circulation areas;
- walkway lighting;
- beam floodlighting;
- local lighting.

EXTERNAL AND LARGE CIRCULATION AREAS

These areas may be illuminated using mounted lamps situated around the perimeter of the site or in the corners of the site at high level. The main objectives of area lighting are to enable staff and machinery to move around the site in safety and to give greater security to the site, as they light a wider area than local lighting. Areas of local danger such as excavations and obstructions should, however, be marked separately with red warning lights or amber flashing lamps. Tungsten filament, mercury vapour or tungsten halogen lamps can be used, and these should be mounted on poles, masts or towers according to the lamp type and wattage. Typical mounting heights for various lamps and wattages are shown in Table 1.3.2.

Large areas are generally illuminated using large, high-mounted lamps, whereas small areas and narrow sites use a greater number of smaller fittings. By mounting the lamps as high as practicable above the working level, glare is reduced; by lighting the site from at least two directions, the formation of dense shadows is also reduced. The spacing of the lamps is also important if under-lit and over-lit areas are to be avoided. Figures 1.3.2 and 1.3.3 show typical lamps and the recommended spacing ratios.

Table 1.3.2 **Mounting heights**

Lamp type	Watts	Minimum height (m)
Tungsten filament	200	4.500
	300	6.000
	750	9.000
Mercury vapour	400	9.000
	1,000	15.000
	2,000	18.000
Tungsten halogen	500	7.500
	1,000	9.000
	2,000	15.000

Most manufacturers provide guidance as to the choice of lamps or combination of lamps. There is a simple method of calculating lamp requirements.

1. Decide on the service illuminance required, and double this figure to obtain the target value.
2. Calculate total lumens required $= \dfrac{\text{area (m}^2) \times \text{target value (lux)}}{0.23}$

3. Choose lamp type.

   Number of lamps required $= \dfrac{\text{total lumens required}}{\text{lumen output of chosen lamp}}$

4. Repeat stage 3 for different lamp types to obtain the most practicable and economic arrangement.
5. Consider possible arrangements, remembering that:
   - larger lamps give more lumens per watt and are generally more economic to run;
   - fewer supports simplify wiring and aid overall economy;
   - corner siting arrangements are possible;
   - clusters of lamps are possible.

BEAM FLOODLIGHTING

Tungsten filament or mercury vapour lamps can be used, but this technique is limited in application on construction sites to supplementing other forms of lighting. Beam floodlights are used to illuminate areas from a great distance. The beam of light is intense, producing high glare, and it should therefore be installed to point downwards towards the working areas. Generally the lamps are selected direct from the manufacturer's catalogue without calculations.

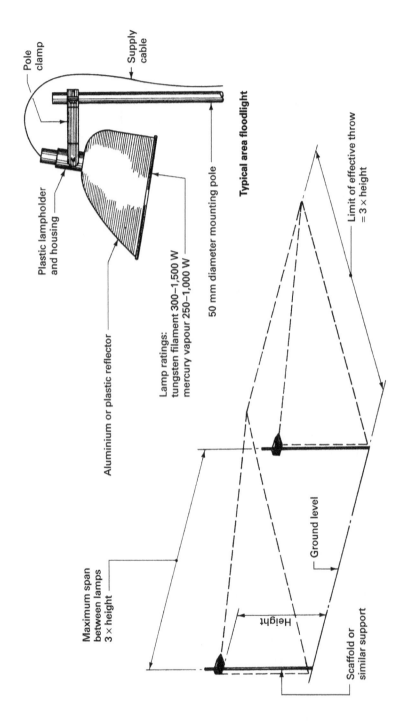

Pole clamp

Supply cable

Plastic lampholder and housing

Aluminium or plastic reflector

Lamp ratings:
tungsten filament 300–1,500 W
mercury vapour 250–1,000 W

50 mm diameter mounting pole

Typical area floodlight

Limit of effective throw
= 3 × height

Maximum span
between lamps
3 × height

Ground level

Height

Scaffold or
similar support

Figure 1.3.2 Area lighting: lamps and spacing ratios: 1

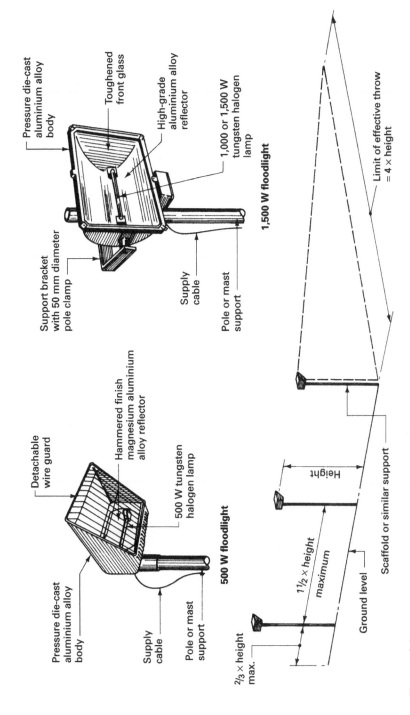

Pressure die-cast aluminium alloy body

Toughened front glass

High-grade aluminium alloy reflector

1,000 or 1,500 W tungsten halogen lamp

1,500 W floodlight

Support bracket with 50 mm diameter pole clamp

Supply cable

Pole or mast support

Detachable wire guard

Hammered finish magnesium aluminium alloy reflector

500 W tungsten halogen lamp

500 W floodlight

Pressure die-cast aluminium alloy body

Supply cable

Pole or mast support

²/₃ × height max.

1½ × height maximum

Height

Ground level

Scaffold or similar support

Limit of effective throw = 4 × height

Figure 1.3.3   Area lighting: lamps and spacing ratios: 2

WALKWAY LIGHTING

Tungsten filament and fluorescent lamps can be used to illuminate access routes such as stairs, corridors and scaffolds. Bulkhead fittings and suspended fluorescent fittings with a reduced voltage of 110 V single-phase should be used. Festoon lighting, in which the ready-wired lampholders are moulded to the cable itself, can also be used. A standard festoon cable would be 100.000 m long, with rainproof lampholders and protective shades or guards at 3.000 m or 5.000 m centres, using low-energy compact fluorescent bulbs for the respective centres (see Fig. 1.3.4).

For lighting to scaffolds of four- or five-board width, 60 W equivalent lamps should be used, placed at not more than 6.000 m centres and preferably at least 2.400 to 3.000 m above the working platform, either to the wall side or centrally over the scaffold.

LOCAL LIGHTING

The use of light fittings on tripod stands can be used to increase the surface illumination at local points, particularly where finishing trades are involved. A series of lights is available from plastering lights, fluorescent floor lights to floodlights. These fittings must be portable so that shadow casting can be reduced or eliminated from the working plane: therefore it is imperative that these lights are operated off a reduced voltage of 110 V single-phase. Typical examples of suitable lamps are shown in Fig. 1.3.5.

As an alternative to a system of static site lighting connected to the site mains, electrical supply mobile lighting sets are available. These consist of a diesel-engine-driven generator and a telescopic tower with a cluster of tungsten iodine lamps. These are generally cheaper to run than lamps operating off a mains supply. Small two-stroke generator sets with a single lamp attachment suitable for small, isolated positions are also available.

Whichever method of illumination is used on a construction site, it is always advisable to remember the axiom: 'A workman can only be safe and work well when he can see where he is going and what he is doing.'

CONSIDERATE CONTRACTOR SCHEME

The self-governing scheme tries to influence the impact of construction activities on the surrounding communities. This will involve pollution, noise, waste and environmental considerations, one of which is light emissions. As well as maintaining site safety and security, contractors should give regard to the amount of light that is directed outside the construction site boundary. Effective shading and turning lights into the site should be considered to prevent glare and over-illumination of associated properties. The level of illumination can be varied to an acceptable level so that it does not disturb any neighbours.

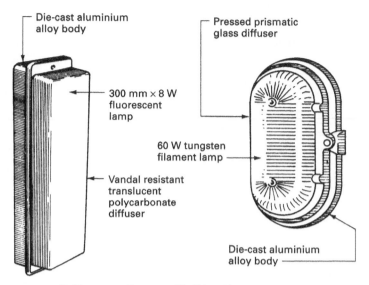

Die-cast aluminium alloy body

Pressed prismatic glass diffuser

300 mm × 8 W fluorescent lamp

60 W tungsten filament lamp

Vandal resistant translucent polycarbonate diffuser

Die-cast aluminium alloy body

Ceiling- or wall-mounted bulkhead lamp fittings

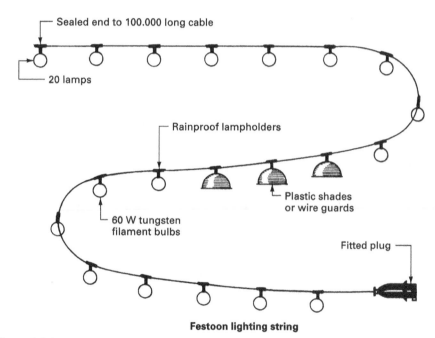

Sealed end to 100.000 long cable

20 lamps

Rainproof lampholders

Plastic shades or wire guards

60 W tungsten filament bulbs

Fitted plug

Festoon lighting string

Figure 1.3.4   Typical walkway lighting fittings

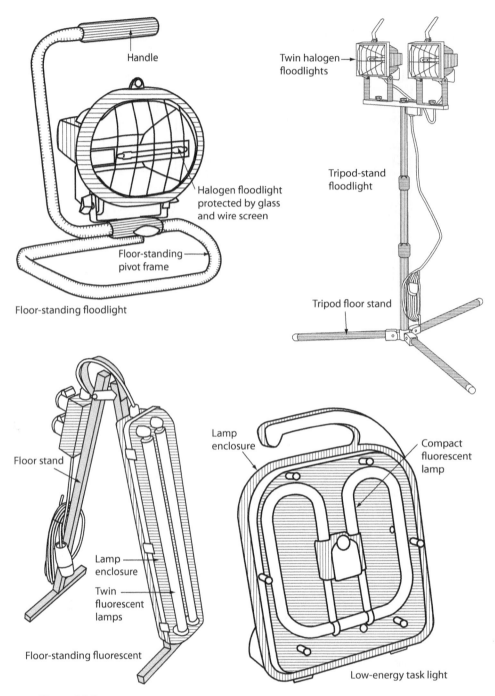

Handle

Twin halogen
floodlights

Halogen floodlight
protected by glass
and wire screen

Tripod-stand
floodlight

Floor-standing
pivot frame

Tripod floor stand

Floor-standing floodlight

Floor stand

Lamp
enclosure

Compact
fluorescent
lamp

Lamp
enclosure

Twin
fluorescent
lamps

Floor-standing fluorescent

Low-energy task light

Figure 1.3.5   Local lighting: suitable lamps and fittings

Winter construction

1.4

Approximately 20 per cent of the workforce of the United Kingdom is employed either directly or indirectly by the construction industry, which creates a large part of our gross domestic product. Therefore any fluctuation in productivity will affect a large number of operatives, contractors and elements of the supply chain. There is a general loss in output in the construction industry in normal circumstances during the winter period of about 10 per cent, due to bad weather, extreme cold and low light levels, which can result in a waste of resources. A severe winter with a long period of sub-zero temperatures can treble this typical loss of output, resulting in loss of cash flow to the main contractor and subcontractors, plus reduced pay to employees. The contractor may be unable through contractual conditions to claim for severe weather delays, which will have a marked effect on their profit earned on the contract. The client may also instruct liquidated damages to be actioned for the overrun on a completion date. The project owner also suffers through the delay in completing the construction project, which could necessitate extending the borrowing period for the capital to finance the project or the loss of a prospective tenant or lease income.

The UK weather cannot be predicted with any certainty, especially with the effects of global warming now influencing our climate. Some degree of predictability on future weather conditions can be obtained from maps, charts and statistical data issued by the Meteorological Office, and these are often useful for long-term planning, whereas in the short term reliance is placed on local knowledge, daily forecasts and the short-term monthly weather forecasts. Due to the unpredictability of the UK climate, many contractors choose not to invest in protective equipment, materials and plant for winter building when this can only be for a short period.

Contractors must therefore assess the total cost of possible delays against the capital outlay required for plant and equipment, to enable them to maintain full or near-full production during the winter period.

Weather conditions that can have a delaying effect on construction activities are:

- heavy rain, causing damage and remediation to unset finishes;
- high-level winds, causing overturning or safety issues;
- low temperatures, causing the freezing of water and damage to finishes;
- snow freezing and obscuring the work area;
- poor daylight levels.

The worst effects obviously occur when more than one of the above conditions occur simultaneously, with the resultant loss of several weeks of production.

HEAVY RAIN

Prolonged rain affects the site ground conditions and eventually the movement around the site due to mud, which in turn increases site hazards, particularly those associated with excavations and earth-moving works. It also causes problems for operatives trying to walk around a site, thus reducing their productivity.

Delays with most external operations, such as bricklaying and concreting, are usually experienced, particularly during periods of heavy rainfall, where the mortar and concrete finish is washed away. Damage can be caused to unprotected materials stored on site and, in many cases, to newly fixed materials or finished surfaces that have not been fully enclosed within the building element. The higher moisture content of the atmosphere will also delay the drying out of buildings that use traditional finishes. If high winds and rain occur together, driving rain penetration and site hazards due to poor visibility are considerably increased.

HIGH-LEVEL WINDS

High winds during external construction operations can be very hazardous, especially during activities such as frame erection and the fixing of sheet cladding. They can also limit the operations that can be carried out by certain items of plant, such as tower cranes, mobile elevated platforms and suspended cradles. Positive and negative wind pressures can also cause damage to partially completed building envelopes, incomplete structures and materials stored on site.

LOW TEMPERATURES

The lowering of air temperatures during winter causes frost and freezing damage as many site activities are slowed down. Examples include bricklaying and concreting, which both use water as a constituent (water freezes at sub-zero temperatures). Poorly serviced mechanical plant can be difficult to start in winter conditions, and stockpiles of materials can become frozen and difficult to move. General movement and circulation around the site becomes hazardous due to ice, creating danger for site personnel. When high winds are experienced with low temperatures, they will aggravate these effects.

SNOW

This is one of the most variable factors in British weather, ranging geographically from an average of five days a year on which snow falls on low ground in the extreme south-west to 35 days in the north-east of Scotland. Snow will also impede the movement of labour, plant and materials, causing hazardous and slippery conditions, as well as creating cold and uncomfortable working conditions. Externally stored materials will become covered with a layer of snow, which will have to be removed prior to use. High winds encountered with falling snow can cause drifting, increasing site hazards and personal discomfort, and decreasing general movement around the site.

Heavy snow may also have a dramatic effect on the supply chain by delaying the movement of materials to the site from suppliers outside the immediate vicinity as the infrastructure is disrupted.

WINTER BUILDING TECHNIQUES

The major aim of any winter building method or technique is to maintain an acceptable rate of productivity such that, combined with the increased production in the summer months, equates to the completion of the contract on time. Inclement weather conditions can very quickly have a marked effect on the movement of vehicles and personnel around the site and, indeed, off the site, which will be impaired or even brought to a complete standstill unless firm access roads or routes are provided, maintained and swept free of snow. If the access roads and hardstandings form part of the contract and are suitable, these could be constructed at an early stage in the contract before the winter period. The wearing course should be left until the project is near completion to avoid any damage to the road that would require repairing at cost to the contractor. If the permanent road system is not suitable in layout for contractual purposes, then compacted hardcore temporary roads are laid and bulk timbers, timber or concrete sleepers are used for road crossings and entrances, to protect any services on footpaths or verges.

Frozen ground can present problems with all excavating activities. Most hydraulic excavating plant can operate in frozen ground up to a depth of 300 mm, but at a reduced rate of output. Therefore if a deep frost is anticipated it is a wise precaution to protect the areas to be excavated by covering with insulating quilts of mineral wool or glass-fibre mats enclosed in a polythene envelope, which insulates the ground from the freezing conditions, enabling excavation work to continue. Similar precautions can be taken in the case of newly excavated areas to prevent them freezing and giving rise to 'frost-heave' conditions. This is where the water within the ground expands and moves in the direction of least resistance, which is towards the excavated face, causing expansion and movement and subsequent excavation to a stable depth once thawed.

Water supplies should be laid below ground at such a depth as to avoid the possibility of freezing. The actual depth will vary according to the locality of the site, with a minimum depth of 750 mm for any area. If the water supply is temporary and above ground, the pipes should be well insulated and not exposed.

Electrical supplies can fail in adverse weather conditions because the vulnerable parts such as contacts become affected by moisture, frost or ice. These components should be fully protected in the manner advised by their manufacturer. Temporary raised plywood structures with sloping felted roofs and padlocked doors offer protection for electrical equipment from the elements.

Items of plant that are normally kept uncovered on site – such as mixers, dumper trucks, bulldozers and generators – should be protected as recommended by the manufacturer to avoid morning starting problems. These precautions will include selecting and using the correct grades of oil, lubricant and antifreeze, as well as covering engines and electrical systems, draining radiators where necessary. The driver's cab on excavation equipment is normally boarded over for security purposes and should remain free from freezing.

Concreting operations during winter periods can be aided by the use of insulation in the form of bubble-wrap covers, which trap the heat from the hydration process within the concrete structure, allowing the curing of the concrete to proceed unaffected by the freezing conditions. However, extreme cold will delay the setting process of concrete and careful planning with regard to the temperatures on the day of pouring must be undertaken. Another method is to use hot water with the mixing of the concrete, heated so its temperature is raised at the time of placing, and then to cover it with insulation for curing. This is a provision that must be agreed with the concrete supplier, as not all batching plants can provide this facility. Insulated formwork systems are available for extreme latitude temperatures where extended periods of freezing may make the additional expenses of insulating the formwork cost-effective during their period of use.

In order to comply with the Construction Design and Management Regulations, operatives will also need protection from adverse winter conditions if an acceptable level of production is to be maintained on a contract. Such protection can be of one or more of the following types:

- temporary shelters;
- framed enclosures;
- air-supported structures;
- other protective measures, including clothing.

TEMPORARY SHELTERS

This is the cheapest and simplest method of providing protection to the working area. Temporary shelters consist of a screen of reinforced polythene sheeting of suitable gauge fixed to the outside of the scaffold to form a windbreak. If the sheeting is attached to an access scaffolding, wind-load calculations may be required for stability. The sheeting must be attached firmly to the scaffold standards so that it does not flap or tear (a suitable method is shown in Fig. 1.4.2). Temporary reinforced polythene sheeting is now available with eyelets sewn in and specialist sheeting toggles for attaching to scaffold structures. The sheeting has reinforcement strips sewn into its structure; fixings attach to these strips, such that all exposed edges are reinforced. The sheeting is also flame-retardant.

FRAMED ENCLOSURES

These consist of a purpose-made frame with a curved or ridged roof, clad with a translucent polythene sheet that is stretched and fastened to the purpose-built frames set at centres along the building; alternatively a frame enclosing the whole of the proposed structure can be constructed from standard tubular scaffolding components (see Fig. 1.4.1). Such structures can be ordered from a supplier to include doors, temporary floors and erection on site. They can be defined as a temporary building. Anchorage to the ground of the entire framing is also of great importance, and this can be achieved using a screw-type ground anchor as shown in Fig. 1.4.2. This type of structure should be considered for works that have a long programme period during the winter months within the structure's enclosure, in order to receive the maximum return for the investment from the temporary protection cost. Additional artificial lighting may also be required, along with personnel and plant access doors.

AIR-SUPPORTED STRUCTURES

Sometimes called **air domes**, these are being increasingly used on building sites as a protective enclosure for works in progress and for covered material storage areas. Two forms are available:

- internally supported dome;
- air rib dome.

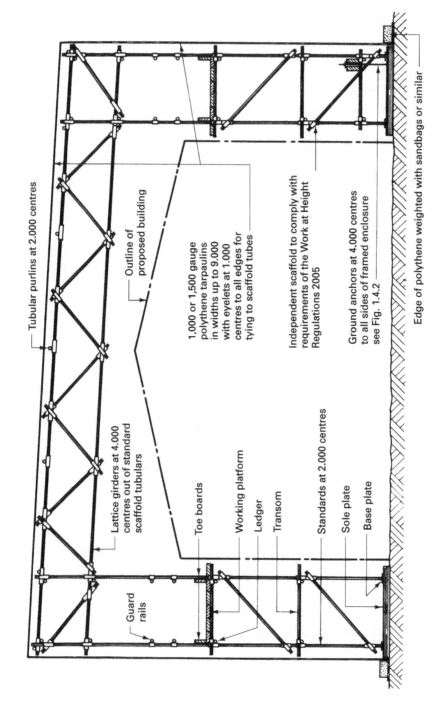

Tubular purlins at 2.000 centres

Outline of proposed building

1,000 or 1,500 gauge polythene tarpaulins in widths up to 9.000 with eyelets at 1.000 centres to all edges for tying to scaffold tubes

Independent scaffold to comply with requirements of the Work at Height Regulations 2005

Ground anchors at 4.000 centres to all sides of framed enclosure see Fig. 1.4.2

Edge of polythene weighted with sandbags or similar

Lattice girders at 4.000 centres out of standard scaffold tubulars

Toe boards

Working platform

Ledger

Transom

Standards at 2.000 centres

Sole plate

Base plate

Guard rails

Figure 1.4.1 Protective screens and enclosures: typical framed enclosure

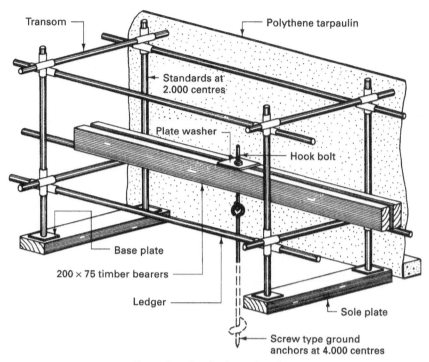

Transom

Polythene tarpaulin

Standards at
2.000 centres

Plate washer

Hook bolt

Base plate

200 × 75 timber bearers

Ledger

Sole plate

Screw type ground
anchors at 4.000 centres

Ground anchor for framed enclosure

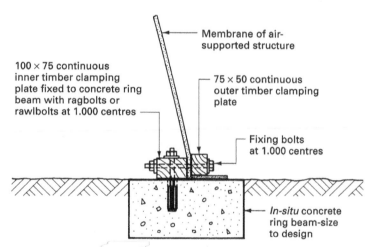

Membrane of air-
supported structure

100 × 75 continuous
inner timber clamping
plate fixed to concrete ring
beam with ragbolts or
rawlbolts at 1.000 centres

75 × 50 continuous
outer timber clamping
plate

Fixing bolts
at 1.000 centres

In-situ concrete
ring beam-size
to design

Ring beam anchorage for air-supported structure

Figure 1.4.2   Protective screens and enclosures: anchorages

The internally supported dome is held up by internal air pressure acting against the covering membrane of some form of PVC-coated nylon or rayon, with access through an airlock or air curtain door, whereas in the air rib dome the membrane is supported by air-inflated ribs to which the covering membrane is attached. The issue with the former is the problems associated with the use of an airlock, which must be maintained and kept closed. The usual shape for an air-supported structure is semi-cylindrical with rounded ends through which daylight can be introduced by the use of a translucent membrane over the structure.

The advantages of air-supported structures are that they are low-cost, light in weight and reusable; only a small amount of labour is required to erect and dismantle them, and with only a low internal pressure (approximately 150 N/m^2 above atmospheric pressure), workers inside are not affected.

Disadvantages are the need to have at least one fan in continuous 24-hour operation to maintain the internal air pressure, the need to provide an airlock or curtain entrance, which will impede or restrict the general site circulation, and the height limitation, which is usually in the region of 45 per cent of the overall span of the structure. The anchorage and sealing of the air-supported structure are also very important, and can be achieved by using a concrete ring beam as shown in Fig. 1.4.2.

OTHER PROTECTIVE MEASURES, INCLUDING CLOTHING

Personal protective clothing/equipment (PPE) is another very important aspect of winter building techniques. It must be provided to satisfy the health and safety objectives of the Construction Design and Management Regulations, which include the protection of operatives from the elements. The ideal protective clothing consists of a suit in the form of jacket and trousers made from a lightweight polyurethane-proofed nylon, with a removable jacket liner for extreme conditions. A strong lightweight safety helmet and a strong pair of steel-toe-capped rubber boots with a good grip tread would complete the protective clothing outfit. The benefits of providing suitable protective clothing are reduced sickness absence, increased productivity in inclement weather conditions, and a warmer and happier workforce.

Heating equipment may be needed on a building site to offset the effects of cold weather on building operations, particularly where the application and drying out of wet trades – for example, plastering – is required to maintain the contract programme. Heating is used indoors and often has to be combined with dehumidification and air moving units to extract moisture from the air during the drying-out process. The main types of heater in use are convectors, radiant heaters and forced-air heaters using electricity, gas or oil as fuel. Electricity is usually too expensive in the context of building operations, except for the operation of pumps and fans included in many of the heating appliances using the main other fuels. Bottled gas is the most common fuel, and must be considered in the context of hire charges, refilling, maintenance and running costs.

Convector heaters using propane or electricity with an output of 3 to 7 kW are suitable only for small volumes, such as site huts and drying rooms. These are normally hard-wired to the site accommodation unit and checked for safety and tested on a regular basis.

Self-contained radiant heaters using propane are an alternative to the convector heater. They have the advantage of being mobile, making them suitable for site-hut heating or for local heating where personnel are working.

Forced-air heaters have high outputs of 30 to 70 kW; they are efficient, versatile and mobile. Two forms of heater are available:

- the direct forced-hot-air heater, which discharges water vapour into the space being heated, which then has to be removed;
- the vented forced-air heater, where the combustion gases are discharged through a flue to the outside air.

Both forms of heater require a fan operated normally by an electrical supply. The usual fuels for the heater are propane or light fuel oil, which produces sufficient heat for general area heating of buildings under construction and for providing heat to several parts of a building using plastic ducting to distribute the hot air.

The natural drying out of buildings after the wet trades is considerably slower in the winter months, due to high humidity and low temperatures, and therefore to meet production targets this drying-out process needs the help of suitable plant. The heaters described above are often used for this purpose, but they are generally very inefficient. Effective drying out or the removal of excess moisture from a building requires the use of a dehumidifier and a vented space heater. The dehumidifier is used in conjunction with a heater: this eliminates the need to heat the cold air that would replace the warm moist air discharged to the outside if only a vented heater was employed. The most common type used in construction is the refrigeration type, where the air is drawn over refrigeration coils that lower the dewpoint, causing the excess moisture in the air to condense and discharge into a container. Depending on the model used, the extraction rate can be from 10 to 150 litres per day. A discharge hose will need to be run from the unit to a drain outside the building. With this process all windows, doors and openings will need to remain shut and air movers will be required to encourage the circulation and extraction of water from the construction work.

CONCRETE

Concrete can be damaged by rain, sleet, snow, freezing temperatures and cooling winds before it has matured, by slowing down the rate at which concrete chemically hydrates and hardens, or by increasing to an unacceptable level the rate at which the water evaporates. If the designed amount of water is removed, the design strength of the concrete will not be reached.

For best results in winter conditions:

- use certified concrete suppliers;
- defrost aggregates prior to mixing;
- have a minimum of delay between mixing and placing;
- make sure the minimum temperature of concrete when placed is 10 °C.

The following additional precautions should ensure that no detrimental effects occur when mixing and placing in winter conditions.

- Newly placed concrete should be kept at a temperature of more than 5 °C for at least three days, because the rate at which concrete sets below this temperature is almost negligible. The use of insulation covers has been examined earlier on page 44. Do not use antifreeze solutions: these are incompatible with concrete mixes.
- Follow the recommendations of BS EN 1992-1-1:2004 *Eurocode 2. Design of concrete structures. General rules and rules for buildings* for guidance.
- When using air-entraining mixtures (see BS ENs 480 and 934: *Admixtures for concrete and grout*) resistance to frost is achieved, as the bubbles act as expansion chambers for ice formations, but concrete design must allow for the effect of porosity, which can be as much as 15 per cent of strength loss.
- If special cements are used, the manufacturer's instructions should be strictly adhered to. Rapid-hardening Portland cement (RHPC) sets at the same rate as ordinary Portland cement (OPC), but because of the extra fineness of particles it has a greater surface area for the hydration reaction. This provides a high rate of heat evolution to help prevent frost damage and effect faster strength gain.

BRICKWORK

Brickwork can be affected by the same climatic conditions as given above for concrete, resulting in damage to the mortar joints and possibly spalling of the bricks over long-term frost-freeze conditions. The following precautions should be taken when bricklaying under winter weather conditions.

- Keep bricks reasonably dry under cover or shrink-wrapped and opened just before use.
- Use a 1:1:6 mortar or a 1:5–6 cement:sand mortar with an air-entraining plasticiser to form microscopic air spaces, which can act as expansion chambers for the minute ice particles that may form at low temperatures.
- Cover up the brickwork on completion with a protective insulating quilt or similar covering for at least three days.
- During periods of heavy frost, use a heated mortar by mixing with heated water (approximately 50 °C) to form a mortar with a temperature of between 15 and 25 °C.

PLASTERING AND RENDERING

Precautions that can be taken when plastering internally during cold periods are simple: close windows and doors or cover the openings with polythene or similar sheeting, maintaining an internal temperature above freezing. Lightweight plasters and aggregates are less susceptible to damage by frost than ordinary gypsum plasters and should be specified whenever possible.

TIMBER

If the correct storage procedure of a covered but ventilated rack has been followed and the moisture content of the fixed joinery is maintained at the correct level, few or no problems should be encountered when using this material during winter weather conditions.

If a contractor's output falls below the planned target figures then fixed charges for overheads, site on-costs and plant costs are being wasted to a proportionate extent. Profits will also be less because of the loss in turnover. Therefore it is usually worthwhile spending an equivalent amount on winter-building precautions and techniques to restore full production.

In general terms, a decrease of about 10 per cent in productivity is experienced by builders during the winter period; by allocating 10 per cent of fixed charges and profit to winter-building techniques, a builder can bring production back to normal. Cost analysis will show that the more exceptional the inclement weather becomes, the greater is the financial reward for money spent on winter-building techniques and precautions.

Shoring 1.5

During demolition works it may be necessary to install shoring as a temporary support applied to a building or structure. The assembly and dismantling of a shoring system must comply with the requirements of the Construction Design and Management Regulations 2007, particularly Regulation 28, which covers stability of structures, and Regulation 29, covering demolition and dismantling. These relate to work on structures where there is a perceived risk of accidental collapse and danger to any person on or adjacent to the site. The operations undertaken must be planned and undertaken in a safe manner, without risk to operatives and the general public. Relevant clauses relating to the avoidance of risk from working at height during demolition also occur in the Work at Height Regulations 2005.

Shoring is a temporary structure but can often remain in situ for a length of time while the structural design is undertaken. Its importance cannot be overlooked as it has to support a structure until the permanent supports are installed. Regardless of the time it is to remain in place, it must be subject to planned safety procedures and competent supervision during assembly, dismantling and any associated demolition, and be subject to an inspection regime.

The following situations may justify the application of temporary shoring:

- to give support to walls that are dangerous or are likely to become unstable because of subsidence, bulging or leaning; if the application is going to be long term, then a structural engineer will need to become involved often through an insurance claim;

- to avoid failure of sound walls caused by the removal of an underlying support, such as where a basement is being constructed near a sound wall, which would possibly undermine the existing foundation of the wall;

- during demolition works to give protective support to an adjacent building or structure in order to avoid any potential claim from an adjacent neighbour;

- to support the upper part of a wall during the formation of a large opening in the section of the wall underneath;

- to give support to a floor or roof to enable a loadbearing wall to be removed and replaced by a supporting beam.

There are various methods of shoring available for the support of structures. Traditionally softwood stress-graded timber was used for raking shores to walls and support to excavated trenches. The modern type of shoring uses a system approach in steel that is tested and certified for various loading conditions. For example, an acrow is a vertical, adjustable prop that is used to support walls above formed openings. Shoring arrangements can also be formed by coupling together groups of scaffold tubulars, which again are load-tested components.

SHORING SYSTEMS

There are three basic shoring systems:

- dead shoring;   - raking shoring;   - flying shoring.

Each shoring system has its own function to perform and is based on the principles of a perfectly symmetrical situation. The simplest example is shown in Fig. 1.5.1. A single raker and short wall plank or binding combine with a sole plate and needle and cleat at the top. The sole plate is set in the ground and inclined at a slightly acute angle to the raker. This allows the raker to be levered and tightened in place with a crowbar before securing with cleats and dogs.

Where flying shores are required, the disposition of buildings does not always lend itself to symmetrical arrangements. Some asymmetrical examples are shown in Fig. 1.5.2. See Figs 1.5.8 and 1.5.9 for details of regularly disposed flying-shore components.

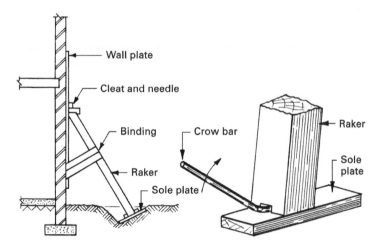

Figure 1.5.1   Single raking shore and base tightening process

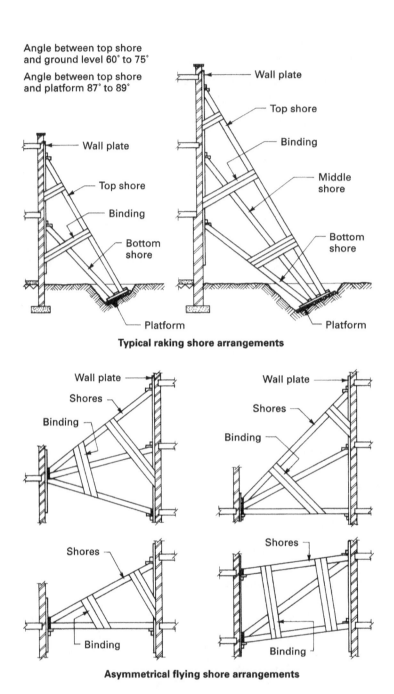

Angle between top shore
and ground level 60° to 75°

Angle between top shore
and platform 87° to 89°

Wall plate

Top shore

Binding

Wall plate

Top shore

Binding

Middle
shore

Bottom
shore

Bottom
shore

Platform

Platform

Typical raking shore arrangements

Wall plate

Shores

Binding

Wall plate

Shores

Binding

Shores

Binding

Shores

Binding

Asymmetrical flying shore arrangements

Figure 1.5.2   Shoring arrangements

DEAD SHORING

This type of shoring is used to support dead loads that act vertically downwards. In its simplest form it consists of a vertical prop or shore leg with a head plate, sole plate and some means of adjustment for tightening and easing the shore. The usual arrangement is to use two shore legs connected over their heads by a horizontal beam or needle. The needles are inserted through small holes through the structure to be supported. The loads are transferred by the needle to the shore legs and hence down to a solid bearing surface. It may be necessary to remove pavings and cut holes in suspended timber floors to reach a suitable bearing surface; if a basement is encountered, a third horizontal member called a transom will be necessary, because it is impractical to manipulate a shore leg through two storeys. A typical example of this situation is shown in Fig. 1.5.3.

The introduction of a strongboy, which is a fitting that sits on top of the adjustable prop, eliminates the need to use needles when supporting light loads. A mortar joint is often cut out and the shoe, then the strongboy is inserted into the joint and the prop is adjusted until the load is supported sufficiently. The strongboys are placed each side of the opening, depending on the wall thickness that needs to be supported. Typical details of the prop and strongboys are illustrated in Fig. 1.5.4.

The following is a typical sequence of operations necessary for a successful dead shoring arrangement for the insertion of a lintel forming an opening on an internal ground-floor block wall below the first floor ceiling level.

1. Carry out a thorough site investigation to determine:
   - number of shores required, by ascertaining possible loadings and window positions;
   - position and direction of floor joists;
   - bearing capacity of soil and floors;
   - location of services that may have to be avoided or bridged.

2. Fix ceiling props between suitable head and sole plates to relieve the wall of floor loads. The struts should be positioned as close to the wall as practicable. This takes some of the internal loading from the floor off the wall and protects the existing floor level from any deflection. The site investigation should establish whether the floor joists are resting on the wall or span over it.

3. Strut all window openings within the vicinity of the shores to prevent movement or distortion of the opening. The usual method is to place timber plates against the external reveals and strut between them.

4. Cut slots through the wall mortar joints slightly larger in size than the strongboy heads.

5. Insert strongboys and the acrow prop into the joint and extend the acrow and tighten to take the load. Ensure that the base of the prop has a suitable sole plate to spread the load.

6. Cut out blockwork, insert pad stones and lintel, wedge and pack into position.

7. Allow the packing and new mortar joints to attain sufficient strength before removing the strongboys, props and sole plates and point up slots.

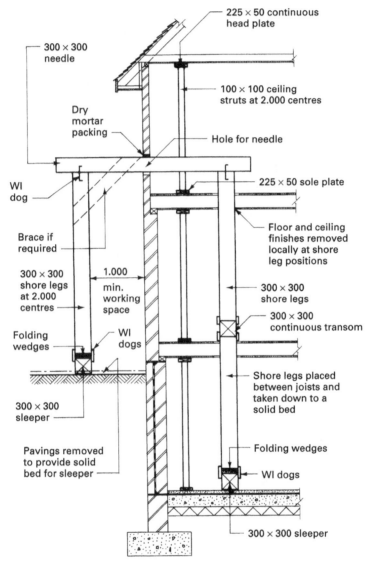

Figure 1.5.3 Dead shoring

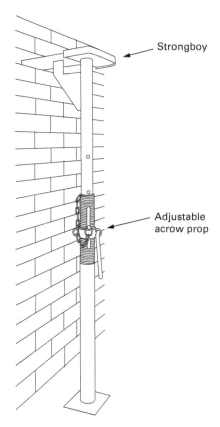

Strongboy

Adjustable
acrow prop

Figure 1.5.4   **Acrow prop and strongboy details**

RAKING SHORING

This shoring arrangement transfers the floor and wall loads to the ground by means of sloping struts or rakers. It is very important that the rakers are positioned correctly so that they are capable of receiving maximum wall and floor loads. The centreline of the raker should intersect with the centrelines of the wall or floor bearing; common situations are detailed in Fig. 1.5.6. One raker for each floor is required and ideally should be at an angle of between 40° and 70° with the horizontal; therefore the number of rakers that can be used is generally limited to three. A four-storey building can be shored by this method if an extra member, called a **rider**, is added (see Fig. 1.5.7).

The use of tested prefabricated strut rakers enables the high strength-to-weight ratio properties of steel to be used for the shoring of vertical structures on a temporary basis. These struts have an adjustable heel and head, which rotates to align with the structure that they are supporting. The raking steel strut can be adjusted for length by a threaded coupling at each end and by the selection of the correct-length prop between the adjusters.

Figure 1.5.5 illustrates the components and set-up required for this type of shore, along with its application horizontally.

The operational sequence for erecting raking shoring is as follows.

1. Carry out site investigation as described for dead shoring.

2. Mark out and cut mortises and housings in wall plate.

3. Set out and cut holes for needles in external wall.

4. Excavate to a firm bearing subsoil and lay grillage platform and sole plate.

5. Cut and erect rakers, commencing with the bottom shore. A notch is cut in the heel so that a crowbar can be used to lever the raker down the sole plate and thus tighten the shore (see Fig. 1.5.1). The angle between sole plate and shores should be at its maximum (about 89°) to ensure that the tangent point is never reached and not so acute that levering is impracticable.

6. Fix cleats, distance blocks, binding and, if necessary, cross-bracing over the backs of the shores.

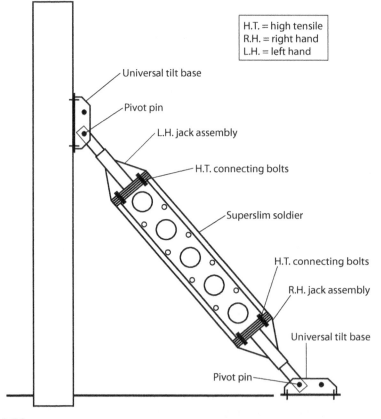

| H.T. = high tensile |
| R.H. = right hand |
| L.H. = left hand |

Universal tilt base

Pivot pin

L.H. jack assembly

H.T. connecting bolts

Superslim soldier

H.T. connecting bolts

R.H. jack assembly

Universal tilt base

Pivot pin

Figure 1.5.5 **Prefabricated 'soldier' shoring arrangement**

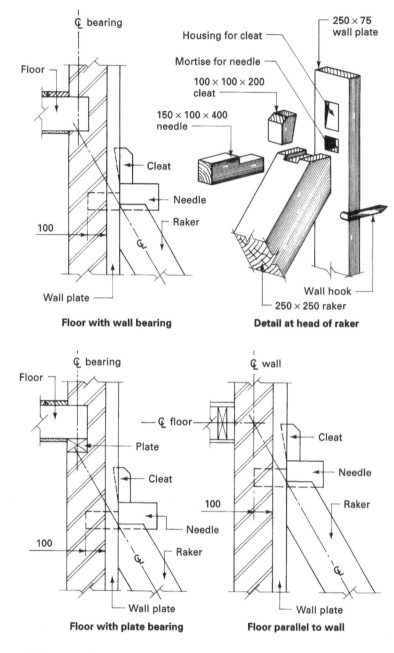

Floor with wall bearing

Detail at head of raker

Floor with plate bearing

Floor parallel to wall

Figure 1.5.6   Raking shore intersections

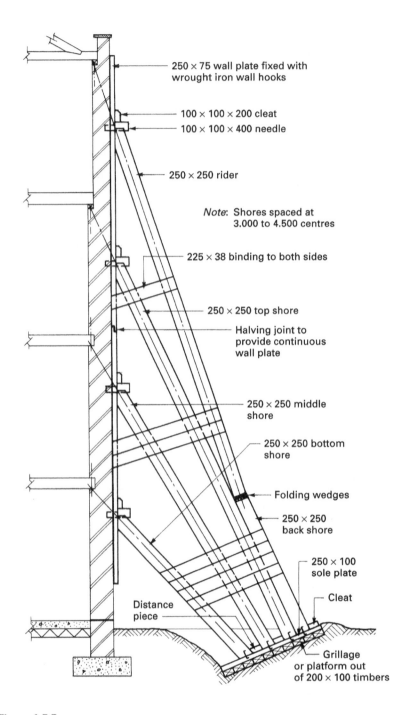

250 × 75 wall plate fixed with wrought iron wall hooks

100 × 100 × 200 cleat
100 × 100 × 400 needle

250 × 250 rider

Note: Shores spaced at 3.000 to 4.500 centres

225 × 38 binding to both sides

250 × 250 top shore

Halving joint to provide continuous wall plate

250 × 250 middle shore

250 × 250 bottom shore

Folding wedges

250 × 250 back shore

250 × 100 sole plate

Cleat

Distance piece

Grillage or platform out of 200 × 100 timbers

Figure 1.5.7 **Typical multiple raking shore**

FLYING SHORING

These shores fulfil the same functions as a raking shore but have the advantage of providing a clear working space under the shoring for the access of equipment and plant. They can be used between any parallel wall surfaces provided the span is not in excess of 12.000 m, when the arrangement would become uneconomic. Short spans up to 9.000 m usually have a single horizontal member, whereas the larger spans require two horizontal shores to keep the section sizes within the timber range commercially available (see Figs 1.5.8 and 1.5.9).

It is possible with all forms of shoring to build up the principal members from smaller sections by using bolts and timber connectors, ensuring all butt joints are well staggered to give adequate rigidity. This is in effect a crude form of laminated timber construction.

The site operations for the setting out and erection of a flying shoring system are similar to those listed for raking shoring.

Again with the introduction of structural steel fabricated sections, a system approach can be taken with flying or horizontal shores. Figure 1.5.5 illustrates a soldier beam horizontal prop that can be used as a flying shore. Consideration must be given to preventing the shore from slipping from its position, and this may involve the use of bracing or a physical fixing at either end.

Flying shores are mainly used within excavations as they can be used in conjunction with hydraulic cylinders to provide a very secure system of temporary support to prevent excavation collapse. A cylindrical propping system is often used as this is the most efficient structural geometrical shape. Several props may be required to span a basement excavation, and careful planning will be required to lower machinery and plant between the network of props.

The use of a specialist scaffolding contractor enables the design and erection of flying shores using standard scaffolding components. They must be designed to take the lateral loads imposed on them by the structures either side that they have to support.

FAÇADE SUPPORT

Many historical listed buildings that have outlived their design life and require refurbishment can retain their façade elevations and have the internal structure of the building demolished entirely and rebuilt to modern methods and specifications: for example, Arsenal's Highbury football ground façade was retained and the stadium demolished and converted into flats as a commercial enterprise.

In order to accomplish this the structure and finishes of the external façade will need to be temporarily supported during the demolition and reconstruction operations. There are a number of ways of achieving this, depending on the severity of the lateral loadings that will need to be restrained and supported. In order to accommodate the overturning nature of the façade wall it is necessary

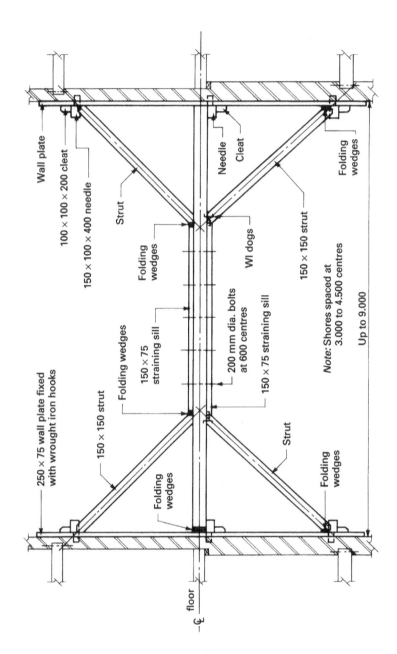

Wall plate

100 × 100 × 200 cleat

150 × 100 × 400 needle

Strut

Needle

Cleat

Folding wedges

Folding wedges

WI dogs

150 × 150 strut

Folding wedges

150 × 75 straining sill

200 mm dia. bolts at 600 centres

150 × 75 straining sill

Note: Shores spaced at 3.000 to 4.500 centres

Up to 9.000

250 × 75 wall plate fixed with wrought iron hooks

150 × 150 strut

Folding wedges

Strut

Folding wedges

₵ floor

Figure 1.5.8 Typical single flying shore

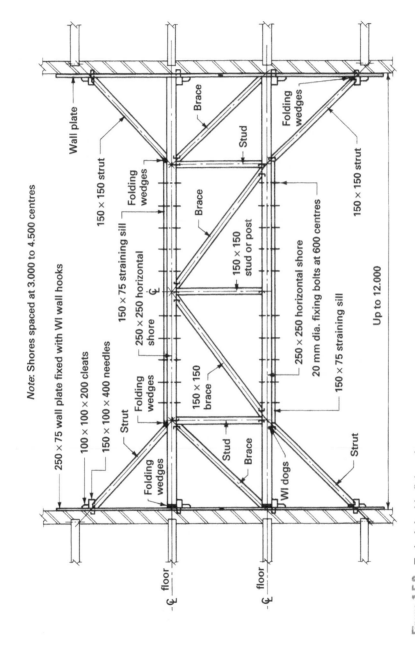

Note: Shores spaced at 3.000 to 4.500 centres

Wall plate

Brace

Folding wedges

Stud

150 × 150 strut

150 × 150 strut

Folding wedges

150 × 75 straining sill

Brace

250 × 250 horizontal shore

150 × 150 stud or post

150 × 250 horizontal shore

20 mm dia. fixing bolts at 600 centres

250 × 75 wall plate fixed with WI wall hooks

100 × 100 × 200 cleats

150 × 100 × 400 needles

Strut

Folding wedges

150 × 150 brace

150 × 75 straining sill

Folding wedges

Stud

Brace

WI dogs

Strut

Up to 12.000

₵ floor

₵ floor

Figure 1.5.9 Typical double flying shore

to increase the width of the supporting structure, such that it can brace against the forces acting on it. Often, additional weight in the form of ballast will need to be added to the supporting structure to balance the weight of the façade wall. See Fig. 1.5.10, which illustrates a light-duty bracing scaffold system and a heavy-duty soldier beam system.

Where openings exist in the façade wall, the fitting of horizontal wallers through an opening will secure the structure to the supporting brace system. In lightly supported structures often a drilled fixing is used to attach the scaffold structure to the wall.

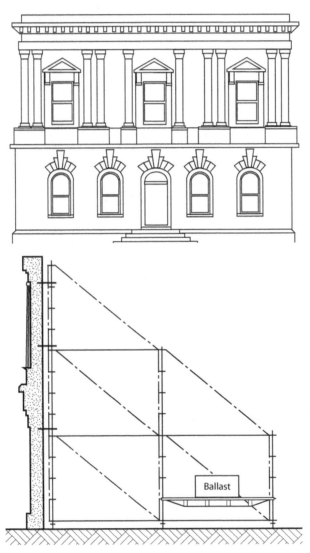

Figure 1.5.10   A light-duty bracing scaffold system and a heavy-duty soldier beam system

In many cases the external façade is adjacent to a pedestrian footpath, on which the access of the general public must be considered. Figure 1.5.11 illustrates a portal beam arrangement, which spans the pedestrian footpath and does not stop access or movement while construction work can commence. This system approach can also be undertaken internally within the boundary of the building, but consideration must be given to the new construction and the removal of the façade supports.

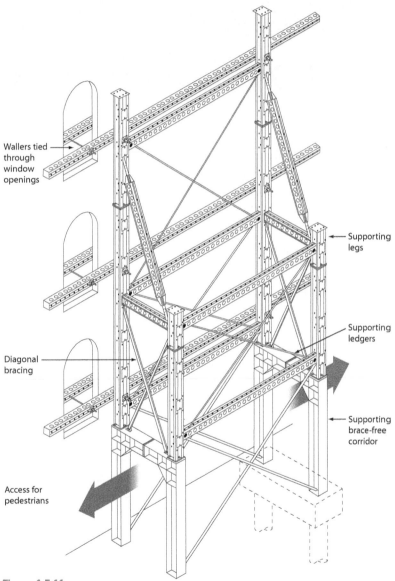

Wallers tied through window openings

Supporting legs

Supporting ledgers

Diagonal bracing

Supporting brace-free corridor

Access for pedestrians

Figure 1.5.11   'Heavy duty' beam supports to a façade

Demolition   1.6

Demolition can be a very dangerous operation to undertake on the pre-contract enabling works; unless the work is of a very small nature, it is best left to a specialist demolition contractor. The National Federation of Demolition Contractors (NFDC) can provide a list of members in the location of the work who are suitably competent and are approved members of the Association. Demolition of a building or structure can be considered under two headings:

■ partial demolition of a structure;

■ complete removal of a structure by demolition.

Before any taking down or demolition is commenced, it is usual to remove carefully all recyclable items such as the valuable metals copper and lead, steel fittings, domestic fittings, windows, doors and frames (for more information on salvaging, see page 68).

Taking down requires a thorough knowledge of building construction and design, so that loadbearing members and walls can be correctly identified and adequately supported by struts, props and suitable shoring. Most partial demolition works will need to be carried out manually using hand tools, such as electric or air-powered breakers. Often a power saw is used to isolate the portion to be removed by cutting into the structure to provide a clean edge. These operations usually relate to the removal of small parts of a building, such as brickwork to form a new structural opening, removal of loadbearing internal walls or roofwork alterations to create dormer windows or loft conversions.

PRELIMINARY CONSIDERATIONS

SURVEY

Before any works of demolition are started, a detailed survey and examination should be made of the building or structure, its surroundings and any boundaries. Photographs of any existing defects on adjacent properties should be taken, witnessed and stored in a safe place in order to provide evidence in any potential insurance claim by an adjacent neighbour. The relationship and condition of adjoining properties that may be affected by the demolition should also be considered and noted, taking into account the existence of any easements, wayleaves, party walls, rights of way and boundary walls. The use of photographic or video evidence of the site and any adjacent buildings or property before any work commences should be undertaken and suitably recorded with a date and time.

- **Roofs and framed structures** Check whether proposed order of demolition will cause unbalanced overturning moments to occur.

- **Walls** Check whether these are loadbearing, party or cross-walls. Examine the condition and thickness of walls to be demolished and of those to be retained.

- **Basements** Careful examination is required to determine whether these extend under public footpaths or beyond boundary of site, and whether a party wall rests on the basement supporting structure.

- **Cantilevers** Check the nature of support to balconies, heavy cornices and stairs.

- **Services** These may have to be isolated, sealed off, protected or removed, and could include any or all of the following:
  - drainage runs;
  - electricity cables;
  - gas mains and service pipes;
  - water mains and service pipes;
  - telephone cables above and below ground level;
  - radio and television relay cables;
  - district heating mains.

All of the service authorities have record location maps of the type, depth and location of the primary service runs. These can be obtained by writing to each service authority (for example, Transco for gas services) and requesting the location map for your development. A careful survey of the whole site is advisable to ensure that any hazardous, flammable or explosive materials, such as oil drums and gas cylinders, are removed before the demolition work commences. The possibility of contamination of the soils at depth or on the surface must also be considered with removal of demolished foundations and materials. Testing would provide information on the type of contaminant and how it must be dealt with, which may require disposal to specialist tips off site. If the method of construction of the existing structure is at all uncertain, all available drawings should be carefully studied and analysed; alternatively, a detailed survey of the building should be conducted under the guidance of an experienced surveyor.

INSURANCE

Insurance companies and their underwriters will regard demolition work as particularly hazardous, not only to the operatives on site but also to the public and adjacent neighbours. Any aspect of demolition is normally contracted out to a specialist demolition contractor, as it may not be included under a general contractor's insurance policy. Even minor work such as removals may not be included. Therefore, where demolition or removals form part of a work contract, general contractors should ensure that adequate insurance is in place to cover all risks, including claims from operatives and other parties to the work. Insurance will also be required for any other third-party risk, including cover for any claim for loss or damage to property, business, public utilities and roads and paving maintained by the local authority. It is often good practice to invite the insurance assessor to view the site before tendering in order to ascertain any additional insurance costs that would need adding into the estimate.

RECYCLING AND RECLAMATION

With the introduction of sustainable methods of construction comes the issue of the disposal of demolition waste and materials from the site clearance. Often these have a value, as they may be metals, hardcore or a material that can be recycled and used as a building material. A substantial amount of a demolished building can be recycled so that a limited amount of waste is taken to landfill. Landfill is now an expensive luxury as the Government levies a landfill tax on the volume that is disposed of in this way. There are a number of materials that can be salvaged and reused. For example:

- roofing slates can be redressed and reused to match existing weathered slate;
- mortar can be cleaned from bricks and the bricks can be relaid to match within conservation areas;
- roof and floor joists can be reused;
- concrete can be crushed and used as hardcore filling;
- any metal can be melted down and reused within the production process;
- glass is broken down into cullet for introduction back into the glass-making process;
- plasterboard can be broken down and the gypsum element reused in manufacturing;
- miscellaneous pieces of timber can be processed into sheet materials such as chipboard or MDF.

With all of these recycling options, there must be a market for them to be sold to. The growth of architectural salvage is now a significant business, responsible for the preservation and reuse of much of the featurework from our old buildings. Feature roof tiles and components, stone lintels, sills and chimney features all have a resale value for clients looking to renovate or replace a damaged feature of their property. Where a market does not exist, then several options may be available to reuse the demolition materials, such as:

- the use of gabion cages – steel cages within which bricks can be placed on the outer faces and the inside filled with demolition rubble. When they are stacked an attractive retaining wall feature is formed;

- the use of an on-site crusher facility that can reduce the demolition rubble down to a graded material that can be sold off site or used as filling on-site and compacted in layers;

- the landscaping of excavated materials into earth mound bunds, which can be grassed and used as a barrier to traffic noise, as a fence or as a boundary feature.

Since 2008 a new set of regulations has been in force regarding the management of construction waste. The Site Waste Management Plan Regulations 2008 cover new build, maintenance, alteration, site clearance and demolition. A requirement of the Regulations is that, for any contract over £300,000, a site waste management plan (SWMP) must be produced. The SWMP must list each party to the contract and the location of the site, and must describe each waste type expected during the course of the project. The action to be taken for each type of waste must be identified. The Regulations were brought in to prevent the occurrence of 'fly tipping' (the illegal dumping of waste) and to protect the environment with correct disposal of waste. A duty of care exists with these Regulations and there must be a provable audit trail from the source of the waste to its eventual disposal.

HOARDINGS

A site survey will determine whether the location justifies special protection for adjoining properties, public highways or other places used by the general public. This may mean the installation of higher hoardings or fan hoardings. Consultation with the local authority will be required to determine their requirements for temporary works, especially the health and safety aspects. There may be requests for footpaths and temporary road closures. The local authority usually requires a formal application, licensing fees and possibly financial deposits against any damages. Some examples of standard boarded and fan hoardings are included under the section on security and protection in Chapter 2.2 of *Construction Technology*.

ASBESTOS SURVEY

Asbestos is a hazard that is present within many of our older buildings. Great care and attention must be taken with the alteration, refurbishment and demolition of such structures as the identification of the presence of asbestos is an essential legal requirement. The precautions that must be taken are understandable as inhaled asbestos fibres may be carcinogenic, in time manifesting as forms of cancer of the lungs known as asbestosis and mesothelioma.

Much of our existing building stock contains large quantities of asbestos. It was originally used in ignorance of its latent effect on personal health, often sprayed onto steelwork as fire protection, used as a pipe insulation and in sheet form as a cladding or lining. Therefore, before altering or demolishing buildings, it is essential that an asbestos appraisal survey is undertaken to assess whether there is a risk to operatives. The Health and Safety at Work etc. Act is the principal statute, under which the Control of Asbestos Regulations 2006 is the current edition.

These regulations cover the management of asbestos on a property other than a dwelling, including:

- the position of the 'duty holder', who is responsible for the control within the premises;
- the identification of asbestos;
- risk assessment of work which may expose employees to asbestos;
- the production of a method statement for planning of the asbestos work;
- the licensing arrangements for working with asbestos;
- notification procedures to the enforcing authority;
- the provision of information, instruction and training on asbestos;
- procedures to prevent exposure to asbestos;
- the use and maintenance of control measures;
- the cleaning and provision of PPE;
- arrangements to deal with any unplanned asbestos release and the reduction of its spread;
- the use of designated areas;
- the provision of air monitoring, testing and site clearance certificates;
- health surveillance;
- storage and labelling of asbestos waste.

These regulations are extensive in the measures that are used to manage and control the removal of asbestos. The consequences of not undertaking this could include enforcement action by the HSE and the shutting down of the project until corrective action has been taken.

HSE Guide HSG264 *Asbestos: The Survey Guide* has been produced to offer specific guidance regarding the management of asbestos, including the new types of asbestos survey that must now be carried out. The following illustrates the surveys and what must be undertaken.

MANAGEMENT SURVEYS

These are surveys that must be undertaken for the normal use and occupancy of a building. Their purpose is to locate the presence and extent of any suspect asbestos containing materials (ACMs) in the building, which could be damaged or disturbed during normal occupancy.

It may be necessary as part of the survey to disturb, in the process of identification, the ACMs; the survey should also contain a report on their condition.

The survey will usually involve sampling and analysis to confirm the presence or absence of ACMs.

REFURBISHMENT AND DEMOLITION SURVEYS

This must be undertaken before any refurbishment or demolition work is carried out. This type of survey is used to locate and describe all ACMs in the area where the refurbishment work will take place, or in the whole building if demolition is planned.

The survey should fully explore and often involve destructive inspection, as necessary, to gain access to all areas, including those that may be difficult to reach. All ACMs will then need to be removed and disposed of safely before any demolition work can proceed. Air testing and monitoring will be required and the building will need to be certified before demolition begins.

Note: The Personal Protective Equipment Regulations 1992 and the Control of Substances Hazardous to Health Regulations 2002 require employers to ensure that their employees are appropriately dressed and protected when working with or in the presence of materials posing a potential health risk. For asbestos surveys and removals, dress requirements will necessitate total isolation by means of disposable outer protective clothing and face-fitted and approved respiratory equipment.

STATUTORY NOTICES

There are some statutory notices that have to be served before the undertaking of any demolition work.

- **The Building Act 1984** The demolition of a building must be notified by the person who intends to carry out the demolition to the local authority Building Control office.

- **The Party Wall Act 1996** Procedures have to be exercised with regard to any work on a wall that is deemed a party wall under this legislation.

- **Town and Country Planning Acts** Buildings that are within a conservation area or are subject to graded listing will be subject to the local authority planning regulations.

- **The HSE** if the work becomes notifiable under the CDM Regulations 2007.

Before commencing any demolition work, the building owner or their agent must notify the public utilities companies, including gas, electricity, water and drainage authorities, the telecommunications services, and the associated companies responsible for other installations, such as radio and cable television relay lines. Note that it is the contractor's responsibility to ensure that all services and other installations have been rendered safe or removed by the authority or company concerned.

SUPERVISION AND SAFETY

Regulation 29 of the Construction Design and Management Regulations 2007 states:

> 'The demolition or dismantling of a structure, or part of a structure, shall be planned and carried out in such a manner as to prevent danger or, where it is not practicable to prevent it, to reduce danger to as low a level as is reasonably practicable. The arrangements for carrying out such demolition or dismantling shall be recorded in writing before the demolition or dismantling work begins.'

This record is a document called a health and safety plan, which must be in place before the work commences.

FURTHER PLANNING AND RISK CONSIDERATIONS

- Method of construction of the building and the structural interdependence and integrity of the supporting elements of the construction.

- Type of materials used, their strengths and weaknesses, their potential as a health and safety hazard (see details of asbestos surveys on page 70) and the possibility for reclamation and reuse (see information on recycling on pages 68–69).

- Whether the demolition is full or partial. Occasionally, façades are retained as part of a preservation order or building listing, and here temporary supports may be required (see Chapter 1.5 on shoring).

- Neighbour considerations. This may include measures to prevent dust and debris spilling onto a public thoroughfare and onto the land of adjacent buildings. Means for protection to be considered: for example, dust sheeting and hoardings.

- Exposure of contaminants. In addition to hazardous materials used in the construction process, there may be contamination of the land from previous occupancies. Consider extraction of soil samples for laboratory analysis.

- Location and isolation of all existing services to the site and informing relevant supplier, including any protection measures if connections are to be reused.

- Local planning authority and Health and Safety Executive requirements and restrictions on hours and days of work. Also, determine acceptable noise and dust levels.

- Handling and disposal methods for debris (see information on recycling, rubble chutes and skips on pages 68–69 and 77–80).

METHODS OF DEMOLITION

There are several methods of demolition, with the choice usually determined by a number of factors:

- **type of structure** – for example, multi-storey framed structure, reinforced concrete chimney;

- **type of construction** – for example, masonry wall, pre-stressed concrete, structural steelwork;

- **location of site** – for example, the proximity of adjacent properties that need protection;

- **time period** – for example, the programmed duration of the demolition works may determine the method to be used.

Each site is unique, and must be assessed with particular regard to selection of demolition technique. Methods vary, and the following procedures below are included for general guidance.

HAND DEMOLITION

The use of manual handling to demolish a structure should only be considered where small volumes of demolition are required. This is with regard to the Control of Vibration at Work Regulations 2005. The amount of time that an operative spends using a mechanical breaker must be controlled and carefully managed to avoid vibration white finger and other hand injuries. Hand demolition involves the progressive demolition of a structure by operatives using hand-held tools; lifting appliances may be used to hoist and lower members or materials once they have been released. Before hand demolition, other methods involving the use of mechanical plant should be considered: the risks associated with using, for example, a mini excavator with a demolition pick fitted are less than those involved in direct hand demolition. Debris should not be allowed to fall freely unless this has been risk-assessed; the use of rubbish chutes with attached skips will reduce the risk of any impact on a third party by falling debris.

HIGH-REACH DEMOLITION

The development of hydraulic technology and powerful, larger excavators enables longer booms to be fitted to a machine, giving a reach of between 15 and 30 metres in height. Coupled with the various machine attachments, this enables a multi-storey structure to be taken down safely while the operator remains protected in an air-conditioned cab.

Figure 1.6.1 illustrates such a machine, which has two booms within its arm; other manufacturers use three, enabling greater distances to be reached in excess of 30 metres. The cab of some machines tilts upwards to enable the operator to see the work area while remaining in a comfortable position.

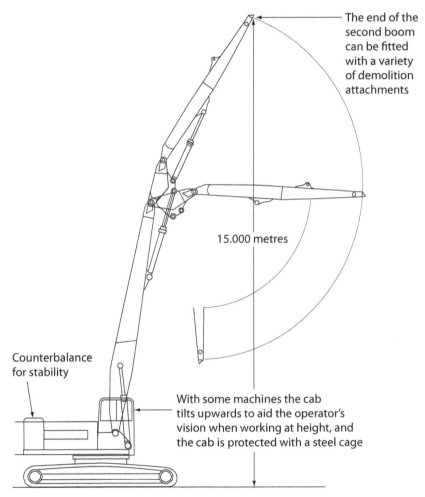

The end of the second boom can be fitted with a variety of demolition attachments

15.000 metres

Counterbalance for stability

With some machines the cab tilts upwards to aid the operator's vision when working at height, and the cab is protected with a steel cage

Figure 1.6.1   High-reach booms fitted to excavator for demolition

Through the use of diesel-powered hydraulics, the operator can safely, powerfully and accurately bring down a structure within a confined area. Dust suppression systems (mainly water sprays) must be used to avoid overspill onto adjacent neighbours.

The end of the boom can be fitted with a number of attachments:

- hydraulic toothed crusher;
- hydraulic shears;
- demolition grab.

Figure 1.6.2 illustrates these attachments, which can be fitted to demolition plant booms to aid the rapid removal of a structure. The first is a demolition crusher, which has high-strength steel teeth capable of crushing concrete by using hydraulic pressure to split the concrete structure apart, using the weaker tension forces within the concrete. The second is a set of demolition shears,

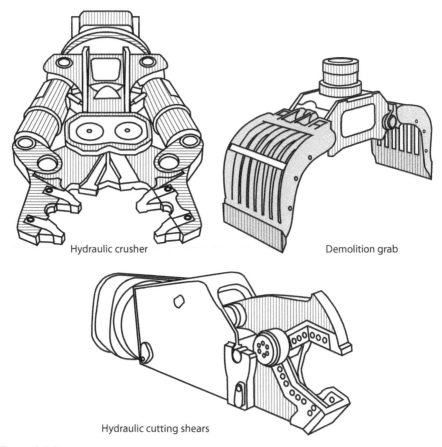

Hydraulic crusher

Demolition grab

Hydraulic cutting shears

Figure 1.6.2   Attachments to excavators for demolition purposes

which have carbon-steel jaws fitted that work by a shearing action against each edge, enabling structural steel components and sections to be snipped in half by a scissor action. The third attachment is a demolition grab, which is used at a later stage for the sorting and disposal of the demolished structure.

DELIBERATE COLLAPSE DEMOLITION

This method involves the removal of key structural members, causing complete collapse of the whole or part of the building. Expert engineering advice should be obtained before this method is used. It should be used only on detached isolated buildings on reasonably level sites so that the safety to personnel (including any machine operator) can be carefully controlled. A complete method statement must be obtained, along with a structural engineer's evaluation of the safest method of undertaking this work. A wide exclusion zone will be required around the building while undertaking this method of removal. It is a better method to use for single- or double-storey demolition projects and not multi-storey ones, where the zone of collapse may be uncertain and cannot be estimated accurately.

DEMOLITION BALL TECHNIQUES

Due to potential difficulties of control over the swinging and slewing of a heavy weight, this method is restricted under UK safety guidance. The issue is the control of the heavy wrecking ball and the uncertainty of the flying debris after a strike. Where it is acceptable, the following techniques could be used:

- vertical drop;
- swinging in line with the jib;
- slewing jib.

Whichever method is used, a skilled operator is essential.

The use of a demolition ball from a normal-duty mobile crane should be confined to the vertical drop technique only. A heavy-duty machine such as a convertible dragline excavator should be used for the other techniques, but in all cases an anti-spin device should be attached to the hoist rope. It is advisable to reduce the length of the crane jib as the demolition work proceeds, but at no time should the jib head be less than 3.000 m above the part of the building being demolished.

Pitched roofs should be removed by hand demolition down to wall-plate level and 50–70 per cent of the internal flooring should be removed to allow for the free fall of debris within the building enclosure. Demolition should then proceed progressively, storey by storey.

Demolition ball techniques should not be used on buildings over 30.000 m high, because the fall of debris is uncontrollable. Attached buildings should be separated from the adjoining structure by hand demolition to leave a space of at least 6.000 m, or half the height of the building, whichever is the greater; a similar clear space is required around the perimeter of the building to give the machine operating space.

DEMOLITION BY EXPLOSIVES

This is a specialist method where charges of explosives are placed within the fabric of the structure and detonated to cause partial or complete collapse. A specialist subcontractor must be used for this method of demolition as the control of high-acceleration explosives is strictly regulated and legislated. Such techniques require a high degree of planning, preparation and management to ensure a successful and safe blowdown of a structure.

OTHER METHODS

Where site conditions are not suitable for the use of explosives, other specialist methods can be considered.

- **Hydraulic hammer** This consists of a high-strength steel point contained within a hydraulic hammer. When the operator starts the breaker, a piston action is exerted on the breaker point, which is driven into the structure, forcing it apart.

- **Thermic lance** This method involves a steel tube, sometimes packed with steel rods, through which oxygen is passed. The tip of the lance is preheated by conventional means to melting point (approximately 1,000 °C) when the supply of oxygen is introduced. This sets up a thermochemical reaction, giving a temperature of around 3,500 °C at the reaction end, which will melt all the materials normally encountered, causing very little damage to surrounding materials. This type of demolition is ideal for cutting through steelwork efficiently where a machine is unable to reach.

The dangers and risks encountered with any demolition works cannot be overemphasised, and all contractors should seek the advice of and employ specialist demolition contractors to carry out all but the simplest demolition tasks.

See also BS 6187: *Code of practice for demolition.*

RUBBLE CHUTES AND SKIPS

RUBBLE CHUTES

A rubble chute consists of a series of plastic or rubber open-ended tubes that are connected together using nylon rope or chains. Once erected in a string, they effectively enclose any rubbish or waste materials that are dropped down them. A skip with a containing hood is placed at the bottom of the chute to contain any overspill or dust. The method is efficient in that it relies on gravity to discharge the waste materials, but dust suppression must be considered around the skip. Care should be taken with the size of materials placed within the chute.

Proprietary chute units typically taper from 510 mm diameter down to 380 mm diameter and have internal ribbing to resist wear and abrasion. Each unit is fitted with two uppermost brackets on opposing sides for chain link retention and fixing of adjacent sections. Figure 1.6.3 shows a typical standard unit with variations having a reinforcing ring (use every sixth section), side entry and hopper attachments. Figure 1.6.4 shows the application to a building, with material discharge to a skip placed at ground level.

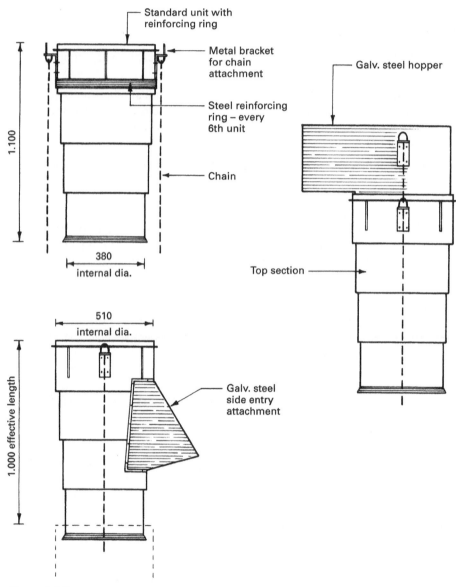

Figure 1.6.3 Typical rubble chute details

SKIPS OR DUMPSTERS

Skips are manufactured from mild-steel sheet with welded box sections and angles to provide ribbed reinforcement and rigidity. They are produced in a wide range of sizes as rubbish and demolition material receptacles to suit all applications, from minor domestic work to major demolition contracts. Skip capacities can be categorised as:

- mini: 1.5 to 2.5 m^3
- midi: 3.0 to 4.0 m^3
- builders: 4.5 to 30 m^3.

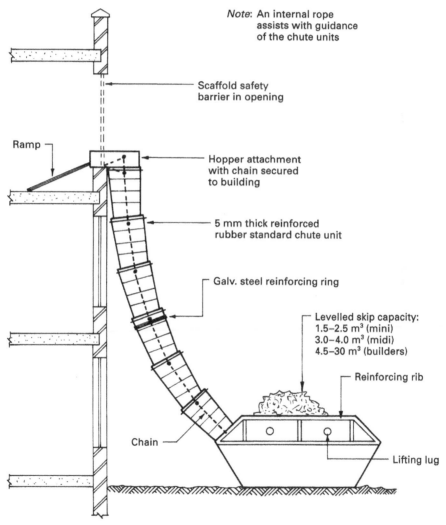

Note: An internal rope assists with guidance of the chute units

Scaffold safety barrier in opening

Ramp

Hopper attachment with chain secured to building

5 mm thick reinforced rubber standard chute unit

Galv. steel reinforcing ring

Levelled skip capacity: 1.5–2.5 m^3 (mini) 3.0–4.0 m^3 (midi) 4.5–30 m^3 (builders)

Reinforcing rib

Chain

Lifting lug

Figure 1.6.4 Rubble chute and skip

Skips are hired out on the basis of a charge for the skip delivered to site, which must be filled and removed within a reasonable time period. The removal charge will reflect the fee levied to the skip contractor for tipping at licensed fill sites, and therefore care should be taken with the type of materials that are placed within the skip.

The local authority highways department and the local police must be consulted if the skip is to occupy all or part of the road or pavement. Lights must be attached to the roadside of any skip left overnight to identify the hazard, and in some cases permission to encroach onto the highway may be required (for example, at road junctions, where a permit may be required along with a fee to the local authority). Provision of temporary traffic control lights may also be necessary.

Contaminated land remediation 1.7

The sustainable reuse of brownfield sites has introduced the problem of how to deal with contaminants associated with the previous use of the land: for example, as a petrol station, gas works or industrial units. The economics of developing on such sites must be balanced with the cost of the remediation of the site under the full weight of the legislation concerned with the process.

The two principal UK regulations that deal with contamination are:

- the Environmental Protection Act 1990 – this is the principal legislation concerning the protection of the environment;
- the Contaminated Land (England) Regulations 2006 – which state that land has to be designated as a special site and remediation proposals notified.

There are several reasons why we have to identify contaminated land and the subsequent decontamination of the land for reuse:

- because the Local Authority Planning Regulations and Conditions make certain requirements regarding development on contaminated land;
- because local authorities have a duty to identify contaminated land and establish the remediation requirements in terms of Part 2A of the Environmental Protection Act 1990;
- on a voluntary basis, due to the environmental policies of a developer or company, when the site has not been identified as contaminated.

Further guidance can be obtained from the SEPA publication *Land Remediation and Waste Management Guidelines*, from which the following subtitles have been taken along with the common methods of dealing with soil contamination.

CONTAMINANTS

The presence of any contamination must be established before any development can commence. Various methods can be employed to establish the presence of any contamination, which are similar to the site investigation techniques, including:

■ desktop survey to establish previous use and industry type;

■ in-situ site testing by taking soil samples for chemical analysis.

The type of contamination will obviously vary according to the industrial processes that were undertaken on the proposed site. Examples of the varied nature of the contamination and its location include:

■ surface dumping of contaminants that have not leached into the soil, such as asbestos sheeting;

■ chemical contaminants that have leached into the surrounding subsoils, such as a ruptured fuel tank on a disused petrol station site that has had an undetected leak for a number of years.

The presence of groundwater in an area of contamination is cause for concern. Water can dissolve the contamination and then move through the subsoil strata. In this way, contamination can migrate well outside the original source area of contamination.

REMEDIATION TECHNIQUES

EXCAVATION AND REMOVAL OF CONTAMINANT

This is the obvious option to undertake. The extent of the contamination will need to be established by observation and testing. Once its chemical composition is known, then the licensed tip that can take the contaminant needs to be identified. Excavation continues until the contaminant is disposed of off-site. The volume removed will then need to be backfilled with suitable materials to bring the site back up to formation level in compacted layers.

There are several variations on the excavation and disposal method:

■ levelling out the site by excavation and covering the contamination with a new hardstanding surface, such as an airport runway or car-parking area, to break the link between the pollutant and any people;

■ using the proposed building and surrounding hardstandings to break the link between the contamination and the occupants, such as by inserting a membrane beneath the foundations;

■ moving contaminated soil under part of the site that is then covered with hardstandings, backfilling the removed soil with suitable materials;

■ raising up site levels with imported fill, breaking any link with the pollutant.

With all of the above methods the presence of groundwater must be carefully considered. This is because the contamination can dissolve within the groundwater, which can move through the soil and cause further problems within the site, should water levels rise or migrate.

CAPPING AND COVER SYSTEMS

This method does not deal with the source or level of contamination: it just ensures that it cannot move from its current position. Suitable membranes and covering systems should be installed in accordance with current guidance so that no contaminant can reach human contact through the capping materials. A break layer is normally provided consisting of granular fill materials covered over the contaminated ground before the geotextile membrane is installed. This also prevents the puncturing of the membrane by any debris within the contaminated layer.

Capping requires the use of a geotextile membrane to contain any pollutants and a clay layer to seal any groundwater within the contaminated zone, as clay is impervious to water in both directions. Consideration must be given to venting any toxic gases from the substrate layers, and the drainage of any groundwater above the clay layer.

The edges of the capping layer must be treated to prevent any movement within the soils strata of any contamination through groundwater movement. This is achieved by the use of slip trenches filled with a drainage material and a vertical membrane barrier. See Figs 1.7.1 and 1.7.2 for details of the capping and drainage systems.

ENCAPSULATION BY SOLIDIFICATION AND STABILISATION

Contaminants can be made immobile using in-situ or ex-situ techniques to immobilise the contamination, preventing it from any migration through the soil structure by the movement of any groundwater.

IN-SITU TECHNIQUES
Slurry walls

This method uses bentonite clay mixed with water, which is placed within an excavated trench taken down to a calculated strata depth where groundwater cannot leach round the slurry wall. When mixed with water, the bentonite forms a semi-solid slurry that is impenetrable to groundwater movement across it. This retains any contamination within groundwater into a designated area. The disadvantage with this method is that no construction work can be

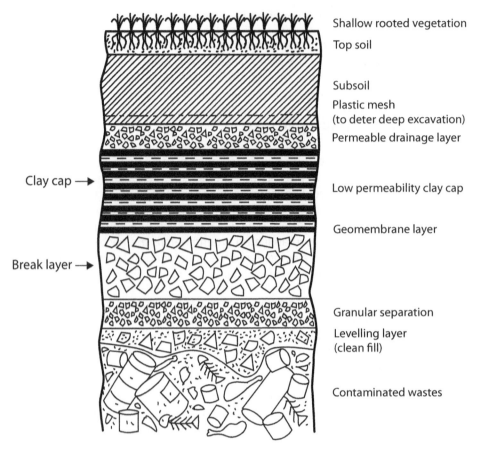

Shallow rooted vegetation

Top soil

Subsoil

Plastic mesh
(to deter deep excavation)

Permeable drainage layer

Clay cap →

Low permeability clay cap

Geomembrane layer

Break layer →

Granular separation

Levelling layer
(clean fill)

Contaminated wastes

Figure 1.7.1 **Capping process for contaminated land**

built onto any of the slurry walls, as they are incapable of supporting any loads. This method could be used outside of the building zone to contain contamination, but would require landscaping over the trenches.

Jet grouting

This process uses a drilling method that forces a grout mixture into the surrounding soil by the use of a hollow drill bit. This is done below the level of contamination and continues until a barrier is built up as each grout drilling combines with the next. When set, the grout forms a permanent barrier to the passage of any groundwater that can carry dissolved contaminants. A large number of drill holes would have to be undertaken to form a horizontal and vertical enclosure of a contaminated area. Systematic details of the system can be seen in Fig. 1.7.3.

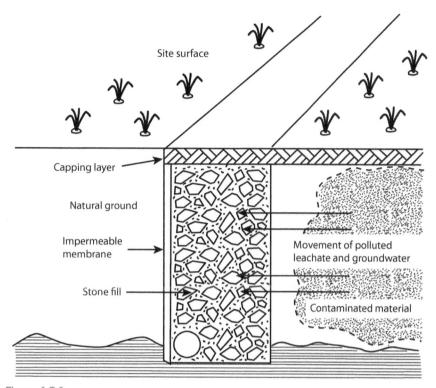

Figure 1.7.2   **Structure of typical perimeter system used with capping layer**

Site surface

Capping layer

Natural ground

Impermeable membrane

Stone fill

Movement of polluted leachate and groundwater

Contaminated material

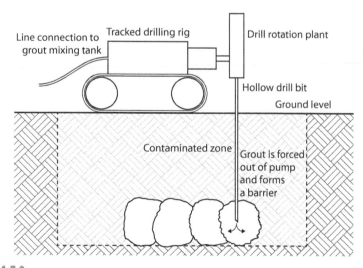

Figure 1.7.3   **Jet grouting to seal in soil contamination**

Line connection to grout mixing tank

Tracked drilling rig

Drill rotation plant

Hollow drill bit

Ground level

Contaminated zone

Grout is forced out of pump and forms a barrier

Cementation

This is the process of mixing cement slurry into a contaminated soil such that, when the cement hydrates and sets over time, the contamination is locked in place and cannot move or migrate through the action of groundwater. The process involves the use of a piling rig fitted with an auger that has mixing paddles on its shaft. The auger is lowered into the ground and cement slurry is pumped down the hollow auger and mixed with the contaminated soil.

EX-SITU TECHNIQUE

Soil washing

This is a method that can be employed with excavation methods in processing contaminated soil through a process to wash out the contaminants, which are then safely disposed of or neutralised. Clean soil is then taken back to the site and compacted in layers. The process involves the removal of larger objects by screening through a rotating screen; these deposits are then disposed to landfill. Chemicals then may be added via high-pressure washing of the soil, which removes and neutralises the contamination. The water is recycled and reused to save resources, and the washed soil and sediment is aerated to actively encourage breakdown of the contamination. The final treated sediment is then taken back to the site and used as filling material.

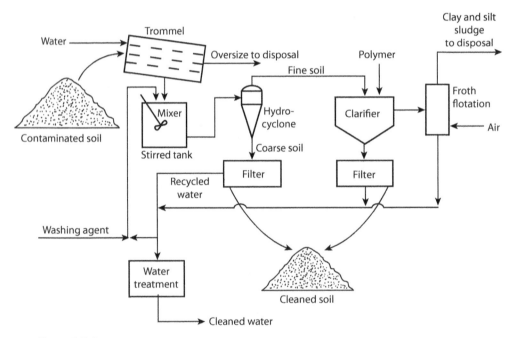

Figure 1.7.4 Soil washing

2

Plant and equipment

Builders' plant 2.1

The general aim of building construction is to produce a structure at reasonable cost to the specified quality and within the agreed time period. To achieve this end the industry has been steadily evolving construction processes that utilise the mechanisation of both off-site and on-site tasks, using sophisticated and efficient construction equipment or plant. This part will concentrate on the broad scope of construction plant and equipment currently used for site-based operations. It will not cover manufacturing equipment used in specialist off-site factory production. For details of manual tools used by individual crafts, the reader is guided to the main trade and craft literature, textbooks and guides that are available elsewhere.

The range of plant now available to construction contractors is extensive, ranging from portable powered hand tools (such as nail guns) through to high-powered mobile machines (such as tracked excavators) to huge, fixed equipment (such as tower cranes). In a text of this nature it is only possible to consider the general classes of plant and their uses; for a full analysis of the many variations, with the different classifications, readers are advised to consult the many textbooks and catalogues devoted entirely to contractors' plant. A good source of information is the Construction Plant-hire Association, for which the website is www.cpa.uk.net. This association was formed in 1941 to represent the plant-hire industry in the United Kingdom, to negotiate general terms and conditions of hiring plant, to give advice and to promote a high standard of efficiency in the services given by its members.

The main reasons for electing to use items of plant are to:

- increase rate of output;
- reduce overall building costs;

- carry out activities that physically cannot be done by manual means;

- reduce the health and safety risks to operatives who would, by using manual methods, be put in harmful situations;

- maintain a planned rate of production where there is a shortage of either skilled or unskilled operatives;

- maintain the high and consistent standards that are required by present-day designs and specifications.

It must not be assumed that the introduction of plant to a contract will always reduce costs. This may well be true with large contracts, but when carrying out small contracts, such as a traditionally built one-off house, it is usually cheaper to carry out the constructional operations by traditional manual methods.

THE SELECTION OF CONSTRUCTION PLANT

The type of plant to be considered for selection will depend on the tasks involved, the time element and the staff available. The person who selects the plant must be competent; the plant operator must be a trained person, to obtain maximum efficiency and for health and safety reasons to ensure safe operation of specialist equipment; the manufacturer's recommended maintenance schedule for the plant must be followed; and, above all, the site layout and organisation must be planned with a knowledge of the capabilities and requirements of the plant.

Furthermore the selection of construction plant relies on:

- technical capacity;

- availability for use;

- financial viability.

TECHNICAL CAPACITY

Technical capacity requires that the plant selected must be 'fit for purpose', in that it can achieve the quality and rate of output needed to meet the client's specification and the contractor's programme.

When deciding on the proposed mechanisation of the tasks and the right plant to choose, a number of factors need to be considered. These will certainly involve:

- the length of the contract programme;

- the amount of work to be done: for example, volume of soil to be excavated for a basement or tonnage of reinforcing steel to be lifted for an r.c. structure;

- the access and egress situations or restrictions: for example, width of entrances or height available under power lines;
- the terrain and topography of the site and the condition and provision of the haul roads: for example, the gradient of the site will affect the speed and traction of vehicles;
- the likely weather conditions: for example, wet clay will foul tyred vehicles and high winds will reduce the number of days a tower crane can operate;
- the proximity to occupied buildings: for example, disruption due to noise, dust and vibration caused by the plant operation.

The soil conditions and modes of access to a site will often influence the choice of plant items that could be considered for a particular task. Congested town sites may severely limit the use of many types of machinery and plant. If the proposed structure occupies the whole of the site, component and material deliveries in modules or prefabricated format will eliminate the need and space for fabrication and assembly plant. Concrete-mixing plant will also be impractical: therefore supplies will be delivered ready mixed. Wet sites usually require plant equipped with caterpillar tracks, whereas dry sites are suitable for tracked and wheeled vehicles or power units. On housing sites, it is common practice to construct the estate road sublayers at an early stage in the contract, to provide firm access routes for mobile plant and hardstanding for static plant, such as cement mixers. The heights and proximity of adjacent structures or buildings may limit the use of a horizontal jib crane and may dictate the use of a crane with a luffing jib – see later.

AVAILABILITY FOR USE

Availability for use is where the identified plant is organised to be available at the right time to meet the demands of the programme. Common types of plant (such as compressors or pumps) are usually procured from the existing plant pool within the company; for specialist equipment or for one-off activities, the plant-hire market is usually very cost-effective; or if the use of the plant is envisaged in further projects, it may be purchased specially for the current project (this is particularly true for companies who do overseas work in less developed countries, where the hire market is small). Thus the choice of buying, hiring or combining a mix of the two for a given construction project needs to be evaluated when deciding what plant to procure.

The advantages of buying plant are:

- plant is available when required;
- the cost of idle time caused by inclement weather, work being behind planned programme or delay in deliveries of materials will generally be less on owned plant than on hired plant;

- builders can apportion the plant costs to the various contracts using the plant, by their own chosen method;
- there is some resale value, although depreciation can seriously affect this, particularly for newly purchased plant.

The advantages of hiring plant are:

- plant can be hired as required and for short periods;
- hire firms are responsible for maintenance, repairs and replacements;
- the contractor is not left with unused expensive plant items after completion of the contract;
- hire rates can include operator, personal protective equipment, fuel and oil.

Within the UK construction industry there has been a gradual increase in the use of specialist plant-hire companies for supplying the majority of fixed and mobile plant or equipment. A number of main contracting companies have established their own commercial plant-hire organisations to benefit from the centralised and planned management of resources and as an income generator in their own right.

FINANCIAL VIABILITY

Financial viability considers how the plant selected must meet the overall financial criteria required by the contractor. With acquisition, the costs of the whole lifecycle of the plant need to be considered so as not to drain the company of funds (for example, loan costs, servicing, hire charges, transport costs, depreciation and so on).

Thus against plant-hire rates and contractual conditions, a contractor will have to compare the cost of buying and owning a similar piece of plant. A comparison of costs can be calculated using either the simple, straight-line method if the capital is available, or by treating the plant as an investment and allowing for loan interest on the capital outlay.

STRAIGHT-LINE METHOD

Capital cost of plant	£45,000.00
Expected useful life	5 years

Yearly working, say 75% of total year's working hours

$$= 50 \text{ weeks} \times 40 \text{ hours} \times \frac{75}{100}$$

$$= 1,500 \text{ hours per year}$$

Assuming a resale value after five years of £7,500.00

Annual depreciation	$= \dfrac{45,000 - 7,500}{5}$	$= £7,500$
Therefore hourly depreciation	$= \dfrac{7,500.00}{1,500.00}$	$= £5.00$
Net cost per hour	$=$	5.00
Add 2% for insurance, etc.	$=$	0.10
Add 10% for maintenance	$=$	0.50
Total cost per hour		£5.60

To the above costs must be added the running costs, which would include fuel, operator's wages and overheads.

INTEREST ON CAPITAL OUTLAY METHOD

Capital cost of plant	£45,000.00
Compound interest on capital @ 6% for 5 years	15,220.00
	£60,220.00
Deduct resale value	7,500.00
	£52,720.00
Add 2% of capital cost for insurance, etc.	900.00
Add 10% of capital cost for maintenance	4,500.00
	£58,120.00
Therefore cost per hour	$= \dfrac{£58,120}{5 \times 1,500} = £7.75$

To the above hourly rate must be added the running costs as given for the straight-line method. This second method gives a more accurate figure for the actual cost of owning an item of plant, and for this reason it is widely used.

To be economic, large items of plant need to be employed continuously and not left idle for considerable periods of time. Careful maintenance of all forms of plant is of the utmost importance: this not only increases the working life of a piece of plant but, if a plant failure occurs on site, it can cause serious delays and disruptions in the programme, and this in turn can affect the company's future planning. To reduce the risk of plant breakdown, a trained and skilled operator should be employed to be responsible for the running, cleaning and daily maintenance of any form of machinery. Time for the machine operator to carry out these tasks must be allowed for in the site programme and the daily work schedules.

For accurate pricing of a bill of quantities, you should consider all plant requirements carefully, and prepare a plant schedule at the pre-tender stage. This should take into account plant types, plant numbers and personnel needed. If the tender is successful, a detailed programme should be prepared in liaison with all those to be concerned in the supervision of the contract, so that the correct sequence of operations is planned and an economic balance of labour and machines is obtained.

MAINTENANCE AND SUPPORT

Whether equipment and plant is supplied by a hire company or is directly owned, full-time skilled mechanics are required to be responsible for running repairs and recommended preventive maintenance. Such tasks would include:

- checking oil levels – daily;
- greasing – daily or after each shift;
- checking engine sump levels – after 100 hours' running time;
- checking gearbox levels – after 1,200 to 1,500 hours' running time;
- checking tyre pressures – daily;
- inspecting chains and ropes – daily;
- checking condition of static equipment, such as formwork and shoring including alignment of locating pins and couplers.

As soon as a particular item of plant has finished its work on site, it should be returned to the company's main plant yard so that it can be reallocated to another contract or, if hired, must be returned to the hire company's depot, to avoid any excess charges that may apply. On its return to the main plant yard or hire depot, an item of plant should be inspected and tested so that any necessary repairs, replacements and maintenance can be carried out before it is re-employed on another site. Also, when returning formwork to a hirer, careful accounting of all the separate pieces needs to be ensured to avoid additional charges. A record of the machine's history should be accurately kept and should accompany the machine wherever it is employed so that the record can be kept up to date.

HEALTH AND SAFETY ISSUES

Apart from the factors previously discussed, you must also consider health and safety when selecting items of plant for use on a particular contract, particularly with regard to two specific issues: noise and vibration.

The aspects of safety that must be legally provided are contained in various Acts of Parliament and Statutory Instruments such as the Health and Safety at Work etc. Act 1974, Control of Pollution Act 1974, Construction (Design and

Management) Regulations 2007, Control of Vibration at Work Regulations 2005 and Noise at Work Regulations 1989. These requirements will be considered in the following sections devoted to various classifications of plant.

Although reaction to noise is basically subjective, excessive noise can damage a person's health and hearing; it can also cause disturbance to working and living environments. Under the Health and Safety at Work etc. Act 1974 and its subsidiary regulations, provision is made for the protection of workers against noise. If the maximum safe daily noise limits for the unprotected human ear are likely to be exceeded, the remedies available are the issue of suitable ear protectors to the workers, housing the plant in a sound-insulated compartment, or the use of quieter plant or processes.

Local authorities have powers under the Control of Pollution Act 1974 to protect the community against noise. Part III of this Act gives the local authority the power to set out its own requirements as to the limit of construction noise acceptable by serving a notice, which may specify:

■ plant or machinery that is, or is not, to be used;

■ hours during which the works may be carried out;

■ level of noise emitted from the premises in question or from any specific point on these premises, or the level of noise that may be emitted during specified hours.

The local authority is also allowed to make provisions in these notices for any change of circumstance.

A contractor will therefore need to know the local authority requirements as to noise restrictions at the pre-tender stage to enable selection of the right plant and/or processes to be employed. Methods for dealing with site noise are given in BS 5228-1:2009 *Noise and vibration control on construction and open sites*. It recommends methods of noise control and provides guidance on methods of predicting and measuring noise and assessing its impact on those exposed to it. It also recommends specific methods of noise control, such as the specification of acoustic barriers, relating to construction and open sites where work construction operations generate significant noise levels. The legislative background to noise control is described and guidance given for the establishment of effective liaison between developers, site operators and local authorities.

The risk of vibration damage to operators of hand-controlled machinery is limited by the requirements of the Control of Vibration at Work Regulations 2005. Several factors combine to define the limits of use, including the extent and degree of exposure and the operating direction or axis of operation. If not properly controlled, Hand Arm Vibration Syndrome (HAVS) is serious and disabling and can cause permanent damage, such as the inability to do fine work and where cold can trigger painful finger blanching attacks. To aid site personnel in preventing HAVS, the Health and Safety Executive publishes a simplified calculator and ready reckoner to establish the maximum time that can be allowed on any one type of vibrating plant. These tools are freely available and published on their website at www.hse.gov.uk/vibration/hav/vibrationcalc.htm. The calculations are based on the equipment manufacturers' measured vibration levels in metres per second squared (m/s^2) and the daily 'trigger' usage time (hours). The exposure action value (EAV) is a daily amount of vibration exposure above which employers are required to take action to control exposure. The greater the exposure level, the greater the risk and the more action employers will need to take to reduce the risk. For hand-arm vibration the current EAV is a daily exposure of 2.5 m/s^2 A(8), which means trigger time as a proportion of an eight-hour working day.

Small powered plant 2.2

A precise definition of builders' small powered plant is not possible, because the term 'small' is relative. For example, it is possible to have small cranes, but these, when compared with a hand-held electric drill, would be considered to be large plant items. Generally, small plant can be considered to be hand-held mobile power tools with a built-in motor and their attendant power sources, such as an air compressor for pneumatic tools, mains electric site voltage, rechargeable batteries or small petrol motors.

Most hand-held power tools are operated by electricity or compressed air, either to rotate the tool or drive it by percussion. Some of these tools are also designed to act as rotary/percussion tools. Generally, the pneumatic tools are used for the heavier work and have the advantage that they will not burn out if a rotary tool stalls under load. However, electrically driven tools are relatively quiet because there is no exhaust noise, and they can be used in confined spaces because there are no exhaust fumes.

ELECTRIC HAND TOOLS

There are many types of portable hand-held electric tools; as they become lighter and more powerful as technology improves, they are becoming increasingly popular. Typical tools of this type are shown in Fig. 2.2.1 and some are described below.

DRILLS

- **Rotary drills** Standard high-speed drills for small holes (< 25 mm diameter) in timber and metals ~ up to 4,000 revolutions per minute (rpm)

- **Percussion drills** Reciprocating hammer drills for holes in masonry and concrete ~ 2,600 rpm and 40,000 blows per minute (bpm)

- **Diamond drills** Slow, durable and robust for cutting cores in hardened concrete up to 100 mm in diameter ~ 1,000 to 2,500 rpm
- **Mixer drills** Fixed with a long-arm stirrer, a heavy-duty high-torque drill for mixing plaster, paints and dry wall compounds ~ 800 rpm
- **Magnetic drills** Used for making high-precision holes in structural steelwork where the drill can be clamped to the steel by a powerful electromagnet ~ clamping force up to 10 kN.

SAWS

Most saws have protecting guards and cut-off switches to ensure the safety of the operator.

- **Reciprocating saws** Its straight, projecting blade cuts with a backwards and forwards motion, and is used for cutting timber-based products and metal, in stud, sheet and pipe form.
- **Jigsaws** Similar operation to reciprocating saw except the blade is at right-angles to direction of motion of saw, which makes it ideal for cutting curves in various types of board material. Can be fitted with a dust extractor to reduce inhalation of harmful particulate material.
- **Circular saws** For all types of timber, plaster or plastic board materials when fitted with a steel blade, or alternatively will cut stone, brick or tile when fitted with an abrasive mineral disc.
- **Alligator saws** Very powerful twin-bladed reciprocating saw that cuts by moving the blades in opposing directions at the same time. Used often for cutting terracotta clay (Poroton®) blocks or aerated concrete blocks.
- **Chain saws** Consist of a short, oblong guide bar around which the cutting chain travels, and is used for tree felling and working on large timber baulks. Due to the fast-moving, exposed blade, this needs considerable skill in handing and the provision of specialist PPE involving Kevlar®-reinforced trousers and jacket as well as gloves and eye protection.

FASTENERS AND FIXING KIT

Fasteners and fixing kits are used to speed up the connection of components and ensure consistency in fixing. Most equipment has soft-handled grips to reduce the vibration for the user.

- **Impact wrench** Used for tightening bolts in steel-frame erection at around 2,200 rpm and 2,700 bpm and can handle bolt diameters up to 20 mm. Some models can achieve a consistent predetermined torque. Often cordless for ease of portability, although most models weigh in the region of 3 kg.

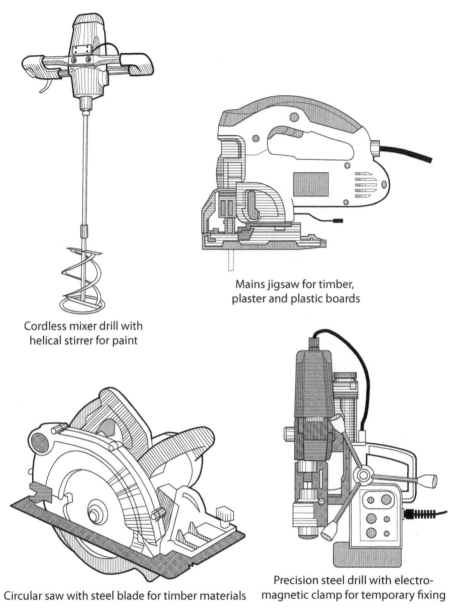

Cordless mixer drill with
helical stirrer for paint

Mains jigsaw for timber,
plaster and plastic boards

Circular saw with steel blade for timber materials

Precision steel drill with electromagnetic clamp for temporary fixing

Figure 2.2.1   Typical portable electric hand tools

- **Screwdriver** Used for securing self-tapping screws when fixing drylining board, claddings or fascias. Often fitted with an automatic depth-sensitive 'set and forget' nosepiece for consistent fastener depth. Also often fitted with a sensitive clutch that will operate only when the screwdriver bit is in contact with the screw head, slipping when a predetermined tension has been reached. Lightweight (~ 1kg) and delivering a consistent pre-determined torque.

- **Stapler** Used to fix sheathing materials in place, such as roofing felt or breather membranes. Generally a cordless device that operates at up to 8 bar pressure and carries a refillable staple cartridge (typically carrying 160–200 staples). Heavier than the screwdriver but still reasonably portable at ~2 kg.

- **Nailer** Similar to the stapler, but with a cartridge for nails. Used for rapid fixing of skirting boards, floorboards and architraves.

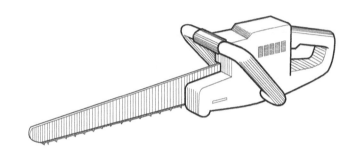

Alligator saw for clay aerated concrete blocks

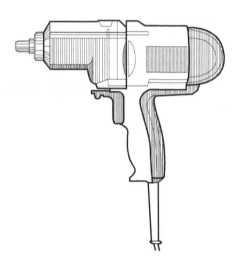

Mains-powered heavy-duty impact wrench with central T-grip for stability

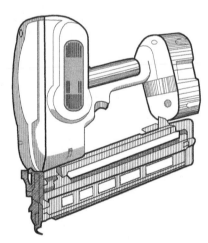

Cordless nail gun with central T-grip for stability

Figure 2.2.1 *Continued*

There has also been considerable development in the use of cordless battery-powered electric hand tools. This is mainly due to the improved battery technology that has developed over the last ten years, which has made cordless hand-held equipment very flexible and productive. Modern cordless tools use lightweight lithium ion (Li-ion) battery packs, which can provide 18 volts and deliver 350 watts of power at the tool interface. Cheaper alternatives are the older nickel cadmium (Ni-Cd) type, which cannot hold their charge for very long, and the nickel metal hydride (Ni-MH) battery packs. The latter are good because they do not exhibit any 'memory effect' (when a battery loses charging capacity after it has been recharged repeatedly), but they still have limited charge-life compared to Li-ion batteries.

Figure 2.2.2 shows the typical features of a mains-powered rotary drill. This is used for boring holes into timber, masonry and metals using twist drills. A wide range of twist-drill capacities is available, with single- or two-speed motors for general-purpose work, the low speed being used for boring into or through timber. Chucks are specified on larger precision drills or diamond drills, where larger holes or cores are required. Dual-purpose electric drills are very versatile: the rotary motion can be combined with or converted into a powerful but rapid percussion motion, making the tool suitable for boring into concrete (provided that special tungsten-carbide-tipped drills are used). Most cordless drills incorporate quick, keyless chuck grips and an LED spotlight for ease of operation and use.

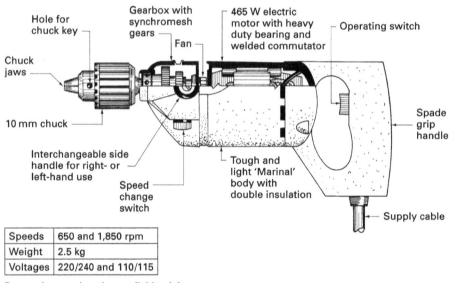

Speeds	650 and 1,850 rpm
Weight	2.5 kg
Voltages	220/240 and 110/115

Percussion version also available giving 18,000 and 31,500 impacts per min.

Figure 2.2.2 Typical corded mains electric drill (courtesy: Stanley–Bridges)

Where electric hand-held tools are mains-corded they should preferably operate off a reduced voltage supply of 110 V and should conform to the recommendations of BS EN 60745: *Hand-held motor-operated electric tools*. Protective guards and any recommended protective clothing, such as goggles and ear protection, should be used as instructed by the manufacturers and as laid down in the Construction (Design and Management) Regulations 2007, the Personal Protective Equipment at Work Regulations 1992 and the Noise at Work Regulations 2005. Electric power tools must never be switched off while under load, because this could cause the motor to become overstrained and burn out. If electrical equipment is being used on a site, the Electricity at Work Regulations and relevant first-aid placards should be displayed in a prominent position.

PNEUMATIC TOOLS

These tools need a supply of compressed air as their power source, and on building sites this is generally in the form of a mobile compressor powered by a diesel, petrol or electric motor, the most common power unit being the diesel engine. Compressors for building works are usually of the piston or reciprocating type, where the air is drawn into a cylinder and compressed by a single stroke of the piston; alternatively two-stage compressors are available where the air is compressed to an intermediate level in a large cylinder before passing to a smaller cylinder, where the final pressure is obtained. Air receivers are usually incorporated with and mounted on the chassis, to provide a constant source of pressure for the air lines and to minimise losses due to pressure fluctuations of the compressor and frictional losses due to the air pulsating through the distribution hoses.

One of the most common pneumatic tools used in building is the **breaker**, which is intended for breaking up hard surfaces such as roads and old foundations. These breakers vary in weight, ranging from 15 to 40 kg with air consumptions of 10 to 20 m^3/min. A variety of breaker points or cutters can be fitted into the end of the breaker tool to tackle different types of surface (Fig. 2.2.3). **Chipping hammers** are a small, lightweight version of the breaker described above, having a low air consumption of 0.5 m^3/min or less. **Backfill tampers** are used to compact the loose spoil returned as backfill in small excavations, and weigh approximately 23 kg, with an air consumption of 10 m^3/min. Compressors to supply air to these tools are usually specified by the number of air hoses that can be attached and by the volume of compressed air that can be delivered per minute. Other equipment that can be operated by compressed air includes vibrators for consolidating concrete, small trench sheeting or sheet pile-driving hammers, concrete-spraying equipment, paint sprayers and hand-held rotary tools, such as drills, grinders and saws.

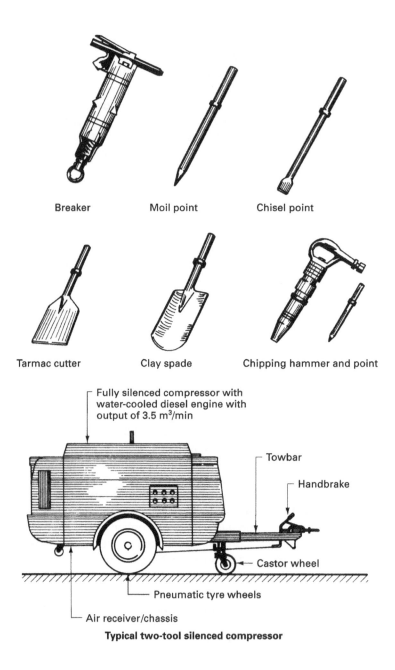

Breaker   Moil point   Chisel point

Tarmac cutter   Clay spade   Chipping hammer and point

Fully silenced compressor with
water-cooled diesel engine with
output of 3.5 m³/min

Towbar

Handbrake

Castor wheel

Pneumatic tyre wheels

Air receiver/chassis

Typical two-tool silenced compressor

Figure 2.2.3   Typical pneumatic tools and compressor

Pneumatic tools can be very noisy. In view of the legal requirements of the Health and Safety at Work etc. Act 1974 and the Control of Pollution Act 1974, these tools are normally fitted with a suitable muffler or silencer. Models are produced with built-in silencers fitted not only on the tool holder, but also on the compressor unit. The mufflers that can be fitted to pneumatic tools such as the breaker have no loss of power effect on the tool, provided the correct muffler is used. There are also issues with regard to user comfort and health with the avoidance of Hand Arm Vibration Syndrome (HAVS) (see the previous section).

CARTRIDGE HAMMERS

Cartridge hammers or guns are used for the quick fixing together of components or for firing into a surface a pin with a threaded head to act as a bolt fixing. The gun actuates a 0.22 cartridge that drives a hardened and tempered steel pin into an unprepared surface. Holding power is basically mechanical, caused by the compression of the material penetrated against the shank of the pin, although with concrete the heat generated by the penetrating force of the pin causes the silicates in the concrete to fuse into glass, giving a chemical bond as well as the mechanical holding power. BS 4078-1:1987 *Powder actuated fixing systems. Code of practice for safe use* specifies the design, construction, safety and performance requirements for cartridge-operated fixing tools, and this standard also defines two basic types of tool:

- **Direct-acting tools**, in which the driving force on the pin comes directly from the compressed gases from the cartridge. These tools are high-velocity guns with high muzzle energy (typical exit velocities are in the range 400–500 m/s).

- **Indirect-acting tools**, in which the driving force is transmitted to the pin by means of an intervening piston with limited axial movement. This common form of cartridge-fixing tool is trigger operated, having a relatively low velocity and muzzle energy. Typical velocities of this type are lower, often less than 100 m/s. A typical example is shown in Fig. 2.2.4.

Hammer-actuated fixing tools working on the piston principle are also available, the tool being hand held and struck with a hammer to fire the cartridge. The cartridges and pins are designed for use with a particular model of gun, and should in no circumstances be interchanged between models or different makes of gun. Cartridges are produced in seven strengths, ranging from extra low to extra high, and for identification purposes are colour-coded in accordance with BS 4078. A low-strength cartridge should be used first for a test firing, gradually increasing the strength until a satisfactory result is obtained. If an over-strength cartridge is used it could cause the pin to pass right through the base material. Fixing pins near to edges can be dangerous because the pin is deflected towards the free edge by following the path of

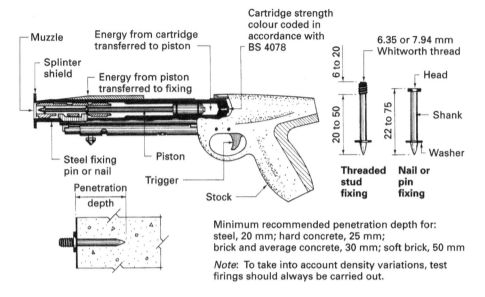

Muzzle

Splinter shield

Energy from cartridge transferred to piston

Energy from piston transferred to fixing

Cartridge strength colour coded in accordance with BS 4078

6 to 20

20 to 50

22 to 75

6.35 or 7.94 mm Whitworth thread

Head

Shank

Washer

Threaded stud fixing

Nail or pin fixing

Steel fixing pin or nail

Piston

Trigger

Stock

Penetration depth

Minimum recommended penetration depth for:
steel, 20 mm; hard concrete, 25 mm;
brick and average concrete, 30 mm; soft brick, 50 mm

Note: To take into account density variations, test firings should always be carried out.

Figure 2.2.4 **Typical cartridge hammer (courtesy: Douglas Kane Group)**

low resistance. Minimum edge distances recommended for concrete are 100 mm using a direct-acting tool and 50 mm using an indirect-acting tool; for fixing into steel the minimum recommended distance is 15 mm using any type of fixing tool.

The safety and maintenance aspects of these fixing tools cannot be overstressed. In particular, consideration must be given to the following points.

- Pins should not be driven into brittle or hard materials such as vitreous-faced bricks, cast iron or marble.

- Pins should not be driven into base materials where there is a danger of the pin passing through.

- Pins should not be fired into existing holes.

- During firing the tool should be held at right angles to the surface with the whole of the splinter guard flush with the surface of the base material. Some tools are designed so that they will not fire if the angle of the axis of the gun to the perpendicular exceeds 7°.

- Operatives under the age of 18 years should not be allowed to use cartridge-operated tools.

- All operators should have a test for colour blindness before being allowed to use these fixing tools.

- No one apart from the operator and an assistant should be in the immediate vicinity of firing to avoid accidents due to ricochet, splintering or re-emergence of pins.

- Operators should receive instructions as to the manufacturer's recommended method of loading, firing and the action to be taken in cases of misfiring.

- Protective items such as goggles and earmuffs are normally supplied by the hire company and should be worn as recommended by the manufacturers.

VIBRATORS

After concrete has been placed it should be consolidated either by hand tamping or by using special vibrators. The power for vibrators can be supplied by a small petrol engine, an electric motor or in some cases by compressed air. Three basic forms of vibrator are used in building works: poker vibrators, vibration tampers and clamp vibrators. **Poker vibrators** are immersed into the wet concrete, and because of their high rate of vibration they induce the concrete to consolidate. The effective radius of a poker vibrator is about 1 m: therefore the poker or pokers should be inserted at approximately 600 mm centres to achieve an overall consolidation of the concrete.

Vibration tampers are small vibrating engines that are fixed to the top of a tamping board for consolidating concrete pavings and slabs. **Clamp vibrators** are similar devices but are attached to the external sides of formwork to vibrate the whole of the form. Care must be taken when using this type of vibrator to ensure that the formwork has sufficient in-built strength to resist the load of the concrete and to withstand the vibrations.

In concrete members that are thin and heavily reinforced, careful vibration will cause the concrete to follow uniformly around the reinforcement, and this increased fluidity due to vibration will occur with mixes that in normal circumstances would be considered too dry for reinforced concrete. Owing to the greater consolidation achieved by vibration, up to 10 per cent more material may be required than with hand-tamped concrete.

Separation of the aggregates can be caused by over-vibrating a mix: therefore vibration should be stopped when the excess water rises to the surface. Vibration of concrete saves time and labour in the placing and consolidating of concrete, but does not always result in a saving in overall costs because of the high formwork costs, extra materials costs and the cost of providing the necessary plant.

POWER FLOATS

Power floats are hand-operated rotary machines powered by a petrol engine or an electric motor that drives the revolving blades or a revolving disc.

The objective of power floats is to produce a smooth, level surface finish to concrete beds and slabs suitable to receive the floor finish without the need for a cement/sand screed. The surface finish that can be obtained is comparable to that achieved by operatives using hand trowels but takes only one-sixth of the time, giving a considerable saving in both time and money. Most power floats can be fitted with either a revolving disc or blade head, and these are generally interchangeable.

The surfacing disc is used for surface planing after the concrete has been vibrated and will erase any transverse tamping line marks left by a vibrator beam as well as filling in any small cavities in the concrete surface. The revolving blades are used after the disc-planing operation to provide the finishing and polishing, which can usually be achieved with two passings. The time at which disc planing can be started is difficult to specify: it depends on factors such as the workability of the concrete, temperature, relative humidity and the weight of the machine to be used. Experience is usually the best judge but, as a guide, if imprints of not more than 2 to 4 mm deep are made in the concrete when walked upon, it is generally suitable for disc planing. If a suitable surface can be produced by a traditional concrete placing method the disc-planing operation prior to rotary blade finishing is often omitted. Generally, blade finishing can be commenced once the surface water has evaporated; a typical power float is shown in Fig. 2.2.5. Power floats can also be used for finishing concrete floors with a granolithic or similar topping.

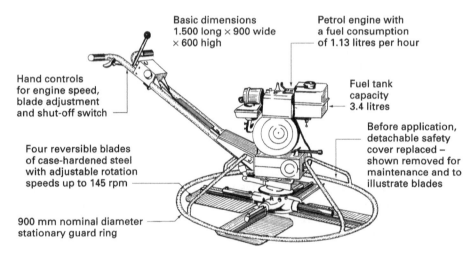

Basic dimensions
1.500 long × 900 wide
× 600 high

Petrol engine with
a fuel consumption
of 1.13 litres per hour

Hand controls
for engine speed,
blade adjustment
and shut-off switch

Fuel tank
capacity
3.4 litres

Four reversible blades
of case-hardened steel
with adjustable rotation
speeds up to 145 rpm

Before application,
detachable safety
cover replaced –
shown removed for
maintenance and to
illustrate blades

900 mm nominal diameter
stationary guard ring

Figure 2.2.5 Typical power float (courtesy: Construction Equipment and Machinery)

Pumps are among the most important items of small plant for the building contractor, because they must be reliable in all conditions, easy to maintain, easily transported and efficient. The basic function of a pump is to move liquids vertically or horizontally, or in a combination of the two directions. Before selecting a pump a builder must consider what task the pump is to perform, and this could be any of the following:

- keeping excavations free from water;

- lowering the water table to a reasonable depth;

- moving large quantities of water such as the dewatering of a cofferdam;

- supplying water for general purposes.

Having defined the task the pump is required to carry out, the next step is to choose a suitable pump, taking into account the following factors:

- volume of water to be moved;

- rate at which water is to be pumped;

- height of pumping, which is the vertical distance from the level of the water to the pump and is usually referred to as **suction lift** or head. It should be the shortest distance practicable to obtain economic pumping;

- height and distance to outfall or discharge point, usually called the **delivery head**;

- loss due to friction in the length of hose or pipe, which increases as the diameter decreases. In many pumps the suction and delivery hoses are marginally larger in diameter than the pump inlet to reduce these frictional losses;

- power source for pump, which can be a petrol engine, diesel engine or electric motor. Pumps powered by compressed air are available, but these are unusual on general building contracts.

Pumps in common use for general building works can be classified under three headings:

- centrifugal;

- displacement;

- submersible.

CENTRIFUGAL PUMPS

These are classed as normal or self-priming. They consist of a rotary impeller that revolves at high speed, forcing the water to the sides of the impeller chamber and thus creating a vortex that sucks air out of the suction hose.

Atmospheric pressure acting on the surface of the water to be pumped causes the water to rise into the pump, initiating the pumping operation. **Normal centrifugal pumps** are easy to maintain, but they require priming with water at the commencement of each pumping operation. Where continuous pumping is required, such as in a basement excavation, a **self-priming pump** should be specified. These pumps have a reserve supply of water in the impeller chamber so that if the pump runs dry the reserve water supply will remain in the chamber to reactivate the pumping sequence if the water level rises in the area being pumped.

DISPLACEMENT PUMPS

These are either reciprocating or diaphragm pumps. **Reciprocating pumps** work by the action of a piston or ram moving within a cylinder. The action of the piston draws water into the cylinder with one stroke and forces it out with the return stroke, resulting in a pulsating delivery. Pumps of this type can have more than one cylinder, forming what is called a duplex (two-cylinder) or triplex (three-cylinder) pump. Some reciprocating pumps draw water into the cylinder in front of the piston and discharge at the rear of the piston: these are called double-acting pumps, as opposed to the single-acting pumps where the water moves in one direction only with the movement of the piston. Although highly efficient and capable of increased capacity with increased engine speed, these pumps have the disadvantage of being unable to handle water containing solids.

Displacement pumps of the **diaphragm** type can, however, handle liquids containing 10–15 per cent of solids, which makes them very popular. They work on the principle of raising and lowering a flexible diaphragm of rubber or rubberised canvas within a cylinder by means of a pump rod connected via a rocker bar to an engine crank. The upward movement of the diaphragm causes water to be sucked into the cylinder through a valve; the downward movement of the diaphragm closes the inlet valve and forces the water out through another valve into the delivery hose. Diaphragm or lift and force pumps are available with two cylinders and two diaphragms, giving greater output and efficiency. Typical pump examples are shown in Fig. 2.2.6.

SUBMERSIBLE PUMPS

These are used for extracting water from deep wells and sumps (see Chapter 3.1), and are suspended in the water to be pumped. The power source is usually an electric motor to drive a centrifugal unit, which is housed in a casing with an annular space to allow the water to rise upwards into the delivery pipe or rising main. Alternatively an electric submersible pump with a diaphragm arrangement can be used where large quantities of water are not involved.

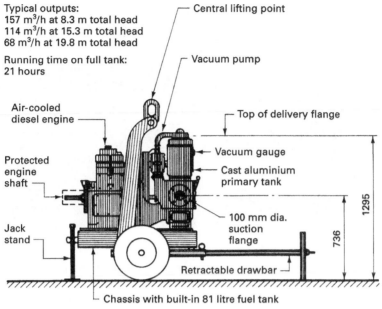

Typical outputs:
157 m³/h at 8.3 m total head
114 m³/h at 15.3 m total head
68 m³/h at 19.8 m total head

Running time on full tank:
21 hours

Central lifting point

Vacuum pump

Air-cooled
diesel engine

Top of delivery flange

Protected
engine
shaft

Vacuum gauge

Cast aluminium
primary tank

Jack
stand

100 mm dia.
suction
flange

1295

736

Retractable drawbar

Chassis with built-in 81 litre fuel tank

Typical self-priming centrifugal pump

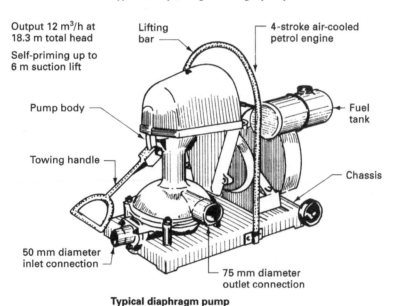

Output 12 m³/h at
18.3 m total head

Self-priming up to
6 m suction lift

Lifting
bar

4-stroke air-cooled
petrol engine

Pump body

Fuel
tank

Towing handle

Chassis

50 mm diameter
inlet connection

75 mm diameter
outlet connection

Typical diaphragm pump

Figure 2.2.6   Typical pumps (courtesy: Sykes Pumps and William R. Selwood)

Earth-moving and excavation plant 2.3

The selection, management and maintenance of builders' plant are particularly important when considered in the context of earth-moving and excavation plant. Before deciding to use any form of plant for these activities, the site conditions and volume of work entailed must be such that it will be an economic venture.

The difference between plant that is classified as earth-moving equipment and excavating machines is very slight, because a piece of plant that is designed primarily to excavate will also be capable of moving the spoil to an attendant transporting vehicle; likewise, machines basically designed to move loose earth will also be capable of carrying out excavation works to some degree.

The choice of excavation plant is usually based on an understanding of the site conditions, type of soil materials present, familiarity with a particular manufacturer's machines, availability or personal preference. There are many good works of reference and manufacturers' catalogues devoted to the analysis of the various machines, to aid the would-be buyer or hirer. In this book we will only consider the general classes of plant, pointing out their intended uses and providing some typical examples of the various types.

COMBINED EXCAVATION AND EARTH-MOVING PLANT

Before any earth-moving work is started, a drawing should be produced indicating the areas and volumes of 'cut' and 'fill' required to enable a programme to be prepared to reduce machine movements to a minimum. For example, in new road or rail contracts, often referred to as permanent way construction, large quantities of earth have to be moved to form a predetermined level or gradient. A useful planning tool for these types of contract is a mass haul diagram indicating the volumes of earth to be moved, the direction of movement, and the need to import more spoil, remove the surplus spoil from site or incorporate it into the landscaping scheme. For a typical example of 'cut' and 'fill' work, see Fig. 2.3.1.

TRACK-TYPE TRACTORS WITH DOZER BLADES

A track-type tractor with dozer blades is primarily a dozer tractor with crawler tracks, fitted with a mould board or blade at the front for stripping and oversite excavations up to a depth of 400 mm (depending on machine specification). In such excavations, the dozer blade pushes the loosened material ahead of the machine. Depending on the blade type it can be fixed to the dozer or fixed at 90° to the direction of travel, while 'angledozers' have the mould board set at an angle, in plan, to the machine's centreline. Most mould boards can be set at an angle in either the vertical or the horizontal plane to act as an angledozer; on some models, the leading cutting edge of the dozer blade can be fitted with teeth for excavating in hard ground. These machines can be very large, with cutting of 1.2 to 4 m in width and 600 mm to 1.2 m in height, with a depth of cut up to 400 mm. Most dozer blades are mounted on crawler tracks, although small bulldozers with a wheeled base are available. The control of the dozer blade is hydraulic, as shown in Fig. 2.3.2. In common with other tracked machines, one of the disadvantages of this arrangement is the need for a special transporting vehicle, such as a low loader, to move the equipment between sites. These machines are very powerful over short distances of up to 150 m for levelling, spreading and grading soil materials. With specialist plough implements fixed to the rear of the machine, called rippers or scarifiers, they can also be used to break up areas of shallow paving in demolition works and for clearing out scrub and root materials on greenfield sites.

GRADERS

Graders are similar machines to bulldozers in that they have an adjustable mould blade, either fitted at the front of the machine or slung under the centre of the machine's body. They are used for finishing – to fine limits, and to the required formation level – large areas of ground that have been scraped or bulldozed, such as in road base or airport runway construction. These machines can be used only to grade the surface, because their low motive power is generally insufficient to enable them to be used for oversite excavation work.

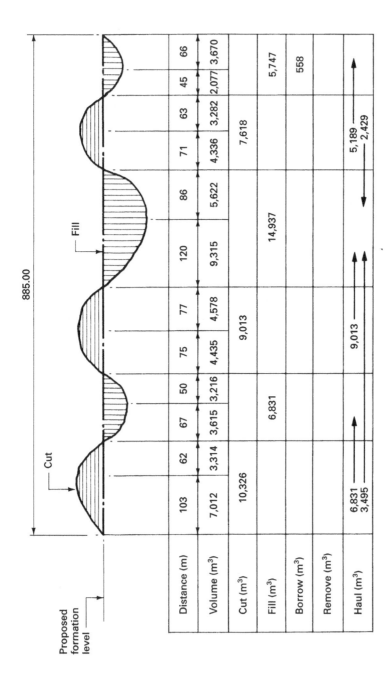

Distance (m)	103	62	67	50	75	77	120	86	71	63	45	66
Volume (m³)	7,012	3,314	3,615	3,216	4,435	4,578	9,315	5,622	4,336	3,282	2,077	3,670
Cut (m³)	10,326			9,013					7,618			
Fill (m³)			6,831				14,937				5,747	
Borrow (m³)											558	
Remove (m³)												
Haul (m³)	6,831 3,495				9,013				5,189 2,429			

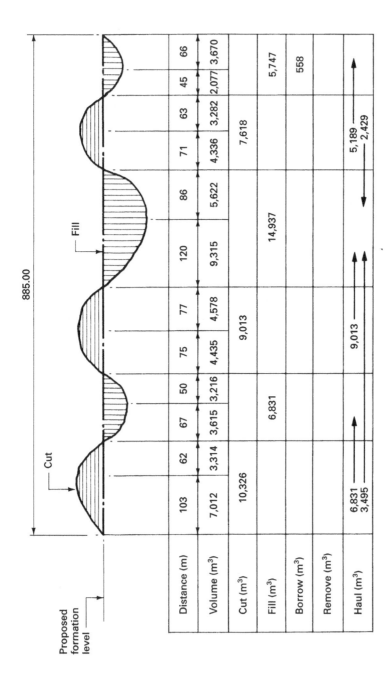

Figure 2.3.1 Typical mass haul diagram

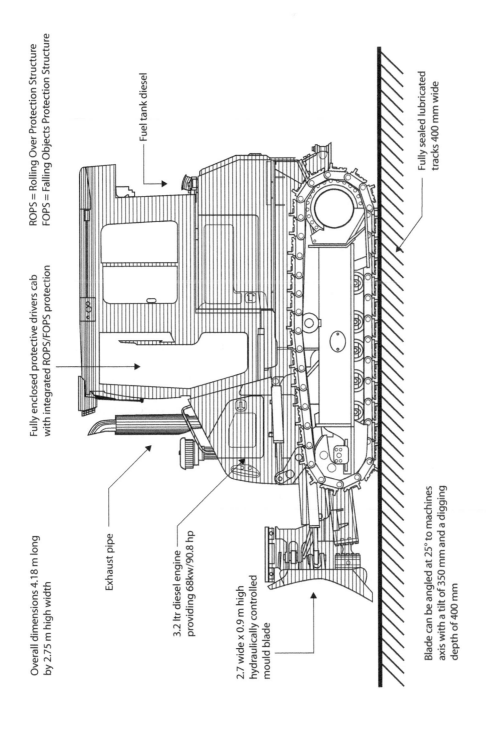

Overall dimensions 4.18 m long
by 2.75 m high width

Fully enclosed protective drivers cab
with integrated ROPS/FOPS protection

ROPS = Rolling Over Protection Structure
FOPS = Falling Objects Protection Structure

Fuel tank diesel

Exhaust pipe

3.2 ltr diesel engine
providing 68kw/90.8 hp

2.7 wide x 0.9 m high
hydraulically controlled
mould blade

Blade can be angled at 25° to machines
axis with a tilt of 350 mm and a digging
depth of 400 mm

Fully sealed lubricated
tracks 400 mm wide

Figure 2.3.2 Typical small size tractor-powered bulldozer details

SCRAPERS

Where projects require excavation and levelling and/or grading of materials up to 3 km, track-type tractors become very inefficient. This type of permanent way construction involves the excavation and transportation of huge volumes of soil, and the most useful piece of contractors' plant to use is the scraper. The scraper consists of a power unit and a scraper bowl. It is a very efficient way of shifting large volumes of soil, and is used to excavate and transport soil where surface stripping, site levelling and cut and fill activities are planned. These earth-moving scrapers are capable of producing a very smooth and accurate formation level, and come in two basic types:

■ a towed scraper;

■ an independent-wheeled tractor-scraper.

The design and basic operation of the scraper bowl is similar in both types. It consists of a shaped bowl with a cutting edge that can be lowered to cut the top surface of the soil up to a depth of 300 mm. As the bowl moves forward, the loosened earth is forced into the container; when full, the cutting edge is raised to seal the bowl. To ensure that a full load is obtained, many contractors use a bulldozer to act as a pusher at the rear. The bowl is emptied by raising the front apron and ejecting the collected spoil with a dozer blade mechanism that raises the rear portion and spreads the collected spoil as the machine moves forwards. To avoid spoil material clogging the cutting edge, some scrapers have an elevator belt or auger fitted within the bowl, which lifts the material off the cutting edge and then dumps it at the back of the bowl. This reduces the rolling resistance, helps eliminate voids in the bowl and improves overall productivity.

The **towed scraper** consists of a wheeled scraper bowl towed behind a tracked tractor or wheeled hauler power unit. The bowl sizes generally range from 5 to 20 m³ with towed scrapers. The speed of operation is governed by the speed of the towing vehicle, which does not normally exceed 8 km/h when hauling and 3 km/h when scraping for a tracked tractor. For this reason this type of towed scraper should be used only on small hauls of up to 300 m. The towed scraper works best with soft clays and silts on flat haul roads and surfaces, where only shallow cuts or light material stripping is required – see Fig. 2.3.3.

The **independent-wheeled tractor-scraper** incorporates a purpose-built power unit, so it has a greater haulage range with a higher top speed, and is generally easier to control. They can deal with a greater variety of soil types, including gravels and some rocks. Some models have two engines working together, one in the front 'tractor' unit and the other in the rear 'scraper' unit; this is known as a tandem machine. They can also be matched with tracked tractor machines to push from the rear if needed to increase traction. The bowl capacities of these types of machine start at around 20 m³ and go up to around 45 m³. Typical examples are shown in Fig. 2.3.3. The torque available

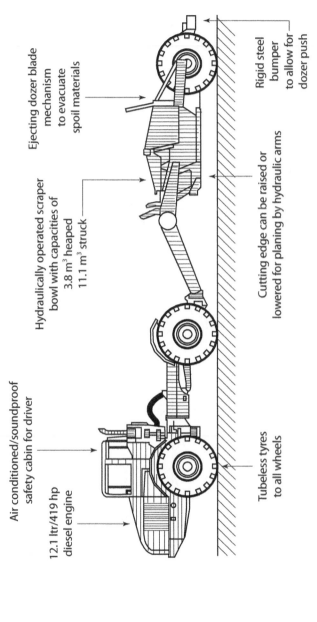

a) Towed scraper

Air conditioned/soundproof
safety cabin for driver

12.1 ltr/419 hp
diesel engine

Hydraulically operated scraper
bowl with capacities of
3.8 m³ heaped
11.1 m³ struck

Ejecting dozer blade
mechanism
to evacuate
spoil materials

Rigid steel
bumper
to allow for
dozer push

Cutting edge can be raised or
lowered for planing by hydraulic arms

Tubeless tyres
to all wheels

Towed scraper
(Volvo T450 D Hauler Unit and Ashland 1-180 T52 Scraper)

Overall size
8.5 m long
3.7 m high

Overall size
9.63 m long
2.5 m high

Figure 2.3.3 Typical towed and independent-wheeled scraper details (courtesy: Volvo and CAT)

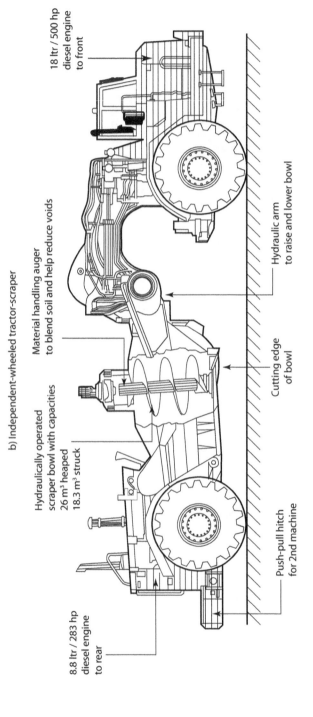

b) Independent-wheeled tractor-scraper

18 ltr / 500 hp diesel engine to front

Material handling auger to blend soil and help reduce voids

Hydraulically operated scraper bowl with capacities 26 m³ heaped 18.3 m³ struck

8.8 ltr / 283 hp diesel engine to rear

Hydraulic arm to raise and lower bowl

Cutting edge of bowl

Push-pull hitch for 2nd machine

CAT 627 Independent Tractor Scraper

(overall size: 14.5 m long 4.28 m high)

Figure 2.3.3 *Continued*

at the interface between the wheels or tracks and the haul surface is known as the 'rimpull'. This is a function of the weight of the loaded machine and the traction available, and is an important factor to compare when deciding on the most appropriate type of scraper to use for a specific project. Another property that is often considered when selecting scrapers or other heavy earth-moving plant is the 'retardation' speed: that is, the maximum safe speed that the brakes can bring under their control in a given situation.

To achieve maximum output and efficiency of scrapers, consider the following.

- When working in hard ground, the surface should be pre-broken by a ripper or scarifier, and a pushing vehicle should assist in cutting. Usually one bulldozer acting as a pusher can assist three scrapers if the cycle of scrape, haul, deposit and return is correctly balanced.

- Where possible, the cutting operation should take place downhill to take full advantage of the weight of the unit.

- Haul roads should be kept smooth to enable the machine to obtain maximum speeds.

- Recommended tyre pressures should be maintained, or extra resistance to forward movement will be encountered.

EXCAVATING MACHINES

There are two basic types of excavating machine: those that are mechanically rigged with steel cables and pullies, and those that are hydraulically fed with oil in sealed pistons. Traditionally mechanically rigged 'universal' machines were used, as they could be rigged with a variety of different booms and buckets to form a variety of excavating configurations. However, with the development of mechanical technology, the more efficient hydraulic controls and systems machines have tended to predominate.

In excavation, the most important factor is the amount of break-out force available at the bucket: this is the amount of digging pull that can be exerted by the machine bucket on the soil being excavated, and is expressed in kilo-Newtons (kN). Most manufacturers provide this information to assist in the correct choice of plant to balance the material to be excavated.

MULTI-PURPOSE EXCAVATORS

Multi-purpose excavators are based on a tyred tractor power unit and are very popular with the small to medium-sized building contractor because of their versatility. The tractor is usually a diesel-powered wheeled vehicle as shown in Fig. 2.3.4, fitted with a hydraulically controlled loading shovel at the front and a hydraulically controlled back-acting bucket or hoe at the rear of the vehicle. The primary function of the **front-loading shovel** is to scoop up loose material in the bucket, raise the loaded spoil and manoeuvre into a

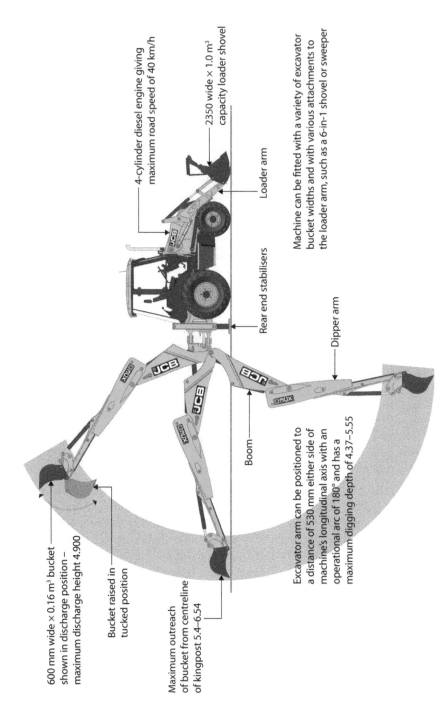

4-cylinder diesel engine giving maximum road speed of 40 km/h

2350 wide × 1.0 m³ capacity loader shovel

Loader arm

Rear end stabilisers

Machine can be fitted with a variety of excavator bucket widths and with various attachments to the loader arm, such as a 6-in-1 shovel or sweeper

Dipper arm

Boom

Excavator arm can be positioned to a distance of 530 mm either side of machine's longitudinal axis with an operational arc of 180° and has a maximum digging depth of 4.37–5.55

600 mm wide × 0.16 m³ bucket shown in discharge position – maximum discharge height 4.900

Bucket raised in tucked position

Maximum outreach of bucket from centreline of kingpost 5.4–6.54

Figure 2.3.4 **Typical backhoe loader details (courtesy: JCB Sales)**

position to discharge its load into an attendant lorry or dumper. The backhoe loader shovel is driven towards the spoil heap with its shovel lowered almost to ground level, and uses its own momentum to force the bucket to bite into the spoil heap, filling the shovel. These tractor-based wheeled machines are highly manoeuvrable, as many models incorporate all-wheel steering. The capacity of this type of shovel ranges from around 1.0 to 1.3 m^3. Another popular version of the backhoe loader shovel is known as the '6-in-1' bucket, which enables the machine to perform the functions of bulldozing, grabbing, grading and loading. Other attachments that can be incorporated or interchanged with the front backhoe loader shovel include forklift attachments and a side-tip shovel for use in areas where space prevents turning the machine.

The rear backacter is similar to the front-loading shovel except that the bucket, dipper arm and boom pivot towards the machine instead of away from it. It also has the capability to slide the backacter to the left or right of the centre of the machine, allowing it excavate on boundary lines and close to existing obstructions. To ensure the stability of the machine, it is essential that the weight of the machine is removed from the axles during a backacting excavation operation. This is achieved by stabiliser legs at the rear of the machine working in conjunction with the bucket at the front of the machine. The backacter is used mainly for digging foundation and service trenches. As the machine excavates, it moves away from the excavation so that the wheels and stabilising jacks do not collapse the newly formed trenches. Once the bucket is full, it can deposit the spoil material to the left or the right, or directly into the back of a waiting lorry or dumper. Standard and deep-profile buckets can be used and are available in widths ranging from 230 mm for service trenches up to 1,100 mm for ditch digging. Deep-profile buckets allow a capacity of around 0.5 m^3 for increased bulk muck shifting.

The backacter can also have a variety of attachments including hydraulic breakers, ground-boring augers and kerb lifters. Many of these attachments can be locked into position by mechanical or hydraulic 'quick-hitch' systems, which speed up attachment changeover times and increase productivity. However, care must be taken to ensure that they are securely fixed with safety pins fully engaged to minimise accidents. The multiple features of these types of machines are illustrated in Fig. 2.3.5.

A common variation of this multi-purpose excavator is the tracked version as shown in Fig. 2.3.6. Although this only has a hydraulic backacter mechanism, it is generally a more powerful machine with a longer reach and a greater digging depth. Also its tracks make it ideal for use in rough or wet terrain. It also can have various types of bucket and other tool attachments.

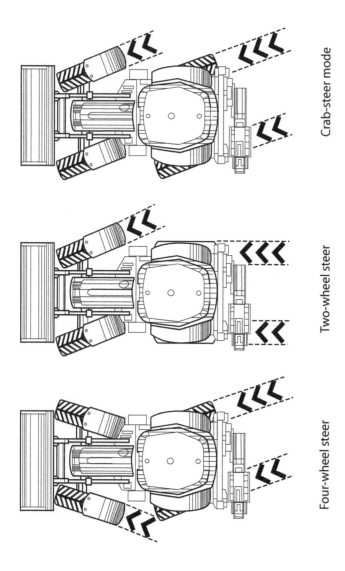

Four-wheel steer

Two-wheel steer

Crab-steer mode

Figure 2.3.5   Features of multi-purpose backhoe loaders

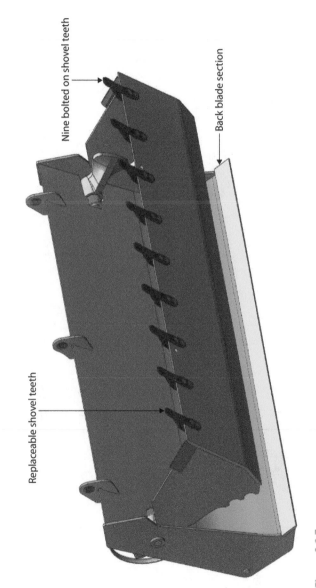

Nine bolted on shovel teeth

Back blade section

Replaceable shovel teeth

Figure 2.3.5 *Continued* Typical 6-in-1 bucket details (courtesy: JCB Sales)

Bucket

Hydraulic breaker

Compactor

Forks fitted to 6-in-1 shovel

Forks

Snow blade

Figure 2.3.5 *Continued* Typical multi-purpose hydraulic attachments (courtesy: JCB Sales)

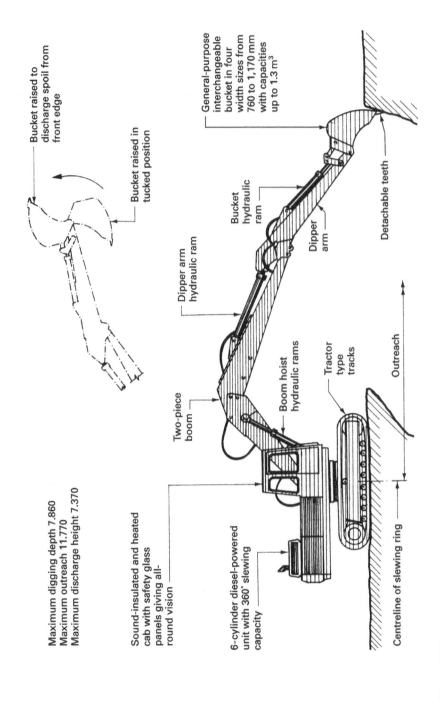

Bucket raised to discharge spoil from front edge

Bucket raised in tucked position

General-purpose interchangeable bucket in four width sizes from 760 to 1,170 mm with capacities up to 1.3 m³

Detachable teeth

Bucket hydraulic ram

Dipper arm

Dipper arm hydraulic ram

Two-piece boom

Boom hoist hydraulic rams

Tractor type tracks

Outreach

Centreline of slewing ring

Maximum digging depth 7.860
Maximum outreach 11.770
Maximum discharge height 7.370

Sound-insulated and heated cab with safety glass panels giving all-round vision

6-cylinder diesel-powered unit with 360° slewing capacity

Figure 2.3.6 Typical tracked hydraulic backacter details (courtesy: Hymac)

MINI- AND MICRO-EXCAVATORS

The development of improved hydraulics and engine capacity has seen the use of mini- and micro-excavator machines rapidly increase, particularly among small contractors. They are ideal to use where access and manoeuvrability of existing multi-purpose machines is restricted. Also, with the Manual Handling Regulations, it has become important to mechanise manual labour to minimise health and safety risks. The mini-excavators are used mainly for service trenching and strip foundations although, with the variety of hydraulic attachments available, they can be used for numerous construction tasks, such as carrying, trenching and demolition works.

The smallest, most compact machines are known as micro-excavators. These are particularly useful in refurbishment projects, for breaking up old solid floors or excavating internal foundations because they are small enough to fit through a standard-sized door. On large contracts these machines, when fitted with breaker attachments, can be easily craned and lifted up for careful, selective demolition of existing concrete-frame structures. The dozer blade on both the mini- and micro-machines is used in a similar way to the jacks on the larger multi-purpose machines, to provide stability to the boom and dipper arms when in use. Examples of these types of machine are shown in Fig. 2.3.7.

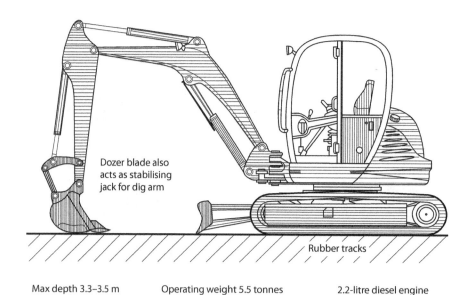

Dozer blade also acts as stabilising jack for dig arm

Rubber tracks

Max depth 3.3–3.5 m
Max reach 5.7 m

Operating weight 5.5 tonnes
O/a width 2.0 m

2.2-litre diesel engine
providing 46 hp

Figure 2.3.7   Hydraulic mini- and micro-excavators (courtesy: JCB)

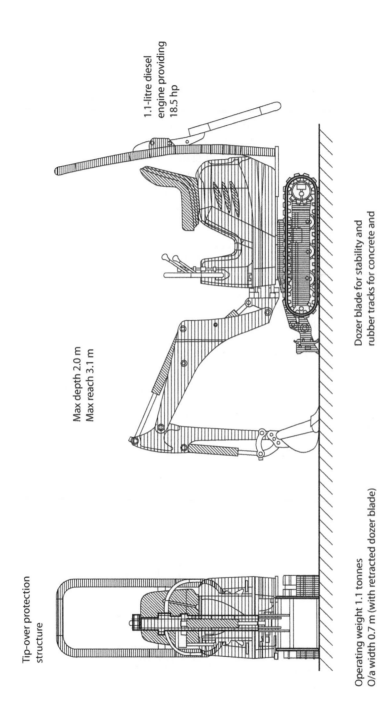

Tip-over protection structure

Max depth 2.0 m
Max reach 3.1 m

1.1-litre diesel engine providing 18.5 hp

Dozer blade for stability and rubber tracks for concrete and tarmacked surfaces

Operating weight 1.1 tonnes
O/a width 0.7 m (with retracted dozer blade)

Figure 2.3.7 *Continued*

FACE SHOVEL

A face shovel can be used as a loading shovel or for excavating into the face of an embankment or berm. Most machines tend to be the tracked mechanical units, although less powerful units can be wheel-based. The bucket sizes vary depending on the material to be loaded and whether it is loose or solid material. The smaller machines are used on construction sites to load or excavate spoil materials, as illustrated in Fig. 2.3.8. They are particularly useful for handling rock and stone that has been previously loosened by other means. These smaller, wheeled or tracked hydraulic machines are used for loading general civil engineering materials, with bucket sizes ranging from 0.5 to 1.5 m³. These machines are limited in the depth to which they can dig below machine level, which is generally within the range 0–300 mm. The discharge operation simply involves the tipping of the shovel about its support pivot at a height of around 3 m into the waiting lorry or dumper.

The larger machines are much more powerful, with bucket capacities of up to 4.5 m³, and work in quarries or open-cast mining. These are versatile machines that can both dig from the face of an embankment and skim the surface of the ground. This is because the parallelogram linkage formed by the stick and boom rams keeps the bucket parallel to the ground during material penetration; also, the master cylinder maintains a level bucket while raising the boom. The soil can be lifted and dumped through the bottom of the bucket at a height of over 7 m, which allows use of large articulated dump trucks. Discharging the soil through the bottom of the bucket involves a clam-shell movement between the front and rear parts. A typical machine is illustrated in Fig. 2.3.9.

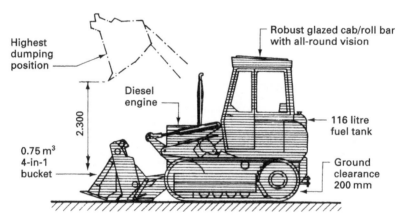

Overall dimensions: length (bucket on ground) 4.300
width (bucket) 1.800

Maximum lifting height 4.675
Top speeds: forward 6.7 km/h, reverse 8.2 km/h
Width of tracks 300 mm

Highest dumping position

Robust glazed cab/roll bar with all-round vision

Diesel engine

2.300

116 litre fuel tank

0.75 m³ 4-in-1 bucket

Ground clearance 200 mm

Figure 2.3.8 Typical tracked face shovel

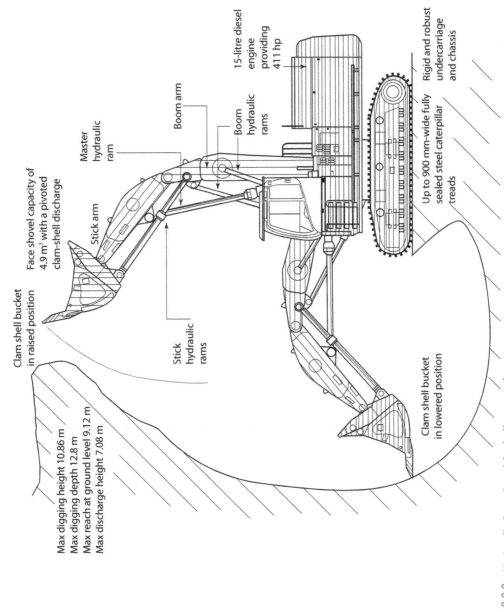

Clam shell bucket
in raised position

Face shovel capacity of
4.9 m³ with a pivoted
clam-shell discharge

Master
hydraulic
ram

Boom arm

15-litre diesel
engine
providing
411 hp

Stick arm

Boom
hydraulic
rams

Rigid and robust
undercarriage
and chassis

Stick
hydraulic
rams

Up to 900 mm-wide fully
sealed steel caterpillar
treads

Max digging height 10.86 m
Max digging depth 12.8 m
Max reach at ground level 9.12 m
Max discharge height 7.08 m

Clam shell bucket
in lowered position

Figure 2.3.9 Hydraulic face shovel for bulk excavation

SKID STEER LOADERS

Skid steer loaders are wheeled while multi-terrain loaders are tracked, but they are both powered and controlled by the same principles. The machine is a small rigid-framed, diesel-engine-powered machine with lift arms used to attach a loading shovel or a wide variety of other attachments. Both types of machine are highly manoeuvrable because the vehicle is turned by generating a differential velocity at opposite sides of the vehicle: that is, to turn right, the left-hand-side wheels turn in preference to the right-hand-side wheels; in this way the vehicle can pirouette on the spot. The power is transmitted to both the wheels (or tracks) and the loading arm by diesel-powered hydraulic pumps; there are no transmission axles. These modern skid loaders have fully enclosed cabs and other features to protect the operator, such as a side access door. These machines are used frequently in controlled top-down demolition of concrete-framed buildings, where materials are moved and dozered down a waste shaft. See Fig. 2.3.10.

DRAGLINE

Dragline machines are used for road and port construction excavations and open-cast mining. They are a remnant of the traditional 'universal' rope and cable machines that could be rigged for a variety of excavator configurations, including a backacter and a skimmer. The development of modern, more efficient and more powerful hydraulic technology has largely seen the demise of 'universal' rope and cable machines, but there is still a vibrant second-hand market in many regions of the developing world.

These larger machines are used in strip-mining operations to move overburden above coal, rock and aggregate, and can weigh over 2,000 metric tonnes. This type of excavator is essentially a crane with a long jib to which is attached a drag bucket for excavating in loose and soft soils below the level of the machine. This machine is for bulk excavation where fine limits are not of paramount importance, because this is beyond the capabilities of the machine's design. Discharge of the collected spoil is similar to that of a backacter, through the open front end of the bucket (see Fig. 2.3.11). A machine rigged as a dragline can be fitted with a grab bucket as an alternative, for excavating in very loose soils below the level of the machine. Outputs of dragline excavators will vary according to operating restrictions, from 30 to 80 bucket loads per hour.

TRENCHERS

These are machines designed to excavate trenches of constant width with considerable accuracy and speed. Widths available range from 200 to 1,500 mm, with depths up to 7.3 m. Most trenchers work on a conveyor principle, having a series of small cutting buckets attached to two endless chains, which are supported by a boom that is lowered into the ground to the

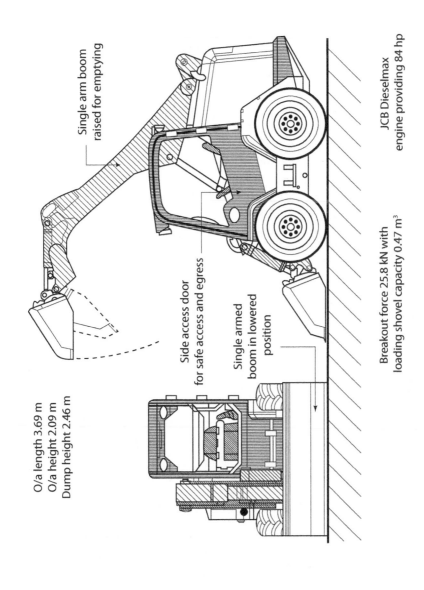

Single arm boom
raised for emptying

Side access door
for safe access and egress

Single armed
boom in lowered
position

O/a length 3.69 m
O/a height 2.09 m
Dump height 2.46 m

JCB Dieselmax
engine providing 84 hp

Breakout force 25.8 kN with
loading shovel capacity 0.47 m³

Figure 2.3.10   Skid steer loader (JCB Skid Steer 260)

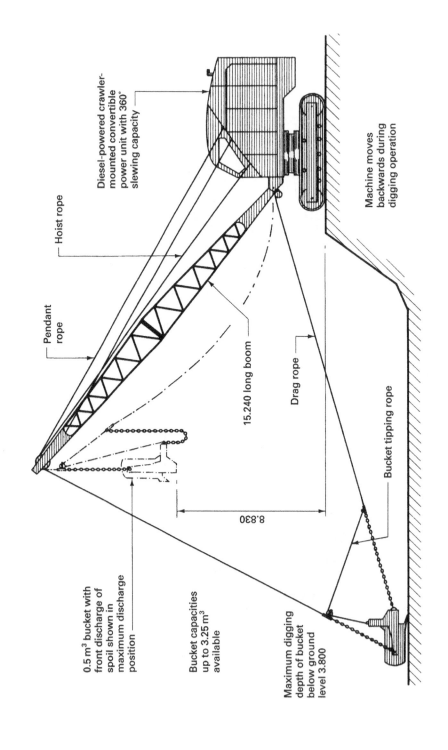

Diesel-powered crawler-mounted convertible power unit with 360° slewing capacity

Hoist rope

Pendant rope

15.240 long boom

Drag rope

8.830

Bucket tipping rope

Machine moves backwards during digging operation

0.5 m³ bucket with front discharge of spoil shown in maximum discharge position

Bucket capacities up to 3.25 m³ available

Maximum digging depth of bucket below ground level 3.800

Figure 2.3.11 Traditional 'universal' rope-rigged dragline (courtesy: Ruston-Bucyrus)

required depth. The spoil is transferred to a cross-conveyor to deposit the spoil alongside the trench being dug; alternatively it can be fed via belt conveyor directly into a waiting truck for disposal. With a depth of dig of some 1.5 m outputs of up to 2 m^3 per minute can be achieved, according to the nature of the subsoil. Some trenchers are fitted with an angled mould blade to enable the machine to carry out the backfilling operation (see Fig. 2.3.12 for a typical example).

COMPACTION PLANT

Compaction is the action of squeezing out air voids between soil grains and fragments. The plant used to do this job are known as rollers (see Fig. 2.3.13). They are designed to compact stone fill materials or surface finishes, such as tarmacadam for pavings. In their simplest form they rely on either deadweight to carry out the compaction operation or vibration.

Deadweight or static rollers are usually diesel-powered and driven by a seated operator within a cab. These machines can be obtained with weights ranging up to 25 tonnes, distributed to the ground through two large-diameter rear wheels and a wider but smaller front steering wheel. Many of these rollers carry water tanks, to add to the dead load and to supply small sparge or sprinkler pipes fixed over the wheels to dampen the surfaces, to prevent the adhesion of tar or similar material when being rolled. These rollers can also be fitted with a scarifier to the rear of the vehicle for ripping up the surfaces of beds or roads.

Vibrating rollers depend mainly on the vibrations produced by a powered eccentric overturning weight mechanism in the front rolling drum. They operate to produce 2 mm-amplitude vibrations at a frequency in the region of 3,000 vibrations per minute. They are similar in operation to the static rollers and, although large and heavy (ranging from 4 to 16 tonnes), they weigh less than their static roller counterparts, and so are quicker and more efficient. The rear pneumatic-tyred wheels support the diesel engine and also provide additional traction. Vibrating rollers are particularly effective for the compaction and consolidation of granular soils. Manufacturers often supply tables to assist in the choice of roller. The tables correlate the type of soil against the maximum depth that can be successfully compacted and the amount of soil that can be compacted per hour.

Also available are much smaller hand-guided vibrating rollers, with weights ranging from 0.5 to 5 tonnes. They can have single or double rollers and are available with or without water sprinkler attachments. These vibrating machines can provide the same degree of compaction as their heavier counterparts but, being lighter and smaller, they can be manoeuvred into buildings for consolidating small areas of hardcore or similar bed material.

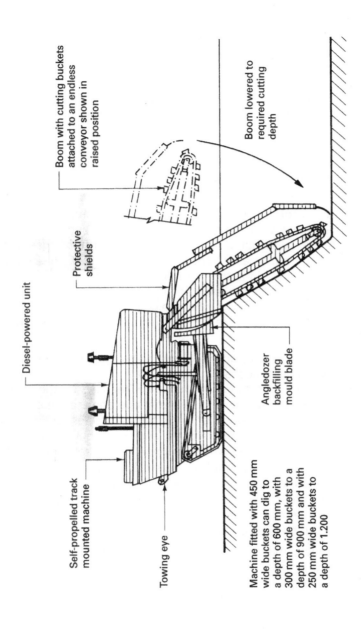

Boom with cutting buckets attached to an endless conveyor shown in raised position

Boom lowered to required cutting depth

Protective shields

Diesel-powered unit

Angledozer backfilling mould blade

Self-propelled track mounted machine

Towing eye

Machine fitted with 450 mm wide buckets can dig to a depth of 600 mm, with 300 mm wide buckets to a depth of 900 mm and with 250 mm wide buckets to a depth of 1.200

Figure 2.3.12   Typical trencher machine (courtesy: Davis Manufacturing)

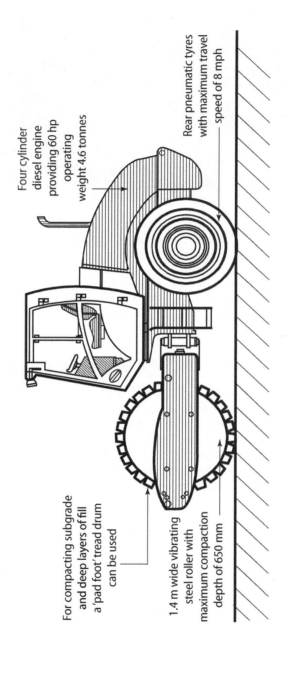

Four cylinder diesel engine providing 60 hp operating weight 4.6 tonnes

Rear pneumatic tyres with maximum travel speed of 8 mph

For compacting subgrade and deep layers of fill a 'pad foot' tread drum can be used

1.4 m wide vibrating steel roller with maximum compaction depth of 650 mm

Figure 2.3.13   Typical single drum vibrating roller

Plant for transportation

2.4

The movement of construction materials, building components, equipment, plant and even personnel is a key factor in the efficient programme of any project. In all cases the method of transportation must match the site location, the spatial access, the environmental constraints, the quantity of materials/component specified, the complexity of the project and the allowable costs. This section will describe the basic forms of transport used in construction practice today.

LORRIES AND TRUCKS

Transportation of machines and materials between sites is usually carried out using suitably equipped or adapted lorries or trucks. These range from the small flat-bed or high-sided pickup trucks weighing less than 800 kg unladen to the very long low-loader lorries, used to convey tracked plant, such as cranes and bulldozers. The small pick-up truck is usually based on the manufacturer's private car or van range. It benefits from lower vehicle excise duty than heavier lorries and the fact that, for trucks of less than 7.5 tonnes gross vehicle weight, the driver does not require a Heavy Goods Vehicle licence. Lorries and pick-up trucks are standard forms of transportation encountered in all aspects of daily living and as such must comply with the regulations made under the Road Traffic Act, Vehicles (Excise) Act and Customs and Excise Act. Most of the statutory requirements will be incorporated into the design and finish of the vehicle as purchased, but there are certain regulations regarding such matters as projection of loads, maximum loading of vehicles and notifications to be given to highway authorities, police forces and the Department for Transport; these matters will be the direct concern of the building contractor. Precise information on these requirements can be found in the Road Traffic (Authorisation of Special Types) (General) Order 2003. To satisfy EU regulations, lorry drivers on public roads carrying a payload of more than 3,500 kg must record their driving hours in a logbook or use tachographs (this only applies where the driver exceeds four hours of driving in any 24-hour period and where travel distances exceed 50 km). Further information can be found in the VOSA briefing document *Rules on drivers' hours and tachographs*.

Most lorries designed and developed for building contractors' use are powered by diesel engines, which are more economical and robust than the petrol engine, although they are heavier and more expensive. Given the mileages in excess of 20,000 (32,000 km) per year experienced by most contractors, diesel engines usually prove to be a worthwhile proposition.

The leading motor manufacturers produce a vast range of lorries to meet a wide range of payloads and power requirements. This is achieved by configuring a variety of engine sizes, drive systems and powered axles to customised, specific body systems, such as a variety of tipper and flat-bed trucks. The most common configuration is a rigid-frame chassis with double axles at the front and double or bogie axles at the rear, with the facility to raise one of these off the ground when empty to reduce rolling resistance and be more fuel-efficient. The rigid chassis lorries often incorporate additional hydraulic self-loading equipment, such as clamshell grabs for tipper lorries carrying loose materials, or pallet grabs for loading palleted loads from a flat-bed chassis. These lorries with integrated hydraulic self-loading equipment are commonly known as 'HIABs' after the Finnish manufacturer Hydrauliska Industri AB that first developed them.

Most lorry manufacturers offer a choice of cabin arrangements depending on the anticipated length of the haul journey. The standard-use cab is for short road hauls; a day cab provides more storage space and a cushioned bench in addition to the driver's cockpit, as well as simple drink preparation facilities; the larger sleeper cabin provides bunk beds and additional catering and ancillary facilities for overnight hauls.

For large civil engineering projects where there is a large quantity of loose soil materials to be moved, rigid-frame trucks or articulated dump trucks are often used. Where there are good haul roads, or where public road haulage is required, rigid-frame trucks are better as they can travel at higher speeds and still handle varying grades of slope. However, where haul roads are poor or not present, or in tight loading areas, articulated trucks are better as the vehicle steers by pivoting on an articulation point behind the cab instead of the wheels turning. Typical projects involving 'off-road' haulage include road and dam construction, sand and gravel pits, mines, rock quarries, tunnel construction and landfill sites. Articulated trucks can handle weak ground conditions, including soft cut and fill areas as the machine's drive-train can be configured in a number of different ways to give all-wheel drive and differential locking axles. There are many combinations of engines and load-carrying capacities; typically in the UK it is not unusual for machines to carry payloads in the region

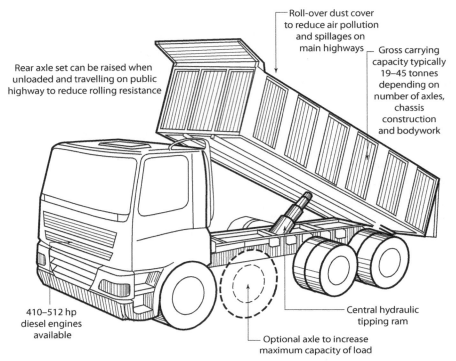

Roll-over dust cover to reduce air pollution and spillages on main highways

Gross carrying capacity typically 19–45 tonnes depending on number of axles, chassis construction and bodywork

Rear axle set can be raised when unloaded and travelling on public highway to reduce rolling resistance

410–512 hp diesel engines available

Optional axle to increase maximum capacity of load

Central hydraulic tipping ram

a) Rear-tipper lorry with central ram for hauling soil and rubble materials

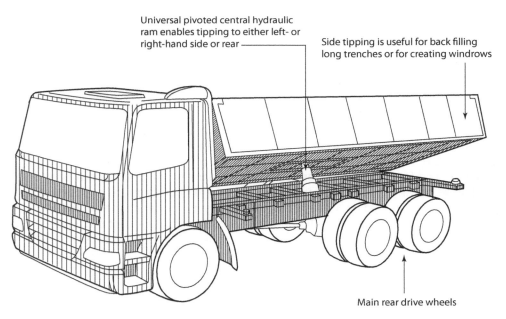

Universal pivoted central hydraulic ram enables tipping to either left- or right-hand side or rear

Side tipping is useful for back filling long trenches or for creating windrows

Main rear drive wheels

b) Tipper lorry with three-way pivoted ram mechanism

Lorry unit based on the DAF FAD XF Range

Figure 2.4.1   **Types of lorry transport used in construction**

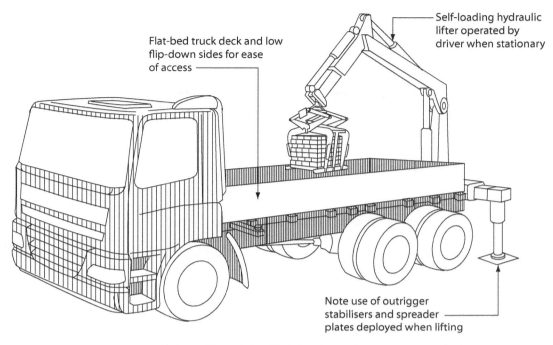

Flat-bed truck deck and low flip-down sides for ease of access

Self-loading hydraulic lifter operated by driver when stationary

Note use of outrigger stabilisers and spreader plates deployed when lifting

c) Flat bed 'HIAB' lorry with self-loading hydraulic equipment

Lorry unit based on the DAF FAD XF Range

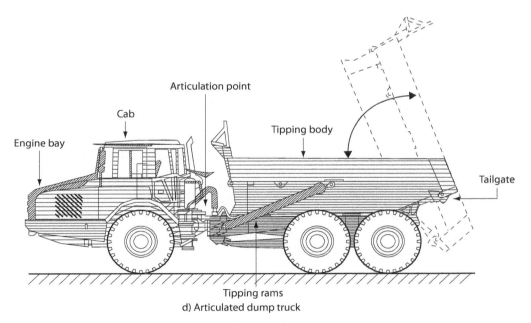

Engine bay

Cab

Articulation point

Tipping body

Tailgate

Tipping rams

d) Articulated dump truck

Based on Volvo A25F/A30F articulated dump truck

Figure 2.4.1   *Continued*

of 40–50 tonnes. These trucks can come with a number of useful features such as exhaust braking which saves the service brakes by restricting the exhaust gases from escaping and lowering engine speed.

For examples of typical lorry types described above, see Fig. 2.4.1.

Table 2.4.1 is provided to help select the most appropriate haulage plant for excavation works. Reference is also made to scrapers and dozers studies in Chapter 2.3, which are also involved in mass haulage of soils.

WHEEL WASHERS

For construction projects in heavily urbanised areas, the needs of the community must be carefully considered. Substructure works result in a lot of excavated materials having to be disposed of in landfill and waste recycling centres off the construction site. This creates a nuisance for the community and road users: earth, dust and materials excavated from the site are often dropped by the lorries and other vehicles used to convey them.

This disturbance can be reduced in several ways:

- by using earth bunding and landscaping on site, so that the material does not need to be disposed of;
- by using piled foundations where mass excavation is not required;
- by using on-site crushing facilities to recycle demolition materials into suitable fill materials.

Where there is bulk excavation, the most efficient and cost-effective method of reducing dirt and mud on public roads is by using wheel washers to clean the lorries before they leave the construction site. To get the maximum benefit, their use must be carefully considered against the volume to be excavated and the time for which they will be required on a construction site. For example, very small excavations can be controlled with a pressure washer and a general operative, who can remove dust and debris and sufficiently clean the wheels before the lorries leave the site.

Before considering the installation of a temporary wheel washer, several factors will need to be taken into account:

- availability of a suitable water supply;
- silt disposal solutions;
- duration required on site;
- location of wheel washer;
- availability of a 415-volt electrical supply or generator;
- suitable lifting equipment for servicing the lorry ramps;
- area for location of the silt settlement tanks.

Table 2.4.1 Haul distances for excavation plant (note: table not to scale)

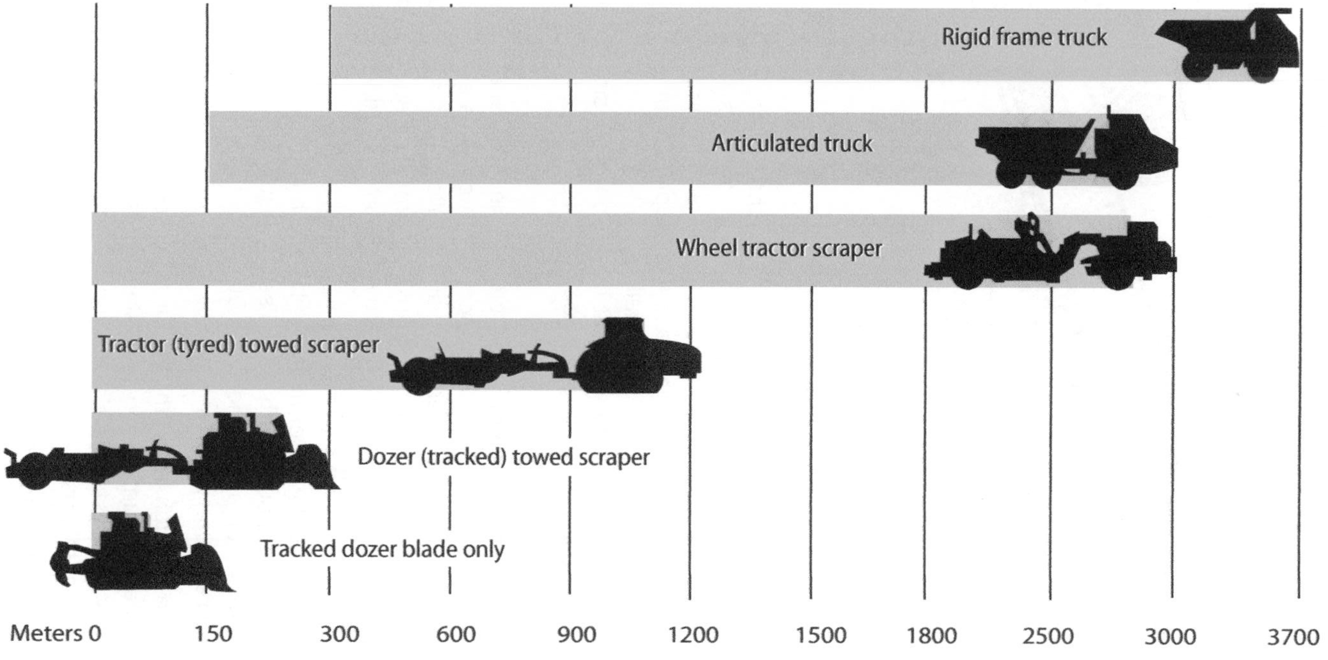

| Meters 0 | 150 | 300 | 600 | 900 | 1200 | 1500 | 1800 | 2500 | 3000 | 3700 |

Rigid frame truck

Articulated truck

Wheel tractor scraper

Tractor (tyred) towed scraper

Dozer (tracked) towed scraper

Tracked dozer blade only

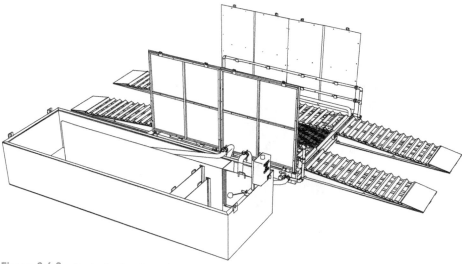

Figure 2.4.2 **Typical wheel washer**

Figure 2.4.2 illustrates a compact arrangement that can be used for light-to-medium washing of the construction site traffic. It consists of a raised platform, which is accessed by heavy-duty removable ramps. The raised middle section houses some high-pressure jet sprays that force water at the underneath of any vehicle. The two parallel side screens contain a series of horizontal water jets that spray water at the wheels and sides of the vehicle. This effectively cleans the lorry of any earth particles that have attached to the wheels, body and sides.

The debris removed by the jets is collected in the tray beneath the ramped and horizontal raised section. This is then pumped to the adjacent settlement tank where any solids and water-borne silt are allowed to settle, allowing the recycling of the water for reuse. The tank does require regular maintenance to remove any build-up of solids. The removable ramps enable any solids to be removed from the wash platform area.

Various sizes of wheel washer are available, with options to allow them to be housed flush in the ground with buried settlement tanks and higher-capacity twin tanks, which can accommodate 1,000 washes per day, for permanent solutions such as quarry working.

There are drive-through baths that require no electrical connections, as they rely on the small amount of debris washing off as the wheel moves through water. A dry-wheel system is also designed to remove debris as the wheels are driven over steel ribs.

Factors involved in selecting the appropriate wheel cleaning plant will normally include the amount of debris expected, the time of year and the nature of the materials to be excavated. A typical arrangement can be delivered and assembled on site and become operational within one day.

DUMPERS

Dumpers are versatile, labour-saving pieces of plant available to the builder for the horizontal movement of general materials, ranging from masonry and aggregates to fluids, such as wet concrete. These diesel-powered vehicles require only one operative, the driver, and can traverse the rough terrain encountered on many building sites. Many sizes and varieties are produced, giving options such as two- or four-wheel drive, hydraulic- or gravity-operated container, side or high-level discharge, self-loading facilities and special adaptations for collecting and transporting crane skips.

A recent plant development of the dumper is the 'dumpster', which operates as a mechanical wheelbarrow. Dumpsters are ideal for small excavation works such as domestic extension foundations with restricted access, or in confined refurbishment projects where plant is required to transport demolition or other loose materials. They are highly manoeuvrable, can fit through standard doorways and small openings in walls, and can discharge into standard-height skips. They are steered by controlling the differential speed of the left- and right-side rubber tracks and can pirouette on the spot just as the skid steer loader – see Chapter 2.3.

Specification for dumpers and dumpsters is usually given by quoting the container capacity in litres for heaped, struck and water levels. See Fig. 2.4.3 for a typical example.

ROUGH-TERRAIN FORKLIFTS AND TELESCOPIC HANDLERS

Rough-terrain forklifts (RTF) are used for vertical loading situations, such as raising pallets of bricks onto scaffolding or floor plates. Their maximum height range depends on the mast of the machine and the integrated hydraulics, but typically they rise to operate at a loading height of about 6.5 m, with a maximum carrying capacity of around 2 tonnes for the largest machines. Rough-terrain forklifts can handle undulating terrain when loaded by use of a tilting mast, which can be inclined backwards by up to 10° when travelling. Some machines also come with the option of four-wheel drive, which also improves traction over rough terrain.

Telescopic handlers, often referred to as 'telehandlers', are similar to RTFs in that they are used for vertical lifting, but they have an added advantage in that they can reach over obstructions by use of their inclined hydraulic boom, which replaces the vertical mast. Besides the use of standard forks that can carry palletted loads, there are a variety of other attachments such as a crane hook, various front shovels, a road sweeper attachment and even a safe working man platform. All of these increase the use and flexibility of the telehandler and make it one of the most useful pieces of construction plant available.

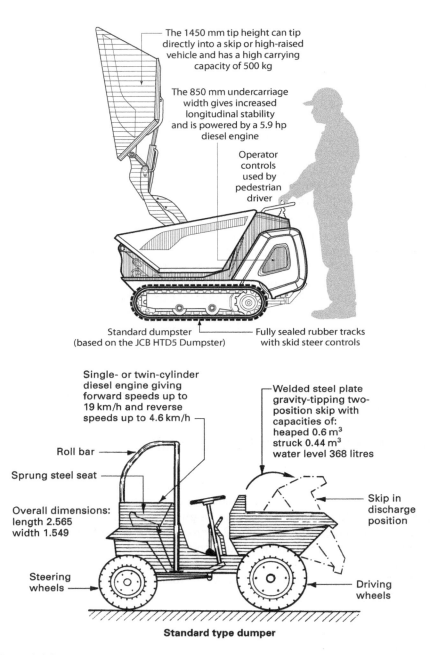

The 1450 mm tip height can tip directly into a skip or high-raised vehicle and has a high carrying capacity of 500 kg

The 850 mm undercarriage width gives increased longitudinal stability and is powered by a 5.9 hp diesel engine

Operator controls used by pedestrian driver

Standard dumpster (based on the JCB HTD5 Dumpster)

Fully sealed rubber tracks with skid steer controls

Single- or twin-cylinder diesel engine giving forward speeds up to 19 km/h and reverse speeds up to 4.6 km/h

Welded steel plate gravity-tipping two-position skip with capacities of:
heaped 0.6 m^3
struck 0.44 m^3
water level 368 litres

Roll bar

Sprung steel seat

Skip in discharge position

Overall dimensions:
length 2.565
width 1.549

Steering wheels

Driving wheels

Standard type dumper

Figure 2.4.3 **Typical diesel dumpers and dumpsters**

Generally all machines offer a four-wheel drive with four-wheel steer for greater manoeuvrability. Typical details are shown in Fig. 2.4.4.

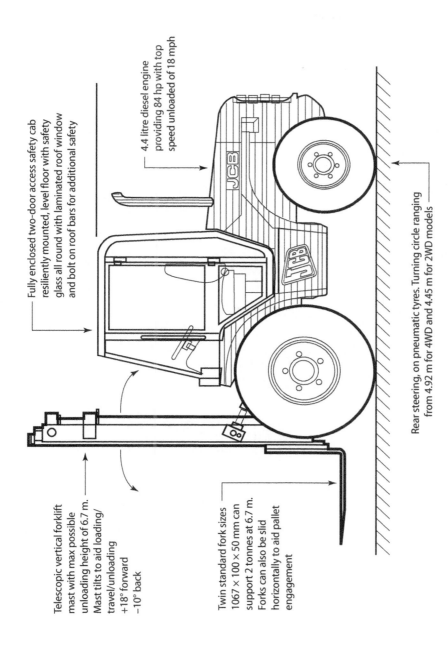

Fully enclosed two-door access safety cab resiliently mounted, level floor with safety glass all round with laminated roof window and bolt on roof bars for additional safety

4.4 litre diesel engine providing 84 hp with top speed unloaded of 18 mph

Telescopic vertical forklift mast with max possible unloading height of 6.7 m. Mast tilts to aid loading/ travel/unloading +18° forward −10° back

Twin standard fork sizes 1067 × 100 × 50 mm can support 2 tonnes at 6.7 m. Forks can also be slid horizontally to aid pallet engagement

Rear steering, on pneumatic tyres. Turning circle ranging from 4.92 m for 4WD and 4.45 m for 2WD models

a) Rough terrain forklift (based on the JCB RTF 940)

Figure 2.4.4   **Rough-terrain forklifts and telescopic handlers**

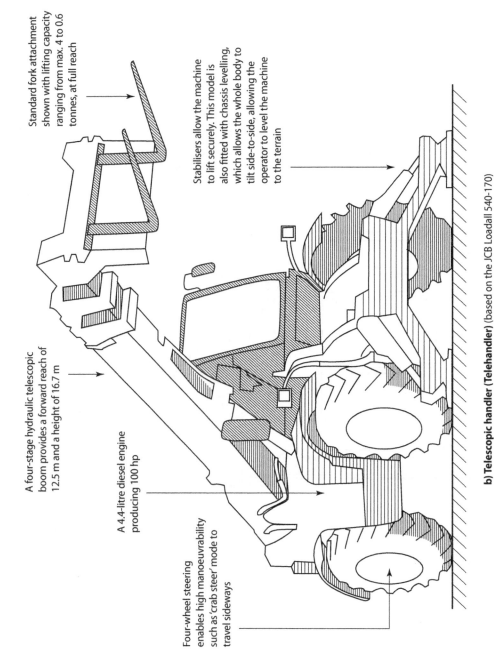

Standard fork attachment shown with lifting capacity ranging from max. 4 to 0.6 tonnes, at full reach

Stabilisers allow the machine to lift securely. This model is also fitted with chassis levelling, which allows the whole body to tilt side-to-side, allowing the operator to level the machine to the terrain

A four-stage hydraulic telescopic boom provides a forward reach of 12.5 m and a height of 16.7 m

A 4.4-litre diesel engine producing 100 hp

Four-wheel steering enables high manoeuvrability such as 'crab steer' mode to travel sideways

b) Telescopic handler (Telehandler) (based on the JCB Loadall 540-170)

Figure 2.4.4 *Continued*

The distinction between an elevator and a conveyor is usually one of direction of movement, in that elevators are considered as those belts moving materials mainly vertically, whereas conveyors are a similar piece of plant moving materials mainly horizontally. Elevators are not common on building sites but, if used wisely, can be economical for such activities as raising bricks or roofing tiles to their fixing position. Most elevators consist of an endless belt with raised transverse strips at suitable spacings, against which the materials to be raised can be placed, usually to a maximum height of 7 m. Conveyors or endless belts are used mainly for transporting loose materials such as aggregates and even concrete. They are generally considered economic only on large sites or quarries where there may be large volumes of aggregate to be stockpiled, or where there are considerable access problems in moving spoil material, such as excavating basements beneath existing buildings.

PASSENGER AND MATERIAL HOISTS

Traditionally materials were hoisted up the building by hand using a rope and pulley system or gin wheel. Except in the smallest repair and maintenance works (usually associated with re-roofing), these have been replaced by electric-powered scaffold hoists, which fix onto scaffolding standards via specially designed gallows brackets and can lift loads of up to 200 kg. For heavier loads of up to 500 kg and where horizontal transport is required, horizontal scaffold gantry hoists can be specified. These have to be designed individually to ensure their stability by the use of ballasted counter-weights incorporated into the gantry steel beam. In more permanent locations or where large heavy loads need to be moved horizontally (for example, in a steel beam and column fabrication workshop), gantry beam hoists may form a permanent fixture within the building. These comprise a heavy universal beam or channel(s) to which a gantry hoist is fixed via a series of electric-powered rollers, which track along the gantry beam. The size of loads that can be carried in these situations varies from 500 kg up to 5 tonnes.

A recent development in lifting is the use of relatively lightweight but strong mobile material hoists. These can be easily manoeuvred by one person to safely jack up structural or ducting components for fixing. Typically these can lift weights of 300 kg up to a height of 6.7 m, or of 450 kg up to 3.5 m, depending on the configuration of the base unit. There is also a range of different forks and boom attachments to suit a variety of specific situations, such as a pipe cradle for lifting circular components. All mobile hoists must be set up on firm, level hard standing or floor by trained operatives. See Fig. 2.4.5.

For raising materials to higher levels a static mast hoist is often used. Static mast hoists can be designed for just materials use or for passengers, although today with the restriction on space on most construction sites, mast hoists are designed to carry both.

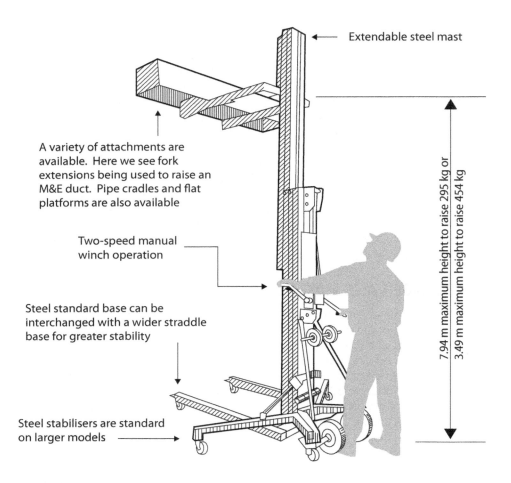

Extendable steel mast

A variety of attachments are available. Here we see fork extensions being used to raise an M&E duct. Pipe cradles and flat platforms are also available

Two-speed manual winch operation

Steel standard base can be interchanged with a wider straddle base for greater stability

Steel stabilisers are standard on larger models

7.94 m maximum height to raise 295 kg or
3.49 m maximum height to raise 454 kg

Mobile materials hoist ('Genie® Superlift™ Advantage' (courtesy: Genie Industries))

Figure 2.4.5   **Materials hoist**

The static materials mast hoist consists of a mast or tower with the lift platform either cantilevered from the small section mast or centrally suspended, with guides on either side, within an enclosing tower. It needs to be plumb and tied to the structure or scaffold at the intervals recommended by the manufacturer to ensure stability. The operation of a materials mast hoist should be entrusted to a trained operator who has a clear view from the operating position.

Passenger hoists, like materials hoists, can be driven by a petrol, diesel or electric motor, and can be of a cantilever or enclosed variety. The cantilever type consists of one or two passenger hoist cages operating on one or both sides of the cantilever tower. Tying-back requirements are similar to those needed for the materials hoist. Passenger and material hoists should conform to the recommendations of BS 7212: *Code of practice for safe use of construction hoists*. Typical multi-use mast hoist details are shown in Fig. 2.4.6.

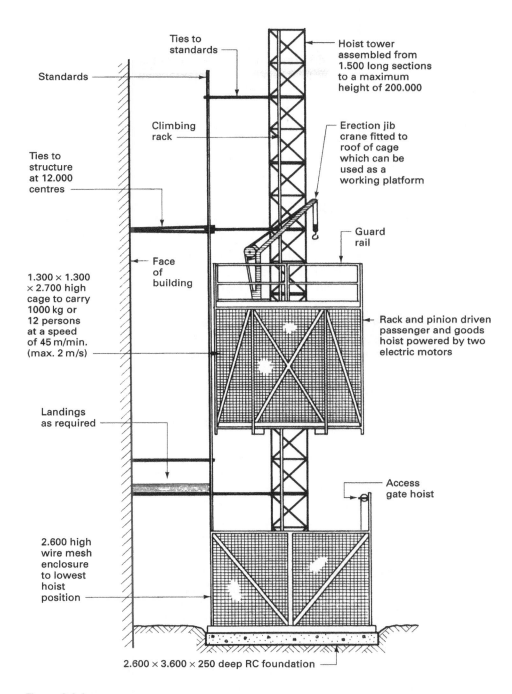

Ties to standards

Standards

Ties to structure at 12.000 centres

Climbing rack

Hoist tower assembled from 1.500 long sections to a maximum height of 200.000

Erection jib crane fitted to roof of cage which can be used as a working platform

Guard rail

Face of building

1.300 × 1.300 × 2.700 high cage to carry 1000 kg or 12 persons at a speed of 45 m/min. (max. 2 m/s)

Rack and pinion driven passenger and goods hoist powered by two electric motors

Landings as required

Access gate hoist

2.600 high wire mesh enclosure to lowest hoist position

2.600 × 3.600 × 250 deep RC foundation

Figure 2.4.6   Typical multi-use passenger and materials hoist

Like all construction lifting equipment, hoists are governed by the Lifting Operations and Lifting Equipment Regulations 1998. These regulations, known by the acronym LOLER, were made under the Health and Safety at Work etc. Act 1974. The regulations provide an extensive checklist with regard to assessment of risk in the use of hoists. Here is a brief summary of the main points.

■ Wherever access can be gained, and wherever anyone at ground level could be struck by the platform or counterweight, enclosures and gates should be at least 2 m high. Access gates must be kept closed at all times except for the necessary loading and unloading of the platform. The platform itself must be fitted with a device capable of supporting a full load in the event of the hoist ropes or hoisting gear failing. Furthermore, the hoist must be fitted with an automatic device to prevent the platform or cage over-running.

■ Hoist operation is to be from one point only at all times if not controlled from the cage itself. The operator must have a clear view of the hoist throughout its entire travel. If this is not possible, auxiliary visibility devices or a signaller or banksman must be located to cover all landings.

■ Winches and carriers must have independent devices to prevent free fall and an automatic braking system that is applied whenever the controls are not in the normal operating position. Also, multiple roping or cylinders (hydraulic hoists) and equipment with high factors of safety should be a standard facility.

■ Safe working loads (SWL) in terms of materials and/or maximum number of passengers to be carried must be displayed on all platforms and associated equipment. Other acceptable terms that may be used include rated capacity or working load limit.

■ Examination and inspection include the engagement of appropriately trained and competent employees, capable of ensuring that a hoist is safe to use. Pre-use checks should be conducted at the beginning of each working day, after alterations to height and after repairs. Thorough and detailed examination of all components will be necessary at periodic intervals determined by the degree of exposure and nature of the work. All examinations and inspections must be recorded and filed.

■ Passenger-carrying hoists specifically require the cages to be constructed in such a manner that passengers cannot fall out, become trapped or be struck by objects falling down the hoistway. Other requirements include the need for gates that will prevent the hoists being activated until they are closed and which can be opened only at landing levels. Over-run devices must also be fitted at the top and bottom of the hoistway.

■ Facilities for the prevention of movement or tipping of materials during transportation by hoist are essential. Security of loads such as loose materials should be provided by suitable containers, such as wheelbarrows that are scotched or otherwise restrained.

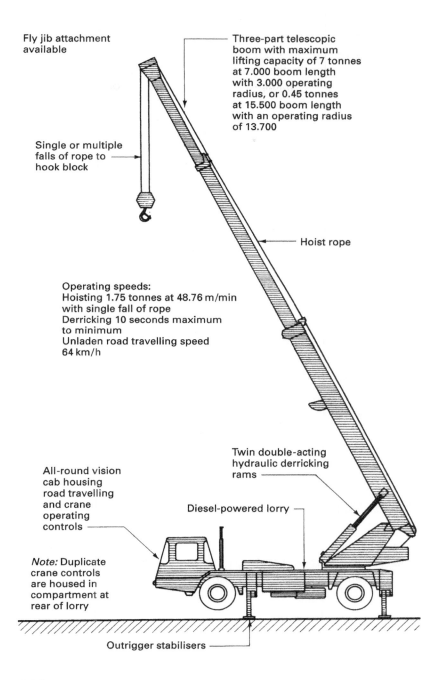

Fly jib attachment available

Three-part telescopic boom with maximum lifting capacity of 7 tonnes at 7.000 boom length with 3.000 operating radius, or 0.45 tonnes at 15.500 boom length with an operating radius of 13.700

Single or multiple falls of rope to hook block

Hoist rope

Operating speeds:
Hoisting 1.75 tonnes at 48.76 m/min with single fall of rope
Derricking 10 seconds maximum to minimum
Unladen road travelling speed 64 km/h

Twin double-acting hydraulic derricking rams

All-round vision cab housing road travelling and crane operating controls

Diesel-powered lorry

Note: Duplicate crane controls are housed in compartment at rear of lorry

Outrigger stabilisers

Figure 2.4.7   Lorry-mounted telescopic crane (courtesy: Coles Cranes)

A crane may be defined as a device or machine for lifting loads by means of a wire rope. With all cranes, the shorter the lifting radius, the greater the lifting capacity and the maximum load denotes the 'capacity' of the crane: for example, a '20-tonne crane' can carry a 20,000 kg load at its shortest reach. (Note that the abbreviation for the metric tonne is 't'.) In most cases cranes are a hire item, as most general contactors do not have the work for a crane to be on permanent standby. The range of cranes available for hire is very wide, so choice must be made on a basis of:

■ weight, dimensions and characteristics of the load;

■ operational speeds, radii, heights of lifts and area of movement;

■ number, frequency and type of lifting operation;

■ length of time for which the crane will be required and sequencing within the overall contract programme;

■ site terrain conditions, including restriction from over-sailing adjacent buildings and interfering with aircraft flight paths;

■ prevailing wind speeds, which can restrict the use of the crane;

■ space available for crane access, erection, operation and dismantling.

CONSTRUCTION CRANES

The subdivision of construction crane types can be very wide and varied, but one simple method of classification is to consider cranes under two general headings:

■ mobile cranes;

■ tower cranes.

MOBILE CRANES

Mobile cranes used on construction sites come in a wide variety of designs and capacities, all generally with a 360° rotation or slewing circle and loading capacities ranging from 10 t to 1,000 t depending on type, configuration and power. They can be grouped into four main types:

■ lorry-mounted cranes;

■ city cranes;

■ rough-terrain cranes;

■ crawler cranes.

Lorry-mounted cranes

Lorry-mounted cranes consist of either a crane mounted on a specially configured commercial lorry body and chassis or a purpose-built lorry with the crane designed as an integral part of the undercarriage – often termed an 'all-terrain' crane. The operator drives the vehicle between sites from a conventional cab but in some cases has to operate the crane engine and controls from a separate crane-operating position. Generally the all-terrain versions have better manoeuvrability and traction as they incorporate independent steering and drive chains for their multiple front and rear axles. The capacity of these cranes ranges from 2 to 100 tonnes in the freestanding position. This capacity is possible through using the jack outrigger stabilisers built into the chassis and a heavy counterweight bolted to the rear of the lorry ranging from 2 to 24 tonnes.

The most popular type of lorry-mounted crane is the telescopic jib crane because of the short time required to prepare the crane for use on arrival on site, making it ideal for short hire periods (see Fig. 2.4.7 for a typical example). Mobile lorry cranes can travel between sites at speeds of up to 90 km/h, which makes them very mobile, but they are not suited to major off-road hauls over rough terrain. They require a firm, level surface from which to operate and will carry timber or steel spreader plates to place under the outriggers before lifting. These cranes are designed for frequent and long-distance road travel, with all the benefits and economies of a commercial vehicle in terms of fuel economy, emissions and low maintenance. They are ideal where there is firm, level ground, good access, space to manoeuvre readily, and where only one operator is required.

City cranes

City cranes are compact wheel-mounted mobile cranes built directly onto the undercarriage. They are generally of relatively low lifting capacities of 25–40 tonnes. They can be distinguished from other mobile cranes by the fact that the driver has only one cab position for both driving and operating the crane. They are extremely mobile, with most models incorporating all-steer wheels enabling sideways 'crab' movement. In addition, to aid their use in cramped urban situations, they are designed with low headroom requirements, some only requiring 3-metre headroom clearance, for improved access. As they are wheeled, they come equipped with wide hydraulic outrigger stabilisers and bearing plates. They require good, firm ground and shallow gradients on which to work, and can be driven on public roads. The small-capacity machines have a fixed boom or jib length, whereas the high-capacity cranes can have sectional lattice jibs or, more commonly, a telescopic boom for increased lifting capacities. See Fig. 2.4.8.

Rough-terrain cranes

Rough-terrain cranes are similar to city cranes in that they are wheel-based, but are designed with higher ground clearance and a more compact wheel base, and incorporate three-mode steering (front-wheel, rear-wheel or all-wheel steering) to clear obstructions and manoeuvre in the most restricted terrain. They are also more highly powered at lower engine speeds, with improved torque to deal with heavier soil traction. The cranes are 'pick and carry' machines, where the outriggers are only used to level and stabilise the crane during hoisting and unloading, and in that respect they are similar to telescopic handlers (see Fig. 2.4.4 (b)). Rough-terrain cranes can be driven on the public road but are slower and less efficient to haul from site to site than lorry-mounted cranes. They generally have telescopic booms designed to lift payloads similar to those lifted by city cranes.

Crawler cranes

This form of mobile crane is a purpose-built, heavy crane supported on a strong undercarriage, turntable chassis slewing ring and wide crawler tracks, making it very stable but also mobile.

With their wide axle base and rear counterweight, crawler cranes have a low centre of gravity and can deal with a variety of rough-terrain situations.

The main advantage over their nearest rival, the wheeled rough-terrain crane, is that not only can they lift very heavy loads and transport them horizontally to a different location, but also they can perform each lift with little set-up, as the crane is stable on its tracks alone, with no outriggers. The main disadvantage of this form of mobile crane is the dismantling and assembly costs and the general need for special low-loading lorries to transport the crane parts between sites. Some crawler cranes can comprise up to 15 separate pieces including the crawler track units, which are removed from the undercarriage of the crane to make it narrow enough to fit on a low loader. Due to the variety of models on the market, the range of loads that crawler cranes can carry is large, typically 30 to 1,000 t. The lower capacity machines up to 60 t use single telescopic jibs or lattice-construction jibs with additional sections and fly jibs to obtain the various lengths and capacities required (see Fig. 2.4.9). For very heavy lifts of up to 1,000 t the lifting jib is usually derricked by additional mast jibs and suspended counterweights, as illustrated in Fig. 2.4.10. Outrigger stabilisers may also be fitted for large static lifts.

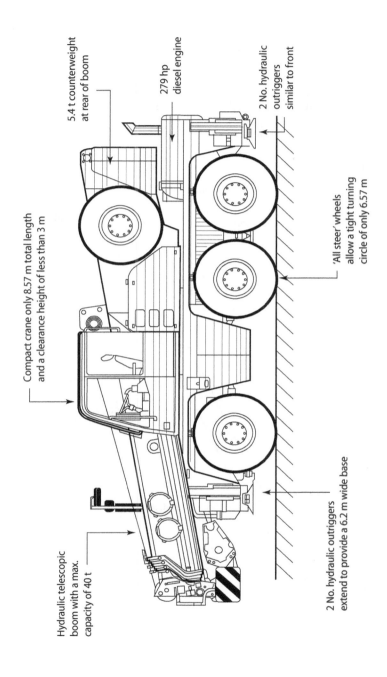

Hydraulic telescopic boom with a max. capacity of 40 t

Compact crane only 8.57 m total length and a clearance height of less than 3 m

5.4 t counterweight at rear of boom

279 hp diesel engine

2 No. hydraulic outriggers similar to front

'All steer' wheels allow a tight turning circle of only 6.57 m

2 No. hydraulic outriggers extend to provide a 6.2 m wide base

Mobile city crane

('AC 40 City' crane) (courtesy: the Terex Corporation)

Figure 2.4.8 **Typical city crane**

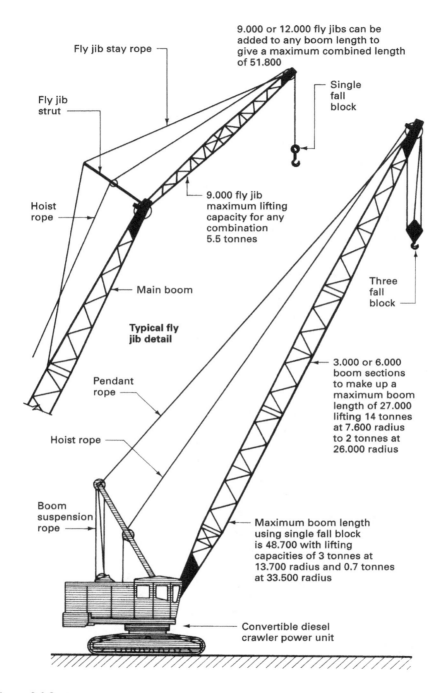

Fly jib stay rope

9.000 or 12.000 fly jibs can be added to any boom length to give a maximum combined length of 51.800

Single fall block

Fly jib strut

9.000 fly jib maximum lifting capacity for any combination 5.5 tonnes

Hoist rope

Three fall block

Main boom

Typical fly jib detail

3.000 or 6.000 boom sections to make up a maximum boom length of 27.000 lifting 14 tonnes at 7.600 radius to 2 tonnes at 26.000 radius

Pendant rope

Hoist rope

Boom suspension rope

Maximum boom length using single fall block is 48.700 with lifting capacities of 3 tonnes at 13.700 radius and 0.7 tonnes at 33.500 radius

Convertible diesel crawler power unit

Figure 2.4.9   Typical small capacity crawler crane (courtesy: Thomas Smith & Sons)

TOWER CRANES

The tower crane has been universally accepted by the building industry as a standard piece of plant required for construction of medium- to high-rise structures. Tower cranes are available in several forms, either with a horizontal jib carrying a saddle or a trolley, or with a luffing or derricking jib with a lifting hook at its far end. Horizontal jibs can bring the load closer to the tower, whereas luffing jibs can be raised to clear obstructions such as adjacent buildings, which is an advantage on confined sites. The basic types of tower crane available are:

- self-supporting static tower cranes;
- self-erecting mobile tower cranes;
- specialist tower cranes:
  - supported static tower cranes;
  - travelling tower cranes;
  - climbing cranes.

Self-supporting static tower cranes

Self-supporting static tower cranes generally have a greater lifting capacity than other types of crane. The mast of the self-supporting tower crane must be firmly anchored at ground level, either to a concrete base with holding-down bolts or to a special mast base section fitted with customised ballast units. They are particularly suitable for confined sites, and should be positioned with a jib of sufficient length to give overall coverage of the new structure and the loading areas. Generally these cranes have a static tower with one of three main types of top slewing jib arrangement:

- 'A' frame or hammer-head jib – where the jib is held horizontal or in a slightly raised position by inclined tie bars to the top of a projecting 'A' frame fixed above the slewing ring (see Fig. 2.4.11);
- flat top jib – where there is no 'A' frame but the jib is more substantial as it acts as a true cantilever (this arrangement is useful to avoid congestion when there are two or more overlapping cranes on site);
- luffing jib – used on congested or hemmed-in sites or to avoid over-sailing as the jib can be raised or lowered to reposition the radii of the load; this arrangement requires a shorter tower but will need two motors: one to move the load and the other to raise the luffing jib.

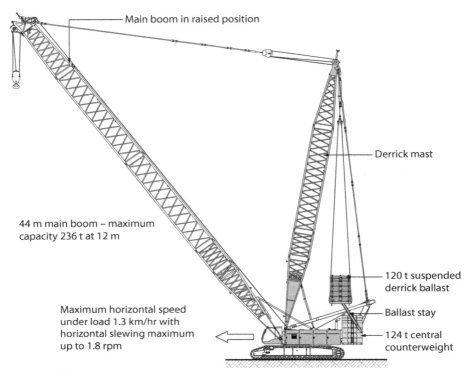

Main boom in raised position

Derrick mast

44 m main boom – maximum
capacity 236 t at 12 m

120 t suspended
derrick ballast

Ballast stay

Maximum horizontal speed
under load 1.3 km/hr with
horizontal slewing maximum
up to 1.8 rpm

124 t central
counterweight

Figure 2.4.10   Typical large capacity crawler crane

Before starting to erect a static tower crane, careful consideration must be given to its position on the site. As for all forms of plant, maximum utilisation is the ultimate aim, so a central position within reach of all storage areas, loading areas and activity areas is required. Generally, output will be in the region of 18 to 20 lifts per hour; the working sequence of the crane needs to be carefully planned and coordinated to take full advantage of the crane's capabilities.

Tower cranes have to be assembled piece by piece on site. The most common method of assembly is by using a second mobile lorry-based crane or an all-terrain crane. The lower tower section is lifted and bolted onto the pre-prepared concrete foundation and bolted into position; subsequent tower sections in three- or six-metre lengths are then bolted together until the desired height is reached. The top slewing ring, cab and counterweight jib are then assembled and lifted up into position; finally the main jib is lifted up and the guy rods or wires are attached.

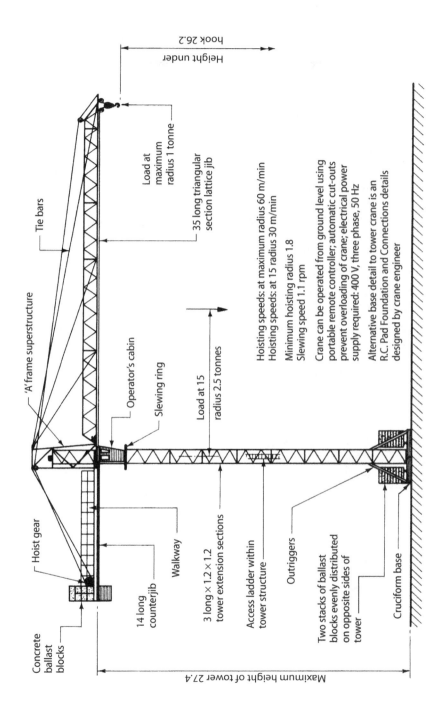

The following labels appear on the figure:

Height under hook 26.2

Load at maximum radius 1 tonne

35 long triangular section lattice jib

Tie bars

'A' frame superstructure

Operator's cabin

Slewing ring

Load at 15 radius 2.5 tonnes

Hoisting speeds: at maximum radius 60 m/min
Hoisting speeds: at 15 radius 30 m/min

Minimum hoisting radius 1.8
Slewing speed 1.1 rpm

Crane can be operated from ground level using portable remote controller; automatic cut-outs prevent overloading of crane; electrical power supply required: 400 V, three phase, 50 Hz

Alternative base detail to tower crane is an R.C. Pad Foundation and Connections details designed by crane engineer

Hoist gear

Concrete ballast blocks

14 long counterjib

Walkway

3 long × 1.2 × 1.2 tower extension sections

Access ladder within tower structure

Two stacks of ballast blocks evenly distributed on opposite sides of tower

Outriggers

Cruciform base

Maximum height of tower 27.4

Figure 2.4.11 Typical self-supporting static tower crane (courtesy: Stothert & Pitt)

An alternative method of assembly and erection is to raise the lower tower section onto a concrete base including an enlarged, sleeved section incorporating an array of four hydraulic jacks. With this in place, you then assemble the jib and counterweight jib and fix these to the tower section with the aid of a mobile crane. Using the hoisting facilities of the jib, further tower sections can be fitted inside the sleeved section and elevated by the hydraulic jacks while supported within the sleeve; this procedure is repeated until the desired height has been reached. The procedure using the sleeved section and hydraulic jacks is shown in Fig. 2.4.12.

Self-erecting mobile tower cranes

Self-erecting mobile tower cranes have become increasingly popular due to their speed of erection and dismantling, the quietness of operation, the reduction of on-site vehicle movements during assembly and deconstruction of crane, and their ability to fold away to avoid over-sailing issues. The smaller, lightweight versions can be transported to site by HIAB or towed, depending on the manufacturer's recommendations. See Fig. 2.4.13.

The main features of self-erecting tower cranes are:

■ the vertical mast is mounted on a ground-level slewing mechanism and held in the vertical position by guy wires and rods;

■ the vertical guys are held taut by ballast units, also fixed to the slewing mechanism;

■ there is a horizontal load path for higher load-handling capacity (unlike low-pivot luffing cranes).

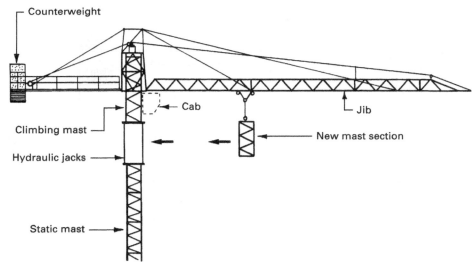

Figure 2.4.12 **Crane mast assembly**

Once the stabilisers are set and the crane is ballasted, it can unfold in about fifteen minutes and fold away in even fewer. The self-erecting tower crane is low in power consumption, with the smaller capacity cranes only requiring single-phase power. It also benefits from a small on-site footprint, making it ideal for cramped, urban developments, particularly on small to medium building projects. It is pedestrian-operated at ground level, mostly by the use of a wired or wireless remote control unit, which helps the operator keep a close watch on the load.

The larger self-erecting tower cranes are based on all-terrain cranes in that they are designed with an integral lorry-based crane carrier. Also, instead of a single inclined jib, when deployed they have a vertical mast and a horizontal boom, all braced with wire ropes or steel rods. Erection takes place with a number of separate assembly winches, which engage the jib and mast assemblies. The whole mast and boom arrangement is mounted on a slewing turntable chassis together with ballasted counterweights, which secure the guying wires and rods (see Fig. 2.4.14).

In the case of a self-erecting all-terrain tower crane, during the lift the outrigger stabilisers raise the crane carrier off the ground. As with all outrigged all-terrain cranes, great care must be taken to distribute the loads safely to the ground through a detailed risk assessment and correct positioning of spreader plates. Another safety feature is that the operator does not need to climb a ladder to reach the cabin, but can enter the cabin at ground level, and the cabin then climbs up the vertical mast to the operating level. This type of cabin, known as an elevating cabin, allows the operator a clear view of the load during slinging and the lift.

SPECIALIST TOWER CRANES

Supported static tower cranes

Supported static tower cranes are similar in construction to self-supporting tower cranes but are used for lifting to a height in excess of that possible with self-supporting or travelling tower cranes. The tower or mast is fixed or tied to the structure using single or double steel stays to provide the required stability. This tying back will induce stresses in the supporting structure, which must therefore be of adequate strength. Supported tower cranes usually have horizontal jibs, because the rotation of a luffing jib mast renders it unsuitable for this application (see Fig. 2.4.15 for a typical example).

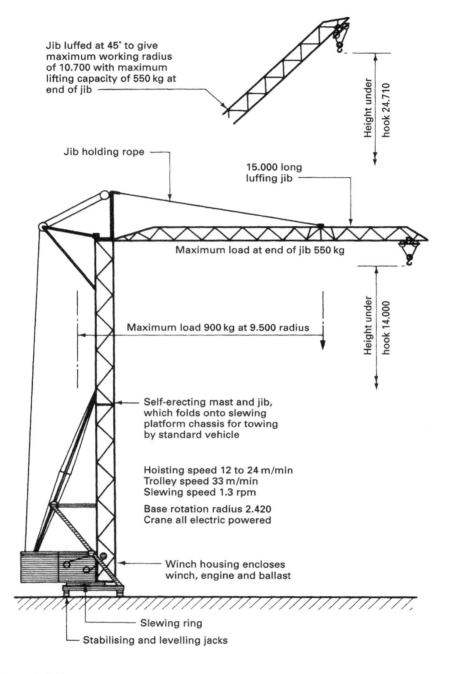

Jib luffed at 45° to give maximum working radius of 10.700 with maximum lifting capacity of 550 kg at end of jib

Height under hook 24.710

Jib holding rope

15.000 long luffing jib

Maximum load at end of jib 550 kg

Height under hook 14.000

Maximum load 900 kg at 9.500 radius

Self-erecting mast and jib, which folds onto slewing platform chassis for towing by standard vehicle

Hoisting speed 12 to 24 m/min
Trolley speed 33 m/min
Slewing speed 1.3 rpm

Base rotation radius 2.420
Crane all electric powered

Winch housing encloses winch, engine and ballast

Slewing ring

Stabilising and levelling jacks

Figure 2.4.13 Typical small self-erecting tower crane, towed version (courtesy: Manitou)

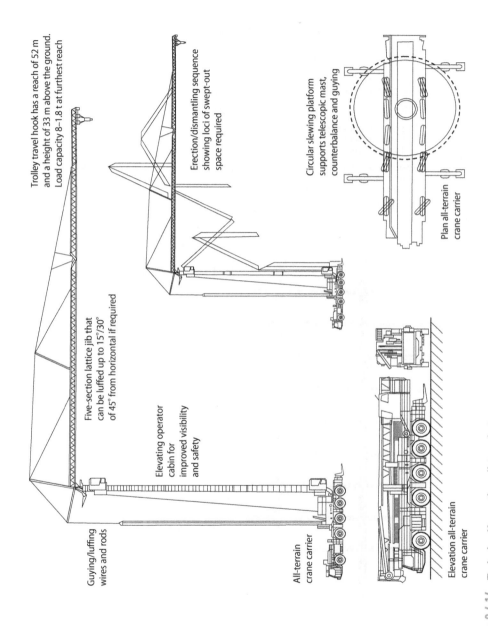

Trolley travel hook has a reach of 52 m and a height of 33 m above the ground. Load capacity 8–1.8 t at furthest reach

Erection/dismantling sequence showing loci of swept-out space required

Circular slewing platform supports telescopic mast, counterbalance and guying

Plan all-terrain crane carrier

Five-section lattice jib that can be luffed up to 15°/30° of 45° from horizontal if required

Elevating operator cabin for improved visibility and safety

Guying/luffing wires and rods

All-terrain crane carrier

Elevation all-terrain crane carrier

Figure 2.4.14   Typical self-erecting all-terrain tower crane

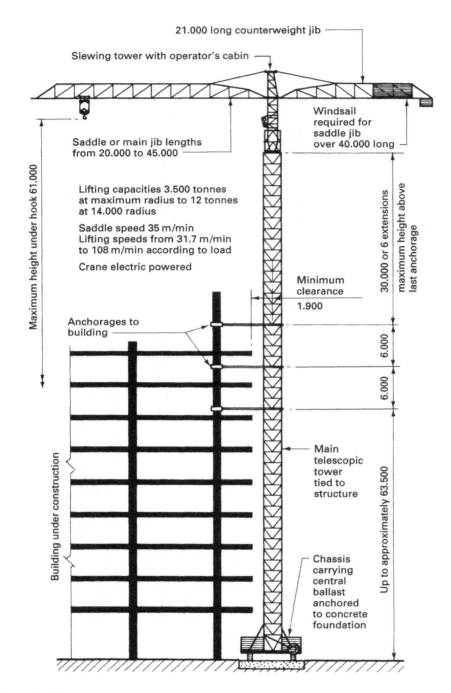

21.000 long counterweight jib

Slewing tower with operator's cabin

Windsail required for saddle jib over 40.000 long

Saddle or main jib lengths from 20.000 to 45.000

Lifting capacities 3.500 tonnes at maximum radius to 12 tonnes at 14.000 radius

Saddle speed 35 m/min
Lifting speeds from 31.7 m/min to 108 m/min according to load

Crane electric powered

Maximum height under hook 61.000

Minimum clearance 1.900

Anchorages to building

30.000 or 6 extensions maximum height above last anchorage

6.000

6.000

Main telescopic tower tied to structure

Building under construction

Up to approximately 63.500

Chassis carrying central ballast anchored to concrete foundation

Figure 2.4.15   Typical supported static tower crane (courtesy: Babcock Weitz)

Travelling tower cranes

To obtain better site coverage with a tower crane, a rail-mounted or travelling crane could be used. This type of crane travels on heavy, wheeled bogies mounted on a wide-gauge (4.2 m) railway track, with gradients not exceeding 1 in 200 and curves not less than 11 m radius, depending on mast height. These are ideal where there is a large, rectilinear area, perhaps of flooring or roofing, to be constructed. To ensure the stability of the crane, it is essential that the base for the railway track sleepers is accurately prepared, well drained and regularly inspected and maintained. The motive power is electricity, the supply of which should be attached to a spring-loaded drum, which will draw in the cable as the crane reverses to reduce the risk of the cable becoming cut or trapped by the wheeled bogies. Travelling cranes can be supplied with similar lifting capacities and jib arrangements as given for static cranes (see Fig. 2.4.16 for a typical example).

Climbing cranes

These are used for very tall buildings, being located within and supported by the structure under construction. The tower mast requires only a small (1.5 to 2 m^2) opening in each floor, which conveniently can be a lift shaft or stairwell in the new building. The crane is supported in collars which surround the tower at two different floor levels, generally about 12 m apart – see Fig. 2.4.17. The lower collar, at the bottom of the tower, takes vertical forces and part of the overturning moment, while the top collar takes the horizontal forces of the remainder of the moment. When the crane is operating the tower is fixed to both collars, allowing the forces generated by the crane to be transferred to the building structure via the collars. To climb the crane up to the next position, an additional higher level collar is assembled around the tower. A climbing support track is hung from what has now become the middle collar, which provides the purchase for the mast to climb up the shaft. The devices fixing the tower to the collars are released and the crane is climbed up the support track powered by a hydraulically operated motor housed in the lower mast section. Once the bottom of the tower has reached the middle collar, the tower is fixed to the middle and top collars, leaving the bottom collar to be dismantled and made available for the next climb.

CRANE SKIPS, SLINGS AND WIRE ROPE

Cranes are required to lift all kinds of materials, ranging from prefabricated components to loose and fluid materials. Various skips or containers have been designed to carry loose or fluid materials (see Fig. 2.4.18). Skips should be of sound construction, easy to attach to the crane hook, easily cleaned, easy to load and unload, and of a suitable capacity.

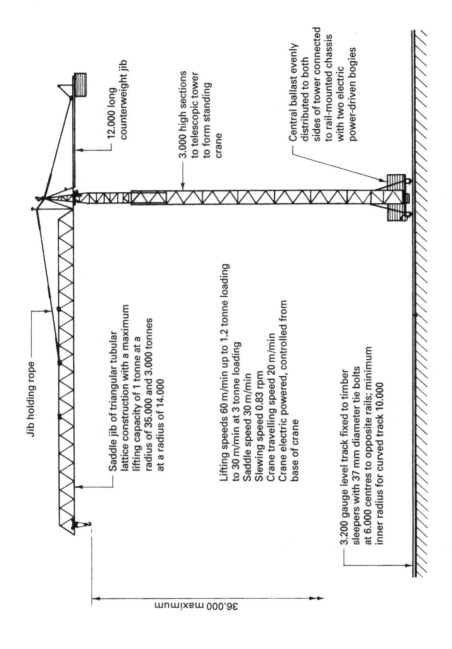

Jib holding rope

12.000 long counterweight jib

3.000 high sections to telescopic tower to form standing crane

Central ballast evenly distributed to both sides of tower connected to rail-mounted chassis with two electric power-driven bogies

Saddle jib of triangular tubular lattice construction with a maximum lifting capacity of 1 tonne at a radius of 35.000 and 3.000 tonnes at a radius of 14.000

Lifting speeds 60 m/min up to 1.2 tonne loading to 30 m/min at 3 tonne loading
Saddle speed 30 m/min
Slewing speed 0.83 rpm
Crane travelling speed 20 m/min
Crane electric powered, controlled from base of crane

3.200 gauge level track fixed to timber sleepers with 37 mm diameter tie bolts at 6.000 centres to opposite rails; minimum inner radius for curved track 10.000

36.000 maximum

Figure 2.4.16   Typical travelling tower crane (courtesy: Richier International)

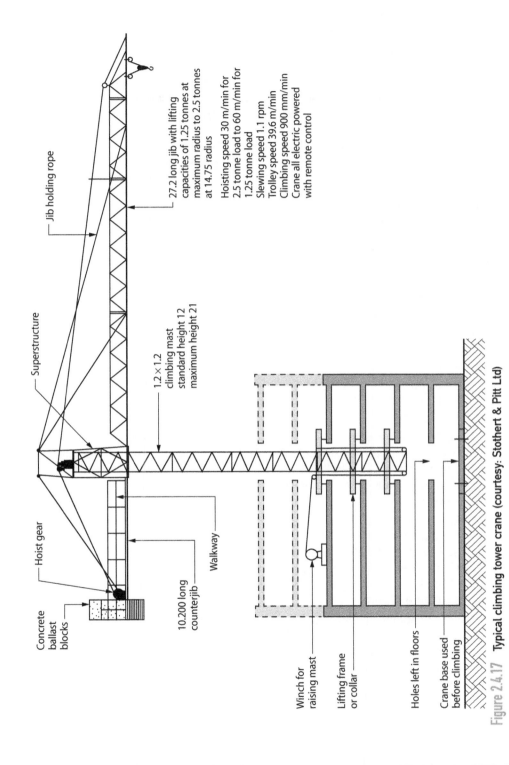

Jib holding rope

27.2 long jib with lifting capacities of 1.25 tonnes at maximum radius to 2.5 tonnes at 14.75 radius

Hoisting speed 30 m/min for 2.5 tonne load to 60 m/min for 1.25 tonne load
Slewing speed 1.1 rpm
Trolley speed 39.6 m/min
Climbing speed 900 mm/min
Crane all electric powered with remote control

Superstructure

1.2 × 1.2 climbing mast standard height 12 maximum height 21

Hoist gear

Concrete ballast blocks

10.200 long counterjib

Walkway

Winch for raising mast

Lifting frame or collar

Holes left in floors

Crane base used before climbing

Figure 2.4.17 Typical climbing tower crane (courtesy: Stothert & Pitt Ltd)

Prefabricated components such as cladding panels are usually hoisted from predetermined lifting points using wire or chain slings (see Fig. 2.4.18). Wire ropes consist of individual wires twisted together to form strands, which are then twisted together around a steel core to form a rope or cable. Ordinary lay ropes are formed by twisting the wires in the individual strands in the opposite direction to the group of strands, whereas in Lang's lay ropes the wires in the individual strands are twisted in the same direction as the groups of strands. Lang's lay ropes generally have better wearing properties thanks to the larger surface area of the external wires, but they have the tendency to spin if the ends of the rope are not fixed. For this reason, ordinary lay ropes with a working life of up to two years are usually preferred for cranes. All wire ropes are lubricated during manufacture, but this does not preclude the need to clean and lubricate wire ropes when exposed to the elements.

LIFTING EQUIPMENT SAFETY CONSIDERATIONS

The Lifting Operations and Lifting Equipment Regulations 1998 detail the minimum requirements for lifting appliances, chains, ropes and lifting gear. Here are some of the main requirements.

- Examination of all forms of lifting gear to ensure sound construction, materials appropriate for the conditions of use, adequate strength for the intended task, retention in good order, and inspection at regular intervals depending on use and exposure as determined by a competent person.

- Adequate support, strength, stability, anchoring, fixing and erection of lifting appliances to include an appropriate factor of safety against failure.

- Travelling and slewing cranes require a 500 mm- or preferably 600 mm-wide minimum clearance wherever practicable between the appliance and fixtures such as a building or access scaffold. If such a clearance cannot be provided, movement between the appliance and fixture should be prevented.

- Safe means of access is required for the crane operator and the signaller. If the access platform is above ground it must be of adequate size, close-boarded or plated, provided with guard rails at least 950 mm above platform level with toe boards at least 150 mm high. Any gap between the toe board and intermediate guard rail or between guard rails must be no more than 470 mm.

- A cab or cabin is required for the crane operator; this must provide an unrestricted view for safe use of the appliance. The cabin must have adequate protection from the weather and harmful substances, with a facility for ventilation and heating. The cabin must also allow access to the machinery for maintenance work.

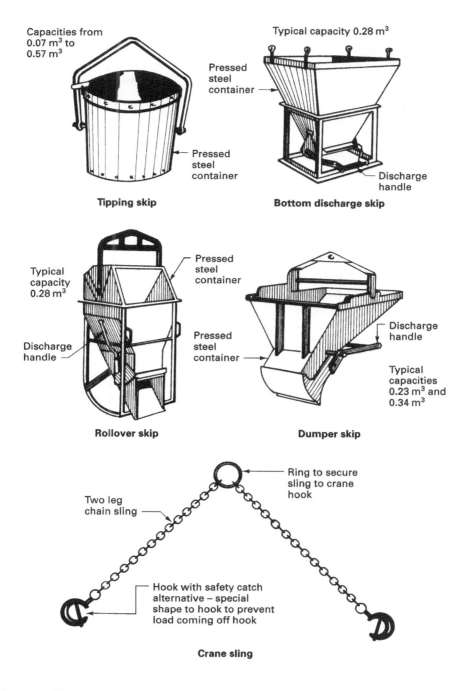

Capacities from
0.07 m^3 to
0.57 m^3

Pressed
steel
container

Tipping skip

Typical capacity 0.28 m^3

Pressed
steel
container

Discharge
handle

Bottom discharge skip

Typical
capacity
0.28 m^3

Pressed
steel
container

Discharge
handle

Rollover skip

Pressed
steel
container

Discharge
handle

Typical
capacities
0.23 m^3 and
0.34 m^3

Dumper skip

Ring to secure
sling to crane
hook

Two leg
chain sling

Hook with safety catch
alternative – special
shape to hook to prevent
load coming off hook

Crane sling

Figure 2.4.18   Typical crane skips and sling

- Equipment that can be adapted for various operating radii and other configurations must be clearly marked with corresponding safe working loads for these variables.

- Brakes, controls and safety devices must be clearly marked to prevent accidental operation or misuse.

- Safe means of access are to be provided for examination, repair and servicing, particularly where there is a possibility of a person falling from any height.

- Strength and fixing requirements for equipment supporting pulley blocks and gin wheels must be ascertained.

- Stability of lifting appliances on soft ground, uneven surfaces and slopes must be considered. Cranes must be either anchored to the ground or to a foundation, or suitably counterweighted or stabilised to prevent overturning.

- Rail-mounted cranes must have a track laid that is secured on a firm foundation to prevent risk of derailment. There must be provision for buffers, effective braking systems and adequate maintenance of both track and equipment.

- Strength requirements are to be assessed for mounting cranes on bogies, trolleys or wheeled carriages.

- Cranes (particularly those used for lifting persons) require hoisting and lowering limiters in addition to capacity indicators and limiters.

- Measures must be taken to prevent a freely suspended load from moving uncontrollably. Devices that could be fitted include: multiple ropes, pawl and ratchet, check valves for hydraulics and safety nets.

- Cranes must be erected under planned conditions and the supervision of a competent person.

- There are specific requirements for persons operating lifting equipment. Both plant operators of lifting appliances and bankspersons or slingers must be trained, experienced and over 18 years of age. This usually requires them to undergo training and a combination of practical and theory tests within the Construction Plant Competence Scheme (CPCS). This is a national scheme run by the Construction Industry Training Board and covers all types of construction plant operators.

- If the banksperson cannot see the whole passage of a lift, an efficient signalling system must be used. A signaller must also be trained and over 18 years of age, and capable of giving clear and distinct communications by hand.

- Testing, examination and inspections are required for all equipment. Chains, slings, ropes, hooks, shackles, eyebolts, pulleys, blocks, gin wheels, sheer legs and other small components are no less important than grabs and winches. All must be tested and thoroughly examined before being put into operation. Further thorough examination is required every 12 months (6 months if for lifting persons), and inspections are to be conducted weekly and documented.

- All cranes must be clearly marked with their safe maximum working loads relevant to lifting radius and maximum operating radius, particularly when fitted with a derricking jib. Lifting equipment not designed for personnel must be clearly marked as such.

- Jib cranes must be fitted with an automatic safe load indicator such as a warning light for the operator and a warning bell for persons nearby.

- Except for testing purposes, the safe working load must not be exceeded.

- When loads are approaching the safe maximum load, the initial lift should be short. A check should then be made to establish safety and stability before proceeding to complete the lift.

Apart from the legal requirements summarised above, common-sense precautions on site must be taken, such as the clear marking of high-voltage electric cables and leaving the crane in an 'out of service' position when unattended or if storm or high wind conditions prevail. On sites with more than one crane in operation, cranes must be fitted with an electronic anti-collision system; however, best practice will also try to ensure the careful zoning of work by a designated crane coordinator. Most tower cranes can operate in wind conditions of up to 60 km/h (16.5 m/s or 38 mph); if wind speeds are any higher, it is recommended that the crane is taken out of service.

The usual 'out of service' position is as follows.

- jibs to be left on free slew and pointed in the direction of the wind on the leeward side of the tower;

- fuel and power supplies switched off;

- load removed;

- hook raised to highest position;

- hook positioned close to tower;

- wheels on rail-mounted cranes chocked or clamped.

Concrete mixers and pumps

2.5

The mixing and transporting of concrete and mortar mixes are important activities on most building sites, from the very small to the very large contract. The choice of method for mixing and transporting the concrete or mortar must be made on the basis of the volume of mixed material required in any given time and also on the horizontal and vertical transportation distances involved. Consideration must also be given to the use of ready-mixed concrete, especially where large quantities are required and/or site space is limited.

CONCRETE MIXERS

Most concrete mixers used on building sites are derived from the minimum recommendations of BS 1305: *Specification for batch type concrete mixers*. Although obsolescent, this British Standard provides useful references for two basic forms: the drum type or free-fall concrete mixer, and the pan type or forced-action concrete mixer.

Drum mixers are subdivided into three distinct forms:

- **tilting drum** (T) – in which the single-compartment drum has an inclinable axis, with loading and discharge through the front opening. This form of mixer is primarily intended for small-batch outputs ranging from 100 T to 200 T litres. Note that mixer output capacities are given in litres for sizes up to 1,000 litres and in cubic metres for outputs over 1,000 litres; a letter suffix designating the type is also included in the title. In common with all drum mixers, tilting mixers have fixed blades inside the revolving drum that lift the mixture and, at a certain point in each revolution, allow the mixture to drop towards the bottom of the drum to recommence the mixing cycle. The complete cycle time for mixing one batch from load to reload is usually specified as 2½ minutes. Typical examples of tilting drum mixers are shown in Fig. 2.5.1.

- **non-tilting drum** (NT) – in which the single-compartment drum has two openings, and rotates on a horizontal axis; output capacities range from 200 NT to 750 NT litres. Loading is through the front opening and discharge through the rear opening by means of a discharge chute collecting the mixture from the top of the drum. The chute should form an angle of not less than 40° with the horizontal axis of the drum.

- **reversing drum** (R) – a version of a mixer with a drum rotating on a horizontal axis that is more popular than the non-tilting drum mixer described above. Capacities of this type of mixer range from 200 R to 500 R litres. Loading is through a front opening and discharge from a rear opening, carried out by reversing the rotation of the drum (see Fig. 2.5.2).

Generally, mixers with an output capacity exceeding 200 litres are fitted with an automatic or manually operated water system, which will deliver a measured volume of water to the drum of the mixer.

Forced action mixers (P) are generally for larger-capacity outputs than the drum mixers described above, and can be obtained within the range from 200 P to 2,000 P litres. The mixing of the concrete is achieved by the relative movements between the mix, pan and blades or paddles. Usually the pan is stationary while the paddles or blades rotate, but rotating-pan models are also available, consisting of a revolving pan and a revolving mixer blade or star, giving a shorter mixing time of 30 seconds with large outputs. In general, pan mixers are not easily transported, so they are usually employed only on large sites where it would be economical to install this form of mixer.

CEMENT STORAGE

Cement for the mixing of mortars or concrete can be supplied in 25 kg bags or in bulk for storage on site before use. Bagged cement requires a dry and damp-free store to prevent air setting taking place (see Chapter 3.2 of *Construction Technology*). If large quantities of cement are required, an alternative storage method is the silo, which will hold cement supplied in bulk under ideal conditions. A typical cement silo consists of an elevated welded-steel cylindrical container supported on four cross-braced legs, with a bottom discharge outlet to the container. Storage capacities range from 12 to 50 tonnes. Silos can be incorporated into an on-site static batching plant, or they can have their own weighing attachments. Some of the advantages of silo storage for large quantities of cement are:

- bulk cement costs less per tonne than bagged cement;

- unloading is by direct pumping from delivery vehicle to silo;

- less site space is required for any given quantity to be stored;

- the first cement delivered is the first to be used, because it is pumped into the top of the silo and extracted from the bottom.

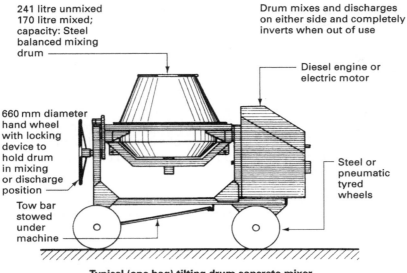

241 litre unmixed
170 litre mixed;
capacity: Steel
balanced mixing
drum

Drum mixes and discharges
on either side and completely
inverts when out of use

Diesel engine or
electric motor

660 mm diameter
hand wheel
with locking
device to
hold drum
in mixing
or discharge
position

Steel or
pneumatic
tyred
wheels

Tow bar
stowed
under
machine

Typical (one bag) tilting drum concrete mixer

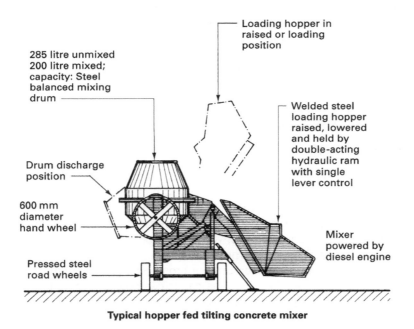

285 litre unmixed
200 litre mixed;
capacity: Steel
balanced mixing
drum

Loading hopper in
raised or loading
position

Welded steel
loading hopper
raised, lowered
and held by
double-acting
hydraulic ram
with single
lever control

Drum discharge
position

600 mm
diameter
hand wheel

Mixer
powered by
diesel engine

Pressed steel
road wheels

Typical hopper fed tilting concrete mixer

Figure 2.5.1   Typical tilting drum mixers (courtesy: Liner Concrete Machinery)

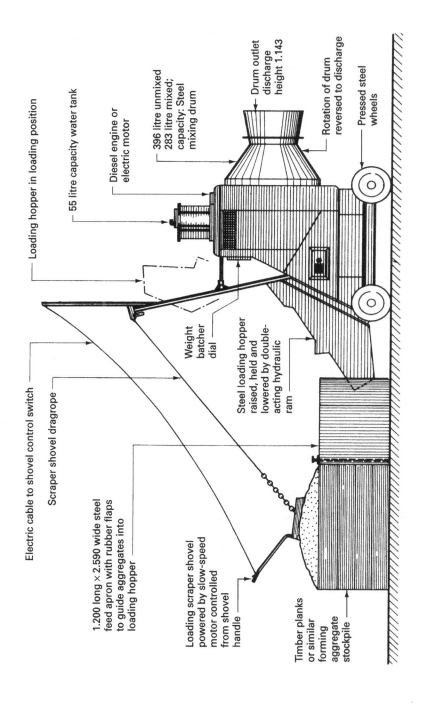

Electric cable to shovel control switch

Scraper shovel dragrope

Loading hopper in loading position

55 litre capacity water tank

Diesel engine or electric motor

396 litre unmixed 283 litre mixed; capacity: Steel mixing drum

Drum outlet discharge height 1.143

Rotation of drum reversed to discharge

Pressed steel wheels

1.200 long × 2.590 wide steel feed apron with rubber flaps to guide aggregates into loading hopper

Weight batcher dial

Steel loading hopper raised, held and lowered by double-acting hydraulic ram

Loading scraper shovel powered by slow-speed motor controlled from shovel handle

Timber planks or similar forming aggregate stockpile

Figure 2.5.2 Typical reversible-drum mixer (courtesy: Liner Concrete Machinery)

READY-MIXED MORTAR

Many sites are congested, particularly in town and city centres where land is at a premium and density of building is high. Space for cement-mixing plant and materials is not always available. Furthermore, labour and safety requirements for mixing and transporting mortar about the site have reduced the cost viability of site preparation. By comparison, safe quality-controlled factory preparation and delivery of mortar offers many advantages, including a guarantee of consistency batch after batch. If required, the manufacturer can also declare the following:

- workability and workable life;*
- chloride content;
- air content;
- compressive strength;
- bond strength;
- water absorption;
- density;
- water vapour permeability;
- thermal conductivity;
- mix proportions.**

* It is usual to incorporate a cement set retarder in the mix to extend the mortar's working life. This is to allow for any delay in transport and the time taken to work through a large batch delivery.

** Mix selection is normally to the National Annex of EN BS 1996 *Design of masonry structures to Eurocode 6* and BS EN 1052-1:1999 *Methods of test for masonry.*

READY-MIXED CONCRETE

The popularity of ready-mixed concrete has increased tremendously. The ready-mixed concrete industry consumes a large proportion of the total cement output of the United Kingdom, supplying many millions of cubic metres of concrete per annum to all parts of the country. The British Ready-Mixed Concrete Association has laid down minimum standards for plant, equipment, personnel and quality control.

Ready-mixed concrete is supplied to sites in specially designed truck mixers, which are basically a mobile mixing drum mounted on a lorry chassis. Truck mixers can be employed in one of three ways.

- Loaded at the depot with dry batched materials plus the correct quantity of water, the truck mixer is used to complete the mixing process at the depot before leaving for the site. During transportation to the site, the mix is kept agitated by the revolving drum. On arrival the contents are remixed before being discharged.

- Fully or partially mixed concrete is loaded into the truck mixer at the depot. During transportation to the site, the mix is agitated by the drum revolving at one to two revolutions per minute. On arrival the mix is finally mixed by increasing the drum's revolutions to between 10 and 15 revolutions per minute for a few minutes before being discharged.

- When the time taken to deliver the mix to the site may be unacceptable, the mixing can take place on site by loading the truck mixer at the depot with dry batched materials and adding the water on arrival on site, before completing the mixing operation and subsequent discharge.

All forms of truck mixer carry a supply of water, which is usually used to wash out the drum after discharging the concrete and before returning to the depot. See Fig. 2.5.3 for typical truck mixer details.

Truck mixers are heavy vehicles, weighing up to 24 tonnes when fully laden, with a turning circle of some 15 m; they require both a firm surface and turning space on site. The site allowance time for unloading is usually 30 minutes, allowing for the discharge of a full load in 10 minutes and leaving 20 minutes of free time to permit for a reasonable degree of flexibility in planning and programming to both the supplier and the user. Truck mixer capacities vary with the different models but 4, 5 and 6 m³ are common sizes. Consideration must be given by the contractor to the best unloading position, because most truck mixers are limited to a maximum discharge height of 1.5 m, using a discharging chute to a semicircular coverage around the rear of the vehicle within a radius of 3 m.

To obtain maximum advantage from the facilities offered by ready-mixed concrete suppliers, building contractors must place a clear order of the exact requirements, which should follow the recommendations given in BS EN 206-1 and BS 8500-1 and 2. The supply instructions should contain the following:

- type of cement;
- types and maximum sizes of aggregate;
- test and strength requirements;
- testing methods;
- slump or workability requirements;
- volume of each separate mix specified;
- delivery programme;
- any special requirements, such as a pumpable mix.

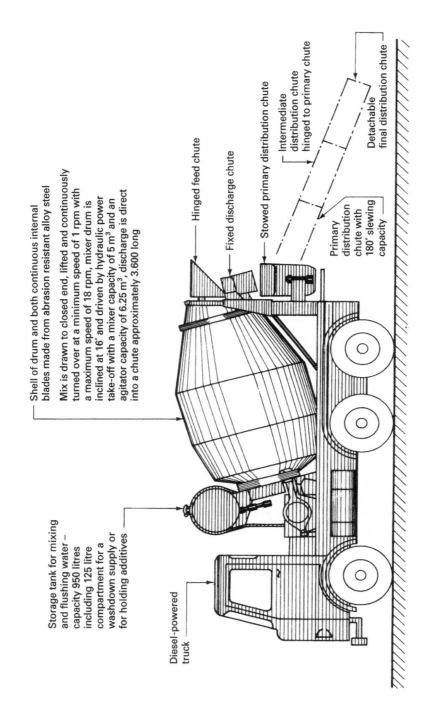

Shell of drum and both continuous internal blades made from abrasion resistant alloy steel

Mix is drawn to closed end, lifted and continuously turned over at a minimum speed of 1 rpm with a maximum speed of 18 rpm, mixer drum is inclined at 16° and driven by hydraulic power take-off with a mixer capacity of 5 m³ and an agitator capacity of 6.25 m³, discharge is direct into a chute approximately 3.600 long

Storage tank for mixing and flushing water – capacity 950 litres including 125 litre compartment for a washdown supply or for holding additives

Diesel-powered truck

Hinged feed chute

Fixed discharge chute

Stowed primary distribution chute

Intermediate distribution chute hinged to primary chute

Detachable final distribution chute

Primary distribution chute with 180° slewing capacity

Figure 2.5.3 Typical ready-mixed concrete truck details

Note: Much of the above can be rationalised by specifying the concrete grade (for example, C30) and the mix category (for example, 'Designed'). For more detail, see BS EN 206-1: *Concrete. Specification, performance, production and conformity* and the complementary British Standards to BS EN 206-1, BS 8500-1: *Concrete. Method of specifying and guidance for the specifier* and BS 8500-2: *Concrete. Specification for constituent materials and concrete*. See also the summary in Chapter 3.2 of *Construction Technology*.

CONCRETE PUMPS

There are several advantages to moving large volumes of concrete by using a pump and pipeline.

- Concrete is transported from the point of supply to the placing position in one continuous operation.
- Faster pours can be achieved with less labour. Typical placing figures are up to 100 m³ per hour using a two-person crew consisting of the pump operator and an operator at the discharge end.
- No segregation of mix is experienced with pumping, and a more consistent placing and compaction is obtained, requiring less vibration.
- Generally, site plant and space requirements are reduced.
- It is the only method available for conveying wet concrete both vertically and horizontally in one operation.
- There is no shock loading of formwork.
- Generally, the net cost of placing concrete is reduced.

Set against the above advantages, there are also some limitations.

- Concrete supply must be consistent and regular, which can usually be achieved by well-planned and organised deliveries of ready-mixed concrete. Note that, under ideal conditions, the discharge rate of each truck mixer can be in the order of 10 minutes.
- Concrete mix must be properly designed and controlled, because not all concrete mixes are pumpable. The concrete is pumped under high pressure, which can cause bleeding and segregation of the mix. The mix must be properly designed to avoid these problems, as well as having good cohesive, plasticity and self-lubricating properties to enable it to be pumped through the system without excessive pressure and without causing blockages.
- More formwork will be needed to receive the high output of the pump, to make it economical.

Most pumps used today are of the twin-cylinder hydraulically driven design, as either a trailer pump or a lorry-mounted pump using a small-bore (100 mm diameter) pipeline capable of pumping concrete 85 m vertically and 200 m horizontally. However, these figures will vary with the actual pump used (see Fig. 2.5.4 for a typical example). The delivery pipes are usually of rigid seamless steel in 3 m lengths except where flexibility is required, as on booms and at the delivery end. Large-radius bends of up to 1 m radius giving 22½°, 45° and 90° turning are available to give flexible layout patterns. Generally, small-diameter pipes of 75 and 100 mm are used for vertical pumping, whereas larger diameters of up to 150 mm are used for horizontal pumping. If a concrete mix with large aggregates is to be pumped, the pipe diameter should be at least three or four times the maximum aggregate size.

The time required on site to set up a pump is approximately 30 to 45 minutes. The pump operator will require a supply of water and grout for the initial coating of the pipeline; this usually requires about two or three bags of cement. A hardstanding should be provided for the pump, with adequate access and turning space for the attendant ready-mixed concrete vehicles. The output of a concrete pump will be affected by the distance the concrete is to be pumped, so the pump should be positioned as close to the discharge point as is practicable. Pours should be planned so that they progress backwards towards the pump, removing the redundant pipe lengths as the work proceeds.

Generally, if the volume of concrete to be placed is sufficient to warrant hiring a pump and operators, it will result in an easier, quicker and usually cheaper operation than placing the concrete by the traditional method of crane and skip, with typical outputs of 15 to 20 m³ per hour as opposed to the 60 to 100 m³ per hour output of the concrete pump. Concrete pumping and placing demands a certain amount of skill and experience, so most pumps in use are hired out and operated by specialist contractors.

IMPORTANT SAFETY ISSUES

Under the COSHH Regulations all construction products must be properly labelled with clear instructions on how they should be safely prepared and used, and also how the waste must be disposed of. All products are supplied with a Materials Safety Data Sheet (MSDS) which carefully explains the potential risks and suggested control measures. This includes paints, solvents, cleaning materials and cement, for example.

Mortar and concrete contain cement. Mortar may also contain lime. Both ingredients can burn, and skin contact with fresh mortar or concrete may result in skin ulceration or dermatitis. Unlike heat burns, cement burns might not be felt until some time after contact with fresh concrete, so there might be no warning of damage occurring. Contact with the skin should be prevented by wearing suitable protective clothing – in particular, reinforced plastic gloves, waterproof overalls and footwear, and protection for the eyes. Face protection may also be appropriate for certain applications. Where skin

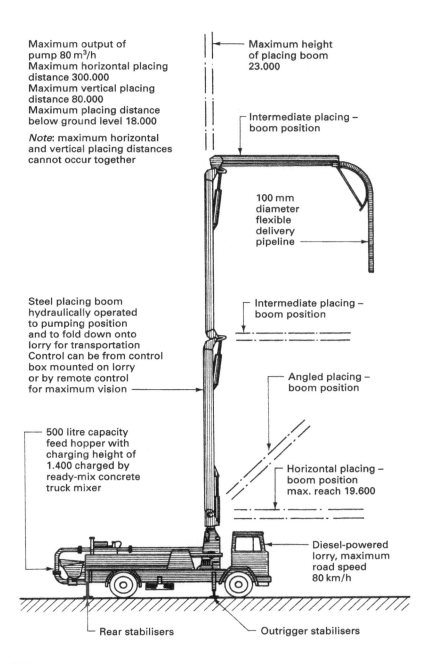

Maximum output of
pump 80 m³/h
Maximum horizontal placing
distance 300.000
Maximum vertical placing
distance 80.000
Maximum placing distance
below ground level 18.000

Note: maximum horizontal
and vertical placing distances
cannot occur together

Maximum height
of placing boom
23.000

Intermediate placing –
boom position

100 mm
diameter
flexible
delivery
pipeline

Steel placing boom
hydraulically operated
to pumping position
and to fold down onto
lorry for transportation
Control can be from control
box mounted on lorry
or by remote control
for maximum vision

Intermediate placing –
boom position

Angled placing –
boom position

500 litre capacity
feed hopper with
charging height of
1.400 charged by
ready-mix concrete
truck mixer

Horizontal placing –
boom position
max. reach 19.600

Diesel-powered
lorry, maximum
road speed
80 km/h

Rear stabilisers

Outrigger stabilisers

Figure 2.5.4   Typical lorry-mounted concrete pump (courtesy: Schwing)

HEALTH AND SAFETY WARNING

- Contains chromium (VI). May produce an allergic reaction.
- Risk of serious damage to eyes.
- Contact with wet cement, wet concrete or wet mortar may cause irritation, dermatitis or burns.
- Contact between cement powder and body fluids (e.g. sweat and eye fluid) may also cause skin and respiratory irritations, dermatitis or burns.
- Avoid eye and skin contact by wearing suitable eye protection, clothing and gloves.
- Avoid breathing dust.
- On contact with eyes or skin, rinse immediately with plenty of clean water. Seek medical advice after eye contact.
- Keep out of the reach of children.

IRRITANT

Figure 2.5.5 **Safety sign on cement materials**

contact occurs, immediate washing with soap and water is recommended. Eye contact requires immediate washing of the affected area with clean water. If cement is swallowed, the mouth should be washed out, followed by drinking plenty of clean water. Operatives should take care to prevent fresh concrete from entering boots and use working methods that do not require them to kneel in fresh concrete. Cement products are packaged with a safety label, similar to that shown in Fig. 2.5.5.

References:

Control of Substances Hazardous to Health Regulations 2002

Personal Protective Equipment at Work Regulations 1992

Advanced temporary access systems 2.6

A scaffold is a temporary frame, usually constructed from steel or aluminium alloy tubes clipped or coupled together to provide a means of access to high-level working areas, as well as providing a safe platform from which to work.

The two basic forms of scaffolding – the putlog scaffold with its single row of uprights or standards set outside the perimeter of the building and partly supported by the structure, and independent scaffolds that have two rows of standards – have been covered in Chapter 2.5 of *Construction Technology*. Here we will look at special scaffolds, such as suspended, truss-out and gantry scaffolds, as well as the easy-to-erect system scaffolds.

Unless a scaffold has a basic configuration as described, for example, by the National Access and Scaffolding Confederation (NASC), the scaffold should be designed by calculation, by a competent person, to ensure it will have adequate strength and stability. It cannot be overemphasised that all scaffolds must comply fully with the minimum requirements set out in the Work at Height Regulations 2005 and BS EN 12811-1: 2003 *Temporary works equipment. Scaffolds. Performance requirements and general design*.

TRUSS-OUT SCAFFOLDS

Truss-out scaffolds are a form of independent tied scaffold that rely entirely on the building for support. They are used where it is impossible or undesirable to erect a conventional scaffold from ground level. The 'truss-out' is the supporting scaffolding structure that projects from the face of the building. Anchorage is provided by adjustable struts fixed internally between the floor and ceiling, from which the cantilever tubes project. Except for securing rakers, only right-angle couplers should be used. The general format for the remainder of the scaffold is as used for conventional independent scaffolds (see Fig. 2.6.1).

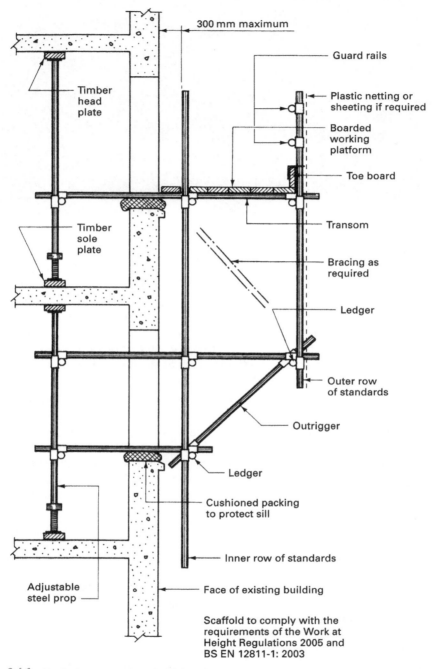

300 mm maximum

Timber
head
plate

Guard rails

Plastic netting or
sheeting if required

Boarded
working
platform

Toe board

Timber
sole
plate

Transom

Bracing as
required

Ledger

Outer row
of standards

Outrigger

Ledger

Cushioned packing
to protect sill

Inner row of standards

Adjustable
steel prop

Face of existing building

Scaffold to comply with the
requirements of the Work at
Height Regulations 2005 and
BS EN 12811-1: 2003

Figure 2.6.1   Typical truss-out scaffold details

SUSPENDED SCAFFOLDS

Suspended scaffolds consist of a working platform suspended from supports, such as outriggers, that cantilever over the upper edge of a building. In this form they give a temporary means of access to the face of a building for cleaning and light maintenance work. Many new, tall structures have suspension tracks incorporated in the fascia or upper edge beam. Alternatively, they can have a cradle suspension track fixed to the upper surface of the flat roof, on which is supported a manual or power trolley with retractable davit arms for supporting the suspended working platform or cradle. All forms of suspended cradles must conform with the minimum requirements set out in the Work at Height Regulations 2005 with regard to platform boards, guard rails and toe boards. Cradles may be single units or grouped together to form a continuous working platform; if grouped together they are connected to one another at their abutment ends with hinges. Figure 2.6.2 shows typical suspended scaffold details.

MOBILE TOWER SCAFFOLDS

Mobile tower scaffolds are used mainly by painters and maintenance staff to gain access to ceilings where it is advantageous to have a working platform that can be readily moved to a new position. The scaffold is basically a square tower constructed from scaffold tubes, mounted on wheels fitted with brakes. Platform access is gained by short, opposing, inclined ladders or one inclined ladder within the tower base area. See Chapter 2.5 of *Construction Technology* for more detail.

BIRDCAGE SCAFFOLDS

Birdcage scaffolds are used to provide a complete working platform at a high level over a large area. They consist of a two-directional arrangement of standards, ledgers and transoms to support a close-boarded working platform at the required height. To ensure adequate stability, standards should be placed at not more than 2.4 m centres in both directions, and the whole arrangement must be adequately braced.

GANTRIES

Gantries are forms of scaffolding used primarily as elevated loading and unloading platforms over a public footpath, where the structure under construction or repair is immediately adjacent to the footpath. As for hoardings, local authority permission is necessary for gantries, and their specific requirements (such as pedestrian gangways, lighting and dimensional restrictions) must be fully met. It may also be necessary to comply with police requirements as to when loading and unloading can take place.

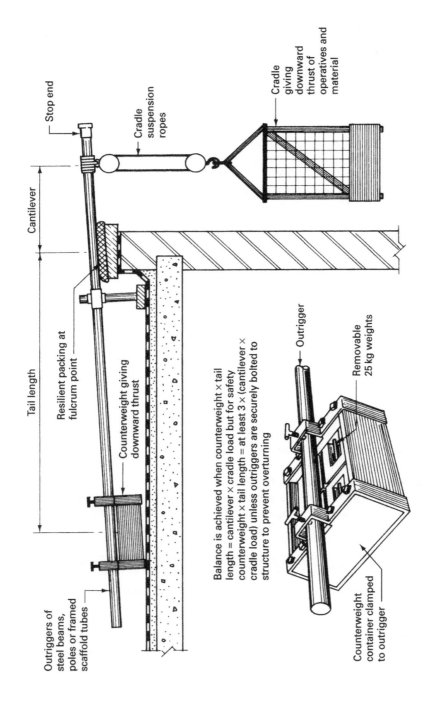

Stop end

Cradle suspension ropes

Cradle giving downward thrust of operatives and material

Cantilever

Tail length

Resilient packing at fulcrum point

Counterweight giving downward thrust

Outriggers of steel beams, poles or framed scaffold tubes

Balance is achieved when counterweight × tail length = cantilever × cradle load but for safety counterweight × tail length = at least 3 × (cantilever × cradle load) unless outriggers are securely bolted to structure to prevent overturning

Outrigger

Removable 25 kg weights

Counterweight container clamped to outrigger

Figure 2.6.2 Typical suspended scaffold details

The gantry platform can also serve as a storage and accommodation area, as well as providing the staging from which a conventional independent scaffold can be erected, to provide access to the face of the building. Gantry scaffolds can be constructed from standard structural steel components as shown in Fig. 2.6.3.

SYSTEM SCAFFOLDS

System scaffolds are based on the traditional independent steel-tube scaffold, but instead of being connected together with a series of loose couplers and clips they usually have integral interlocking connections. They are easy to erect, adaptable, and can generally be assembled and dismantled by semi-skilled operatives. The design of these systems is such that the correct position of handrails, lift heights and all other aspects of the Work at Height Regulations 2005 are automatically met.

Another advantage found in most of these system scaffolds is the elimination of internal cross-bracing, giving a clear walk-through space at all levels; façade-bracing, however, may still be required. Figure 2.6.4 shows details of a typical system scaffold and is intended only to be representative of the many scaffolding systems available.

MECHANICAL ACCESS PLANT

The use of mechanical access plant is increasing as the benefits for productivity and safety are recognised. This equipment is acknowledged by many to be the safest and most efficient means of providing temporary access to height for many work activities. It is becoming common on most construction sites due to the need for workers to have safe access to their place of work, as required by the Work at Height Regulations 2005, if there is no other way to complete the work.

However, like any piece of mechanical plant, access plant must be used responsibly and only after an appropriate risk assessment has been made. The Provision and Use of Work Equipment Regulations 1998 (PUWER) require a risk assessment to be undertaken when planning and selecting mechanical access plant. The Lifting Operations and Lifting Equipment Regulations 1998 (LOLER) require that any equipment used for lifting or lowering loads or people should have been designed to a recognised standard and thoroughly examined at regular intervals.

SCISSOR LIFT

A scissor lift is an electrical powered four-wheeled platform that enables safe vertical movement when stationary. To achieve this, it uses linked, folding supports in a criss-cross 'scissor-like' pattern. The working platform may also have a slide-out 'extension' to allow closer access to the work area, because of the inherent limits of vertical only movement.

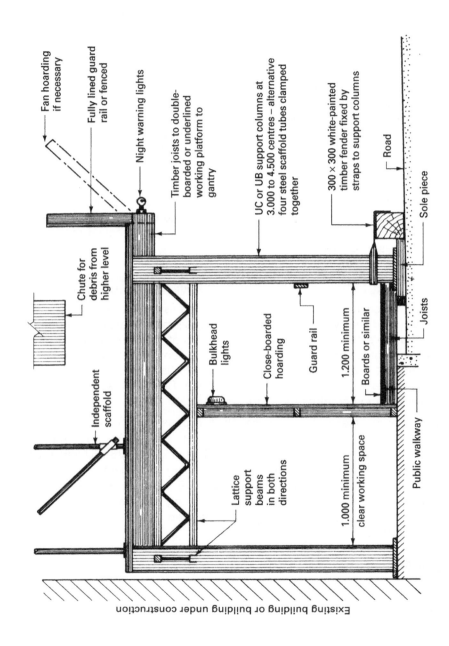

Fan hoarding if necessary

Fully lined guard rail or fenced

Night warning lights

Timber joists to double-boarded or underlined working platform to gantry

UC or UB support columns at 3.000 to 4.500 centres – alternative four steel scaffold tubes clamped together

300 × 300 white-painted timber fender fixed by straps to support columns

Road

Sole piece

Chute for debris from higher level

Bulkhead lights

Close-boarded hoarding

Guard rail

1.200 minimum

Boards or similar

Joists

Independent scaffold

Lattice support beams in both directions

1.000 minimum clear working space

Public walkway

Existing building or building under construction

Figure 2.6.3 Typical gantry scaffold details

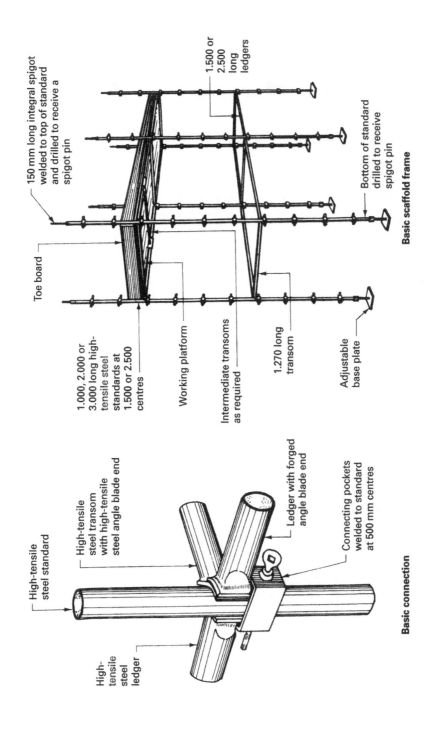

150 mm long integral spigot welded to top of standard and drilled to receive a spigot pin

1.500 or 2.500 long ledgers

Basic scaffold frame

Bottom of standard drilled to receive spigot pin

Toe board

1.000, 2.000 or 3.000 long high-tensile steel standards at 1.500 or 2.500 centres

Working platform

Intermediate transoms as required

1.270 long transom

Adjustable base plate

High-tensile steel standard

High-tensile steel transom with high-tensile steel angle blade end

Ledger with forged angle blade end

Connecting pockets welded to standard at 500 mm centres

High-tensile steel ledger

Basic connection

Figure 2.6.4 Typical system scaffold (courtesy: SGB Anglok Scaffolding)

Scissor lifts are used where there is a solid level floor surface, mostly for installing, for example, suspended ceilings, duct work and mechanical and electrical services. The maximum height of the working platform varies depending on the type of scissor lift, but heights of 12 m for the larger ones are not uncommon, and as with most mechanical access plant they are controlled by the operator who stands on the elevating platform.

The movement of the scissor action can be hydraulic, pneumatic or mechanical means depending on the power system employed, but for safety reasons they require no power to enter the 'descent' mode, just the release of a manual fail-safe valve (see Fig.2.6.5).

MOBILE ELEVATED WORK PLATFORM

Mobile elevated work platforms (MEWPs) comprise a basket formed by a working platform and guard rails, with either an inclined telescopic or an articulated hydraulic boom supported on a wheel-based electrical, LPG or diesel power unit. MEWPs can be self-driven, vehicle-mounted or trailer-mounted, with a range of rated capacities, working heights and outreaches. Most MEWPs work within the height range of 10–25 m, with a maximum of 40 m height achievable with the largest telescopic models. Most are designed to carry the weight of two people.

The major risks associated with MEWPs concern minimising overturning and stability control: for example, telescopic booms driven without proper training have a tendency to 'bounce' the operator. This is particularly relevant when considering the ground conditions that the MEWP will cross and be used on. Some of the MEWPs with a longer wheel base are specifically designed as rough-terrain vehicles with four-wheel drive and smooth, integrated transmission drive systems. Generally MEWPs have lower lifting capacities than scissor lifts, but their main advantage is that they can reach over obstacles to get close to the work area. They are particularly popular with steel fixers for steel frame erection. See Fig. 2.6.6.

MAST-CLIMBING WORK PLATFORMS

Mast-climbing work platforms are a relatively recent addition to site access plant and are mainly used for fixing cladding systems or façade maintenance on medium- and high-rise buildings. They are similar in fabrication to passenger and materials hoists – see Chapter 2.4 *Plant for transportation* on page 134 – and are often made of the same modular components. They are essentially a vertical steel mast connected to a horizontal work platform by a rack and pinion mechanism, powered by an electric motor to provide lift. This configuration allows access to the face of the building at any intermediate height from ground level up to the full height of the building. This is particularly useful when a construction project is too high or too complex for traditional scaffolding.

The stability of the mast and platform relies on it being tied at regular intervals to the building's structure at approximately 5–7.5 m intervals. Provided the mast is correctly tied, there is no maximum height to which a mast climber cannot be used. The working platform has continuous guard rails and some models have horizontal extensions that close the gap between the working platform and the building, making it safer. For single mast systems the maximum load is typically 1.5 t over a maximum platform length of 3 m; with dual-mast systems this can increase to up to 5 t over a platform length of typically 30 m.

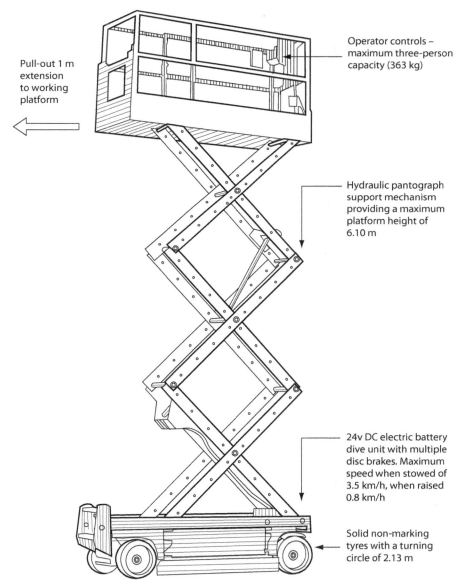

Pull-out 1 m extension to working platform

Operator controls – maximum three-person capacity (363 kg)

Hydraulic pantograph support mechanism providing a maximum platform height of 6.10 m

24v DC electric battery dive unit with multiple disc brakes. Maximum speed when stowed of 3.5 km/h, when raised 0.8 km/h

Solid non-marking tyres with a turning circle of 2.13 m

Figure 2.6.5   Scissor lift (Genie GS-2032 Scissor Lift, courtesy: Genie Industries)

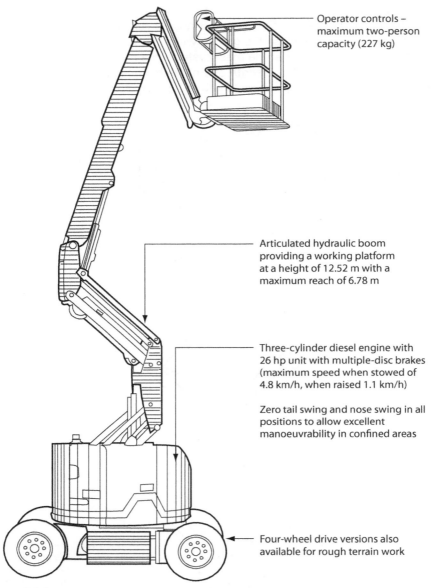

Operator controls –
maximum two-person
capacity (227 kg)

Articulated hydraulic boom
providing a working platform
at a height of 12.52 m with a
maximum reach of 6.78 m

Three-cylinder diesel engine with
26 hp unit with multiple-disc brakes
(maximum speed when stowed of
4.8 km/h, when raised 1.1 km/h)

Zero tail swing and nose swing in all
positions to allow excellent
manoeuvrability in confined areas

Four-wheel drive versions also
available for rough terrain work

Figure 2.6.6. **Mobile elevated work platform (MEWP) (Genie Z22/34 Articulated Boom Lift, courtesy: Genie Industries)**

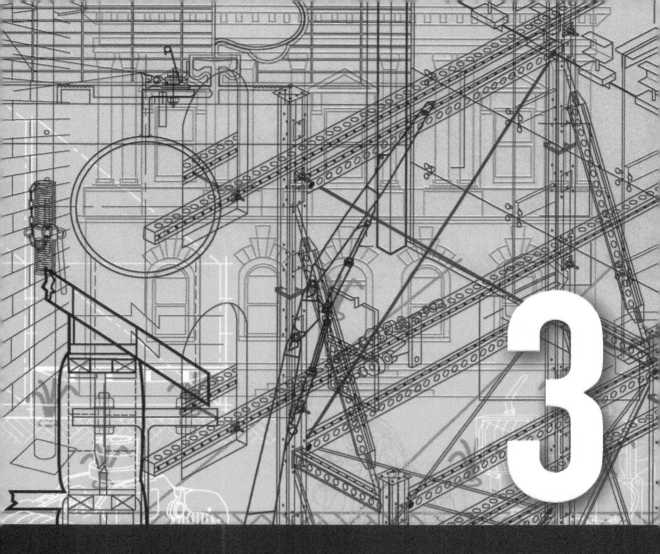

3

Substructure

Subsoil analysis and groundwater control 3.1

SUBSOIL ANALYSIS

Loadings in buildings consist of the combined dead weight of the structure plus the imposed loading from people, furnishings and the effect of snow and wind. These exert a downward pressure on the soil on which the structure is founded, and this in turn promotes a reactive force in the form of an upward pressure from the soil. The structure is, in effect, sandwiched between these opposite pressures, and the design of the building must be able to resist the resultant stresses set up within the structural members and the general building fabric. The supporting subsoil must be able to develop sufficient reactive force to give stability to the structure to prevent failure due to unequal settlement and to prevent failure of the subsoil due to shear. For designers to select, design and detail a suitable foundation, they must have adequate data on the nature of the soil on which the structure will be founded. This is normally obtained from a planned soil investigation programme.

SOIL INVESTIGATION

Soil investigation is specific in its requirements, whereas site investigation is all-embracing, taking into account such factors as topography, location of existing services, means of access and any local restrictions. Soil investigation is a means of obtaining data regarding the properties and characteristics of subsoils by studying geological maps and previous soil reports local to the area, as well as providing samples for testing or providing a means of access for visual inspection. Disturbed samples are useful for classifying and describing the soils through a visual inspection and sieve analysis, but more complex laboratory analysis, such as determining the shear strength of the soil, requires undisturbed samples to be taken. These try to capture the sample in its 'natural' state in terms of soil compaction, water content, volume and density. The actual data required and the amount of capital that can be reasonably expended on any soil investigation programme will depend on the type of structure proposed and on how much previous knowledge the designer has of a particular region or site.

The main methods of soil investigation are:

- trial pits – for small contracts where foundation depths are not likely to exceed 3 m deep;
- window sampling – for small to medium contracts with foundations up to 10 m deep;
- boreholes – for medium to large contracts with foundations up to 30 m deep.

TRIAL PITS

Trial pits are a relatively cheap method of obtaining soil data by enabling easy visual inspection of the soil stratum in its natural condition. The pits are usually machine-excavated to a plan size of 1.50 m × 0.45 m, and located over the site to suit the scope of the investigation. The number of pits required depends on the size of the site and the previous knowledge the soil engineer has of the site, although a series of pits set out on a 20 m grid would give a reasonable coverage of most sites. The pits need to be positioned so that the data obtained are truly representative of the actual conditions, but not in such a position where their presence could have a detrimental effect on the proposed foundations. In very loose soils, or soils having a high water table, trial pits can prove to be uneconomical because of the need for pumps and/or trench support to keep the pits dry and accessible. The spoil removed will provide disturbed samples for testing purposes, whereas undisturbed samples are difficult to obtain from trial pits.

WINDOW SAMPLING

Window samplers are lightweight sampling devices designed for taking disturbed soil samples from depths of up to 8 m. They fill the gap between lightweight hand augers and the bigger track or vehicle-mounted drilling rigs used for boreholes. The window samplers are cylindrical tubes of varying diameters, manufactured from high-tensile steel tubes complete with a replaceable cutting shoe. They are driven into the ground with a hydraulic breaker hammer or purpose-designed portable soil sampling rig. After driving the sample tube into the ground, a disturbed soil sample is removed from one of the windows for analysis.

BOREHOLES

These enable disturbed or undisturbed samples to be removed for analysis and testing. The core diameter of the samples obtained varies from 100 to 200 mm according to the method employed in extracting the sample. Disturbed samples can be obtained by using a rotary flight auger or by percussion boring in a manner similar to the formation of small-diameter bored piles using a tripod or shear leg rig. Undisturbed samples can be obtained from cohesive soils using 450 mm long × 100 mm diameter sampling tubes, which are driven into the soil to collect the sample; upon removal the tube is capped, labelled and sent off to a laboratory for testing. Undisturbed rock samples can be obtained by core drilling with diamond-tipped drills where necessary.

SOIL CHARACTERISTICS

Understanding and working out the characteristics of the soil is crucial for any construction project.

CLASSIFICATION OF SOILS

Soils may be classified by any of the following methods:

- physical properties;
- chemical composition;
- geological origin;
- particle size.

The physical properties of soils can be closely associated with their particle size, both of which are important to the foundation engineer, architect or designer. All soils can be defined as being coarse-grained or fine-grained, each resulting in different properties.

COARSE-GRAINED SOILS

Coarse-grained soils include sands and gravels with a low proportion of voids, negligible cohesion when dry, high permeability and slight compressibility, which takes place almost immediately on the application of load.

FINE-GRAINED SOILS

Fine-grained soils include the cohesive silts and clays with a high proportion of voids, high cohesion, very low permeability and high compressibility, which takes place slowly over a long period of time.

There are of course soils that can be classified in between these two extremes. BS 1377: *Methods of test for soils for civil engineering purposes* divides particle sizes as follows:

- clay particles: less than 0.002 mm;
- silt particles: between 0.002 and 0.06 mm;
- sand particles: between 0.06 and 2 mm;
- gravel particles: between 2 and 60 mm;
- cobbles: between 60 and 200 mm.

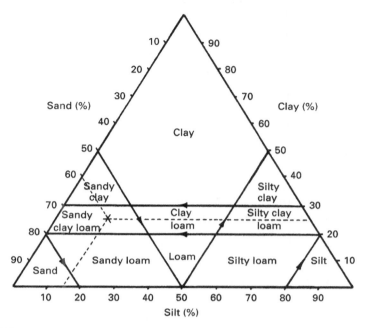

Figure 3.1.1 **Triangular chart of soil classification**

Soils composed mainly of clay, silt and sand can be classified on the triangular chart shown in Figure 3.1.1. Given a soil sample analysis of 60 per cent sand, 15 per cent silt and 25 per cent clay, it can be identifed as sandy clay loam.

Note: Silt is fine particles of sand, easily suspended in water. Loam is very fine particles of clay, easily dissolved in water.

The silt, sand and gravel particles are also further subdivided into fine, medium and coarse, with particle sizes lying between the extremes quoted above.

Very fine-grained soils such as clays are difficult to classify positively by their particle size distribution alone, so use is made of the reaction of these soils to a change in their moisture content. If the moisture content is high, the volume is large and the soil is basically a suspension of clay particles in water. As the moisture content decreases, the soil passes the liquid limit and becomes plastic. The liquid limit of a soil is defined as 'the moisture content at which a soil passes from the plastic state to the liquid state', and this limit can be determined by the test set out in BS 1377.

Further lowering of the moisture content will enable the soil to pass the plastic limit and begin to become a solid. The plastic limit of a soil is reached when a 20 g sample just fails to roll into a 3 mm diameter thread when rolled between the palm of the hand and a glass plate. When soils of this nature reach the solid state, the volume tends to remain constant, and any further decrease in moisture content will only alter the appearance and colour of the sample. It must be clearly understood that the change from one definable state to another is a gradual process and not a sudden change.

SHEAR STRENGTH OF SOILS

The resistance that can be offered by a soil to the sliding of one portion over another, or its shear strength, is of importance to the designer because it can be used to calculate the bearing capacity of a soil and the pressure it can exert on such members as timbering in excavations. Resistance to shear in a soil under load depends mainly on its particle composition. If a soil is granular in form, the frictional resistance between the particles increases with the load applied, and consequently its shear strength also increases with the magnitude of the applied load. Conversely, clay particles, being small, develop no frictional resistance, so clay's shear strength will remain constant whatever the magnitude of the applied load. Intermediate soils such as sandy clays normally give only a slight increase in shear strength as the load is applied.

To ascertain the shear strength of a particular soil sample, the triaxial compression test (described in BS 1377) is usually employed for cohesive soils. Non-cohesive soils can be tested in a shear box: a split box into which the sample is placed and subjected to a standard vertical load while a horizontal load is applied to the lower half of the box until the sample shears.

COMPRESSIBILITY

Another important property of soils that must be ascertained before a final choice of foundation type and design can be made is compressibility. Two factors must be taken into account:

- the rate at which compression takes place;
- the total amount of compression when a full load is applied.

When dealing with non-cohesive soils, such as sands and gravels, the rate of compression will keep pace with the construction of the building; when the structure is complete, there should be no further settlement if the soil remains in the same state. A soil is compressed when loaded by the expulsion of air and/or water from the voids, and by the natural rearrangement of the particles. In cohesive soils, the voids are often completely saturated with water, which in itself is nearly incompressible, so compression of the soil can take place only by the water moving out of the voids, allowing settlement of the particles. Expulsion of water from the voids within cohesive soils *can* occur, but only at a very slow rate, mainly because of the resistance offered by the plate-like particles of the soil through which it must flow. This gradual compressive movement of a soil is called consolidation. Uniform settlement will not normally cause undue damage to a structure, but uneven settlement can cause differential movement resulting in progressive structural damage.

STRESSES AND PRESSURES

The comments on shear strength and compressibility clearly indicate that cohesive soils present the most serious problems when considering foundation choice and design. The two major conditions to be considered are:

- shearing stresses;
- vertical pressures.

SHEARING STRESS

The maximum stress under a typical foundation carrying a uniformly distributed load will occur on a semicircle the radius of which is equal to half the width of the foundation; the isoshear line value will be equal to about one-third of the applied pressure (see Fig. 3.1.2). The magnitude of this maximum pressure should not exceed the shearing resistance value of the soil.

VERTICAL PRESSURE

Vertical pressure acts within the mass of the soil on which the structure is founded, and should not be of such a magnitude as to cause unacceptable settlement of the structure. Vertical pressures can be represented on a drawing by connecting together points that have the same value, forming what are termed pressure bulbs. Most pressure bulbs are plotted up to a value of 0.2 of the pressure per unit area, which is considered to be the limit of pressure that could influence settlement of the structure. Typical pressure bulbs are shown in Figs 3.1.2 and 3.1.3. A comparison of these typical pressure bulbs will show that vertical pressure generally decreases with depth; the 0.2 value will occur at a lower level under strip foundations than under rafts, isolated bases and bases in close proximity to one another, which form combined pressure bulbs. The pressure bulbs illustrated in Figs 3.1.2 and 3.1.3 are based on the soil being homogeneous throughout the depth under consideration. In reality this is not always the case, so it is important that soil investigation is carried out at least to the depth of the theoretical pressure bulb. Great care must be taken where underlying strata of highly compressible soil are encountered to ensure that these are not overstressed if cut by the anticipated pressure bulb.

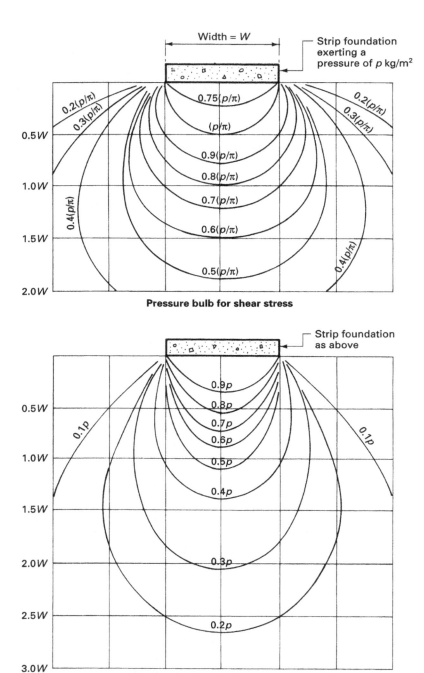

Figure 3.1.2   Strip foundations: typical pressure bulbs

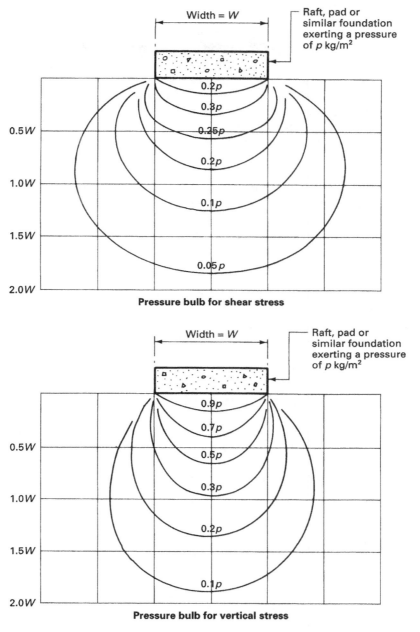

Width = W

Raft, pad or
similar foundation
exerting a pressure
of p kg/m²

0.2p
0.3p
0.5W   0.25p
         0.2p
1.0W   0.1p

1.5W
         0.05p
2.0W

Pressure bulb for shear stress

Width = W

Raft, pad or
similar foundation
exerting a pressure
of p kg/m²

0.9p
0.7p
0.5W   0.5p
         0.3p
1.0W   0.2p

1.5W
         0.1p
2.0W

Pressure bulb for vertical stress

Figure 3.1.3   Raft or similar foundations: typical pressure bulbs

CONTACT PRESSURE

It is often incorrectly assumed that a foundation that is uniformly loaded will result in a uniform contact pressure under the foundation. This would only be true if the foundation was completely flexible, such as the bases to a pin-jointed frame. The actual contact pressure under a foundation will be governed by the nature of the soil and the rigidity of the foundation and, because in practice most large structures have a rigid foundation, the contact pressure distribution is not uniform. In cohesive soils there is a tendency for high stresses to occur at the edges; this is usually reduced slightly by the yielding of the clay soil. Non-cohesive soils give rise to a parabolic contact pressure distribution, with increasing edge pressures as the depth below ground level of the foundation increases. When selecting the basic foundation format, consideration must be given to the concentration of the major loads over the position where the theoretical contact pressures are at a minimum to obtain a balanced distribution of contact pressure (see Fig. 3.1.4).

PLASTIC FAILURE

Plastic failure can occur in cohesive soils if the ultimate bearing capacity of the soil is reached or exceeded. As the load on a foundation is increased, the stresses within the soil also increase until all resistance to settlement has been overcome. Plastic failure, which can be related to the shear strength of the soil, occurs when the lateral pressure being exerted by the wedge of relatively undisturbed soil immediately below the foundation causes a plastic shear failure to develop, resulting in a heaving of the soil at the sides of the foundation moving along a slip circle or plane. In practice this movement tends to occur on one side of the building, causing it to tilt and settle (see Fig. 3.1.5). Plastic failure is likely to happen when the pressure applied by the foundation is approximately six times the shear strength of the soil.

APPROXIMATE BEARING CAPACITIES FOR DIFFERENT SUBSOILS

CLAY

Very stiff boulder clays and hard clays	420–650 kN/m^2
Stiff and sandy clays	220–420 kN/m^2
Firm and sandy clays	110–220 kN/m^2
Soft clays	55–110 kN/m^2
Very soft clays	<55 kN/m^2

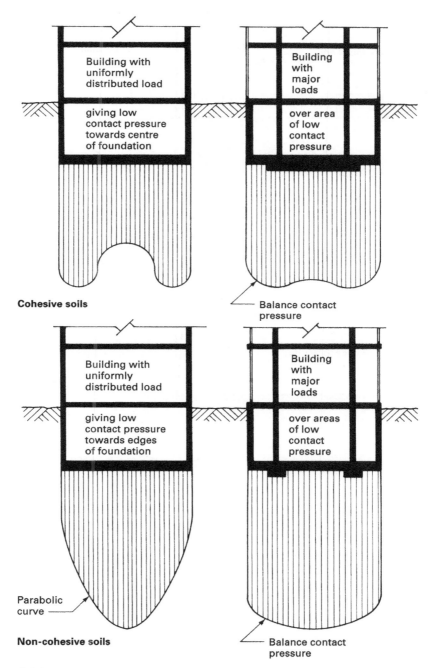

Cohesive soils

Building with uniformly distributed load

giving low contact pressure towards centre of foundation

Building with major loads

over area of low contact pressure

Balance contact pressure

Non-cohesive soils

Building with uniformly distributed load

giving low contact pressure towards edges of foundation

Building with major loads

over areas of low contact pressure

Parabolic curve

Balance contact pressure

Figure 3.1.4   Typical contact pressures

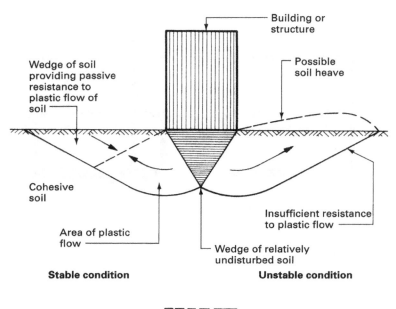

Building or structure

Wedge of soil providing passive resistance to plastic flow of soil

Possible soil heave

Cohesive soil

Insufficient resistance to plastic flow

Area of plastic flow

Wedge of relatively undisturbed soil

Stable condition   **Unstable condition**

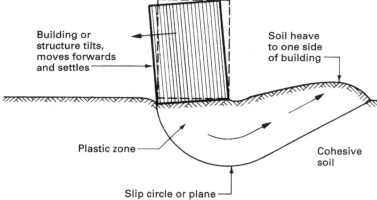

Building or structure tilts, moves forwards and settles

Soil heave to one side of building

Plastic zone

Cohesive soil

Slip circle or plane

Note: Failure is more usual on one side only than on both sides of the building or structure. Failure can occur if pressure applied is about six times the shear stress of the soil.

Figure 3.1.5 **Plastic failure of foundations**

SAND

Compact graded sands and gravels	430–650 kN/m²
Loose graded sands and gravel	220–430 kN/m²
Compact sands of consistent grade	220–430 kN/m²
Loose sands of consistent grade	110–220 kN/m²
Silts	55–110 kN/m²

Note: Figures given are for dry sands. If saturated, the bearing capacity is about one half of that given.

ROCKS

Sound igneous and gneissic rocks	10,700 kN/m²
Dense lime and sandstones	4,300 kN/m²
Slate and schists	3,200 kN/m²
Dense shale, mudstones and soft sandstone	2,200 kN/m²
Clay shales	1,100 kN/m²
Dense chalk	650 kN/m²
Thin bedded limestone and sandstone	*
Broken or shattered rocks and soft chalk	*

* To be assessed in situ, as may be unstable.

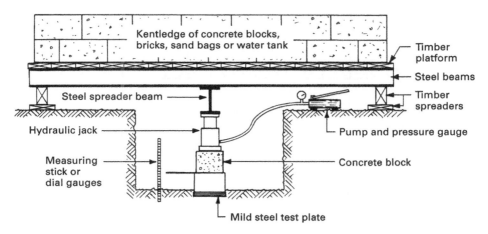

Figure 3.1.6   Plate bearing test: principles

PLATE BEARING TEST

In addition to subjecting subsoil to these site and laboratory tests, trial pit investigation provides an opportunity to load the soil physically in positions where it is anticipated that the foundations will be. The equipment shown in Fig. 3.1.6 replicates the design situation at a proportional reduction: a proportionally smaller foundation bearing area is loaded proportionally less. Data are recorded as measured in settlement (mm) and soil bearing capacity (kN/m^2).

The procedure is to jack between the steel test plate and an immovable load or kentledge hydraulically until the required pressure or resistance is achieved. Thereafter, pressure increase and corresponding settlement are measured, usually at 12- and 24-hour intervals. Loading is applied in increments of about one-fifth of the anticipated ultimate load per 24-hour interval. The results are plotted graphically, as shown in Fig. 3.1.7. The safe design load should be the ultimate load divided by a factor of safety of at least 3.

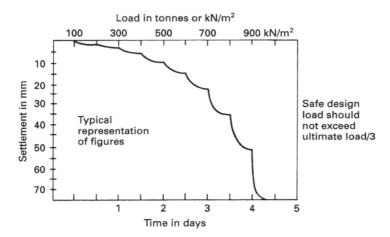

Figure 3.1.7 **Load-settlement graph**

EXAMPLE

If the foundation design is to be a pad of base dimensions 1,400 mm × 1,400 mm loaded to 40 tonnes, a proportionally reduced test plate of 700 mm × 700 mm (1/4 area) may be used to carry 10 tonnes (40 × 1/4). Therefore:

10 tonnes × safety factor 3 = 30 tonnes (approx. 300 kN) ultimate load to be applied.

GROUNDWATER CONTROL

When rainwater collects in surface depressions, it is called surface water; when it percolates down into the permeable subsoil and collects in tiny soil voids, it is called groundwater; the upper surface of the groundwater is called the water table.

Both surface and groundwater can be effectively controlled by a variety of methods, designed either to exclude the water from a particular area or merely to lower the water table to give reasonably dry working conditions, especially for excavation activities. This is important as working in waterlogged conditions can delay substructure work considerably and can cause safety problems, such as flooding and the collapse of the sides of excavations.

The extent to which water affects the stability and bearing capacity of a subsoil will depend on the physical characteristics of the soil, in particular the particle size, which ranges from the fine particles of clay soils to the larger particles or boulders of some granular soils. The effect of the water on these particles is that of a lubricant, enabling them to move when subjected to a force, such as a foundation loading, or simply causing them to flow by movement of the groundwater. The number and disposition of the particles, together with the amount of water present, will determine the amount of movement that can take place. The finer particles will be displaced more easily than the larger particles; this can create voids and encourage the larger particles to settle.

The voids caused by excavation works encourage water to flow, because the opposition to the groundwater movement provided by the soil has been removed. In cases where flow of water is likely, the civil engineering contractor must introduce mechanical or geotechnical processes to restrict this flow to keep the excavations to safe. These processes can be broadly classified into two groups:

- permanent exclusion of groundwater;
- temporary exclusion of groundwater by lowering the water table.

PERMANENT EXCLUSION OF GROUNDWATER

SHEET PILING

Sheet piling is suitable for all types of soil except boulder beds, and is used to form a barrier or cut-off wall to the flow of groundwater. The sheet piling can be of a permanent nature, designed to act as a retaining wall, or it can be a temporary enclosure to excavation works in the form of a cofferdam (for details see Chapter 3.2). To ensure true watertightness, the clutch connections between individual piles need to be welded or grouted. Vibration and noise due to the driving process may render this method unacceptable, and the capital costs can be high unless they can be apportioned over several contracts on a use and reuse basis.

A recent development is the use of vinyl or composite sheet piles which, unlike their steel counterparts, do not rust – always a consideration when specifying the gauge (thickness) of steel sheet piles. Composite sheets piles are manufactured from recycled materials and provide a rust-free, chemically resistant and maintenance-free support and water cut-off solution, costing less than steel alternatives. Although not as strong as steel, they can retain soils and water up to 2 m high, with typical section modules ranging from 300 cm^3 to 2,000 cm^3 per metre length. Ideally suited to installation in soft clays or loose granular soils, the vinyl piles can be installed using a range of vibratory hammers.

DIAPHRAGM WALLS

Diaphragm walls are suitable for all types of soil, and are usually of in-situ reinforced concrete installed using the bentonite slurry method (see Chapter 3.2 for details). This form of diaphragm wall has several advantages:

- low installation noise and vibration;
- can be used in restricted spaces;
- can be installed close to existing foundations.

However, this method is generally uneconomic unless the diaphragm wall forms part of the permanent structure.

SLURRY TRENCH CUT-OFF WALLS

Slurry trench cut-off walls are non-structural, thin, diaphragm walls cast in situ from bentonite slurry and not reinforced. They are suitable for subsoils of silts, sands and gravels. They can be used on sites where there is sufficient space to enclose the excavation area with a cut-off wall of this nature, sited so that there is sufficient earth remaining between the wall and the excavation to give the screen or diaphragm wall support. Provided adequate support is given, these walls are rapidly installed and are cheaper than the structural version.

THIN-GROUTED MEMBRANE WALLS

Thin-grouted membrane walls are an alternative method to the slurry trench cut-off wall when used in silt and sand subsoils. They are also suitable for installation in very permeable soils and made-up ground where bentonite methods are unsuitable. As with slurry trench cut-off walls, ample earth support is required for this non-structural cut-off wall. The common method of formation is to drive into the ground a series of touching universal beam or column sections, sheet pile sections or small steel box sections to the required depth. A grout injection pipe is fixed to the web or face of the section, and this is connected by means of a flexible pipe to a grout pump at ground level. As the sections are withdrawn, the void created is filled with cement grout to form the thin membrane (see Fig. 3.1.8).

SECANT PILING

Secant piling is an alternative method to the reinforced concrete diaphragm wall, consisting of a series of interlocking reinforced concrete-bored piles (see *Construction Technology*, Chapter 3.6, where these were mentioned in the context of basement construction). The formation of the secant bored piles can be carried out as described in Chapter 3.4, ensuring that the piles interlock for their entire length. This will require special cutting tools to form the key in the alternate piles for the interlocking intermediate piles. The pile diameter selected will be determined by the strength required after completion of the excavations to one side of the wall. The usual range of diameters used is between 300 and 600 mm. The piling can be faced with a reinforced rendering or covered with a mesh reinforcement, sprayed with concrete to give a smooth finish (called shotcrete or gunite). An alternative method is to cast, in front of the piles, a reinforced wall terminating in a capping beam to the piles (see Fig. 3.1.8).

GROUT INJECTION

Grouts of all kinds are usually injected into the subsoil by pumping in the mixture at high pressure through tubes placed at the appropriate centres, according to the solution being used and/or the soil type. Soil investigation techniques will reveal the information required to enable the engineer to decide on the pattern and spacing of the grout holes, which can be drilled with pneumatic tools or tipped drills. The pressure needed to ensure satisfactory penetration of the subsoil will depend on the soil conditions and results required, but is usually within the range of 1 N/mm^2 for fine soils to 7 N/mm^2 for cement grouting in fissured and jointed rock strata.

There are many types of grouting systems depending on the soil type present and the pressure of the groundwater. These are usually designed and applied by companies that specialise in this form of permanent groundwater exclusion. Here are some typical examples.

Cement grouts

In common with all grouting methods, cement grouts are used to form a 'curtain' in soils that have high permeability, making temporary exclusion pumping methods economic. Cement grouts are used in fissured and jointed rock strata, and are injected into the ground through a series of grouting holes bored into the ground in lines, with secondary intermediate borehole lines if necessary. The grout can be a mixture of neat cement and water, cement and sand up to a ratio of 1:4, or PFA (pulverised fuel ash) and cement in the ratio of 1:1 with 2 parts of water by weight. The usual practice is to start with a thin grout and gradually reduce the water:cement ratio as the process proceeds, to increase the viscosity of the mixture. To be effective, this form of treatment needs to be extensive.

An elevation on a thin cement grouted curtain

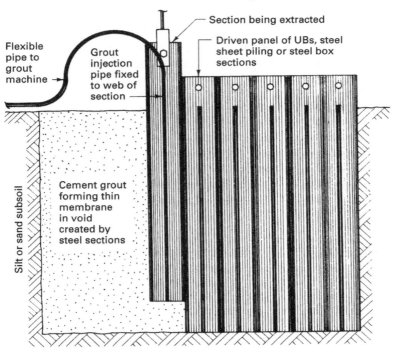

Flexible pipe to grout machine

Grout injection pipe fixed to web of section

Section being extracted

Driven panel of UBs, steel sheet piling or steel box sections

Silt or sand subsoil

Cement grout forming thin membrane in void created by steel sections

Section through a secant pile wall with an in-situ concrete facing

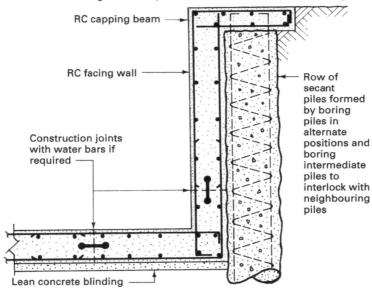

RC capping beam

RC facing wall

Construction joints with water bars if required

Lean concrete blinding

Row of secant piles formed by boring piles in alternate positions and boring intermediate piles to interlock with neighbouring piles

Figure 3.1.8 **Thin grouted membrane and secant piling**

Clay/cement grouts

Clay/cement grouting is suitable for sands and gravels where the soil particles are too small for cement grout treatment. The grout is introduced by means of a sleeve grout pipe that limits its spread and, as with cement grouting, the equipment is simple and can be used in a confined space. The clay/cement grout is basically bentonite with additives such as Portland cement or soluble silicates to form the permanent barrier. One disadvantage of this method is that at least 4 m of natural cover is required to provide support for the non-structural barrier.

Chemical and resin grouts

Chemical and resin grouting is suitable for use in sands and gravels to stabilise the soil, and can also be used for underpinning works below the water-table level. The chemicals are usually mixed before injection into the ground through injection pipes inserted at 600 mm centres. The chemicals form a permanent gel in the earth, which increases the strength of the soil and reduces its permeability. This method is called the one-shot method – a liquid base diluted with water is mixed with a catalyst to control the gel-setting time before being injected into the ground. An alternative, two-shot method can be used – the first chemical (usually sodium silicate) is injected into the ground followed immediately by the second chemical (calcium chloride) to form a silica gel. The reaction of the two chemicals is immediate, whereas in the one-shot method the reaction of the chemicals can be delayed to allow for full penetration of the subsoil, which will in turn allow a wider spacing of the boreholes.

One main disadvantage of chemical grouting is the need for at least 2 m of natural cover. Resin grouts are similar in application to the chemical grouts, but have a lower viscosity, which enables them to penetrate the fine sands that are unsuitable for chemical grouting applications.

Bituminous grouts

Bituminous grouts are suitable for injection into fine sands to decrease the permeability of the soil, but they will not increase the strength of the soil, and are therefore unsuitable for underpinning work.

FREEZING

Freezing is suitable for all types of subsoil with a moisture content in excess of 8 per cent of the voids. The basic principle is to insert freezing tubes into the ground and circulate a freezing solution around the tubes to form ice in the voids, creating a wall of ice to act as the impermeable barrier. This method will give the soil temporary extra mechanical strength, but there is a slight risk of ground heave, particularly when operating in clays and silts. The circulating solution can be a brine of magnesium chloride or calcium chloride at a temperature of between -15 and -25 °C, which would take between 10 and 17 days to produce an ice wall 1 m thick, depending on the type of subsoil. For works of short duration where quick freezing is required, the more expensive liquid nitrogen can be used as the circulating medium. A typical freezing arrangement is shown in Fig. 3.1.9. Freezing methods of soil stabilisation are especially suitable for excavating deep shafts and driving tunnels.

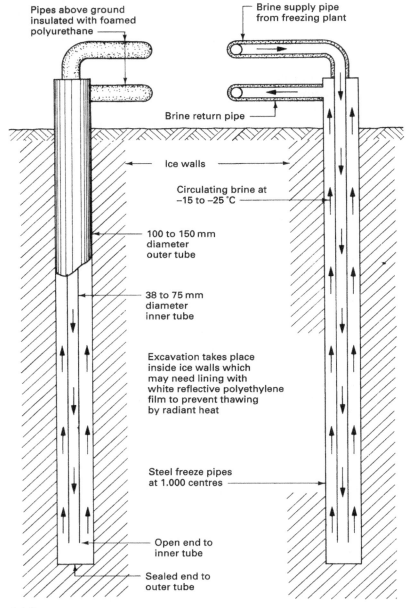

Pipes above ground insulated with foamed polyurethane

Brine supply pipe from freezing plant

Brine return pipe

Ice walls

Circulating brine at −15 to −25 °C

100 to 150 mm diameter outer tube

38 to 75 mm diameter inner tube

Excavation takes place inside ice walls which may need lining with white reflective polyethylene film to prevent thawing by radiant heat

Steel freeze pipes at 1.000 centres

Open end to inner tube

Sealed end to outer tube

Figure 3.1.9   Exclusion of groundwater by freezing

SUMP PUMPING

Sump pumping is suitable for most subsoils and, in particular, gravels and coarse sands when working in open shallow excavations. The sump or water collection pit should be excavated below the formation level of the excavation and preferably sited in a corner position to reduce to a minimum the soil movement due to settlement, which is a possibility with this method. Open sump pumping is usually limited to a maximum depth of 7.5 m because of the suction lift limitations of most pumps. An alternative method to the open sump pumping is the jetted sump, which will achieve the same objective and will also prevent the soil movement. In this method a metal tube is jetted into the ground and the void created is filled with a sand medium, a disposable hose and a strainer, as shown in Fig. 3.1.10.

WELLPOINT SYSTEMS

Wellpoint systems are popular for water lowering in non-cohesive soils up to a depth of between 5 and 6 m. To dewater an area beyond this depth requires a multi-stage installation (see Fig. 3.1.11). The basic principle is to water-jet into the ground a number of small diameter wells, which are connected to a header pipe attached to a vacuum pump (see Fig. 3.1.12).

Wellpoint systems can be installed with the header pipe acting as a ring main enclosing the area to be excavated. The header pipe should be connected to two pumps: the first for actual pumping operations, and the second as a standby pump, because it is essential to keep the system fully operational to avoid collapse of the excavation should a pump failure occur. The alternative system is the progressive line arrangement, where the header pipe is placed alongside a trench or similar excavation to one side or both sides, depending on the width of the excavation. A pump is connected to a predetermined length of header pipe, and further wellpoints are jetted in ahead of the excavation works. As the work including backfilling is completed, the redundant wellpoints are removed and the header pipe is moved forwards.

SHALLOW-BORED WELLS

Shallow-bored wells are suitable for sandy gravels and water-bearing rocks. The action is similar in principle to wellpoint pumping, but is more appropriate for installations that have to be pumped for several months, because running costs are generally lower. This method is subject to the same lift restrictions as wellpoint systems and can be arranged as a multi-stage system if the depth of lowering exceeds 5 m.

DEEP-BORED WELLS

Deep-bored wells can be used as an alternative to a multi-stage wellpoint installation where the groundwater needs to be lowered to a depth greater than 9 m. The wells are formed by sinking a 300 to 600 mm diameter steel lining tube into the ground to the required depth and at spacings to suit the subsoil being dewatered. This borehole allows a perforated well liner to be installed with an electro-submersible pump to extract the water. The annular space is filled with a suitable medium, such as sand and gravel, as the outer steel lining tube is removed (see Fig. 3.1.13).

ELECTRO-OSMOSIS

Electro-osmosis is an uncommon and costly method, which can be used for dewatering cohesive soils such as silts and clays where other pumping methods would not be adequate. It works on the principle that soil particles carry a negative charge, which attracts the positively charged ends of the water molecules, creating a balanced state; if this balance is disturbed, the water will flow. The disturbance of this natural balance is created by inserting two electrodes into the ground and passing an electric charge between them. The positive electrode can be of steel rods or sheet piling, which will act as the anode; a wellpoint is installed to act as the cathode or negative electrode. When an electric current is passed between the anode and cathode, the positively charged water molecules flow to the wellpoint (cathode), where the water is collected and pumped away to a discharge point. The power consumption for this method can vary from 1 kW/m^3 for large excavations up to 12 kW/m^3 of soil dewatered for small excavations, which will generally make this method uneconomic on running costs alone.

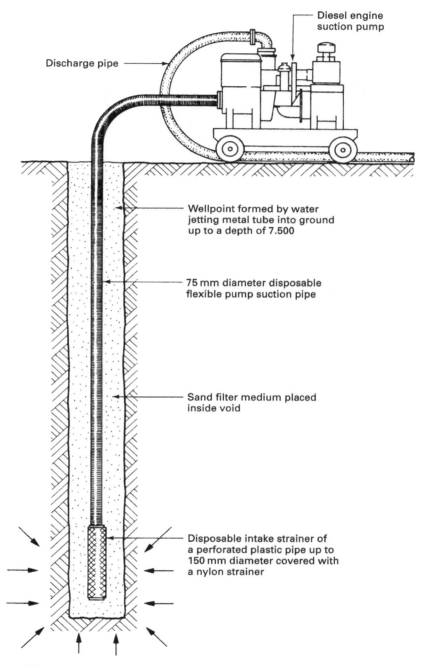

Discharge pipe

Diesel engine
suction pump

Wellpoint formed by water
jetting metal tube into ground
up to a depth of 7.500

75 mm diameter disposable
flexible pump suction pipe

Sand filter medium placed
inside void

Disposable intake strainer of
a perforated plastic pipe up to
150 mm diameter covered with
a nylon strainer

Figure 3.1.10   Typical jetted sump detail

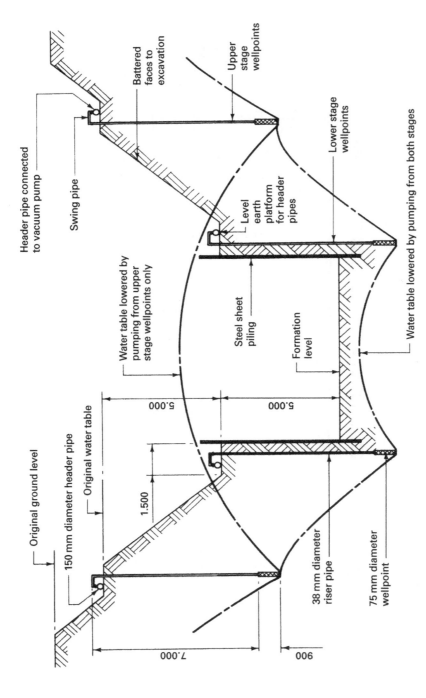

Header pipe connected to vacuum pump

Swing pipe

Battered faces to excavation

Upper stage wellpoints

Lower stage wellpoints

Level earth platform for header pipes

Water table lowered by pumping from both stages

Original ground level

150 mm diameter header pipe

Original water table

Water table lowered by pumping from upper stage wellpoints only

Steel sheet piling

Formation level

5.000

5.000

1.500

38 mm diameter riser pipe

75 mm diameter wellpoint

7.000

900

Figure 3.1.11   Typical example of a multi-stage wellpoint dewatering installation

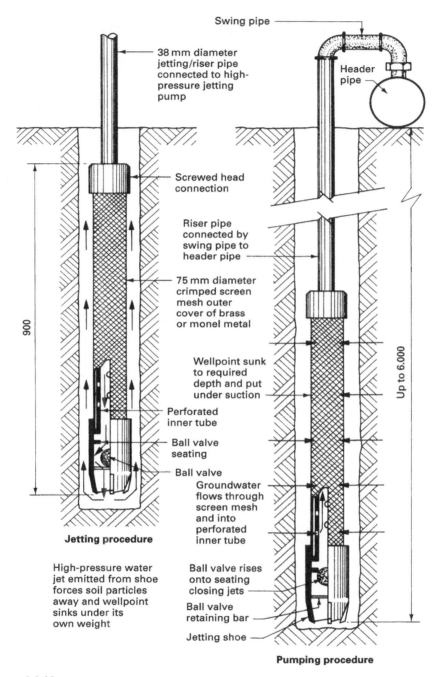

38 mm diameter jetting/riser pipe connected to high-pressure jetting pump

Swing pipe

Header pipe

Screwed head connection

Riser pipe connected by swing pipe to header pipe

75 mm diameter crimped screen mesh outer cover of brass or monel metal

900

Up to 6.000

Wellpoint sunk to required depth and put under suction

Perforated inner tube

Ball valve seating

Ball valve

Groundwater flows through screen mesh and into perforated inner tube

Jetting procedure

High-pressure water jet emitted from shoe forces soil particles away and wellpoint sinks under its own weight

Ball valve rises onto seating closing jets

Ball valve retaining bar

Jetting shoe

Pumping procedure

Figure 3.1.12 Typical wellpoint installation details

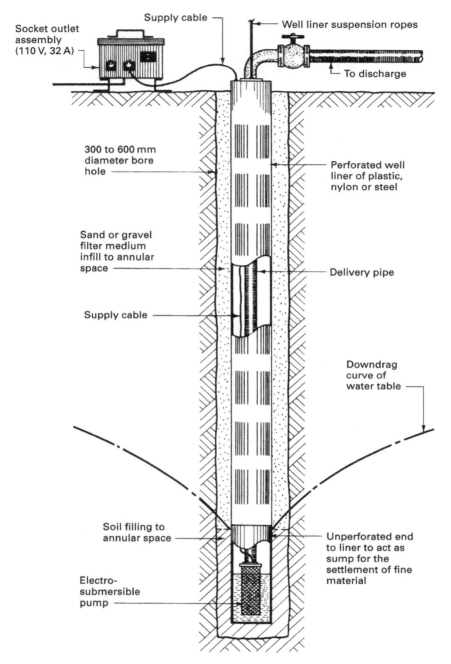

Socket outlet
assembly
(110 V, 32 A)

Supply cable

Well liner suspension ropes

To discharge

300 to 600 mm
diameter bore
hole

Perforated well
liner of plastic,
nylon or steel

Sand or gravel
filter medium
infill to annular
space

Delivery pipe

Supply cable

Downdrag
curve of
water table

Soil filling to
annular space

Unperforated end
to liner to act as
sump for the
settlement of fine
material

Electro-
submersible
pump

Figure 3.1.13 Typical deep-bored well details

As explained, the exclusion of groundwater can be achieved using a variety of methods depending on the type of soil, its compaction and any artesian water pressure that may be present. Table 3.1.1 provides an easy reference to compare the methods previously discussed. Shown for completeness are the methods of groundwater exclusion that also provide structural ground support for substructures such as basements, which are covered later in Chapter 3.5.

Table 3.1.1 Comparison of groundwater reduction methods

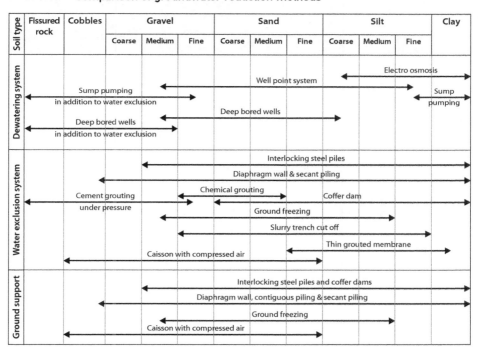

Deep substructure excavations   3.2

The Construction (Design And Management) Regulations 2007, the Confined Spaces Regulations 1997 and The Work at Height Regulations 2005 are all concerned with the safety aspects of deep excavations. The construction of deep trenches is to be avoided wherever possible through choosing alternative techniques to reduce the risk of harm to operatives. These methods include, for example, trenchless pipe systems such as pipe jacking (see page 252) or specifying the use of steel trench boxes. Details of modern trench support systems can be found in *Construction Technology*, Chapters 2.4 and 3.1.

Where deep substructure excavations cannot be avoided, the methods that can be used to support the sides can be considered under five headings:

1. raking struts;
2. steel sheet piling;
3. diaphragm walls;
4. cofferdams;
5. caissons.

RAKING STRUTS

Raking struts are used where it is possible to excavate the basement area back to the perimeter line without the need for shoring – so firm subsoils must be present. The perimeter is trimmed and the timbering or steel sheet piling placed in position and strutted, using either raking timber or adjustable steel struts converging on a common grillage platform, similar to raking shoring. Alternatively, each raker is taken down to a separate sole plate and the whole arrangement is adequately braced. Another method is to excavate back to the perimeter line on the subsoil's natural angle of repose and then cast the base slab, to protect the excavation bottom from undue drying out and subsequent shrinkage. The perimeter trimming can then be carried out, timbered and strutted using the base slab as the abutment (see Fig. 3.2.1). The retaining wall is cast in stages as with the perimeter trench method, and the strutting is transferred to the stem as the work proceeds.

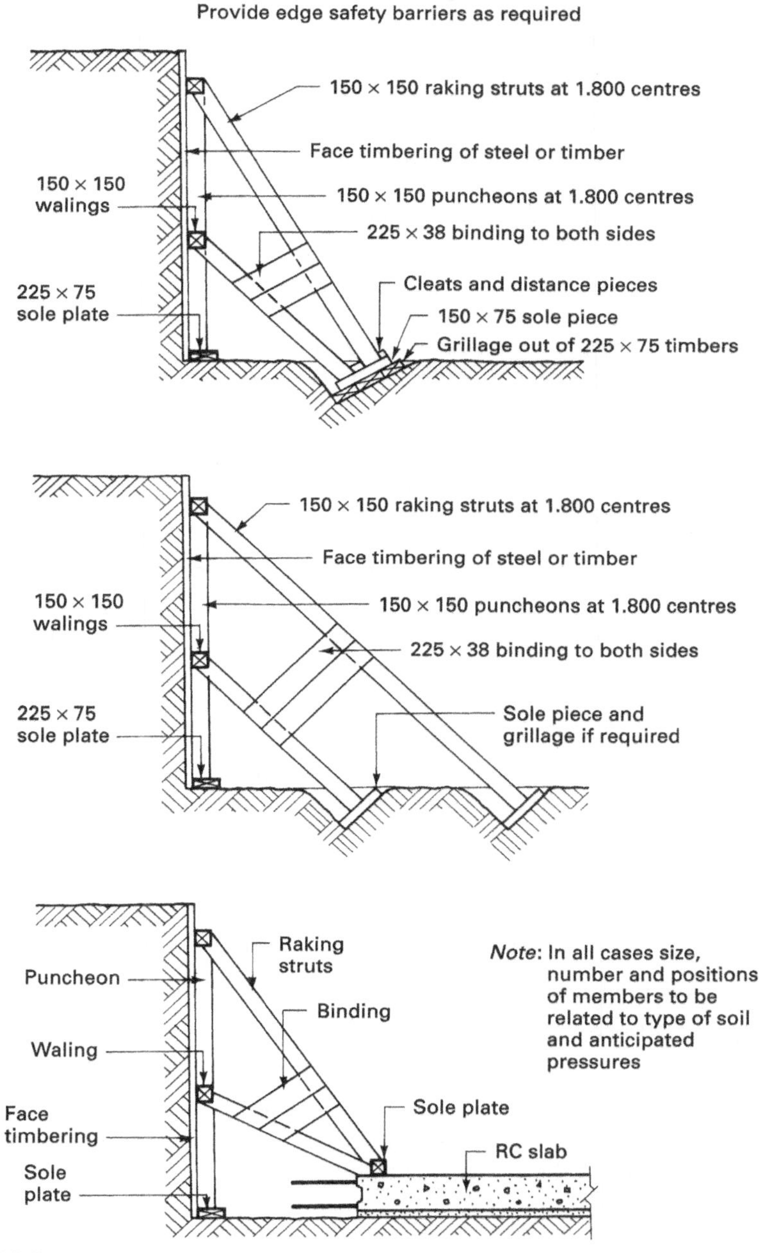

Figure 3.2.1　Basement timbering: raking struts

Steel sheet piling is the most common form of sheet piling, although timber is also common, and in recent years the use of composite vinyl panels made from a variety of recycled plastics has also become popular. Sheet piling can be used in temporary works, such as timbering to excavations in soft and/or waterlogged soils, and in the construction of cofferdams. This material can also be used to form permanent retaining walls, especially those used for riverbank strengthening, and in the construction of jetties. Four common forms of steel sheet pile are the Larssen U-shaped, Frodingham Z-shaped, straight-web piles and tubular piles, all of which have an interlocking joint, known as a clutch, to form a water seal. This may need caulking or grouting where high water pressures are encountered. Straight-web sheet piles are used to provide temporary support to deep service trenches. Tubular piles are used where there are large lateral surcharge and soil loads, because of their increased rigidity and section modulus. Larssen and Frodingham sheet piles are suitable for all uses, and can be obtained in lengths up to 30 m according to the particular section chosen – typical sections are shown in Fig. 3.2.2.

Erecting and installing a series of sheet piles and keeping them vertical in all directions usually requires a guide frame or trestle constructed from large-section timbers or hot-rolled steel sections. The piles are pitched or lifted by crane, using the lifting holes sited near the top of each length, and positioned between the guide walings of the trestle (see Figs 3.2.3 and 3.2.4). When sheet piles are being driven using a percussion or vibratory hammer (see later), there is a tendency for them to creep or lean in the direction of driving. Correct driving methods will help to eliminate this tendency, and the generally accepted method is to install the piles in panels according to these steps:

1. Pitch a pair of piles and drive them until approximately one-third of their length remains above ground level: these act as anchor piles to stop the remainder of the piles in the panel from leaning or creeping while being driven. It is essential that this first pair of piles are driven accurately and plumb in all directions.

2. Pitch a series of piles in pairs adjacent to the anchor piles to form a panel of 10 to 12 pairs of piles.

3. Partially drive the last pair of piles to the same depth as the anchor piles.

4. Drive the remaining piles in pairs, including the anchor piles to their final set.

5. The last pair of piles remain projecting for about one-third of their length above the ground level, to act as guide piles to the next panel.

To facilitate accurate and easy driving, there should be about a 6 mm clearance between the pile faces and the guide walings, and you should use timber spacer blocks in the troughs of the piles (see Fig. 3.2.3).

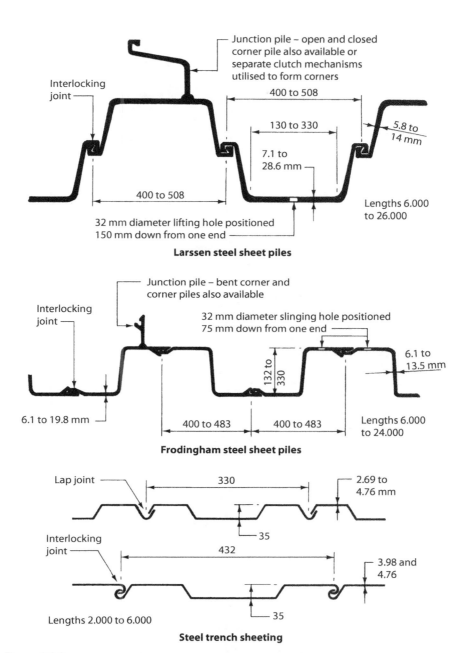

Interlocking joint

Junction pile – open and closed corner pile also available or separate clutch mechanisms utilised to form corners

400 to 508

130 to 330

5.8 to 14 mm

7.1 to 28.6 mm

400 to 508

Lengths 6.000 to 26.000

32 mm diameter lifting hole positioned 150 mm down from one end

Larssen steel sheet piles

Interlocking joint

Junction pile – bent corner and corner piles also available

32 mm diameter slinging hole positioned 75 mm down from one end

6.1 to 13.5 mm

132 to 330

6.1 to 19.8 mm

400 to 483

400 to 483

Lengths 6.000 to 24.000

Frodingham steel sheet piles

Lap joint

330

2.69 to 4.76 mm

35

Interlocking joint

432

3.98 and 4.76

Lengths 2.000 to 6.000

35

Steel trench sheeting

Figure 3.2.2 Typical steel sheet pile and trench sheeting foundations

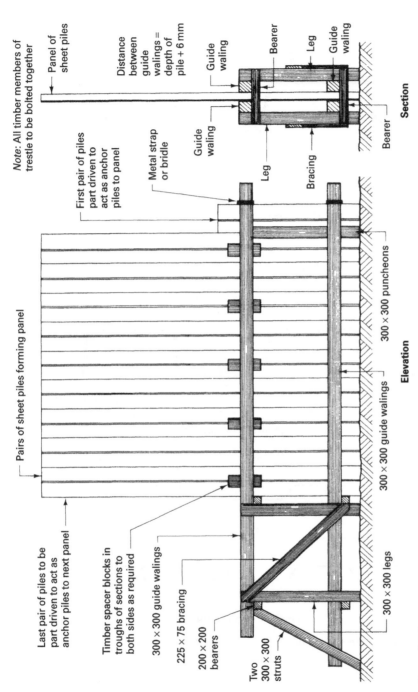

Note: All timber members of trestle to be bolted together

Panel of sheet piles

Distance between guide walings = depth of pile + 6 mm

Guide waling

Bearer

Leg

Guide waling

Guide waling

Bearer

Section

First pair of piles part driven to act as anchor piles to panel

Metal strap or bridle

Guide waling

Leg

Bracing

Pairs of sheet piles forming panel

300 × 300 puncheons

300 × 300 guide walings

Elevation

Last pair of piles to be part driven to act as anchor piles to next panel

Timber spacer blocks in troughs of sections to both sides as required

300 × 300 guide walings

225 × 75 bracing

200 × 200 bearers

Two 300 × 300 struts

300 × 300 legs

Figure 3.2.3 Conventional timber trestle for installing steel sheet piles using a percussion or vibratory hammer

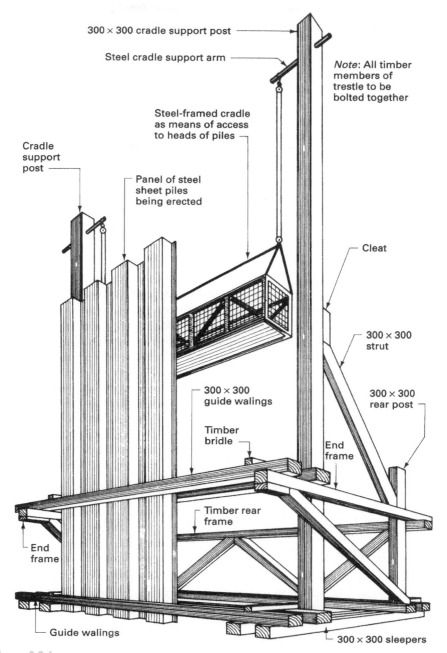

300 × 300 cradle support post

Steel cradle support arm

Note: All timber members of trestle to be bolted together

Steel-framed cradle as means of access to heads of piles

Cradle support post

Panel of steel sheet piles being erected

Cleat

300 × 300 strut

300 × 300 rear post

300 × 300 guide walings

Timber bridle

End frame

Timber rear frame

End frame

300 × 300 sleepers

Guide walings

Figure 3.2.4 Typical timber trestle for large steel sheet piles

Steel sheet piles may be driven to the required set using percussion hammers, vibratory systems or hydraulic drivers.

- **Percussion hammers** are activated by compressed air or diesel power. They are usually equipped with leg grips bolted to the hammer and fitted with inserts to grip the face of the pile, to ensure that the hammer is held in line with the axis of the pile. Wide, flat driving caps are also required to prevent the head of the pile being damaged by impact from the hammer. These tend to be fast and effective but very noisy, so are not suitable for residential or built-up areas.

- **Vibratory systems** use an oscillating mechanism clamped to the top of the pile that vibrates at a high frequency, causing the ground below the sheet pile to liquefy. The pile then sinks under the combined weight of the clamp and its self-weight. The system is less noisy than the percussion hammer, but takes longer to achieve. It works best in silt or sandy soil.

- **Hydraulic drivers** can be used to push the piles into suitable subsoils, such as clays, silts and fine granular soils. These driving systems are vibrationless and almost silent, making them ideal for installing sheet piling close to other buildings. The driving head usually consists of a power pack crosshead containing eight hydraulic rams, two of them connected to the top of each pile with special pile connectors. Piles are driven incrementally in batches of four piles. One pair of rams pushes one pile using the pulling resistance from the adjacent three pairs of rams, fitted to piles that have previously been partly vibrated into position. Each ram has a short stroke of 750 mm; this process is repeated on each of the three other piles until the desired penetration of all four piles is reached. This is by far the quietest, most accurate and most effective of all three methods and can be applied to all sheet piles, whether U-shaped, Z-shaped or circular rolled hollow sections.

Piles used in temporary works should have well-greased interlocking clutches to enable them to be extracted with ease, using specially designed sheet pile extractors. The extractors usually consist of a compressed-air-activated ram giving between 120 and 200 upward blows per minute, which causes the jaws at the lower end to grip the pile and force it out of the ground.

AIDS TO DRIVING SHEET PILES

In soft and silty subsoils, the installation of steel sheet piles can be assisted using high-pressure water jetting to the sides and toes of the piles. The jets should be positioned in the troughs of the piles and to both sides of the section. In very soft subsoils, this method can sometimes be so effective that the assistance of a driving hammer is not required. To ensure that the pile finally penetrates into an undisturbed layer of subsoil, the last few metres or so of driving should be carried out without jetting. In sands and gravels, or where there is hard ground, some systems incorporate a rotary drill fixed to the driving mast to break up the ground just below the pile being sunk.

The Giken™ system for driving tubular piles has a rotary 'gyro-piler', which is a coring attachment fitted to the tubular pile, allowing it to bore through buried structures such as old foundations. This is particularly useful when existing retaining walls or bridge abutments are being replaced such as in a typical road widening scheme.

ACCESS

Access in the form of a working platform to the heads of sheet piles is often necessary during installation to locate adjacent piles, to position and attach the driving hammer, and to carry out inspections of the work in progress. Suitable means of access are:

- independent scaffolding;
- suspended cradle (see Fig. 3.2.4);
- mobile platform mounted on a hydraulic arm;
- working platform in the form of a cage suspended from a mobile crane, which is then hooked on to the pile heads.

DIAPHRAGM WALLS

A diaphragm is a dividing membrane, and in the context of building a diaphragm wall can be used as a retaining wall to form the perimeter wall of a basement structure, to act as a cut-off wall for river or similar embankments and to retain large masses of soil, such as the side wall of a road underpass.

IN-SITU CONCRETE METHOD

In-situ concrete diaphragm walls are being used widely in modern construction work and have a number of advantages.

- The final wall can be designed and constructed as the required structural wall.
- Diaphragm walls can be constructed before the bulk excavation takes place, eliminating the need for temporary works, such as timbering.
- Methods that can be employed to construct the wall are relatively quiet and have little or no vibration.
- Work can be carried out immediately adjacent to an existing structure.
- They can be designed to resist vertical and/or horizontal forces.
- Walls are watertight when constructed.
- Virtually any plan shape is possible.
- Overall, they are an economic method for the construction of basement or retaining walls.

There are two methods by which a cast in-situ diaphragm wall may be constructed:

- touching (contiguous) or interlocking (secant) bored piles;
- excavation of a deep trench in panels using the bentonite slurry method.

The formation of bored piles is fully described in Chapter 3.4 on piling, and the application of contiguous and secant piles were described in *Construction Technology*, Chapter 3.6. These types of substructure wall require a reinforced tie beam over the heads of the piles and a facing to take out the irregularities of the surface. This facing usually takes the form of a cement rendering, lightly reinforced with a steel-welded fabric mesh.

The general method used to construct diaphragm walls is the bentonite slurry system. Bentonite is manufactured from a montmorillonite clay that is commonly called fuller's earth because of its historical use by fullers in the textile industry to absorb grease from newly woven cloth. When mixed with the correct amount of water, bentonite shows thixotropic properties, giving a liquid behaviour when agitated and a gel structure when undisturbed.

The basic procedure is to replace the excavated spoil with the bentonite slurry as the work proceeds. The slurry forms a soft gel or 'filter cake' at the interface of the excavation sides with slight penetration into the subsoil. Hydrostatic pressure caused by the bentonite slurry thrusting on the filter cake cushion is sufficient to hold back the subsoil and any groundwater that may be present. This alleviates the need for timbering and/or pumping, and can be successfully employed up to 36 m deep.

Diaphragm walls constructed by this method are executed in alternate panels from 4.5 m to 7 m long with widths ranging from 500 to 900 mm using a specially designed hydraulic grab attached to a standard crane, or by using a continuous cutting and recirculating machine. Before the general excavation commences, a guide trench about 1 m deep is excavated and lined with lightly reinforced walls. These walls act as a guideline for the excavating machinery and provide a reservoir for the slurry, enabling pavings and underground services to be broken out ahead of the excavation.

To form an interlocking and watertight joint at each end of the panel, circular stop end pipes are placed in the bentonite-filled excavation before the concrete is placed. The continuous operation of concreting the panel is carried out using a tremie pipe and a concrete mix designed to have good flow properties without the tendency to segregate. This will require a concrete with a high slump of about 200 mm but with high-strength properties ranging from 20 to 40 N/mm^2. Generally the rate of pour is in the region of 15 to 20 m^3 per hour, and as the concrete is introduced into the excavated panel it will displace the bentonite slurry, which is less dense than the concrete, which can be stored for reuse or transferred to the next panel being excavated. The ideal situation is to have the two operations acting simultaneously and in complete unison (see Fig. 3.2.5).

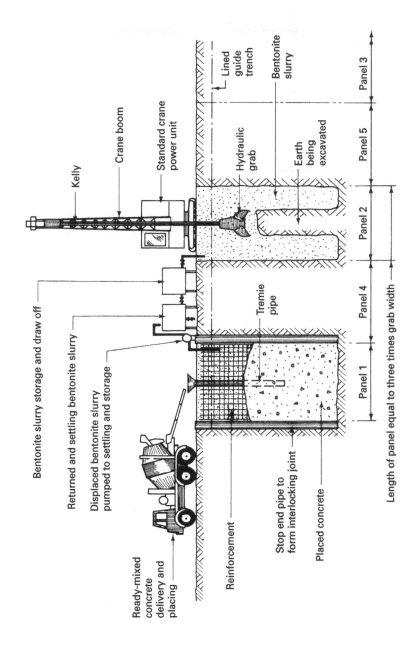

Kelly

Crane boom

Standard crane power unit

Lined guide trench

Bentonite slurry

Hydraulic grab

Earth being excavated

Bentonite slurry storage and draw off

Returned and settling bentonite slurry

Displaced bentonite slurry pumped to settling and storage

Tremie pipe

Ready-mixed concrete delivery and placing

Reinforcement

Stop end pipe to form interlocking joint

Placed concrete

Panel 3

Panel 5

Panel 2

Panel 4

Panel 1

Length of panel equal to three times grab width

Figure 3.2.5   Diaphragm wall construction: bentonite slurry method

Before the concrete is placed, reinforcement cages of high yield steel bars are fabricated on site in one or two lengths. Single length cages of up to 20 m are possible; where cages in excess of this are required, they are usually spot-welded together when the first cage is projecting about 1 m above the slurry level. The usual recommended minimum cover is 100 mm, which is maintained by having spacing blocks or rings attached to the outer bars of the cage.

If diaphragm walls are used merely as a cut-off trench for groundwater control then reinforcement is not always necessary as it is not acting as a retaining wall. Upon completion of the concreting, the bentonite slurry must be removed from the site either by tanker or by diluting so that it can be discharged into the surface water sewers by agreement with the local authority.

PRECAST CONCRETE METHOD

The main concept of precast concrete diaphragm walls is based on the principles of in-situ reinforced concrete walling installed using a bentonite slurry-filled trench but with the advantages obtained by using factory-produced components. The wall is constructed by forming a bentonite slurry-filled trench of suitable width and depth, and inserting into this the precast concrete panel or posts and panels according to the system being employed. As a normal bentonite slurry mix would not effectively seal the joints between the precast concrete components, a special mixture of bentonite and cement with a special retarder additive to control setting time is used. The precast units are placed into position within this mixture, which will set sufficiently within a few days to enable excavation to take place within the basement area right up to the face of the diaphragm wall. To ensure that a clean wall face is exposed upon excavation it is usual practice to coat the proposed exposed wall faces with a special compound to reduce adhesion between the faces of the precast concrete units and the slurry mix.

The usual formats for precast concrete diaphragm walls are either simple tongue and groove jointed panels or a series of vertical posts with precast concrete infill panels, as shown in Fig. 3.2.6. If the subsoil is suitable, it is possible to use a combination of precast concrete and in-situ reinforced concrete to form a diaphragm wall by installing precast concrete vertical posts or beams at suitable centres and linking these together with an in-situ reinforced concrete wall constructed in 2 m-deep stages as shown in Fig. 3.2.7. The main advantage of this composite walling is the greater flexibility in design.

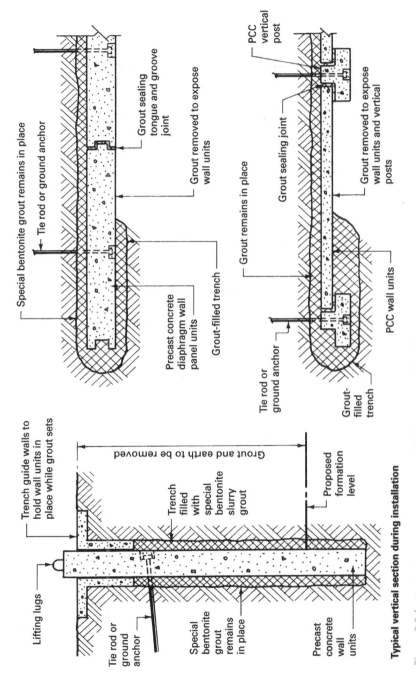

Grout sealing tongue and groove joint

Grout removed to expose wall units

Special bentonite grout remains in place

Tie rod or ground anchor

Precast concrete diaphragm wall panel units

Grout-filled trench

PCC vertical post

Grout sealing joint

Grout removed to expose wall units and vertical posts

Grout remains in place

Tie rod or ground anchor

PCC wall units

Grout-filled trench

Trench guide walls to hold wall units in place while grout sets

Grout and earth to be removed

Trench filled with special bentonite slurry grout

Proposed formation level

Lifting lugs

Tie rod or ground anchor

Special bentonite grout remains in place

Precast concrete wall units

Typical vertical section during installation

Figure 3.2.6   Typical precast concrete panel diaphragm wall details

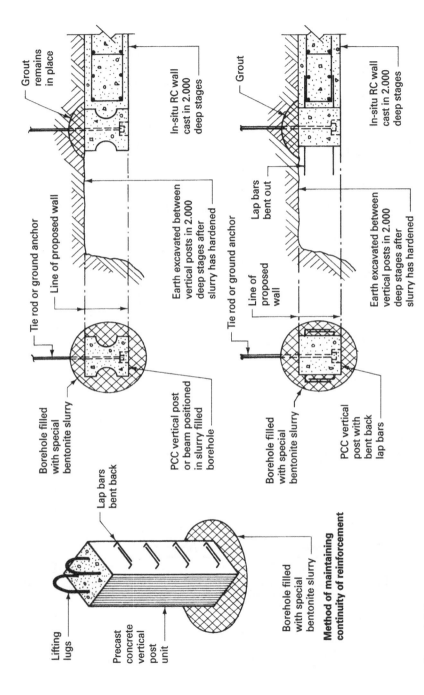

Grout remains in place

In-situ RC wall cast in 2.000 deep stages

Tie rod or ground anchor

Line of proposed wall

Earth excavated between vertical posts in 2.000 deep stages after slurry has hardened

Borehole filled with special bentonite slurry

PCC vertical post or beam positioned in slurry filled borehole

Grout

In-situ RC wall cast in 2.000 deep stages

Lap bars bent out

Tie rod or ground anchor

Line of proposed wall

Earth excavated between vertical posts in 2.000 deep stages after slurry has hardened

Borehole filled with special bentonite slurry

PCC vertical post with bent back lap bars

Lifting lugs

Lap bars bent back

Precast concrete vertical post unit

Borehole filled with special bentonite slurry

Method of maintaining continuity of reinforcement

Figure 3.2.7 Typical precast and in-situ concrete diaphragm wall details

The term 'cofferdam' comes from the French word *coffre* meaning a box, which is an apt description as a cofferdam consists of a watertight perimeter enclosure constructed in the ground or in water. It is usually created to enable the formation of foundations to be carried out in safe working conditions. It is common practice to use interlocking steel trench sheeting or steel sheet piling to form the cofferdam, but any material that will fulfil the same function can be used, including timber piles, precast concrete piles, vinyl piles, earth-filled crib walls and banks of soil and rock. It must be clearly understood that, to be safe, economic and effective, cofferdams must be the subject of structural design. Regulations 32 of the Construction (Design and Management) Regulations 2007 specifically require cofferdams and caissons to be properly designed, constructed and maintained. Of equal importance are Regulations 26 and 35. Regulation 26 requires a safe place to work and safe means of access to and egress from that place of work. Regulation 35 *Prevention of drowning* includes provision for water transport and rescue equipment for operatives working where water levels are a perceived danger. These regulations are complemented by the Confined Spaces Regulations 1997, which include specific reference to chambers and pits (with implications for cofferdams and caissons). Under Regulation 40 *Emergency routes and exits*, there must be preparation of suitable contingency plans in the event of an emergency rescue and availability of resuscitation equipment. Under Regulation 4 of the Work at Height Regulations 2005, specific reference is made to adequate supervision, planning and safety procedures. This includes planning for emergencies and rescue.

SHEET PILE COFFERDAMS

Cofferdams constructed from steel sheet piles or steel trench sheeting can be considered under three headings:

- single-skin cofferdams;
- double-skin cofferdams.
- tubular press-in cofferdams

SINGLE-SKIN COFFERDAMS

These consist of a suitably supported single enclosing row of trench sheeting or sheet piles forming an almost completely watertight box. Trench sheeting could be considered for light loadings up to an excavation depth of 3 m below the existing soil or water level, whereas U-shaped or Z-shaped sheet piles are usually suitable for excavation depths of up to 15 m. The small amount of seepage that will occur through the interlocking joints must not be in excess of that which can be comfortably controlled by a pump; alternatively the joints can be sealed by caulking with suitable mastics, silicone sealants, or by suitable welding techniques. Any seepage of water through the cofferdam can normally be handled by small pumps. The sheet piles can be braced and strutted by using a system of raking struts, horizontal struts or tie rods and ground anchors (see Fig. 3.2.8).

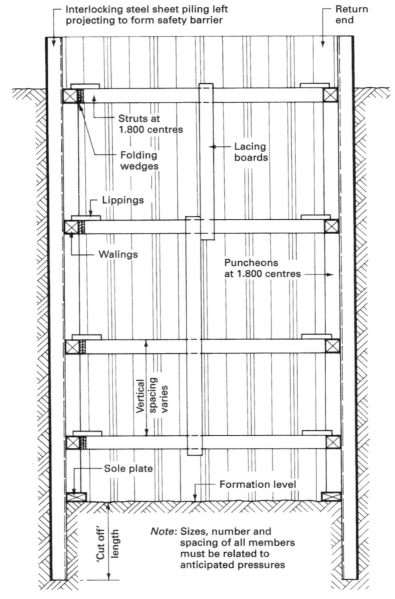

Figure 3.2.8 **Cofferdams: basic principles**

Single-skin cofferdams constructed to act as cantilevers are possible in all soils, but the maximum amount of excavation height will be low relative to the required penetration of the toe of the pile, and this is particularly true in cohesive soils. Most cofferdams are therefore either braced and strutted or anchored using tie rods or ground anchors. Standard structural steel sections or structural timber members can be used to form the support system, but generally timber is economically suitable only for low loadings. The total amount of timber required to brace a cofferdam adequately would be in the region of 0.25 to 0.3 m³ per tonne of steel sheet piling used, whereas the total weight of steel bracing would be in the region of 30–60 per cent of the total weight of sheet piling used to form the cofferdam. Typical cofferdam support arrangements are shown in Fig. 3.2.9. Single-skin cofferdams that are circular in plan can also be constructed using ring beams of concrete or steel to act as bracing without the need for strutting. Diameters up to 36 m are economically possible using this method.

Cofferdams constructed in water, particularly those being erected in tidal waters, should be fitted with sluice gates to act as a precaution against unanticipated weaknesses in the arrangement and, in the case of tidal waters, to enable the water levels on both sides of the dam to be equalised during construction and before final closure. Special piles with an integral sluice gate forming a 200 mm wide × 400 mm deep opening are available. Alternatively a suitable gate can be formed by cutting a pair of piles and fitting them with a top-operated screw gear so that they can be raised to form an opening of any desired depth.

DOUBLE-SKIN COFFERDAMS

These are self-supporting gravity structures constructed by using two parallel rows of piles with a filling material placed in the void created. Gravity-type cofferdams can also be formed by using straight-web sheet pile sections arranged as a cellular construction (see Fig. 3.2.10).

The stability of these forms of cofferdam depends upon the design and arrangement of the sheet piling and upon the nature of the filling material. The inner wall of a double-skin cofferdam is designed as a retaining wall which is suitably driven into the substrata whereas the outer wall acts primarily as an anchor wall. The two parallel rows of piles are tied together with one or two rows of tie rods acting against external steel walings. Inner walls should have a series of low-level weepholes to relieve the filling material of high water pressures and thus increase its shear resistance. For this reason the filling material selected should be capable of being well drained. Therefore materials such as sand, hardcore and broken stone are suitable, whereas cohesive soils such as clay are unsuitable. Note that the weepholes should include a filter mechanisim to prevent the outwashing of fine aggregates and thus ensure the stability of the fill materials. The width-to-height ratio shown in Fig. 3.2.10 of $0.7:0.8H$ can in some cases be reduced by giving external support to the inner wall by means of an earth embankment or berm.

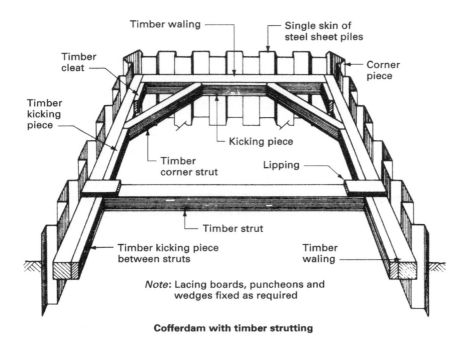

Cofferdam with timber strutting

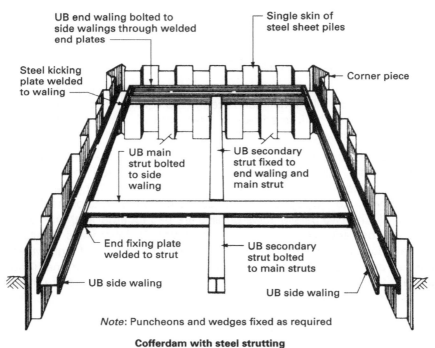

Cofferdam with steel strutting

Figure 3.2.9   Typical cofferdam strutting arrangements

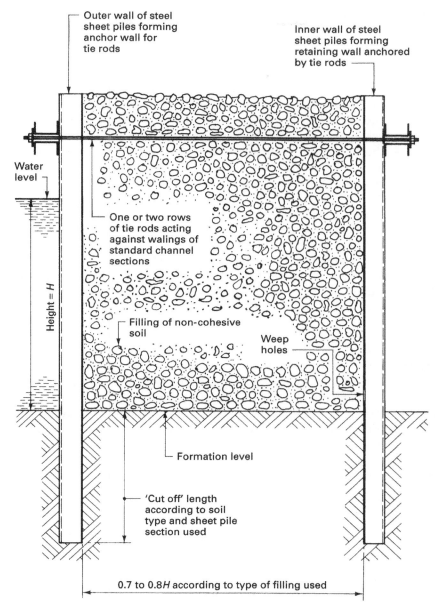

Outer wall of steel sheet piles forming anchor wall for tie rods

Inner wall of steel sheet piles forming retaining wall anchored by tie rods

Water level

One or two rows of tie rods acting against walings of standard channel sections

Height = H

Filling of non-cohesive soil

Weep holes

Formation level

'Cut off' length according to soil type and sheet pile section used

0.7 to 0.8H according to type of filling used

Figure 3.2.10   Typical double-skin cofferdam details

TUBULAR SHEET PILE PRESS-IN METHOD

This method has been developed by Giken™ to hydraulically sink the full range of steel sheet pile cross-sections. The tubular sheet pile section system is particular useful for creating robust cofferdams without the need for using double skin techniques and infilling as described above. The basic configuration is shown in Fig. 3.2.11. The principal method of operation is that a system of hydraulic jacks contained within the 'silent piler' anchors itself to three previously pressed-in piles, then drives in a fourth pile. It then moves its driving mast horizontally to accurately position and push in a fifth pile, which has been delivered to it by the pile runner and clamp crane. Once this new pile is fully pressed in it releases the three jacks and moves them one pile closer to the driving mast clamping them to the next set of three piles; and so the sequence continues. The system is fully automated and can be wireless-controlled from a safe vantage point. The configuration of the tubular piles and their clutch mechanisms for ensuring a watertight fit are shown in Fig. 3.2.11. If additional lateral strength is required, or if the tubular pile wall is to form a permanent part of the structure, such as a basement wall, it is common for a reinforced augured pile to be constructed with the tubular pile once it has been fully driven.

CAISSONS

Caissons are box-like structures that can be sunk through ground or water to install foundations or similar structures below the water line or table. They differ from cofferdams in that they usually become part of the finished foundation or structure, and should be considered as an alternative to the temporary cofferdam if the working depth below the water level exceeds 18 m. The design and installation of the various types of caisson are usually the tasks of a specialist organisation, but building contractors should have a fundamental knowledge of the different types and their uses.

There are four basic types of caisson in general use:

1. box caissons;
2. open caissons;
3. monolithic caissons;
4. pneumatic or compressed-air caissons.

BOX CAISSONS

Box caissons are prefabricated precast concrete boxes that are open at the top and closed at the bottom. They are usually constructed on land, and are designed to be launched and floated to the desired position, where they are sunk on to a previously prepared dredged or rock blanket foundation

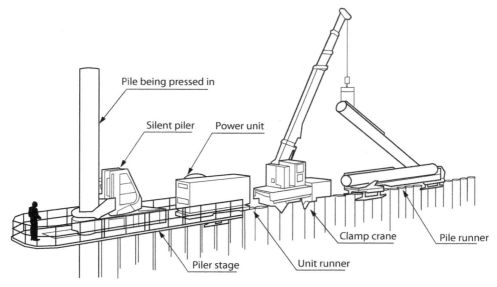

Pile being pressed in

Silent piler   Power unit

Clamp crane   Pile runner

Piler stage   Unit runner

a) Standard components of the GRB system

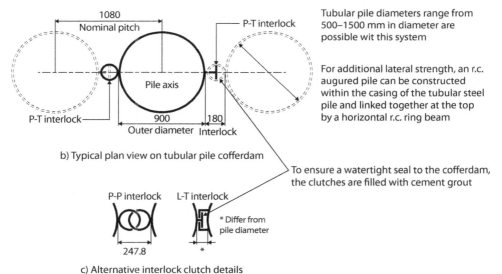

1080
Nominal pitch

P-T interlock

Tubular pile diameters range from 500–1500 mm in diameter are possible wit this system

Pile axis

For additional lateral strength, an r.c. augured pile can be constructed within the casing of the tubular steel pile and linked together at the top by a horizontal r.c. ring beam

P-T interlock

900   180
Outer diameter   Interlock

b) Typical plan view on tubular pile cofferdam

To ensure a watertight seal to the cofferdam, the clutches are filled with cement grout

P-P interlock   L-T interlock

* Differ from pile diameter

247.8   *

c) Alternative interlock clutch details

Figure 3.2.11   Giken Reaction Base (GRB) System

(see Fig. 3.2.12). If the bed stratum is unsuitable for the above preparations, it may be necessary to lay a concrete raft, by using traditional cofferdam techniques, on to which the caisson can be sunk. During installation it is essential that precautions are taken to overcome the problems of flotation by flooding the void with water or adding kentledge to the caisson walls. The sides of the caisson will extend above the water line after it has been finally positioned, providing a suitable shell for such structures as bridge piers, breakwaters and jetties. The void is filled with in-situ concrete placed by pump, tremie pipe or crane and skip. Box caissons are suitable for situations where the bed conditions are such that it is not necessary to sink the caisson below the prepared bed level.

OPEN CAISSONS

Open caissons are sometimes referred to as cylinder caissons because of their usual plan shape. These are of precast concrete and open at both the top and bottom ends, with a cutting edge to the bottom rim. They are suitable for installation in soft subsoils where the excavation can be carried out by conventional grabs, enabling the caisson to sink under its own weight as the excavation proceeds. These caissons can be completely or partially pre-formed; in the latter case further sections can be added or cast on as the structure sinks to the required depth. When the desired depth has been reached a concrete plug in the form of a slab is placed in the bottom by tremie pipe to prevent further ingress of water. The cell void can now be pumped dry and filled with crushed rocks or similar material if necessary to overcome flotation during further construction works.

Open caissons can also be installed in land if the subsoil conditions are suitable. The shoe or cutting edge is formed so that it is wider than the wall above to create an annular space some 75 to 100 mm wide into which a bentonite slurry can be pumped to act as a lubricant and thus reduce the skin friction to a minimum. Excavation is carried out by traditional means within the caisson void, the caisson sinking under its own weight. The excavation operation is usually carried out simultaneously with the construction of the caisson walls above ground level (see Fig. 3.2.13). This method is also used for creating vertical shafts as described later in Chapter 3.3.

MONOLITHIC CAISSONS

Monolithic caissons are usually rectangular in plan and are divided into a number of voids or wells through which the excavation is carried out. They are similar to open caissons but have greater self-weight and wall thickness, making them suitable for structures such as quays, which may have to resist considerable impact forces in their final condition.

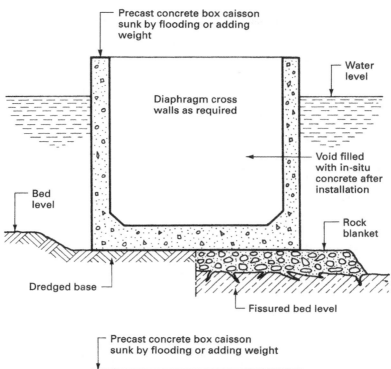

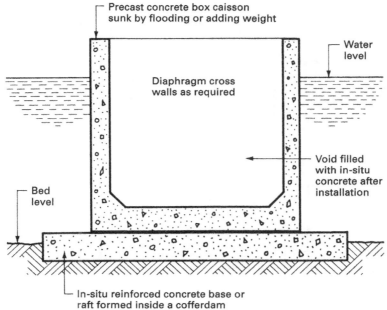

Figure 3.2.12   Typical box caisson details

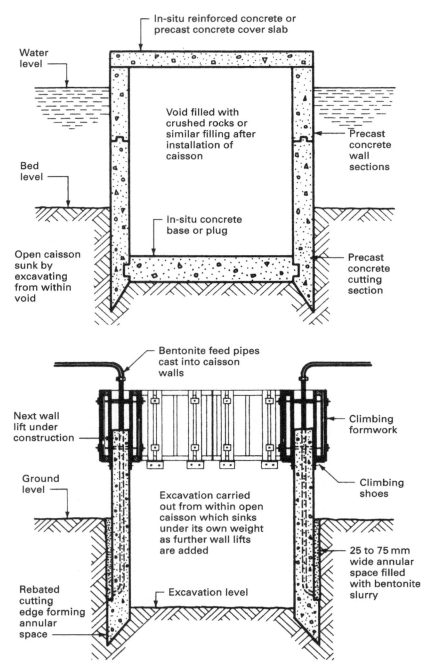

Labels in figure (top diagram):

In-situ reinforced concrete or precast concrete cover slab

Water level

Void filled with crushed rocks or similar filling after installation of caisson

Precast concrete wall sections

Bed level

In-situ concrete base or plug

Open caisson sunk by excavating from within void

Precast concrete cutting section

Labels in figure (bottom diagram):

Bentonite feed pipes cast into caisson walls

Next wall lift under construction

Climbing formwork

Ground level

Climbing shoes

Excavation carried out from within open caisson which sinks under its own weight as further wall lifts are added

25 to 75 mm wide annular space filled with bentonite slurry

Rebated cutting edge forming annular space

Excavation level

Figure 3.2.13 Typical open caisson details

PNEUMATIC OR COMPRESSED-AIR CAISSONS

Pneumatic or compressed-air caissons are similar to open caissons except that there is an airtight working chamber some 3 m high at the cutting edge. They are used where difficult subsoils exist and where hand excavation in dry working conditions is necessary. The working chamber must be pressurised sufficiently to control the inflow of water and/or soil, and at the same time provide safe working conditions for the operatives. The maximum safe working pressure is usually specified as 310 kN/m^2, which will limit the working depth of this type of caisson to about 28 m. When the required depth has been reached, the floor of the working chamber can be sealed over with a 600 mm-thick layer of well-vibrated concrete. This is followed by further well-vibrated layers of concrete until only a small space remains, which is pressure-grouted to finally seal the working chamber. The access shafts are finally sealed with concrete some three to four days after sealing off the working chamber. For typical details, see Fig. 3.2.14.

There are particular health and safety precautions to be taken when constructing compressed-air caissons to avoid decompression sickness and related symptoms. These are outlined in the Work in Compressed Air Regulations 1996 and include notification to the HSE and local emergency services. The detailed compression and decompression procedures are specified in Regulation 11, where there are strict limits to the maximum time that can be spent working in compressed air, the times needed to acclimatise to that pressure and the rest time needed in a 24-hour period. Also the Confined Spaces Regulations 1997 may need to be consulted where constructing any enclosed caisson as there may be limited means of access/egress and restricted ventilation.

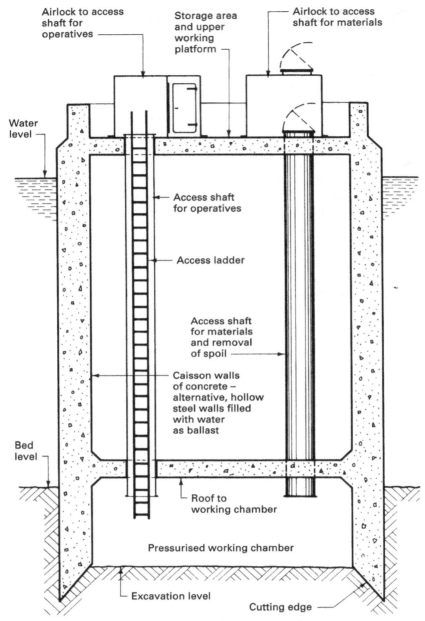

Airlock to access shaft for operatives

Storage area and upper working platform

Airlock to access shaft for materials

Water level

Access shaft for operatives

Access ladder

Access shaft for materials and removal of spoil

Caisson walls of concrete – alternative, hollow steel walls filled with water as ballast

Bed level

Roof to working chamber

Pressurised working chamber

Excavation level

Cutting edge

Excavation carried out within pressurised working chamber and caisson sinks under its own weight

Figure 3.2.14   Typical pneumatic or compressed-air caisson details

Shafts, tunnelling, culverts and pipe jacking 3.3

Techniques for underground construction include the installation of buried services and culverts, transport tunnels in urban areas and basements within buildings. This section will examine the construction of service and transportation tunnels, including the construction of shafts, tunnels, culverts and pipe runs.

Open excavations for the installation of these facilities is generally undertaken for depths between 3 m and 6 m, using sheet piling or similar methods as outlined in *Construction Technology*, Chapters 2.4 and 3.1. When the depth of a projected excavation approaches 6 m, you need to consider the alternative of using underground construction techniques, such as tunnelling or pipe jacking. Other factors that may influence your choice of using underground construction methods are:

- nature of subsoils – this can affect the amount of supports that will be required to keep workers safe in deep trench excavations;

- existing services – in urban areas, with open deep-trench excavations, buried services can be a problem as they may have to be diverted temporarily or permanently, but this can generally be avoided by tunnelling or pipe jacking techniques;

- carriageways – it may be deemed necessary to tunnel or pipe jack under busy roads to avoid disturbing the flow of traffic.

Here are the regulations that particularly apply to tunnelling and pipe jacking.

- **Construction (Design and Management) 2007** This sets out the need for a safe place of work for the protection of operatives working in excavations, shafts and tunnels, covering such aspects as temporary timbering, supervision of works, and means of egress in case of flooding. It also requires that that there is adequate ventilation of excavations and provision of adequate lighting.

- **Confined Spaces Regulations 1997** Regulation 1 includes reference to any 'chamber, tank, vat, silo, pit, trench, pipe, sewer, flue, well or other similar place' as enclosed situations with foreseeable risks. Tunnels and culverts can apply to many of these categories, so Regulations 3 (Duties), 4 (Work in confined spaces) and 5 (Emergency arrangements) are applicable.
- **Work at Height Regulations 2005** Work at height means work in any place, including a workplace at or below ground – so tunnels and culverts. Access to or egress from that place of work by suitable means is a specific requirement.

SHAFTS

Shafts are by definition underground vertical passages but, in the context of building and civil engineering operations, they can also be used to form the excavation for deep, isolated base foundations, service access points or holding tanks for liquid. In common with all excavations, the extent and nature of the temporary support or timbering required will depend on:

- subsoil conditions encountered;
- anticipated ground and hydrostatic pressures;
- materials used to provide temporary support;
- plan size and depth of excavation.

In loose subsoils, a system of sheet piling could be used by driving the piles ahead of the excavation operation. This provides additional lateral support and reduces the inflow of water into the shaft. This form of temporary support is called a cofferdam, and is explained in Chapter 3.2.

Traditional methods used in the past for constructing access shafts involved complex timbering by skilled shoring carpenters, which was relatively expensive. A typical timbered installation detail is shown in Fig. 3.3.1. It also put these workers under considerable risk, both in terms of the shaft collapsing during construction and in the hand-arm vibration caused by hand digging. Although timbering still has its applications, it is now seldom used in underground construction.

The modern alternative is to use precast segmental concrete rings to form the vertical shaft. This is a common method of permanent lining for both deep shafts and tunnels but can also be used for temporary access works (see Fig. 3.3.2). Constructing circular shafts is usually undertaken by specialist subcontractors, as with most tunnelling works, using either of two basic methods:

- jacked caissons;
- underpinning.

In both methods, the circular rings interact with the surrounding soil in an arching effect, so that ground loading is transferred into compressive stress. This can be efficiently resisted by the precast concrete shaft segments, which are first bolted then grouted together. Using these methods, shafts can be sunk up to 40 m deep, depending on ground conditions. Although not as elegant as the timbering solutions of the past, these methods provide a quick and safe solution for sinking a shaft.

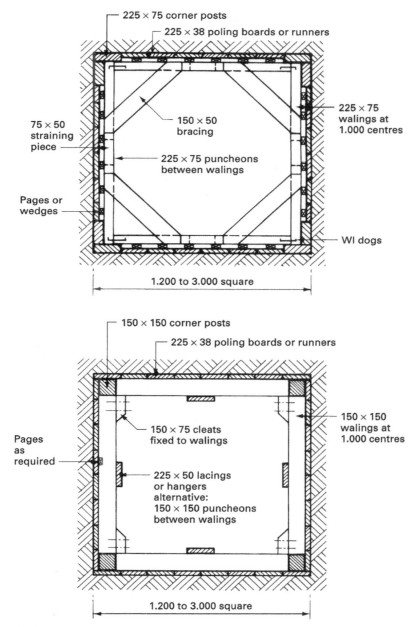

Figure 3.3.1 Traditional shaft timbering: typical plans

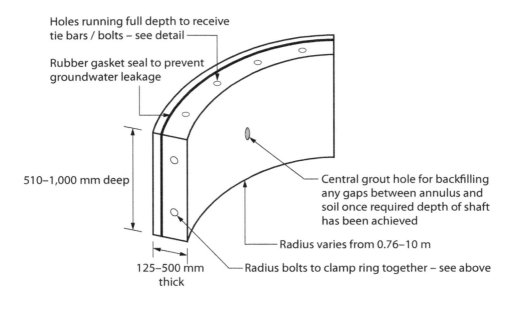

Holes running full depth to receive tie bars / bolts – see detail

Rubber gasket seal to prevent groundwater leakage

510–1,000 mm deep

125–500 mm thick

Central grout hole for backfilling any gaps between annulus and soil once required depth of shaft has been achieved

Radius varies from 0.76–10 m

Radius bolts to clamp ring together – see above

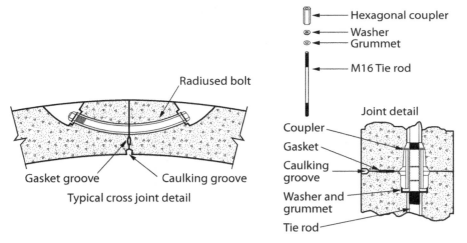

Radiused bolt

Gasket groove

Caulking groove

Typical cross joint detail

Hexagonal coupler

Washer

Grummet

M16 Tie rod

Joint detail

Coupler

Gasket

Caulking groove

Washer and grummet

Tie rod

Fixing connection between horizontal rings

a) Reinforced precast concrete shaft or tunnel segment lining.

Figure 3.3.2   Modern shaft construction methods

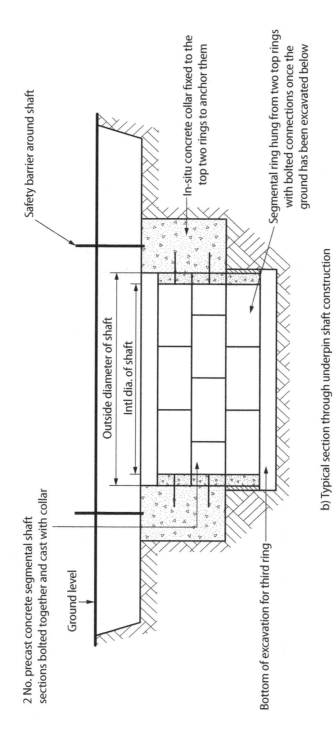

2 No. precast concrete segmental shaft sections bolted together and cast with collar

Safety barrier around shaft

In-situ concrete collar fixed to the top two rings to anchor them

Segmental ring hung from two top rings with bolted connections once the ground has been excavated below

Ground level

Outside diameter of shaft

Intl dia. of shaft

Bottom of excavation for third ring

b) Typical section through underpin shaft construction

Figure 3.3.2 Modern shaft construction methods – *continued*

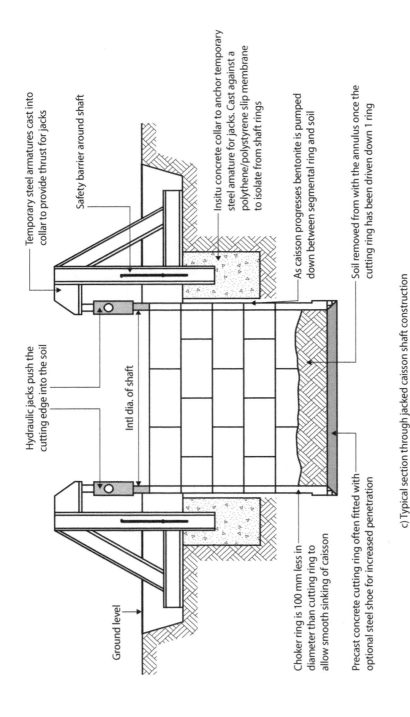

Temporary steel armatures cast into collar to provide thrust for jacks

Safety barrier around shaft

Insitu concrete collar to anchor temporary steel amature for jacks. Cast against a polythene/polystyrene slip membrane to isolate from shaft rings

As caisson progresses bentonite is pumped down between segmental ring and soil

Soil removed from with the annulus once the cutting ring has been driven down 1 ring

Hydraulic jacks push the cutting edge into the soil

Intl dia. of shaft

Choker ring is 100 mm less in diameter than cutting ring to allow smooth sinking of caisson

Precast concrete cutting ring often fitted with optional steel shoe for increased penetration

Ground level

c) Typical section through jacked caisson shaft construction

Figure 3.3.2 **Modern shaft construction methods – *continued***

JACKED CAISSON SYSTEM

Jacked caissons generally range from 2.40 m to 20.00 m in diameter. They comprise a series of segmental precast concrete sections bolted together to form a ring mounted on a steel cutting edge. This whole assembly is then pushed into the ground by a series of jacks located at the ground surface around the circumference of the shaft. The jacks need to be securely anchored to an in-situ circular reinforced concrete collar constructed at ground level around the shaft. Further rings are then bolted in place above the ring that has been sunk to form the next tier. The jacks are then reapplied and the shaft progresses downwards. The sinking is often aided by the introduction of bentonite clay in the gap between the soil and the external face of the shaft rings to act as a lubricant. Depending on the size of the jacks and the rings, up to three tiers of shaft segments can be sunk in one push. The soil inside the shaft is then excavated by long dipper arm backacters or, in larger shafts mini-excavatiors are lowered into the shaft to remove the soil into crane skips. Caisson shaft sinking is the preferred method generally for smaller-diameter shafts, especially in poor ground conditions, such as sands and gravels with a high water table (see Fig. 3.3.2).

UNDERPINNING SYSTEM

Generally for shafts larger than 20.00 m in diameter the underpinning construction method is used. However, ground must be suitable, such as firm clays or rock, and preferably dry.

The preliminary works involve a shallow excavation where the top segmental precast concrete rings are set out. A large in-situ concrete collar is then cast around the top of the shaft with a cross-section of approximately 1.2 m × 1.2 m, and is tied into the segmental ring by short dowel bars projecting from the segments into the collar. The soil beneath the top ring is then excavated down to a depth of 1.5 times the height of the ring. The top ring does not slip as the collar restrains it, which allows operatives to hang new segmental rings below the top ring as if to 'underpin' it. Care must be taken to seal the gap between the outside ring and the soil to minimise water ingress. This system progresses until the desired depth is reached (see Fig. 3.3.2).

TUNNELS

The type of tunnelling and boring method depends on a variety of factors, including:

- ground conditions and temporary support required;
- size of the tunnel required;
- amount of groundwater present;
- availability of plant and equipment;
- number and proximity of other buried services and obstacles;
- safe access and egress of the workers.

For the construction of larger-diameter tunnel/shaft junctions and short lengths of tunnels, the sprayed concrete lining (SCL) method has replaced most traditional hand excavation forms of construction. This is because, as stated above, the risk of hand arm vibration (HAV) has made non-mechanised hand tunnelling very undesirable and therefore impractical in most circumstances.

The fully automated use of tunnel boring machines (TBMs) has also increased. TBMs not only excavate the tunnel, but also use mechanical means to drive the cutting edge forward, remove the spoil material and erect the precast segmental linings. All these actions are contained safely within a shield to protect workers and keep out groundwater.

Table 3.3.1 shows a comparison of a variety of tunnelling methods and the health and safety risks associated with them.

PIPES

For the installation of underground services there has been a huge development in the use of trenchless technology systems. These fall into two main categories of small pipe technology and pipe jacking systems both where minimal surface excavation is needed, thus reducing cost, congestion and inconvenience to the public. These methods are also generally quicker and safer and are classified according to diameter ranges as follows.

- **small pipes** <200 to 900 mm diameter, for thrust, auger-bored or pipe-bursting techniques;
- **medium pipes** 900 to 1,800 mm diameter, for pipe-jacking techniques;
- **large pipes** 1,800 to 3,600 mm diameter, for pipe-jacking techniques.

Two materials are in common use for the pipes: concrete and steel. Spun-concrete pipes are designed with thick walls, and have a rubber joint, making them especially suitable for sewers without the need for extra strengthening. Larger-diameter pipes for pedestrian subway constructions are usually made of cast concrete and can have special bolted connections, making the joints watertight, which also renders them suitable for use as sewer pipes. Steel pipes have a wall thickness relative to their diameter, and usually have welded joints to give high tensile strength, the alternative being a flanged and bolted joint. They are obtainable with various coatings and linings to meet special requirements, such as corrosive-bearing effluents.

SMALL-DIAMETER PIPES

Three methods are available for the installation of small pipes up to 200 mm diameter.

- **Thrust boring** A bullet-shaped solid metal head is fixed to the leading end of the pipe to be installed and is pushed or jacked into the ground, displacing the earth.

Table 3.3.1 Comparison of drive lengths between shafts for pipe jacking and tunnelling methods
Adapted from 'Tunnelling and Pipejacking for Designers', published jointly by the Health and Safety Executive, British Tunnelling Society and the Pipe Jacking Association.

Excavation technique	Nominal internal diameter of pipeline or tunnel lining							
	<0.9 m	0.9 m	1.0 m	1.2 m	1.35 m	1.5 m	1.8 m	>1.8 m
Pipe jack – machine; remote operation from surface	Drive length limited only by capacity of jacking system							
	Man entry not acceptable		Avoid man entry	250 m		400 m	>500 m	>500 m
Pipe jack – machine; operator controlled below ground	Not acceptable			125 m	200 m	300 m	500 m	>500 m
Pipe jack – hand dig	Not acceptable			25 m – 2 drive lengths	50 m – 2 drive lengths	75 m – 2 drive lengths	100 m – 1 drive length. Plan to use minidigger if over 2.1m dia	100 m – 1 drive length. Plan to use minidigger if over 2.1m dia
Tunnel – machine operator controlled + mechanical erector	Not acceptable					250 m	500 m	>500 m
Tunnel – hand dig + mechanical erector	Not acceptable					50 m – 1 drive length	100 m – 1 drive length. Plan to use minidigger if over 2.1m dia	100 m – 1 drive length. Plan to use minidigger if over 2.1m dia
Timber heading – hand dig	Not acceptable			25 m – 2 drive lengths. Minimum cross section inside frames 1.2 m high × 1.0 m wide				

Not acceptable

Avoid

Acceptable

ACCEPTABLE – designers should undertake an assessment of the risks normally associated with small-size pipe jacking/tunnelling and specify the appropriate mitigation measures.

AVOID – designers should undertake a robust technical assessment and risk assessment to justify their decisions to deviate from 'acceptable' criteria.

NOT ACCEPTABLE – designers should not specify the use of pipe jacking/tunnelling of this size and construction method. An alternative design solution should be sought.

- **Auger boring** This is carried out with a horizontal auger boring tool operating from a working pit having at least 2.4 m long × 1.5 m wide clear dimensions between any temporary supporting members. The boring operation can be carried out without casings, but where the objective is the installation of services, concrete or steel casings are usually employed. The auger removes the spoil by working within the bore of the casing, which is being continuously rammed or jacked into position. It is possible to use this method for diameters of up to 1 m.

- **Pipe bursting** This is the replacement of old service pipes using a powerful head device which uses the old service pipe as a 'pilot' hole and requires very little surface excavation. During pneumatic pipe bursting, the pipe-bursting tool is guided through the old or 'host' pipe by a constant-tension winch. As the tool travels through the pipe, its percussive action effectively breaks apart the old pipe and displaces the fragments into the surrounding soil. At the same time, a new high-density polyethylene (HDPE) pipe is pulled in place behind this head device. This method is quick and relatively unobtrusive, as it can use existing inspection chambers or small pits to set up the winch mechanism in the receiving pit. This method can be used to replace most pipes including ceramic, cast iron, precast concrete and plastic pipes.

PIPE JACKING

Pipe jacking can be used for the installation of pipes from 150 mm to 3.6 m diameter, but is mainly employed on the larger diameters of over 1 m. The basic procedure is to force the pipes into the subsoil by means of a series of hydraulic jacks, and excavate, as the driving proceeds, from within the pipes by hand or machine, according to site conditions. The leading pipe is usually fitted with a steel shield or hood to aid the driving process. This is a very safe method, because the excavation work is carried out from within the casing or liner, and the danger of collapsing excavations is eliminated; there is also no disruption of the surface or underground services, and it is a practical method for most types of subsoil.

The most common method is to work from a jacking or working pit, which is formed in a similar manner to traditional shafts except that a framed thrust pad is needed from which to operate the hydraulic jacks. The working pit must be large enough for the jacks to be extended and to allow for new pipe sections to be lowered into the working bay at the bottom (see Fig. 3.3.3).

Pipe jacking can also be carried out from ground level, and is particularly suitable for driving pipes through an embankment to form a pedestrian subway. A series of 300 mm-diameter lined augered boreholes is driven through the embankment to accommodate tie bars, which are anchored to a bulkhead frame on the opposite side of the embankment. The reactions from the ramming jacks are transferred through the tie bars to the bulkhead frame, and the driving action becomes one of pushing and pulling (see Fig. 3.3.4). In firm soils the rate of bore by this method is approximately 3 m per day.

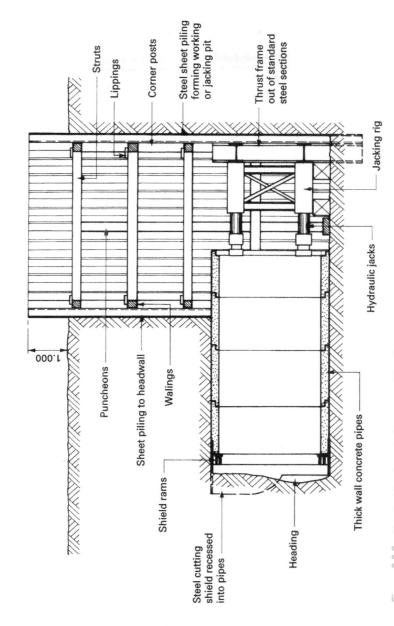

Struts

Lippings

Corner posts

Steel sheet piling forming working or jacking pit

Thrust frame out of standard steel sections

Jacking rig

Hydraulic jacks

Thick wall concrete pipes

Heading

Steel cutting shield recessed into pipes

Shield rams

Walings

Sheet piling to headwall

Puncheons

1.000

Figure 3.3.3 Pipe jacking from below ground level

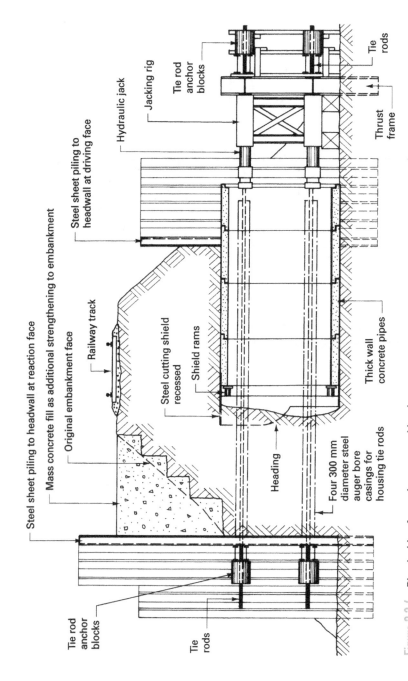

Steel sheet piling to headwall at reaction face

Mass concrete fill as additional strengthening to embankment

Original embankment face

Railway track

Steel cutting shield recessed

Shield rams

Heading

Four 300 mm diameter steel auger bore casings for housing tie rods

Tie rod anchor blocks

Tie rods

Steel sheet piling to headwall at driving face

Hydraulic jack

Jacking rig

Tie rod anchor blocks

Tie rods

Thrust frame

Thick wall concrete pipes

Figure 3.3.4 Pipe jacking from above ground level

Pipes can also be jacked, from ground level, into the earth at a gradient of up to 1:12 using a jacking block attached to a row of tension piles sited below the commencing level.

CULVERTS

A culvert is a box, conduit or pipe of large cross-section usually used to convey pedestrians, water (foul or surface) or services in general under a road, canal or some other elevated construction. Figure 3.3.4 illustrates the concept, using precast concrete pipes under a railway embankment. In the true principle of culvert construction, a void for the pipe would be excavated and covered after the pipes were placed, thus differentiating the process from tunnelling. In practice this may not always be convenient or cost-effective, hence the preference for tunnelling in some situations.

The 'cut and cover' technique involves considerable excavation, with sides usually temporarily restrained by interlocking sheet piles or steel trench sheeting. Groundwater controls may also be required, particularly if in-situ concrete is specified as the culvert material.

CONSTRUCTION

There are three options for culvert construction:

1. in-situ concrete;

2. in-situ concrete base with precast concrete crown;

3. precast concrete.

See Fig. 3.3.5.

PRECAST CONCRETE BOX CULVERTS

Precast concrete box culverts have proved the most popular specification for culverts, with manufacturers producing them in a wide range of sizes from about 1 to 4 m in width, 0.5 to 2.5 m in depth and in standard lengths of 1, 1.5 and 2 m to suit a variety of applications. These include subways, stream crossings, animal (badger) crossings, sewers and sea outfalls. They are not limited to horizontal use, and may also be used as vertical access shafts. Concrete is normally from ordinary Portland cement but, if aggressive soils are encountered, sulphate-resisting cement can be specified. High yield steel reinforcement with a minimum concrete cover of 30 mm is standard.

APPLICATION

Site delivery and installation should be carefully planned to coordinate hire of suitable plant and operatives. Following excavation, a bed or foundation must be prepared from well-compacted selected granules of about 225 mm total

SPRAYED CONCRETE LINING (SCL) METHOD

The use of sprayed concrete as a tunnel lining was originally pioneered for the driving of tunnels for Alpine road and rail tunnels in the 1960s and was known as the New Austrian Tunnel Method (NATM). It has since been developed and used in soft rock-like clays in the UK on a number of successful projects, such as the Jubilee Line and Dockland Light Railway extensions. There was a collapse on the Heathrow Express tunnel using this method, but lessons on the monitoring and management of SCL were learned and it has now become established as a safe and effective method to advance and support tunnel headings in soft soil conditions.

The advantages of using sprayed concrete linings are that it:

- is a structural material that can be used to provide both temporary immediate ground support and a permanent tunnel lining;
- can adapt to a wide variety of tunnel geometries, particularly where they intersect with shafts or inclined ducts;
- is soft when sprayed but rapidly increases in strength and stiffness, providing an increasing amount of support to the ground. Increased support is obtained when used in conjunction with steel mesh, steel ribs and ground anchors.

In SCL tunnels the excavation face is subdivided into small work areas and shorter advance lengths to take account of the condition of the ground and ensure its stability. The order of the excavation by mechanical excavators is firstly the crown, followed by the bench and finally the invert. These small sections (as shown in Fig. 3.3.7) are excavated in turn and stabilised with sprayed shotcrete. The shotcrete lining layer can be between 200 mm and 400 mm thick, and is reinforced with steel mesh and curved circular steel ribs. Ground anchors, bolts and dowels may be driven ahead of the face of the tunnel at an inclined angle, to help support the crown of the tunnel. The nozzle operative needs considerable skill to gauge the consistency of the shotcrete to avoid rebound, where the material bounces off the surface and fails to adhere.

Details of the typical plant used is spraying concrete linings are shown in Fig. 3.3.7.

The excavation is then progressively enlarged to its full dimensions and the activity repeated to advance the tunnel. The sprayed concrete tunnel lining is completed to form a near-circular concrete-lined tunnel. A key requirement is to apply the sprayed concrete lining as quickly as possible to minimise ground movements.

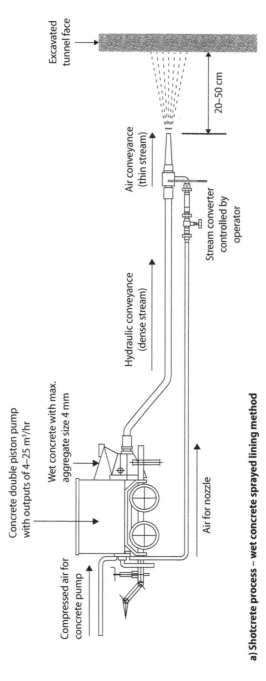

Excavated tunnel face

20–50 cm

Air conveyance
(thin stream)

Hydraulic conveyance
(dense stream)

Stream converter
controlled by
operator

Concrete double piston pump
with outputs of 4–25 m³/hr

Wet concrete with max.
aggregate size 4 mm

Air for nozzle

Compressed air for
concrete pump

a) Shotcrete process – wet concrete sprayed lining method

Figure 3.3.7   Sprayed concrete lining technology for tunnels

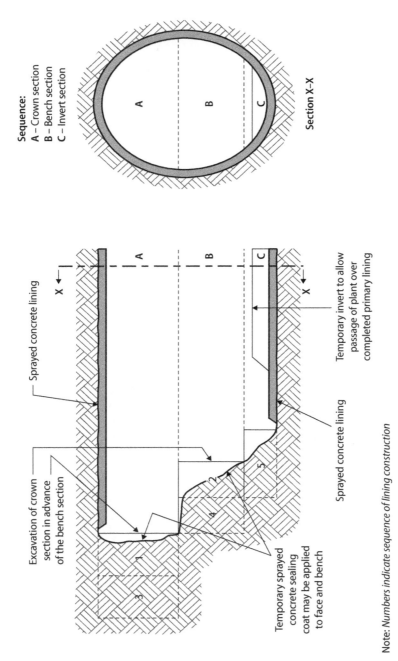

Sequence:
A – Crown section
B – Bench section
C – Invert section

Section X–X

Sprayed concrete lining

Excavation of crown section in advance of the bench section

Sprayed concrete lining

Temporary sprayed concrete sealing coat may be applied to face and bench

Temporary invert to allow passage of plant over completed primary lining

Note: *Numbers indicate sequence of lining construction*

b) General arrangement for tunnelling using sprayed concrete linings in tunnels up to 6 m in diameter

Figure 3.3.7   **Sprayed concrete lining technology for tunnels –** *continued*

The most common method for constructing large-diameter modern tunnels is the earth pressure balance tunnel boring method, in which the cutting mechanism, muck-away and segment erectors are protected by an annular steel shield. This method works well in relatively soft sedimentary engineering soils, such as silty clays and sands, and can operate below the water table. It has been used extensively in some recent major tunnelling projects in the UK (see Fig. 3.3.8). The tunnel boring machine (TBM) has a rotary cutting head, which is steered by hydraulic rams that push off the previous set of segmental rings installed in the tunnel.

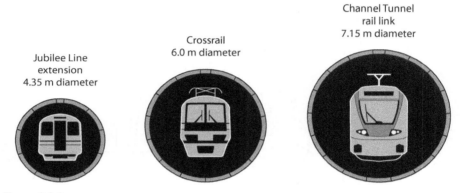

Figure 3.3.8 **Examples of bored precast concrete lined tunnels**

The spoil material from the cutting face is contained in the excavation chamber, which is then raised on to a muck-away conveyer by a helical screw. This arrangement ensures that there is always control over removal of spoil from the face, maintaining face pressure to 'balance' the earth pressure. It helps to maintain the stability of the face as it is being cut, and prevents excessive groundwater leakage. The segmental linings are brought to the mechanical erector by a second conveyer, which lifts and rotates them into the correct position. The segments are then grouted with cement grout.

The basic elements of the TBM are shown in Fig. 3.3.9.

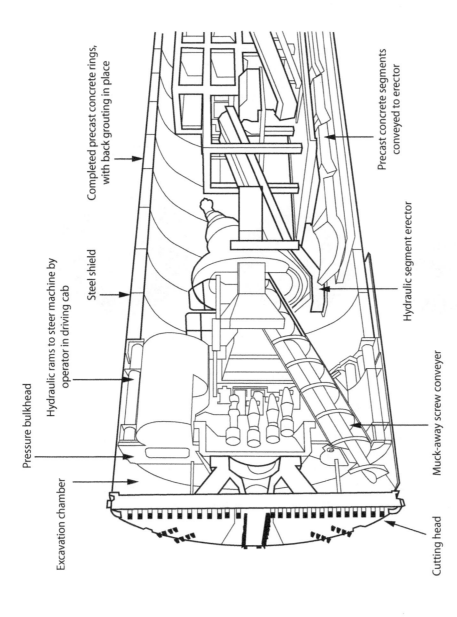

Pressure bulkhead

Hydraulic rams to steer machine by operator in driving cab

Steel shield

Completed precast concrete rings, with back grouting in place

Precast concrete segments conveyed to erector

Hydraulic segment erector

Excavation chamber

Cutting head

Muck-away screw conveyer

A typical earth pressure balance tunnel boring machine

Figure 3.3.9 Basic elements of a tunnel boring machine (TBM)

Specialist pad and piled foundations 3.4

There are many ways in which foundations can be classified, but one of the most common methods is by form, resulting in four basic types.

1. **Strip foundations** To support either light continuous loadings, particularly in domestic buildings, or heavier loadings when suitably reinforced with steel reinforcement cages. These have been described and illustrated in Chapter 3.3 of *Construction Technology*.

2. **Raft foundations** For light loadings, average loadings on soils with low bearing capacities and structures having a basement storey (see Chapter 3.3 of *Construction Technology*).

3. **Pad or isolated foundations** A common method of providing the foundation for columns of framed structures and for the supporting members of portal frames (see Chapter 3.3 of *Construction Technology*, and the summary below).

4. **Piled foundations** A method for structures where the loads have to be transmitted to a point at some distance below the general ground level. Short-bored pile systems for domestic use have been described and illustrated in Chapter 3.3 of *Construction Technology*; this chapter covers the more advanced types of pile foundation available for larger projects.

PAD FOUNDATIONS

Pad foundations are usually square on plan, unless obstructions or limitations such as the site boundary impede the spread in one direction. A rectangular base is generally acceptable in these situations, although the foundation area and amount of reinforcement may need to increase to accommodate the irregular transfer of loads. Pad foundations can be of mass concrete, but are usually reinforced to limit the amount of excavation and concrete fill, and to increase the load-carrying potential. Pads are used mainly to support columns and piers or as a base for heavy manufacturing machinery. Reinforcement in the pad is tied to projecting vertical starter bars for column reinforcement (see Chapter 3.3 of *Construction Technology*).

The area of pad required is obtained by dividing the load on the subsoil (column imposed and dead load + the foundation weight) by the safe bearing capacity of the subsoil.

EXAMPLE

A 40-tonne column and foundation load on a subsoil of safe bearing capacity of 200 kN/m².

40 tonnes is approximately 400 kN (40 × 9.81 = 392.4 kN).

Therefore 400 ÷ 200 = 2 m² base area.

If square, the foundation base will be √2 = 1.4 m × 1.4 m.

On sites where space is restricted and column loadings are high, pad foundations can be sufficiently large to be economically constructed in combination with an adjacent pad. In some situations it may be more practical and economic to combine several pads into a continuous foundation, as described in the next subsection.

The thickness of an individual pad foundation depends on the shear stress (strength) of the concrete and the perimeter of the supporting column. If it is too shallow, the column can 'punch' through the pad and fail in shear. The basic geometry of this punching shear failure is shown in Fig. 3.4.1.

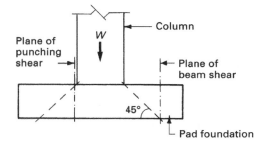

Figure 3.4.1 **Proximity of shear in foundation pads**

When designing a foundation, the area of the base must be large enough to ensure that the load per unit area does not exceed the bearing capacity of the soil on which it rests. When considering column foundations, the ideal situation is to have a square base of the appropriate area, with the column positioned centrally. Unfortunately this is not always possible. When developing sites in urban areas, the proposed structure is often required to abut an existing perimeter wall, with the proposed perimeter columns in close proximity to the existing wall. Such a situation would result in an eccentric column loading if conventional isolated bases were employed. One method of overcoming this problem is to place the perimeter columns on a reinforced concrete continuous column foundation in the form of a strip. The strip is designed as a beam with the columns acting as point loads, which will result in a negative bending moment between the columns, requiring top tensile reinforcement in the strip.

If sufficient plan area cannot be achieved using a continuous column foundation, a combined column foundation could be considered. This consists of a reinforced concrete slab placed at right angles to the line of columns, linking together an outer or perimeter column to an inner column. The base slab must have sufficient area to ensure that the load per unit area will not cause overstressing of the soil, leading to unacceptable settlement. To alleviate the possibility of eccentric loading, the centres of gravity of the columns and slab should be designed to coincide (see Fig. 3.4.2). Combined foundations of this type are suitable for a pair of equally loaded columns, or where the outer column carries the lightest load. The effect of the column loadings on the slab foundation is similar to that for continuous column foundations, where a negative bending moment will occur between the columns.

Where the length of the slab foundation is restricted or the outer column carries the heavier load, the slab plan shape can be in the form of a trapezium, with the widest end nearest to the more heavily loaded column (see Fig. 3.4.2). As with the rectangular base, the trapezoidal base will have negative bending moments between the columns, and to eliminate eccentric loading the centres of gravity of columns and slab should coincide.

Alternative column foundations to the combined and trapezoidal forms are the balanced base foundations. These foundations are usually specified where it is necessary to ensure that no pressure is imposed on an existing substructure or adjacent service, such as a drain or sewer.

Balanced base foundations can be designed in one of two basic forms:

- cantilever foundations;
- balanced base foundations.

A cantilever foundation can take two forms, both of which use a cantilever beam. In one version, the beam acts about a short fulcrum column positioned near the existing structure; in the alternative version, the beam cantilevers beyond a slab base (see Fig. 3.4.3).

A balanced base foundation can be considered where it is possible to place the inner columns directly on to a base foundation. These columns are usually eccentric on the inner base, causing a tendency to rotate within the base. This movement must be resisted or balanced by the inner column base and the connecting beam (see Fig. 3.4.4).

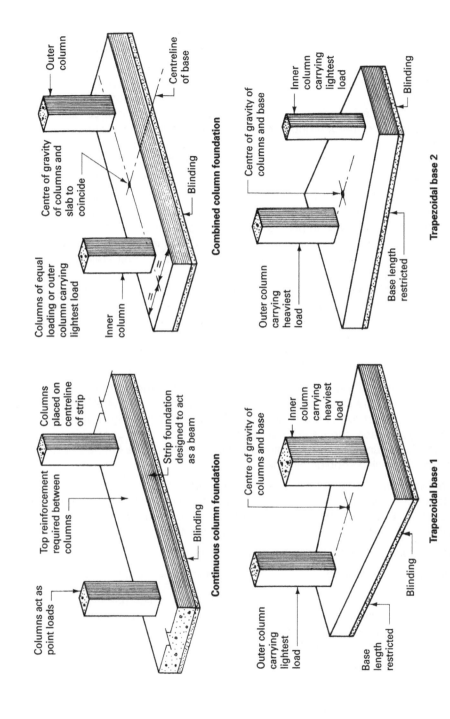

Continuous column foundation

Columns placed on centreline of strip

Top reinforcement required between columns

Strip foundation designed to act as a beam

Blinding

Columns act as point loads

Combined column foundation

Outer column

Centreline of base

Centre of gravity of columns and slab to coincide

Blinding

Columns of equal loading or outer column carrying lightest load

Inner column

Trapezoidal base 1

Inner column carrying heaviest load

Centre of gravity of columns and base

Outer column carrying lightest load

Base length restricted

Blinding

Trapezoidal base 2

Inner column carrying lightest load

Centre of gravity of columns and base

Blinding

Outer column carrying heaviest load

Base length restricted

Figure 3.4.2 Typical combined foundations

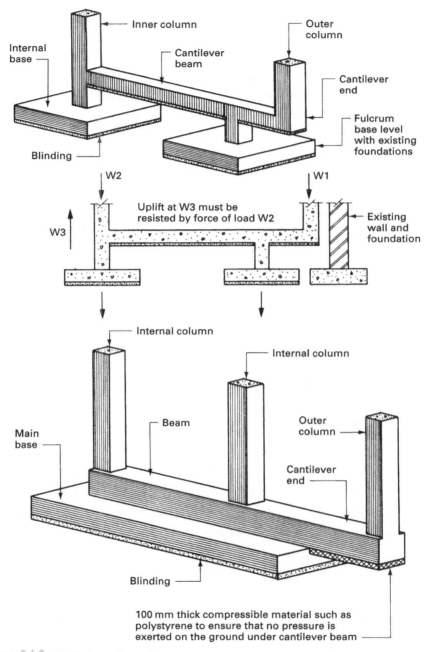

Labels in figure:

Inner column

Outer column

Internal base

Cantilever beam

Cantilever end

Fulcrum base level with existing foundations

Blinding

W2

W1

W3

Uplift at W3 must be resisted by force of load W2

Existing wall and foundation

Internal column

Internal column

Main base

Beam

Outer column

Cantilever end

Blinding

100 mm thick compressible material such as polystyrene to ensure that no pressure is exerted on the ground under cantilever beam

Figure 3.4.3   Typical cantilever bases

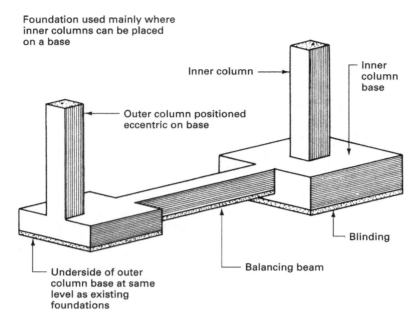

Foundation used mainly where inner columns can be placed on a base

Inner column

Inner column base

Outer column positioned eccentric on base

Blinding

Underside of outer column base at same level as existing foundations

Balancing beam

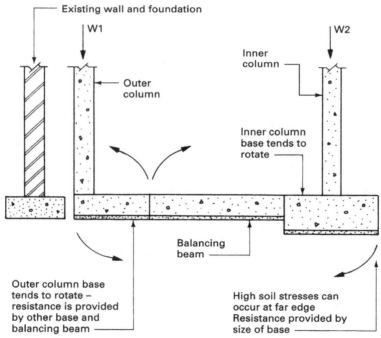

Existing wall and foundation

W1

W2

Inner column

Outer column

Inner column base tends to rotate

Outer column base tends to rotate – resistance is provided by other base and balancing beam

Balancing beam

High soil stresses can occur at far edge Resistance provided by size of base

Figure 3.4.4   Typical balanced base foundation

A pile can be loosely defined as a column inserted in the ground to transmit the structural loads to a lower level of subsoil. Piled foundations have been used in this context for hundreds of years and until the twentieth century were invariably of driven timber. Today a wide variety of materials and methods are available to solve most of the problems encountered when deep foundations are needed. Remember that piled foundations are not necessarily the answer to all awkward foundation problems; they should be considered as an alternative to other techniques when suitable bearing-capacity soil is not found near the lowest floor of the structure.

The unsuitability of the upper regions of a subsoil may be caused by:

- low bearing capacity of the subsoil;
- heavy point loads of the structure exceeding the soil bearing capacity;
- presence of highly compressible soils near the surface, such as filled ground and underlying peat strata;
- subsoils such as clay, which may be capable of moisture movement or plastic failure;
- high water table.

CLASSIFICATION OF PILED FOUNDATIONS

Piled foundations may be classified by the way in which they transmit their loads to the subsoils, or by the way they are formed. Loading to a lower level may be transmitted in a number of ways.

- **End bearing** The shafts of the piles act as columns, carrying the loads through the overlying weak subsoils to firm strata, into which the pile toe has penetrated. This can be a rock stratum or a layer of firm sand or gravel that has been compacted by the displacement and vibration encountered during the driving.

  The design of end-bearing piled foundations can be determined by establishing the safe bearing capacity of the subsoil, by sampling and laboratory analysis, and relating this to the load distribution on each pile. For example, the wall shown in Fig. 3.4.5 carries a load of 100 kN/m on to piles bearing on subgrade of safe bearing capacity of 300 kN/m². Using piles of, say, 0.500 m diameter, their spacing will be:

  - Pile base area = $3.142 \times 0.25^2 = 0.20$ m²
  - Each pile carries: $300 \times 0.2 = 60$ kN
  - The extent of wall on one pile = $100/60 = 1.670$ m
  - Therefore pile spacing is not more than 1.670 m.

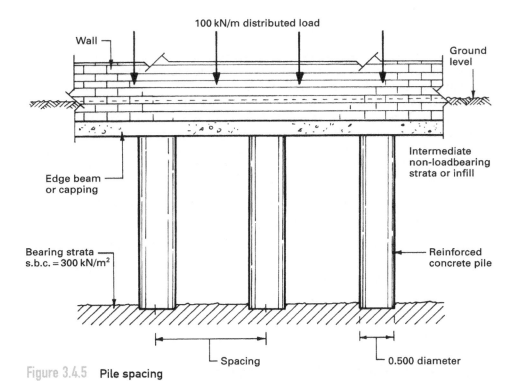

100 kN/m distributed load

Wall

Ground level

Edge beam or capping

Intermediate non-loadbearing strata or infill

Bearing strata
s.b.c. = 300 kN/m^2

Reinforced concrete pile

Spacing

0.500 diameter

Figure 3.4.5 **Pile spacing**

■ **Friction** Any foundation imposes on the ground a pressure that spreads out to form a bulb of pressure, as shown in Fig. 3.4.6. If a suitable load-bearing stratum cannot be found at an acceptable level, particularly in stiff clay soils, it is possible to use a pile to carry this bulb of pressure to a lower level where a higher bearing capacity is found. The friction pile is supported mainly by the adhesion or friction action of the soil around the perimeter of the pile shaft. The distribution of pressure from the pile load as it disperses through the soil from the surface of the vertical pile is given in Fig. 3.4.6 by the K factor coefficient.

In most situations, piles work on a combination of the two principles outlined above, but the major design consideration identifies the pile class.

Piles may be preformed and driven, displacing the soil through which they pass, and are therefore classified as displacement piles. Alternatively, the soil can be bored out and subsequently replaced by a pile shaft, and such piles are classified as replacement piles.

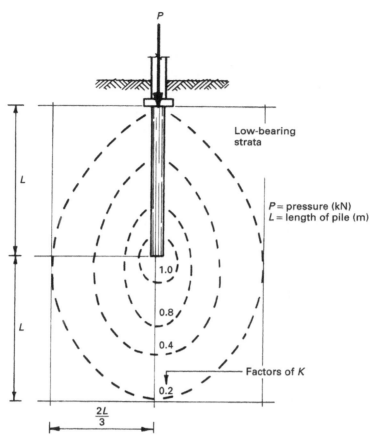

P

Low-bearing
strata

P = pressure (kN)
L = length of pile (m)

1.0

0.8

0.4

Factors of K

0.2

L

L

$\dfrac{2L}{3}$

Figure 3.4.6   Pressure bulb for a friction pile

PILE DOWNDRAG

When piles are driven through weak soils, such as alluvial or silty clay, peat layers or reclaimed land, the long-term settlement and consolidation of the ground can cause downdrag of the piles. In simple terms, this is a transference of the downdrag load of consolidating soil on to a pile shaft. It is called negative skin friction because the resistance to this transfer of load is the positive skin friction of the pile shaft. Downdrag loads have been known to equal the design bearing capacity of the pile.

There are three methods to counteract the effects of downdrag.

1. Increase the number of piles being used to carry the structural loads to accommodate the anticipated downdrag loads: this method may necessitate the use of twice the number of piles required to carry only the structural loadings.

2. Increase the size of the piles designed for structural loads only: this can be expensive, because the cost of forming piles does not increase pro rata with its size.

3. Design for the structural loads only and prevent the transference of downdrag loads by coating the external face of the shaft with a slip layer, reducing the negative skin friction.

Use ground improvement methods such as vibro-compaction to increase the density and compaction of soil before piling. This will increase the bearing capacity and is likely to reduce the number and size of piles required.

DISPLACEMENT PILES

This is a general term applied to piles that are driven, thus displacing the soil, and includes those piles that are preformed, partially preformed or are driven in-situ piles.

TIMBER PILES

Timber piles are usually of square, sawn hardwood or softwood in lengths up to 12 m, with section sizes ranging from 225 mm × 225 mm to 600 mm × 600 mm. They are easy to handle, and can be driven by percussion with the minimum of experience. Most timber piles are fitted with an iron or steel driving shoe, and have an iron ring around the head to prevent splitting due to impact. Although not particularly common, they are used in sea defences, such as groynes, and sometimes as guide piles for large trestles, in conjunction with steel sheet piling. Loadbearing capacities can be up to 350 kN per pile, depending on section size and/or species.

PRECAST CONCRETE PILES

Precast concrete piles are used on medium to large contracts where soft soils overlying a firm stratum are encountered or where there is a high water table. They are also more economical on large projects where at least 100 piles will be required. Generally the economical option for these conditions is to use lengths up to 18 m with section sizes ranging from 250 mm × 250 mm to 450 mm × 450 mm and carrying loadings of up to 1,000 kN. The precast concrete driven pile has little frictional bearing strength because the driving operation moulds the cohesive soils around the shaft, which reduces the positive frictional resistance.

Problems can be encountered when using this form of pile in urban areas due to:

■ transportation of the complete length of pile through narrow or congested streets;

■ the driving process, which is generally percussion, and can set up unacceptable noise and/or vibrations;

■ the fact that many urban sites are in themselves restricted or congested, making it difficult to manoeuvre the long pile lengths around the site.

PREFORMED CONCRETE PILES

Preformed concrete piles are available as reinforced precast concrete or prestressed concrete piles. However, because of the problems listed above and potential site difficulties in splicing or lengthening preformed piles, their use has diminished considerably in recent years in favour of precast piles formed in segments or the partially preformed types of pile. Typical examples of the segmental type are Simplex Westpile's 'Hardrive' and 'Segmental' piles. The Hardrive pile is composed of standard interchangeable precast concrete units of 10, 5 and 2.5 m lengths, designed to carry working loads up to 800 kN. The pile lengths are locked together with four H-section pins located at the corners of an aligning sleeve (see Fig. 3.4.7). The Segmental pile is designed for lighter loading conditions of up to 300 kN, and is formed by joining the 1 m standard lengths together with spigot and socket joints up to a maximum length of 15 m. Special half-length head units are available to reduce wastage to a minimum.

STEEL PREFORMED PILES

Steel preformed piles are used mainly in conjunction with marine structures and where overlying soils are very weak. Two forms are encountered: the hollow box pile, which is made from welded rolled-steel sections to BS EN 10210-2: *Hot finished structural hollow sections of non-alloy and fine grain structural steels. Tolerances, dimensions and sectional properties*, and the universal bearing pile of conventional 'H' section to BS 4-1: *Structural steel sections. Specification for hot-rolled sections*. These piles are relatively light, so are easy to handle and drive. Splicing can be carried out by site welding to form piles up to 15 m long, with loadbearing capacities up to 600 kN. Always apply a protective coating to the pile to guard against corrosion.

COMPOSITE PILES

Composite piles are sometimes called partially preformed piles, and are formed by a method that combines the use of precast and in-situ concrete or steel and in-situ concrete. They are used mainly on medium to large contracts where the presence of running water or very loose soils would render the use of bored or preformed piles unsuitable. Composite piles provide a pile of readily variable length made from easily transported short lengths. Typical examples are Bullivant's Combi pile, Simplex Westpile's Shell and cased piles.

Bullivant's Combi pile is ideal where there are tough ground conditions to be overcome, such as buried boulders or old foundations. The combi pile comprises a 3 m-long steel tubular lead section with sacrificial driving foot, which is driven through the uncertain or obstructive ground and then connected to a standard 250 mm x 250 mm precast concrete pile cross-section. Typically these piles can support up to 425 kN and combine the advantages of a robust steel tubular pile with the economies of precast concrete pile.

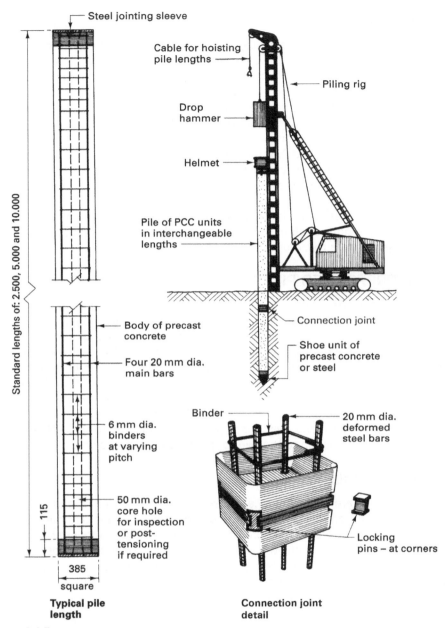

Steel jointing sleeve

Standard lengths of: 2.500, 5.000 and 10.000

Cable for hoisting
pile lengths

Piling rig

Drop
hammer

Helmet

Pile of PCC units
in interchangeable
lengths

Body of precast
concrete

Four 20 mm dia.
main bars

6 mm dia.
binders
at varying
pitch

50 mm dia.
core hole
for inspection
or post-
tensioning
if required

115

385
square

Connection joint

Shoe unit of
precast concrete
or steel

Binder

20 mm dia.
deformed
steel bars

Locking
pins – at corners

**Typical pile
length**

**Connection joint
detail**

Figure 3.4.7   Simplex Westpile's precast modular piles

Simplex Westpile's Shell pile is a driven or displacement pile consisting of a series of precast shells threaded onto a mandrel and top-driven to the required set. After driving and removing the mandrel, the hollow core can be inspected, a cage of reinforcement inserted, and the void filled with in-situ concrete. Lengths up to 60 m with bearing capacities within the range 500 to 1,200 kN are possible with this method. The precast concrete shells are reinforced by a patent system using fibrillated polypropylene film as a substitute for the traditional welded steel fabric (see Fig. 3.4.8).

Piles formed in this manner solve many of the problems encountered with waterlogged and soft substrata, because:

- they are readily adaptable in length;
- the shaft can be inspected internally before the in-situ concrete is introduced;
- the flow of water or soil into the pile is eliminated;
- the presence of corrosive conditions in the soil can be overcome by using special cements in the shell construction.

DRIVEN IN-SITU OR CAST-IN-PLACE PILES

Driven in-situ or cast-in-place piles are an alternative to preformed displacement piles. They are usually applied on medium to large contracts where subsoil conditions or loadings are likely to create variations in the lengths of pile required. They can be formed economically in diameters of 300 to 600 mm with lengths up to 18 m, designed to carry loads of up to 1,300 kN. They generally require heavy piling rigs and an open, level site where noise is unrestricted.

In some systems, the tube used to form the pile shaft is top driven; in others, such as the Franki system, driving is carried out by means of an internal drop hammer working on a plug of dry concrete. In-situ concrete for the core is introduced into the lined shaft through a hopper or skip, and the concrete can be consolidated by impact of the internal drop hammer or by vibration of the tube as it is withdrawn (see Figs 3.4.9 and 3.4.10). A problem that can be encountered with this form of pile is necking due to groundwater movement washing away some of the concrete, thus reducing the effective diameter of the pile shaft and consequently the cover of concrete over the reinforcement. If groundwater is present the other forms of displacement pile previously described should be considered.

PILEDRIVING

Displacement piles are generally driven into the ground by holding them in the correct position against the piling frame and applying hammer blows to the head of the pile. The piling frame can be purpose-made or an adaptation of a standard crane power unit. The basic components of any piling frame are the vertical member that houses the leaders or guides, which in turn support the pile and guide the hammer on to the head of the pile (see Fig. 3.4.11). Pile hammers come in a variety of types and sizes powered by gravity, compressed air or diesel.

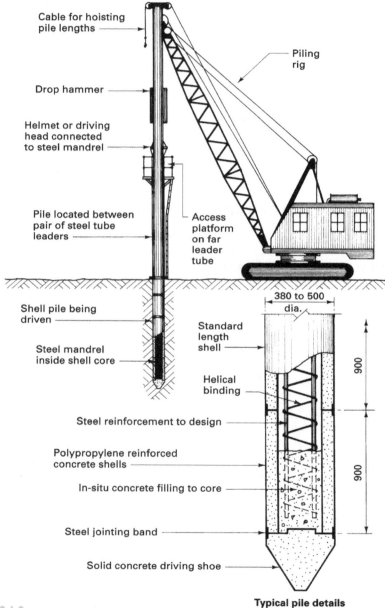

Figure 3.4.8 Simplex Westpile's Shell piles

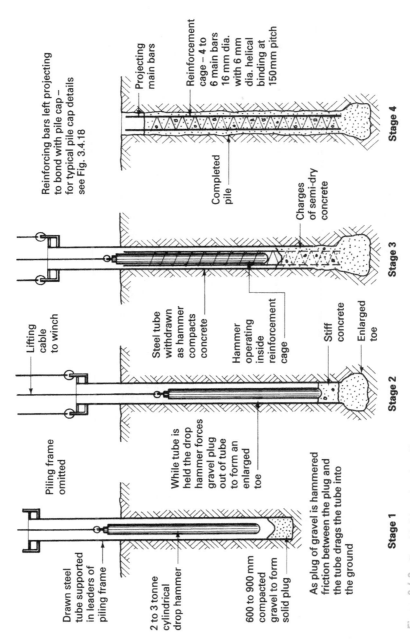

Piling frame omitted

Drawn steel tube supported in leaders of piling frame

2 to 3 tonne cylindrical drop hammer

600 to 900 mm compacted gravel to form solid plug

As plug of gravel is hammered friction between the plug and the tube drags the tube into the ground

Stage 1

While tube is held the drop hammer forces gravel plug out of tube to form an enlarged toe

Lifting cable to winch

Stage 2

Steel tube withdrawn as hammer compacts concrete

Hammer operating inside reinforcement cage

Stiff concrete

Enlarged toe

Stage 3

Charges of semi-dry concrete

Completed pile

Stage 4

Reinforcing bars left projecting to bond with pile cap – for typical pile cap details see Fig. 3.4.18

Projecting main bars

Reinforcement cage – 4 to 6 main bars 16 mm dia. with 6 mm dia. helical binding at 150mm pitch

Figure 3.4.9   Franki driven in-situ piles

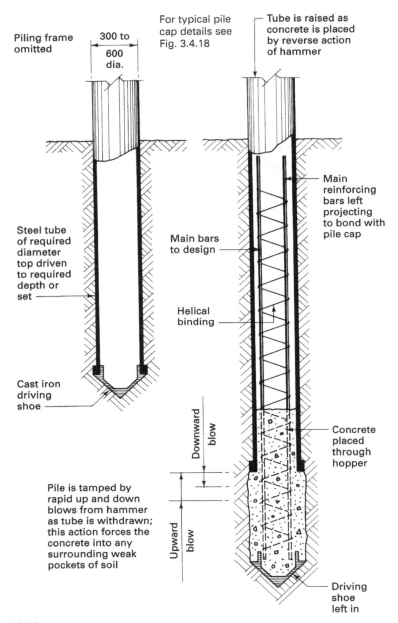

Piling frame omitted

300 to 600 dia.

For typical pile cap details see Fig. 3.4.18

Tube is raised as concrete is placed by reverse action of hammer

Steel tube of required diameter top driven to required depth or set

Cast iron driving shoe

Main bars to design

Helical binding

Downward blow

Upward blow

Pile is tamped by rapid up and down blows from hammer as tube is withdrawn; this action forces the concrete into any surrounding weak pockets of soil

Main reinforcing bars left projecting to bond with pile cap

Concrete placed through hopper

Driving shoe left in

Figure 3.4.10 Vibro cast in-situ pile

DROP HAMMERS

Drop hammers are blocks of cast iron or steel with a mass range of 1,500 to 8,000 kg and are raised by a cable attached to a winch. The hammer, which is sometimes called a monkey or ram, is allowed to fall freely by gravity on to the pile head. The free-fall distance is controllable, but generally a distance of about 1.2 m is employed. Drop hammers are slower than the following power hammers and may inflict more damage to the pile caps.

SINGLE-ACTING HAMMERS

Activated by compressed air, single-acting hammers have much the same effect as drop hammers, in that the hammer falls freely by gravity through a distance of about 1.5 m. Two types are available: in one case, the hammer is lifted by a piston rod; in the other, the piston is static and the cylinder is raised and allowed to fall freely (see Fig. 3.4.11). Both forms of hammer deliver a very powerful blow.

DIESEL HAMMERS

Diesel hammers have been designed to give a reliable and economic method of piledriving. Various sizes giving different energy outputs per blow are available, but most deliver between 46 and 52 blows per minute. The hammer can be suspended from a crane or mounted in the leaders of a piling frame. A measured amount of liquid fuel is fed into a cup formed in the base of the cylinder. The air being compressed by the falling ram is trapped between the ram and the anvil, which applies a preloading force to the pile. The displaced fuel, at the precise moment of impact, results in an explosion that applies a downward force on the pile and an upward force on the ram, which returns to its starting position to recommence the complete cycle. The movement of the ram within the cylinder activates the fuel supply, and opens and closes the exhaust ports (see Fig. 3.4.12). Water- or air-cooled variations exist for popular application to vertical and inclined driving.

VIBRATION TECHNIQUES

Vibration techniques can be used in driving displacement piles where soft clays, sands and gravels are encountered. They are notably efficient, with preformed steel pipe, 'H' and sheet pile sections. The equipment consists of a vibrating unit mounted on the pile head, transmitting vibrations of the required frequency and amplitude down the length of the pile shaft. It achieves this with two eccentric rotors propelled in opposite directions to generate vertical vibrations in the pile. These are, in turn, transmitted to the surrounding soil, reducing its shear strength and enabling the pile to sink into the subsoil under its own weight and also that of the vibrator unit. Figure 3.4.13 shows the operating principles of the unit attached to the head of a sheet steel pile.

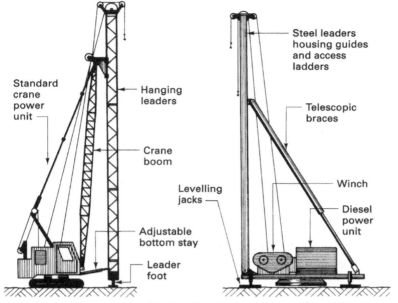

Typical piling frames

- Standard crane power unit
- Hanging leaders
- Crane boom
- Levelling jacks
- Adjustable bottom stay
- Leader foot

- Steel leaders housing guides and access ladders
- Telescopic braces
- Winch
- Diesel power unit

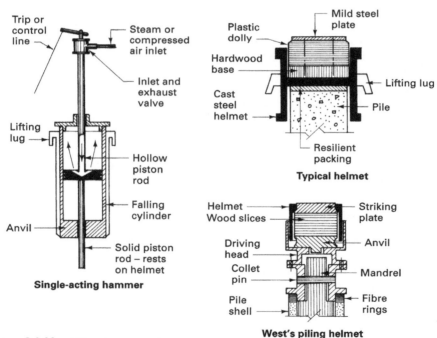

Single-acting hammer

- Trip or control line
- Steam or compressed air inlet
- Inlet and exhaust valve
- Lifting lug
- Hollow piston rod
- Falling cylinder
- Anvil
- Solid piston rod – rests on helmet

Typical helmet

- Mild steel plate
- Plastic dolly
- Hardwood base
- Cast steel helmet
- Lifting lug
- Pile
- Resilient packing

West's piling helmet

- Helmet
- Wood slices
- Driving head
- Collet pin
- Pile shell
- Striking plate
- Anvil
- Mandrel
- Fibre rings

Figure 3.4.11   Piling frames and equipment

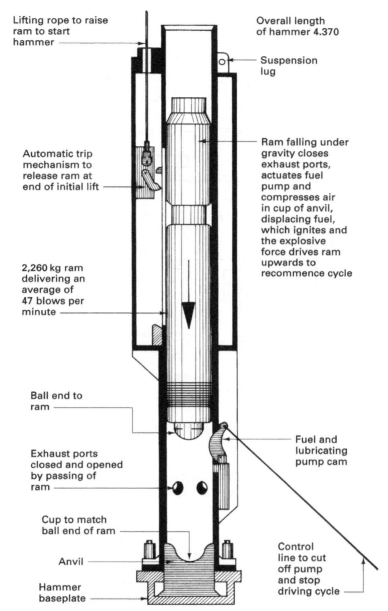

Lifting rope to raise ram to start hammer

Overall length of hammer 4.370

Suspension lug

Automatic trip mechanism to release ram at end of initial lift

Ram falling under gravity closes exhaust ports, actuates fuel pump and compresses air in cup of anvil, displacing fuel, which ignites and the explosive force drives ram upwards to recommence cycle

2,260 kg ram delivering an average of 47 blows per minute

Ball end to ram

Fuel and lubricating pump cam

Exhaust ports closed and opened by passing of ram

Cup to match ball end of ram

Control line to cut off pump and stop driving cycle

Anvil

Hammer baseplate

Figure 3.4.12   Typical diesel hammer details

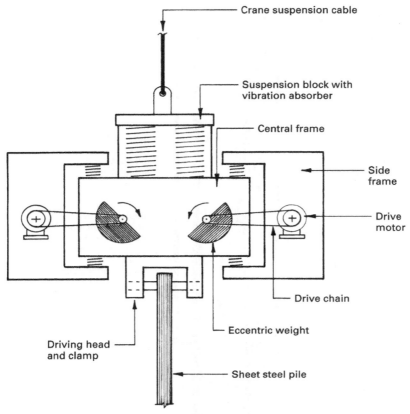

Crane suspension cable

Suspension block with
vibration absorber

Central frame

Side
frame

Drive
motor

Drive chain

Eccentric weight

Driving head
and clamp

Sheet steel pile

Figure 3.4.13   **Operating principles of a vibrating hammer**

To aid the driving process and to reduce the risk of damage to the pile during driving, water-jetting techniques can be used. Water is directed at the soil around the toe of the pile to loosen it and ease the driving process. The water pipes are usually attached to the perimeter of the pile shaft, and are taken down with the pile as it is being driven. The water-jetting operation is stopped before the pile reaches its final depth so that the toe of the pile is finally embedded in undisturbed soil. The main advantage is the relatively silent operation. Additionally there is no damage to the pile cap, and high-voltage electricity may be used instead of combustible fuels.

To protect the heads of preformed piles from damage due to impact from hammers, various types of protective cushioned helmet are used. These can be either of a general nature, as shown in Fig. 3.4.11, or special purpose for a particular system, such as those used for Simplex Westpile's piling.

Sometimes referred to as bored piles, replacement piles are formed by removing a column of soil and replacing it with in-situ concrete or, as in the case of composite piles, with precast and in-situ concrete. Replacement or bored piles are considered for sites where piling is being carried out in close proximity to existing buildings or where vibration and/or noise is restricted. The formation of this type of pile can be considered under two general classifications:

1. percussion bored piles;
2. rotary bored piles;

PERCUSSION BORED PILES

Percussion bored piles are suitable for small and medium-sized contracts of up to 300 piles in clay or gravel subsoils. Pile diameters are usually from 300 to 950 mm, designed to carry loads up to 1,500 kN. Apart from the common factor with all replacement piles that the strata penetrated can be fully explored, these piles can be formed by using a shear leg or tripod rig requiring as little as 1.800 m headroom.

A steel tube made up from lengths (1.000 to 1.400 m) screwed together is sunk by extracting the soil from within the tube liner using percussion cutters or balers, depending on the nature of the subsoil to be penetrated. The steel lining tube will usually sink under its own weight, but it can be driven in with slight pressure, normally applied by means of hydraulic jacks. When the correct depth has been reached, a cage of reinforcement is placed within the liner and the concrete is introduced. Tamping is carried out as the liner is extracted using a winch or hydraulic jack operating against a clamping collar fixed to the top of the steel tube lining. An internal drop hammer can be used to tamp and consolidate the concrete, but usually compressed air is used.

If waterlogged soil is encountered, a pressure pile is usually formed by fixing to the head of the steel liner an airlock hopper, through which the concrete can be introduced and consolidated, while the borehole remains under pressure in excess of the hydrostatic pressure of the groundwater (see Fig. 3.4.14).

ROTARY BORED PILES

Rotary bored piles are the most common and versatile type of bored pile. They range from the short-bored pile used in domestic dwellings (see *Construction Technology*, Chapter 3.3) to the very large-diameter piles used for concentrated loads in multi-storey buildings and bridge construction. The rotary bored pile is suitable for most cohesive soils, such as clay, and is formed using an auger, which may be operated in conjunction with the steel tube liner according to the subsoil conditions encountered.

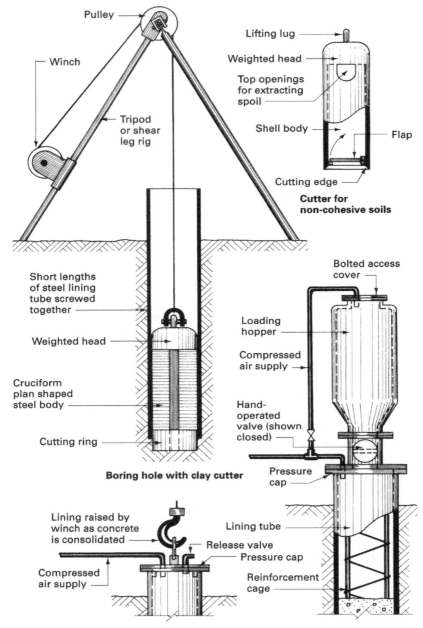

Cutter for non-cohesive soils

Pulley

Winch

Tripod or shear leg rig

Lifting lug

Weighted head

Top openings for extracting spoil

Shell body

Flap

Cutting edge

Short lengths of steel lining tube screwed together

Weighted head

Cruciform plan shaped steel body

Cutting ring

Boring hole with clay cutter

Bolted access cover

Loading hopper

Compressed air supply

Hand-operated valve (shown closed)

Pressure cap

Lining tube

Reinforcement cage

Lining raised by winch as concrete is consolidated

Release valve

Pressure cap

Compressed air supply

Alternative methods of forming pressure piles

Figure 3.4.14   Typical percussion bored pile details

Two common augers are in use. One is the Cheshire auger, which has 1½ to 2 helix turns at the cutting end and is usually mounted on a lorry or tractor. The turning shaft or kelly bar is generally up to 7.500 m long and is either telescopic or extendable. The soil is cut by the auger, raised to the surface, and spun off the helix to the side of the borehole, from where it is removed from site (see Fig. 3.4.15). Alternatively, a continuous flight auger (CFA) can be used, where the spiral motion brings the spoil to the surface for removal from site. Flight augers are usually mounted on a purpose-built power unit similar to a crawler crane.

A development of CFA replacement piling uses a modified auger with an open-ended hollow central core. The borehole is excavated as described for a conventional auger but, on reaching the required depth, concrete is pumped into the hollow core as the auger is withdrawn. This practice is time-efficient, as there is no need to case the pile or temporarily support the excavation. There is also little disturbance to the surrounding subsoil. When the auger is removed, a reinforcement cage is vibrated into the concrete to the required depth. To ease pumping and reinforcement placement, the concrete design may specify rounded aggregates and a higher-than-normal water:cement ratio. Therefore pile diameter and quantity of piles may be greater than with more conventional methods. This technique of placing the concrete as the auger is withdrawing is also known as grout injection piling (see Fig. 3.4.16).

Large-diameter bored piles are usually considered to be those over 750 mm, and can be formed with diameters up to 2.6 m with lengths ranging from 24 to 60 m to carry loadings from 2,500 to 8,000 kN. They are suitable for use in stiff clays for structures having highly concentrated loadings, and can be lined or partially lined with steel tubes as required. The base or toe of the pile can be enlarged or under-reamed up to three times the shaft diameter to increase the bearing capacity of the pile. Reinforcement is not always required, and the need for specialists' knowledge at the design stage cannot be overemphasised. Compaction of the concrete, which is usually placed by a tremie pipe, is generally by gravitational force (see Fig. 3.4.17). Test loading of large-diameter bored piles can be very expensive, and if the local authority insists on test loading, it can render this method uneconomic.

PILE TESTING

The main objective of forming a test pile is to confirm that the design and formation of the chosen pile type is adequate. It is always advisable to form at least one test pile on any piling contract and, indeed, many local authorities will insist on this being carried out. The test pile must not be used as part of the finished foundations, and should be formed and tested in such a position that it will not interfere with the actual contract but is nevertheless truly representative of site conditions.

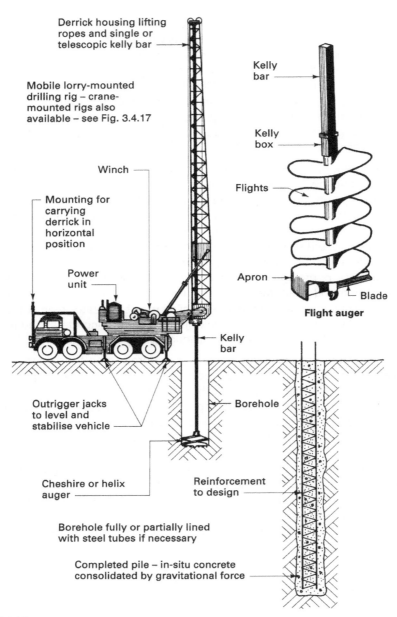

Derrick housing lifting ropes and single or telescopic kelly bar

Mobile lorry-mounted drilling rig – crane-mounted rigs also available – see Fig. 3.4.17

Winch

Mounting for carrying derrick in horizontal position

Power unit

Outrigger jacks to level and stabilise vehicle

Cheshire or helix auger

Borehole fully or partially lined with steel tubes if necessary

Completed pile – in-situ concrete consolidated by gravitational force

Kelly bar

Borehole

Reinforcement to design

Kelly bar

Kelly box

Flights

Apron

Blade

Flight auger

Figure 3.4.15   Typical rotary bored pile details

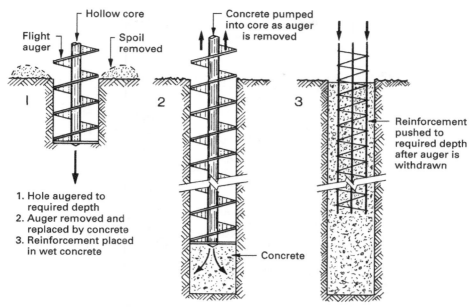

Flight auger — **Hollow core** — **Spoil removed**

Concrete pumped into core as auger is removed

Reinforcement pushed to required depth after auger is withdrawn

1. Hole augered to required depth
2. Auger removed and replaced by concrete
3. Reinforcement placed in wet concrete

Concrete

Figure 3.4.16   **CFA piling using grout injection**

Test piles are usually overloaded by at least 50 per cent of the design working load to near failure or to actual failure, depending on the test data required. Any loading less than total failure loading should remain in place for at least 24 hours. The test pile is bored to the required depth or driven to the required set (which is a predetermined penetration per blow or series of blows), after which it can be tested by one of the following methods.

- A grillage or platform of steel or timber is formed over the top of the test pile and loaded with a quantity of kentledge composed of pig iron or precast concrete blocks. A hydraulic jack is placed between the kentledge and the head of the pile, and the test load gradually applied.

- Three piles are formed, and the outer two piles are tied across their heads with a steel or concrete beam. The object is to jack down the centre or test pile against the uplift of the outer piles. Unless two outer piles are available for the test, this method can be uneconomic if special outer piles have to be bored or driven.

- Anchor-stressing wires are secured into rock or by some other means to provide the anchorage and uplift the resistance to a cross-beam of steel or concrete by passing the wires over the ends of the beam. The test load is applied by a hydraulic jack placed between the cross-member and the pile head, as described for the previous method.

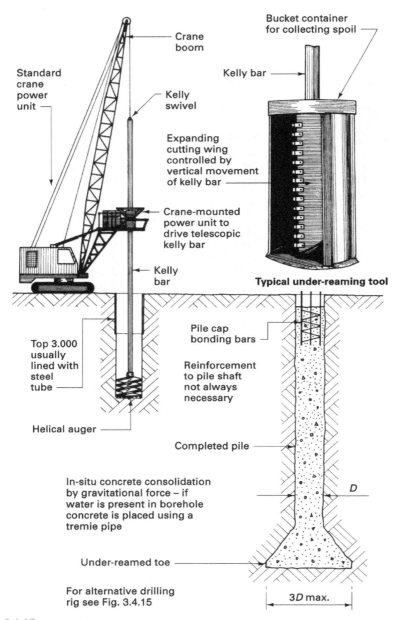

Crane boom

Standard crane power unit

Kelly swivel

Expanding cutting wing controlled by vertical movement of kelly bar

Crane-mounted power unit to drive telescopic kelly bar

Kelly bar

Top 3.000 usually lined with steel tube

Helical auger

Bucket container for collecting spoil

Kelly bar

Typical under-reaming tool

Pile cap bonding bars

Reinforcement to pile shaft not always necessary

Completed pile

In-situ concrete consolidation by gravitational force – if water is present in borehole concrete is placed using a tremie pipe

Under-reamed toe

For alternative drilling rig see Fig. 3.4.15

D

$3D$ max.

Figure 3.4.17 Typical large-diameter bored pile details

- The constant rate of penetration test is a method whereby the pile under test is made to penetrate the ground at a constant rate, generally in the order of about 0.8 mm per minute. The load needed to produce this rate of penetration is plotted against a deflection or timebase. When no further load is required to maintain this constant rate of penetration, the ultimate bearing capacity has been reached.

Note that the safe load for a single pile cannot necessarily be multiplied by the number of piles in a cluster to obtain the safe load of a group of piles.

PILE CAPS

Apart from the simple situations such as domestic dwellings, or where large-diameter piles are employed, piles are not usually used singly but are formed into a group or cluster. The load is distributed over the heads of the piles in the group by means of a reinforced cast in-situ concrete pile cap. To provide structural continuity the reinforcement in the piles is bonded into the pile cap; this may necessitate the breaking out of the concrete from the heads of the piles to expose the reinforcement. The heads of piles also penetrate the base of the pile cap some 100 to 150 mm to ensure continuity of the members.

Piles should be spaced at such a distance that the group is economically formed and at the same time any interaction between adjacent piles is prevented. Actual spacings must be selected on subsoil conditions, but the usual minimum spacings are:

- **friction piles**: three pile diameters or 1.000 m, whichever is greater;
- **end-bearing piles**: two pile diameters or 750 mm, whichever is greater.

The plan shape of the pile cap should be as conservative as possible, and this is usually achieved by having an overhang of 150 mm. The Federation of Piling Specialists has issued the following guide table as to suitable pile cap depths, having regard to both design and cost requirements.

Pile diameter (mm)	300	350	400	450	500	550	600	750
Depth of cap (mm)	700	800	900	1,000	1,100	1,200	1,400	1,800

The main reinforcement is two-directional, formed in bands over the pile heads to spread the loads, and usually takes the form of a U-shaped bar suitably bound to give a degree of resistance to surface cracking of the faces of the pile cap (see typical details shown in Fig. 3.4.18).

In many piling schemes, especially where capped single piles are used, the pile caps are tied together with reinforced concrete tie beams and should be restrained by these beams in two different directions. The beams can be used to carry loadings such as walls to the pile foundations (see Fig. 3.4.18).

The formation of piled foundations is a specialist's task, so a piling contractor may be engaged to carry out the work under one of three arrangements:

1. nominated subcontractor;

2. direct subcontractor;

3. main contractor under a separate contract.

Piling companies normally supply the complete service of advising, designing and carrying out the complete site works, including setting out if required. When seeking a piling tender, the builder should supply the following information:

- site investigation report, giving full details of subsoil investigations, adjacent structures, topography of site and any restrictions regarding headroom, noise or vibration limitations;

- site layout drawings, indicating levels, proposed pile positions, structural loadings and existing buried or overhead services that are present;

- contract data regarding tender dates, contract period, completion dates, a detailed specification and details of any special or unusual contract clauses.

The appraisal of piling tenders is not an easy task, and the builder should take into account not only costs but also any special conditions attached to the submitted tender and the acceptance by the local authority of the proposed scheme and system.

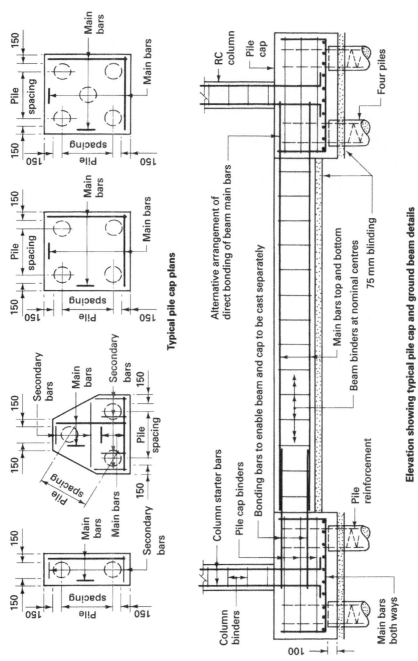

Typical pile cap plans

150 | Main bars
Pile spacing
Pile spacing
150 | 150 | 150 | Main bars | Main bars

150 | Main bars
Pile spacing
Pile spacing
150 | 150 | 150 | Main bars | Main bars

Secondary bars | Main bars | Secondary bars
150 | 150
Pile spacing
Pile spacing
150 | 150 | Main bars | Secondary bars

150 | Main bars | Main bars
Pile spacing
150 | 150 | 150 | Secondary bars

Elevation showing typical pile cap and ground beam details

RC column
Pile cap
Four piles

Alternative arrangement of direct bonding of beam main bars

Bonding bars to enable beam and cap to be cast separately

Main bars top and bottom

Beam binders at nominal centres

75 mm blinding

Column starter bars
Pile cap binders

Pile reinforcement

Column binders

Main bars both ways

100

Figure 3.4.18 Typical pile cap and ground beam details

Deep basements 3.5

The construction of shallow single-storey basements and methods of waterproofing are considered in Chapter 3.6 of *Construction Technology* and Chapter 3.2 of this volume.

One of the main concerns of the contractor and designer when constructing deep and/or multi-storey basements is the elimination, as far as is practicable, of the need for temporary works such as timbering, which in this context would be extensive, elaborate and costly and would put operatives' health and safety at risk. The solution is to be found in the use of piled or diaphragm walling as the structural perimeter wall, constructing this ahead of the main excavation activity, or by using a reinforced concrete land-sunk open caisson if the subsoil conditions are favourable (see Fig. 3.2.12).

Diaphragm or piled walls can be designed to act as pure cantilevers in the first instance, but this is an expensive and usually unnecessary method. The major decision whether to provide temporary support until the floors, which offer lateral restraint, can be constructed, or whether to provide permanent support, if the deep basement is to be constructed free of intermediate floors. Diaphragm walls can take the form of in-situ or precast reinforced concrete wall panels installed using the bentonite slurry trench system (see Fig 3.2.6), while the piled alternative comprises either secant or contiguous augured piles (see Fig 3.1.8) – refer to Chapter 3.2 for details.

MONITORING MOVEMENT

When undertaking deep basement works it is important to set up a monitoring system, whereby any movement in the basement walls or adjacent existing structures can be measured and recorded. This is usually undertaken to reassure the client and other stakeholders that the works are safe and stable as the work proceeds, and that their assets are secure. Sophisticated electro-distance measurement survey stations can be set up using reflectored or reflectorless targets, which record any movement to the nearest 1 mm. These can be manually undertaken by a surveyor on a daily basis, or set up as a fully automated system, which can record up to 1,000 points per hour, 24 hours per day. If the preset trigger movement is exceeded, the engineers can be alerted by text or email so that they can take immediate corrective action.

CONSTRUCTION METHODS

Four basic methods can be used in the construction of deep basements without the need for timbering or excessive subsoil support:

1. propped wall construction method;
2. anchored wall construction method;
3. top-down construction method;
4. berm buttress construction method:

PROPPED WALL CONSTRUCTION METHOD

A series of tube or lattice steel props is installed so that they span between the tops of opposite diaphragm walls, enabling the walls to act initially as propped cantilevers, receiving their final lateral restraint when the internal floors and wall linings have been constructed later (see Fig. 3.5.1). This method is also known as 'bottom-up' construction. The positioning of the props needs to take into account how excavation plant can access the bottom of the excavation to dig out the basement and move other materials in and out. Typically excavation is undertaken by long dipper arm hydraulic backacter machines.

ANCHORED WALL CONSTRUCTION METHOD

The diaphragm walls are exposed by carrying out the excavation in stages, and ground anchors are used to stabilise the walls as the work proceeds. This is a very popular method where no lateral restraint in the form of floors is to be provided, such as in a roadway underpass or where open access is needed to excavate or lift in components (see Fig. 3.5.1). The technique and details of ground anchor installations are given in conjunction with prestressing systems in Chapter 10.1.

TOP-DOWN CONSTRUCTION METHOD

After the perimeter diaphragm walls have been constructed, the ground floor slab and beams are cast, providing top edge lateral restraint to the walls. An opening is left in this initial floor slab through which operatives, materials and excavation plant can pass to excavate the next stage and cast the floor slab

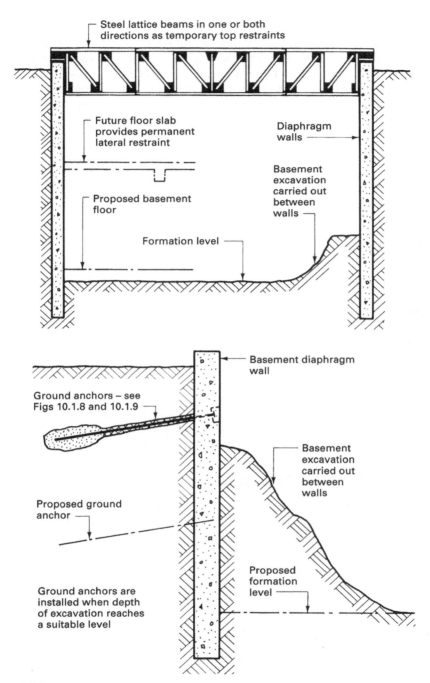

Steel lattice beams in one or both directions as temporary top restraints

Future floor slab provides permanent lateral restraint

Diaphragm walls

Proposed basement floor

Basement excavation carried out between walls

Formation level

Basement diaphragm wall

Ground anchors – see Figs 10.1.8 and 10.1.9

Basement excavation carried out between walls

Proposed ground anchor

Proposed formation level

Ground anchors are installed when depth of excavation reaches a suitable level

Figure 3.5.1   **Deep basement construction methods: 1**

and beams. This method can be repeated until the required depth has been reached (see Fig. 3.5.2). Excavation below the cast floor slabs is undertaken by small tracked mini-diggers with face shovels. They move the excavated material to the access holes, where it is lifted out by long arm dipper excavators. Depending on the size of the basement, this method often requires a temporary system of forced ventilation to evacuate engine exhaust fumes to ensure operator safety. Where internal steel columns are needed to support the floor spans, these are pre-installed in rotary bored pile holes before the ground slab is cast (see Fig. 3.5.3). These columns are known as plunge columns and form the permanent structure of the basement and upper floors. This method allows the creation of basements with large open plan areas, needed for car parking and storage space.

BERM BUTTRESS CONSTRUCTION METHOD

The centre area between the diaphragm walls can be excavated, leaving an earth mound or berm around the perimeter to support the walls while the lowest basement floor is constructed. Slots to accommodate raking struts acting between the wall face and the floor slab are cut into the berm; final excavation and construction of the remainder of the basement can take place working around the raking struts (see Fig. 3.5.2). This method can be scaled up to support deep basements by substituting the adjustable props as shown for the use of large-diameter prefabricated steel tubes (up to 900 mm in diameter) fixed to horizontal steel UB walings.

WATERPROOFING BASEMENTS

Standard methods for waterproofing basements, such as the application of membranes, using dense monolithic concrete structural walls and tanking techniques using mastic asphalt, are covered in the context of single-storey basements (see Chapter 3.6 of *Construction Technology*). Another method that can be used for waterproofing basements is the drained or cavity tanking system, using special floor tiles produced by Atlas Stone Products Ltd.

DRAINED CAVITY SYSTEM

The drained cavity system of waterproofing is suitable for basements of any depth where there are no intermediate floors, or where such floors are so constructed that they would not bridge the cavity. The basic concept of this system is very simple. It accepts that it is possible to have a small amount of water seepage or moisture penetration through a dense concrete structural perimeter wall, and the best method of dealing with any such penetration is to allow it to be collected and drained away without it reaching the interior of the building. This is achieved by constructing an inner non-loadbearing wall to create a cavity, and laying a floor consisting of precast concrete Dryangle tiles

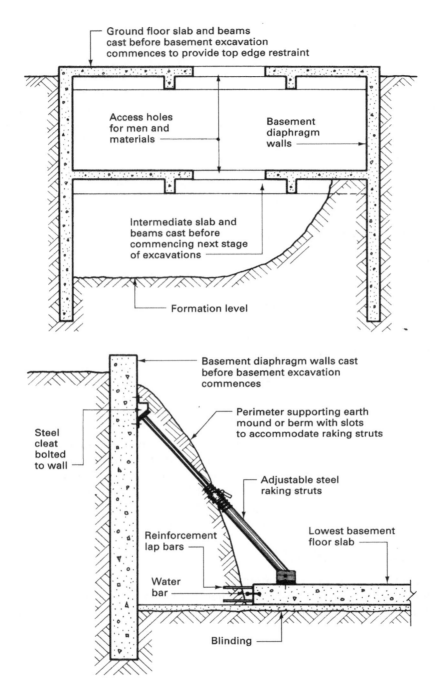

Figure 3.5.2   Deep basement construction methods: 2

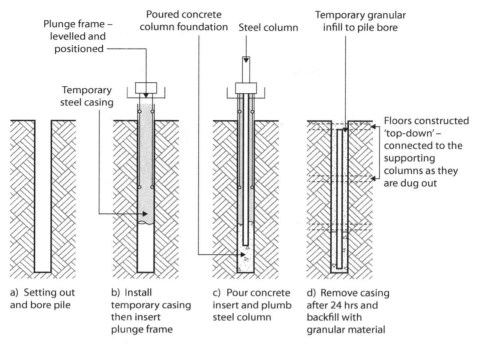

Figure 3.5.3   **Construction of plunge columns used in top-down construction**

over the structural floor of the basement. This will allow any water trickling down the inside face of the outer wall to flow beneath the floor tiles, where it can be discharged into the surface water drains or, alternatively, pumped into the drains if these are at a higher level than the invert of the sump. Typical details are shown in Fig. 3.5.4.

The concrete used to form the structural basement wall should be designed as if it were specified for waterproof concrete, which, it must be remembered, is not necessarily vapourproof. All joints should be carefully designed, detailed and constructed, including the fixing of suitable water bars (see Fig. 3.5.7 in *Construction Technology*). It must be emphasised that the drained cavity system is designed to deal only with seepage that may occur, so the use of a dense concrete perimeter wall is an essential component in the system. The dense concrete used in the outer wall needs to be well vibrated to obtain maximum consolidation, and this can usually be achieved by using poker vibrators.

The in-situ concrete used in any form of basement construction can be placed using chutes, pumps or tremie pipes. The placing of in-situ concrete using pumps is covered in Chapter 2.5, dealing with contractors' plant, but the placing of concrete by means of a tremie pipe is considered in the section on placing concrete below ground.

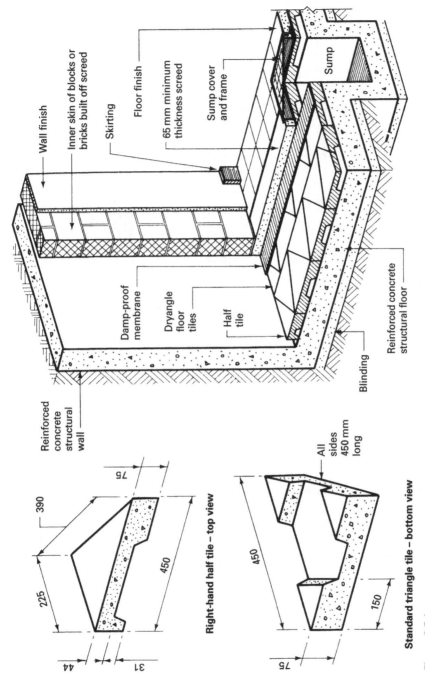

Wall finish

Inner skin of blocks or bricks built off screed

Skirting

Floor finish

65 mm minimum thickness screed

Sump cover and frame

Sump

Damp-proof membrane

Dryangle floor tiles

Half tile

Reinforced concrete structural floor

Blinding

Reinforced concrete structural wall

390

75

225

450

44

31

Right-hand half tile – top view

All sides 450 mm long

450

150

75

Standard triangle tile – bottom view

Figure 3.5.4   Drained cavity system of waterproofing basements

BUOYANCY OR TANKED BASEMENT

Where site space permits, the building dead and imposed loading may be distributed over sufficient basement raft foundation base area to prevent undue settlement. Multi-storey buildings on relatively small sites of restricted foundation bearing area can still be constructed, even where the subsoil composition is unsuited to piling techniques, or where for practical reasons piling is not viable. In these situations it may be possible to design the basement as a huge tank on which the building effectively 'floats'.

In principle, the concept is to counterbalance the mass of subsoil removed for the basement with the mass of the building in its place. Very careful design calculations will be required to ensure the balance, to avoid excessive and eccentric loadings, and to allow for extraneous imposed loading, including wind and snow. Generally, the deeper the excavation, the more compressed the subsoil will be and the greater its bearing capacity. The opportunity to excavate at depth will normally enhance the loading potential considerably, although when exposing clay subsoil there can be a problem of it swelling as the soil above is removed. A substantial loading of in-situ concrete will control this, the weight of which must be incorporated into the balancing calculations. Finally, in large basement construction where a high water table is present, the additional buoyancy caused may need to be counterbalanced by additional tension piles sunk and connected into the basement ground slab to resist the uplift.

PLACING CONCRETE BELOW GROUND

A tremie pipe can be used to place in-situ concrete below the ground or water level where segregation of the mix must be avoided, and can be used in the construction of piled foundations, basements and diaphragm walls. For work below ground level, a rigid tube of plastic or metal can be used; alternatively a flexible PVC tube can be employed, often called elephant trunking. In all cases, the discharge end of the tremie pipe is kept below the upper level of the concrete being placed and the upper end of the pipe is fitted with a feed hopper attachment to receive the charges of concrete. As the level of the placed concrete rises, the tube is shortened by removing lengths of pipe as required. The tremie pipe and its attached hopper head must be supported as necessary throughout the entire concrete-placing operation.

Placing concrete below water level by means of a tremie pipe, very often suspended from a crane, requires more care and skill than placing concrete below ground level. The pipe should be of adequate diameter for the aggregates being used, common tube diameters being 150 and 200 mm. It should be watertight and have a smooth bore. The operational procedure is as follows.

1. The tremie pipe is positioned over the area to be concreted, and lowered until the discharge end rests on the formation level.

2. A travelling plug of plastic or cement bags is inserted into the top of the pipe to act as a barrier between the water and concrete. The weight of the first charge concrete will force the plug out of the discharge end of the tube.

3. When filled with concrete, the tremie pipe is raised so that the discharge end is just above the formation level to allow the plug to be completely displaced, enabling the concrete to flow.

4. Further concrete charges are introduced into the pipe and allowed to discharge within the concrete mass already placed, the rate of flow being controlled by raising and lowering the tremie pipe.

5. As the depth of the placed concrete increases, the pipe is shortened by removing tube sections as necessary.

When placing concrete below the water level by the tremie pipe method, great care must be taken to ensure that the discharge end of the pipe is not withdrawn clear of the concrete being placed, as this could lead to a weak concrete mix by water mixing with the concrete. One tremie pipe will cover an area of approximately 30 m^2 and, if more than one tremie pipe is being used, simultaneous placing is usually recommended. A problem that can occur when using concrete below ground level is the deterioration of the set cement due to sulphate attack by naturally occurring sulphates such as calcium, magnesium and sodium sulphates, which are found particularly in clay soils. The main factors that influence the degree of attack are:

- amount and nature of sulphate in the soil;
- level of water table and amount of groundwater movement;
- form of construction;
- type and quality of concrete.

Details of sulphate-resisting cements that can be used to reduce these effects are given in *Construction Technology*, Chapter 3.2.

Underpinning 3.6

The main objective of underpinning is to transfer the load carried by an existing foundation from its present bearing level to a new level at a lower depth. It can also be used to replace an existing weak foundation. Underpinning may be necessary for one or more of the following reasons:

- as a remedy for:
  - uneven loading;
  - unequal resistance of the subsoil;
  - action of tree roots;
  - action of subsoil water;
  - cohesive soil settlement;
  - deterioration of foundation concrete;
- to increase the loadbearing capacity of a foundation, which may be required to enable an extra storey to be added to the existing structure or if a change of use would increase the imposed loadings;
- as a preliminary operation to lowering the adjacent ground level when constructing a basement at a lower level than the existing foundations of the adjoining structure or when laying deep services near to or below the existing foundations.

Before any underpinning is started, you need to carry out the following surveying and preliminary work.

1. Serve notice to the adjoining owners, often in the form of a Party Wall Agreement, setting out in detail the intention to proceed with the proposed works and giving full details of any proposed temporary supports, such as shoring.

2. Make a detailed survey of the building to be underpinned, recording any defects or cracks, supplemented by photographs and agreed with the building owner where possible.

3. Fix glass slips or 'telltales' across any vertical or lateral cracks to give a visual indication of any further movement taking place.

4. Establish a detailed monitoring regime for the movement of the building and its neighbours. Either take a series of check levels against a reliable datum, or fix metal studs to the external wall with their levels noted. Check these levels periodically as the work proceeds to enable any movements to be recorded and the necessary remedial action taken.

5. Get permission from the adjoining owner to stop up all flues and fireplaces, to prevent the nuisance and damage that can be caused by falling soot.

6. If underpinning is required to counteract unacceptable settlement of the existing foundations, investigate the subsoil to determine the cause and to forecast any future movement so that the underpinning design will be adequate.

7. Reduce the loading on the structure as much as possible by removing imposed floor loads and installing any necessary shoring.

8. Undertake a detailed risk assessment to ensure workers are not put at risk and effective control measures are in place, such as edge protection and support to excavations.

WALL UNDERPINNING

Traditional underpinning to walls is carried out by excavating in stages alongside and underneath the existing foundation, casting a new foundation and building up to the underside of the existing foundation in brickwork or concrete. In modern underpinning, concrete usually takes precedence over brickwork as it is quicker and safer, and requires less shoring and temporary support. Pinning between the old and new work is a rich dry mortar known as a 'dry-pack', which ensures low shrinkage and a good structural bearing between old and new work.

To prevent fracture or settlement, the underpinning stages or bays should be kept short and formed to a definite sequence, so that no two bays are worked consecutively. This will enable the existing foundation and wall to arch or span the void created underneath before underpinning. The number and length of the bays will depend on the:

- total length of wall to be underpinned;
- width of existing foundation;
- general condition of existing substructure;
- superimposed loading of existing foundation;
- estimated spanning ability of existing foundation;
- subsoil conditions encountered.

The generally specified maximum length for bays used in the underpinning of traditional wall construction is 1.5 m, with the proviso that at no time should the sum total of unsupported lengths exceed 25 per cent of the total wall length.

Bays are excavated and timbered as necessary, after which the bottom of the excavation is prepared to receive the new foundation. To give the new foundation strip continuity, dowel bars are inserted at the end of each bay. In in-situ concrete underpinning, splice bars or dowels project into the adjacent pins to provide the continuity. Concrete used in underpinning is usually specified as a 20 mm aggregate mix with a minimum strength of 30 N/mm^2 (C30), using rapid-hardening cement. The final dry-pack mix should consist of 1 part rapid-hardening cement to 3 parts well-graded fine aggregate from 10 mm down to fine sand, with a water/cement ratio of 0.35. In both methods the projection of the existing foundation is cut back to the external wall line, so that the loads are transmitted to the new foundation and not partially dissipated through the original foundation strip on to the backfill material (see Fig. 3.6.1).

JACK OR MIGA PILE UNDERPINNING

Jack or miga pile underpinning can be used where:

- suitable bearing capacity subsoil is too deep to make traditional wall underpinning practical or economic;
- a system giving no vibration is required – note that that this method is also practically noiseless;
- a system of variable depth is required;
- the existing foundation is structurally sound.

The system consists of short, precast concrete pile lengths jacked into the ground until a suitable subsoil is reached (see Fig. 3.6.2). When the jack pile has reached the required depth, the space between the top of the pile and the underside of the existing foundation is filled with a pinned concrete cap. The existing foundation must be in a good condition, because in the final context it will act as a beam spanning over the piles. The condition and spanning ability of the existing strip foundation will also determine the spacing of the piles.

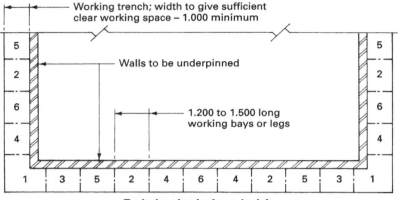

Working trench; width to give sufficient
clear working space – 1.000 minimum

Walls to be underpinned

1.200 to 1.500 long
working bays or legs

5		5
2		2
6		6
4		4
1	3	5	2	4	6	4	2	5	3	1

Typical underpinning schedule

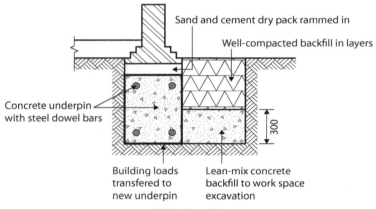

Sand and cement dry pack rammed in

Well-compacted backfill in layers

Concrete underpin
with steel dowel bars

300

Building loads
transfered to
new underpin

Lean-mix concrete
backfill to work space
excavation

Typical section

Figure 3.6.1   **Traditional concrete underpinning**

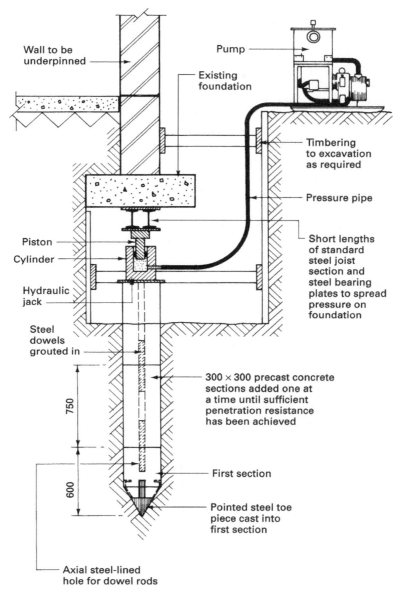

Wall to be
underpinned

Existing
foundation

Pump

Timbering
to excavation
as required

Pressure pipe

Piston

Cylinder

Hydraulic
jack

Short lengths
of standard
steel joist
section and
steel bearing
plates to spread
pressure on
foundation

Steel
dowels
grouted in

300 × 300 precast concrete
sections added one at
a time until sufficient
penetration resistance
has been achieved

750

600

First section

Pointed steel toe
piece cast into
first section

Axial steel-lined
hole for dowel rods

Figure 3.6.2   Jack or miga pile underpinning

NEEDLE AND PILES

If the wall to be underpinned has a weak foundation that is considered unsuitable for spanning over the heads of jack piles, there is an alternative method that gives the same degree of flexibility. This method uses pairs of jacks or (more usually) bored piles in conjunction with an in-situ reinforced concrete beam or needle placed above the existing foundation. The system works on the same principle as a dead shoring arrangement, relying on the arching effect of bonded brickwork. If water is encountered when using bored piles, a pressure pile can be used as an alternative. The formation of both types of pile is described in Chapter 3.4 on page 283. Typical arrangements to enable the work to be carried out, from both sides of the wall or from the external face only, are shown in Fig. 3.6.3. Where the old foundation failed through clay shrinkage and subsidence, it is usual to construct the new concrete needle beams on a void former to allow for swelling of the clay. Similarly the top 3 m of pile is usually sleeved in polythene to de-bond the clay and the pile, and prevent uplift.

PYNFORD STOOLING METHOD

The Pynford method of underpinning enables walls to be underpinned in continuous runs, without the use of needles or raking shoring. The procedure is to cut away portions of brickwork, above the existing foundation, to enable precast concrete or steel stools to be inserted and pinned. The intervening brickwork can be removed, leaving the structure supported entirely on the stools. Reinforcing bars are threaded between and around the stools and caged to form the ring beam reinforcement. After the formwork has been placed and the beam cast, final pinning can be carried out using a well-rammed dry mortar mix (see Fig. 3.6.4). This method replaces the existing foundation strip with a reinforced concrete ring beam, which can be cast in-situ, with a combined reinforced concrete ground slab over the whole footprint of the building. This complete 'raft underpin' is used where complex structural solutions are needed: for example, when retrofitting a large, open-plan basement below an existing building, or when masonry buildings are to be physically jacked and rolled to another location.

HOOPSAFE METHOD

The hoopsafe method is most appropriate as a remedial treatment where differential settlement can be identified, as it places a rigid corset around the foundation and stiffens it up, so that the building moves as one element. A limited amount of external excavation is needed to expose the substructural wall to a depth just above foundation level. Here, an in-situ concrete beam with purpose-made longitudinal voids created with plastic conduits is cast around the building's periphery. The small-diameter voids accommodate steel stressing tendons for post-tensioning, to bind the walls into a solid unit.

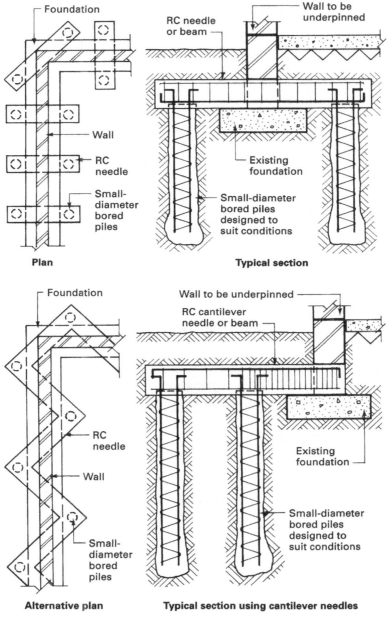

Plan

Typical section

- Foundation
- Wall to be underpinned
- RC needle or beam
- Wall
- RC needle
- Small-diameter bored piles
- Existing foundation
- Small-diameter bored piles designed to suit conditions

Alternative plan

Typical section using cantilever needles

- Foundation
- Wall to be underpinned
- RC cantilever needle or beam
- RC needle
- Wall
- Small-diameter bored piles
- Existing foundation
- Small-diameter bored piles designed to suit conditions

Figure 3.6.3 Needle and pile underpinning

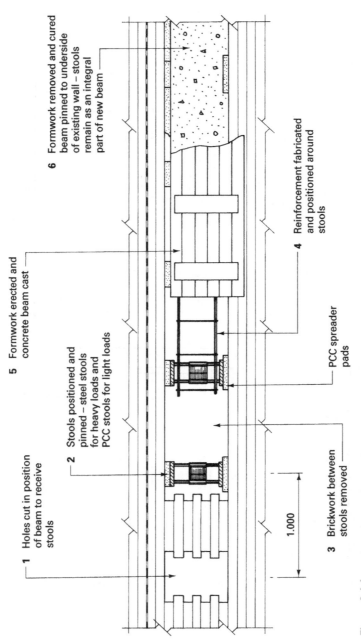

1 Holes cut in position of beam to receive stools

2 Stools positioned and pinned – steel stools for heavy loads and PCC stools for light loads

3 Brickwork between stools removed

4 Reinforcement fabricated and positioned around stools

5 Formwork erected and concrete beam cast

6 Formwork removed and cured beam pinned to underside of existing wall – stools remain as an integral part of new beam

PCC spreader pads

1.000

Figure 3.6.4 Pynford stooling method of underpinning

Continuity provided by the post-tensioned beams integrates the substructural wall and compensates for weaker areas of subsoil. However, where areas or pockets of weakness in the ground can be identified, these should be stabilised by granular consolidation to complement the remedial treatment. This method is relatively fast when applied to regular plan shapes, such as a square or rectangle. Offsets and extensions to buildings are harder to peripherate. Figure 3.6.5 shows the principles of application.

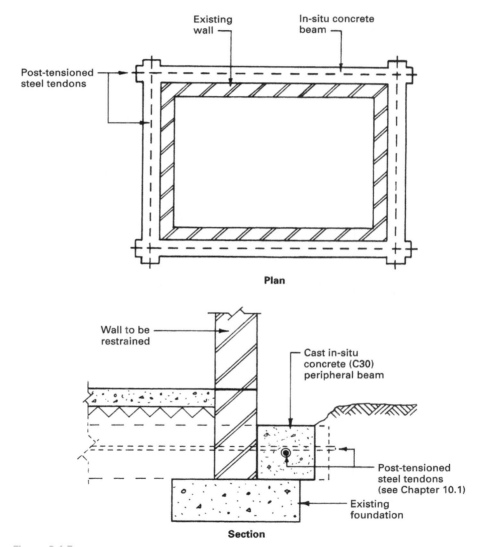

Plan

Section

Figure 3.6.5   **Hoopsafe foundation support**

CASED SCISSOR PILING

Cased scissor piling is another low-cost modern alternative to traditional underpinning for stabilising existing substructural walls and foundations. Figure 3.6.6 shows the installation of reinforced concrete angle piles by air-flushed rotary percussion drill from inside and outside the building. The voids are lined with steel casings, cut off at the surface before lowering reinforcement and placing concrete. Spacings and depth of borings will depend on site conditions, such as the occurrence of solid bearing strata and the extent of structural damage. The inclination of the piles can be varied so that both piles can be bored from the outside, minimising disruption to the building's interior. The relatively short timescale and minimal excavation make for an economic process.

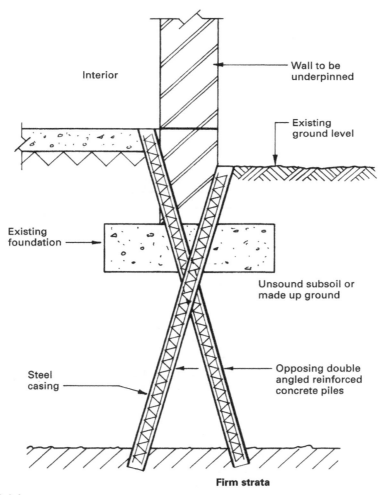

Figure 3.6.6   **Cased scissor pile underpinning**

FINAL DRY-PACKING

Dry-packing or final pinning is usually carried out by ramming a stiff, dry cement mortar mix into the space between the new underpinning work and the existing structure. However, alternative methods are available.

- **Flat jacks** These are circular or rectangular hollow plates of various sizes made from thin sheet metal, which can be inflated with high-pressure water for temporary pinning or, if work is to be permanent, with a strong cement grout. The increase in thickness of the flat jacks is approximately 25 mm.

- **Wedge bricks** These are special bricks of engineering quality, of standard face length but with a one-brick width and a depth equal to two courses. The brick is made in two parts, the lower section having a wide sloping channel in its top bed surface to receive the wedge-shaped and narrower top section. Both parts have a vertical slot through which the bedding grout passes to key the two sections together.

UNDERPINNING COLUMNS

Underpinning a column is a more difficult task than underpinning a wall. It can be carried out on brick or stone columns by inserting a series of stools, casting a reinforced concrete base, and then underpinning by the methods already described.

Structural steel or reinforced concrete columns must be relieved of their loading before any underpinning can take place. This can be achieved by variations of one basic method. A collar of steel or precast concrete members is fixed around the perimeter of the column. Steel collars are usually welded to the structural member, whereas concrete columns are usually chased to a depth of 25 to 50 mm to receive the support collar. The column loading is transferred from the collar to cross-beams or needles, which in turn transmit the loads to the ground at a safe distance from the proposed underpinning excavations. Cantilever techniques that transfer the loadings to one side of the structural member are possible, provided sufficient kentledge and anchorage can be obtained (see Fig. 3.6.7). The underpinning of the column foundation can now be carried out by the means previously described.

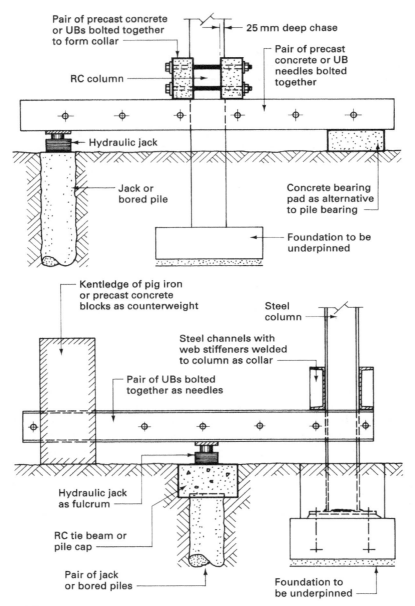

Pair of precast concrete
or UBs bolted together
to form collar

25 mm deep chase

RC column

Pair of precast
concrete or UB
needles bolted
together

Hydraulic jack

Jack or
bored pile

Concrete bearing
pad as alternative
to pile bearing

Foundation to be
underpinned

Kentledge of pig iron
or precast concrete
blocks as counterweight

Steel
column

Steel channels with
web stiffeners welded
to column as collar

Pair of UBs bolted
together as needles

Hydraulic jack
as fulcrum

RC tie beam or
pile cap

Pair of jack
or bored piles

Foundation to
be underpinned

Figure 3.6.7   Typical column underpinning arrangements

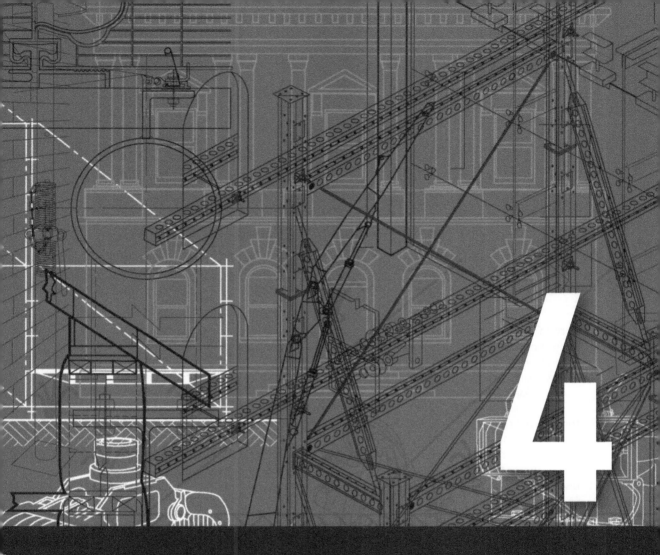

Portal frames

4

Portal frame theory 4.1

Portal frames are used in the construction of medium to large industrial, leisure or retail buildings. These structures offer large, open floor areas with high internal floor-to-roof clearances which can be used, for example, for storage of materials, undertaking indoor sports or manufacturing processes, or can provide high-volume retail spaces. In the main portal frames tend to be single-storey, although intermediate mezzanine floors may often be inserted for internal office space. The basic terminology and elements used in portal frame buildings have been covered in Chapter 10.6 of *Construction Technology*, with an emphasis on steel portal frames, as these are the most common type. This part outlines the detailed structural operation of portal frames and gives examples of types of portal frame that are currently used or can be found in existing structures.

A portal frame may be defined as a continuous or rigid frame that has the basic characteristic of a rigid or restrained joint between the supporting member or column and the spanning member or beam. The continuity of the portal frame reduces the bending moment and resulting bending stresses in the spanning member by allowing the frame to act as one structural entity, thus distributing the stresses throughout the frame.

If a conventional, simply supported beam was used over a large span, an excessive bending moment would occur at mid-span, which would necessitate a deep, heavy beam, a beam shaped to give a large cross-section at mid-span, or a deep cross-member of lattice struts and ties. Fortunately, using a portal frame eliminates the need for these heavier and more costly systems and, particularly in comparison with a lattice truss of struts and ties within the roof space, gives a greater usable volume to the structure as well as a more pleasing, uncluttered internal appearance.

The transfer of stresses from the beam to the column in rigid frames means special care is needed when designing the joint between the members; similarly the horizontal thrust and/or rotational movement at the foundation connection needs careful consideration. Methods used to overcome excessive forces at the foundation are:

- reliance on the passive pressure of the soil surrounding the foundation;

- inclined or eccentric foundations so that the curve of pressure is normal to the upper surface, thus tending to induce only compressive forces;

- a tie bar or beam between opposite foundations;

- introducing a hinge or pin joint where the column connects to the foundation.

HINGES

Portal frames of moderate height and span are usually connected directly to their foundation bases, forming rigid or unrestrained joints. The rotational movement caused by wind pressures tending to move the frames and horizontal thrusts of the frame loadings are generally resisted by the size of the base and the passive earth pressures. When the frames start to exceed 4 m in height and 15 m in span, you should consider introducing a hinged or pin joint at the base connection.

A hinge is a device that will allow free rotation at the fixing point but, at the same time, will transmit both load and shear from one member to another. Hinges are sometimes called pin joints, unrestrained joints or non-rigid joints. Because no bending moment is transmitted through a hinged joint, the design is simplified by the structural connection becoming statically determinate. In practice, it is not always necessary to provide a true 'pivot' where a hinge is included; all that is needed is to provide just enough movement to ensure that the rigidity at the connection is low enough to allow rotation.

Hinges can be introduced into a portal frame design at the base connections and at the centre or apex of the spanning member, giving three basic forms of portal frame.

- **Fixed or rigid portal frame** Here all connections between frame members are rigid. This will give bending moments of lower magnitude, more evenly distributed than with other forms. This form is used for small- to medium-size frames where the moments transferred to the foundations will not be excessive.

- **Two-pin portal frame** In this form of frame, hinges are used at the base connections to eliminate the tendency of the base to rotate. The bending moments resisted by the supporting members will be greater than those encountered in the rigid portal frame. The main use is where high base moments and weak ground conditions are encountered.

- **Three-pin portal frame** This form of frame has hinged joints at the base connections and at the centre of the spanning member. The effect of the third hinge is to reduce the bending moments in the spanning member but to increase deflection. To overcome this latter disadvantage, a deeper beam must be used, or the spanning member must be given a moderate pitch to raise the apex well above eaves level. Two other advantages of the three-pin portal frame are that the design is simplified because the frame is statically determinate and, on site, they are easier to erect, particularly when preformed in sections.

Figure 4.1.1 shows a comparison of the bending moment diagrams for roof loads of the three forms of portal frame with a simply supported beam.

Most portal frames are made under factory-controlled conditions off site, which gives good dimensional and quality control but can create transportation problems. Splices are often used to lessen this problem and to aid site erection. These can be positioned at the points of contraflexure (see Fig. 4.1.1) at the junction between spanning and supporting members and at the crown or apex of the beam. Most hinges or pin joints provide a point at which a splice can be introduced as a pin-jointed connection. See the examples in Chapters 4.2–4.4.

Portal frames constructed of steel, concrete or timber can take the form of the usual roof profiles used for single- or multi-span buildings, such as flat, dual-pitched, mono-pitched, northlight, monitor and arch. The frames are generally connected over the spanning members with purlins designed to carry and accept the fixing of lightweight insulated roof coverings or deckings. The walls can be of similar material fixed to sheeting rails attached to the supporting members or clad with brick or infill panels. Further details of the roofing and cladding systems are provided in Chapter 6.

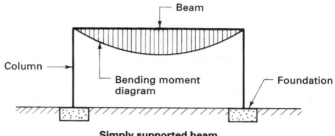

Simply supported beam

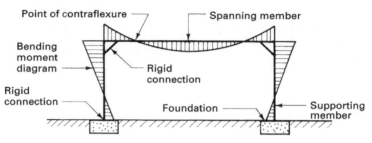

Rigid or fixed portal frame

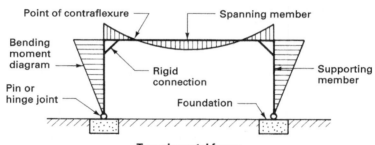

Two-pin portal frame

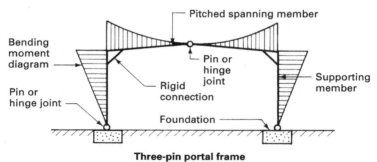

Three-pin portal frame

Figure 4.1.1   Portal frames: comparison of bending moments

Concrete portal frames

4.2

Concrete portal frames are invariably manufactured from high-quality precast concrete that has been suitably reinforced. In the main, the use of precast concrete portal frames is confined to low-pitch (4° to 22.5°) single-span frames, but two-storey and multi-span frames are available, giving a wide range of designs from only a few standard components. Precast concrete portal frames are not as popular as the lighter-weight steel and timber; their main use today is for the construction of durable agricultural and industrial storage buildings.

These frames are generally designed to carry a lightweight roof sheeting or decking (35 kg/m^2 maximum) fixed to precast concrete purlins. Most designs have an allowance for snow loading of up to 150 kg/m^2, an addition to that allowed for the dead load of the roof covering. Wall finishes can be varied and intermixed because they are non-loadbearing; this means that they only have to provide the degree of resistance required for fire, thermal and sound insulation, act as a barrier to the elements, and resist positive and negative wind pressures. Sheet claddings are fixed in the traditional manner, using hook bolts and purlins; sheet wall claddings are fixed in a similar manner to sheeting rails of precast concrete or steel spanning between or over the supporting members. Brick or block wall panels, of solid or cavity construction, can be built off a ground beam constructed between the foundation bases; alternatively they can be built off the reinforced concrete ground floor slab. Remember that all such claddings must comply with any relevant Building Regulations, which will vary depending on whether the end use is for industrial, commercial, storage or agricultural purposes.

FOUNDATIONS AND FIXINGS

The foundations for a precast concrete portal frame usually consist of a reinforced concrete isolated base or pad designed to suit loading and ground-bearing conditions. The frame can be connected to the foundations by a variety of methods.

- **Pocket connection** The foot of the supporting member is located and housed in a void or pocket formed in the base so that there is an all-round clearance of 25 mm to allow for plumbing and final adjustment before the column is grouted into the foundation base (see Fig. 4.2.1).

- **Baseplate connection** A steel baseplate is welded to the main reinforcement of the supporting member, or could be cast into the column using fixing lugs welded to the back of the baseplate. Holding-down bolts are cast into the foundation base; the erection and fixing procedure follows that described for structural steelwork (see Chapter 10.4 of *Construction Technology*).

- **Pin joint or hinge connection** A special base or bearing plate is bolted to the foundation, and the mechanical connection is made when the frames are erected (see Fig. 4.2.4).

The choice of connection method depends largely on the degree of fixity required and the method adopted by the manufacturer of the particular system.

ADVANTAGES OF PRECAST CONCRETE PORTAL FRAMES

- Factory production will result in accurate and predictable components because the criteria for design, quality and workmanship recommended in Eurocode 2: BS EN 1992 *Design of Concrete Structures* can be more accurately controlled under factory conditions than when casting components in situ.

- Most manufacturers produce a standard range of interchangeable components that, within the limitations of their systems, gives a well-balanced and flexible design range covering most roof profiles, single-span frames, multi-span frames and lean-to roof attachments. By adopting this limited range of members, the producers of precast portal frames can offer their products at competitive rates coupled with reasonable delivery periods.

- Maintenance of precast concrete frames is not usually required unless the building owner chooses to paint or clad the frames.

- Precast dense concrete products have an inherent resistance to fire, so it is not usually necessary to provide further fire-resistant treatment. However, the amount of reinforcement concrete cover will vary depending on the fire resistance required – see Building Regulations Approved Document B to ascertain the purpose grouping. For industrial premises the fire resistance is unlikely to be less than two hours, which will justify a minimum of 35 and 50 mm concrete cover for column and beam components respectively.

- The wind resistance of precast concrete portal frames to both positive and negative pressures is such that wind bracing is not usually required.

- Where members of the frame are joined or spliced together, the connections are generally mechanical (nut and bolt), so the erection and jointing can be carried out by semi-skilled operatives.

- In most cases, the foundation design, setting out and construction can be carried out by the portal frame supplier or their nominated subcontractor.

Typical details of single-span frames, multi-span frames, cladding supports, splicing and hinges are shown in Figs 4.2.1 to 4.2.4.

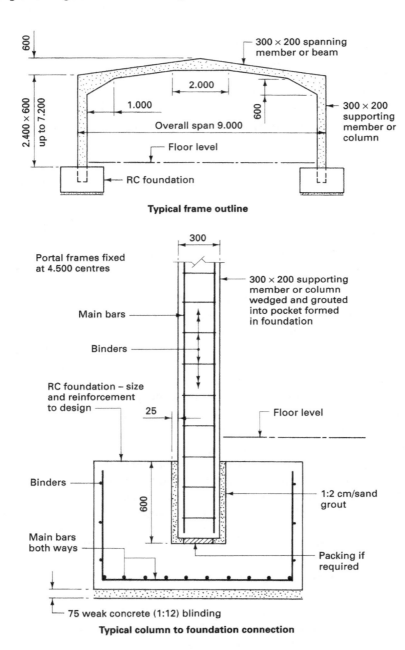

Typical frame outline

Typical column to foundation connection

Figure 4.2.1   Typical single-span precast concrete frame

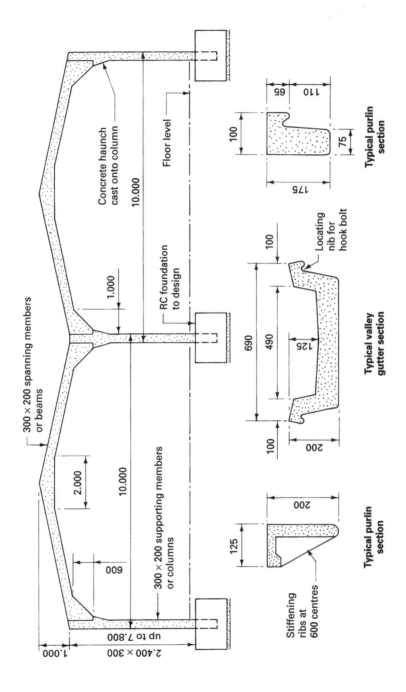

300 × 200 spanning members
or beams

Concrete haunch
cast onto column

Floor level

10.000

1.000

RC foundation
to design

2.000

10.000

300 × 200 supporting members
or columns

600

up to 7.800

2.400 × 300

1.000

**Typical purlin
section**

100

65

110

175

75

**Typical valley
gutter section**

100

690

490

125

125

100

200

Locating
nib for
hook bolt

**Typical purlin
section**

125

200

Stiffening
ribs at
600 centres

Figure 4.2.2 Typical multi-span precast concrete portal frame

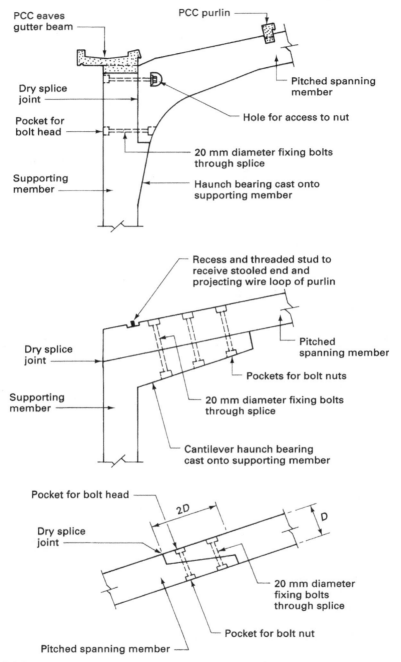

PCC eaves
gutter beam

PCC purlin

Dry splice
joint

Pocket for
bolt head

Supporting
member

Pitched spanning
member

Hole for access to nut

20 mm diameter fixing bolts
through splice

Haunch bearing cast onto
supporting member

Recess and threaded stud to
receive stooled end and
projecting wire loop of purlin

Dry splice
joint

Supporting
member

Pitched
spanning member

Pockets for bolt nuts

20 mm diameter fixing bolts
through splice

Cantilever haunch bearing
cast onto supporting member

Pocket for bolt head

2D

D

Dry splice
joint

20 mm diameter
fixing bolts
through splice

Pocket for bolt nut

Pitched spanning member

Figure 4.2.3   Typical splice details for precast concrete portal frames

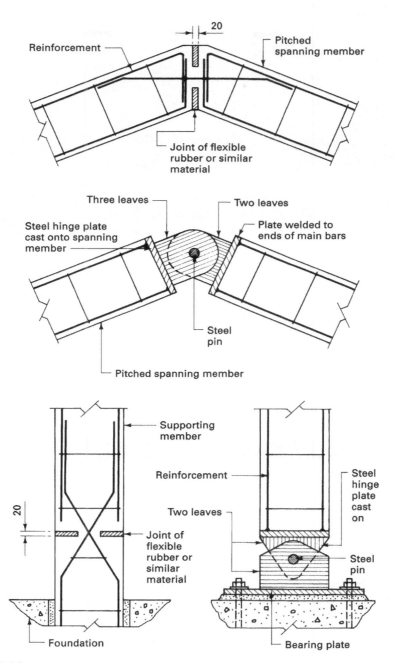

20

Reinforcement

Pitched spanning member

Joint of flexible rubber or similar material

Three leaves

Two leaves

Steel hinge plate cast onto spanning member

Plate welded to ends of main bars

Steel pin

Pitched spanning member

Supporting member

Reinforcement

Steel hinge plate cast on

20

Two leaves

Steel pin

Joint of flexible rubber or similar material

Bearing plate

Foundation

Figure 4.2.4   Typical hinge details for precast concrete portal frames

Steel portal frames 4.3

The basic steel portal terminology and components are described in Chapter 10.6 of *Construction Technology*. Most modern steel portal frames are fabricated from standard universal beam, column and box sections. In temporary and lightweight structures an alternative lattice portal is also produced, constructed of flats, angles or tubular steel. Most forms of roof profile can be designed and constructed, giving a competitive range when compared with other materials used in portal frame construction. Most systems employ welding techniques for the fabrication of components in the fabrication shop, which are joined together on site, usually using bolted connections.

STRUCTURAL FORM

Currently the design of single-span steel structures is covered by BS 5950 – 1. However, structural engineers now more often use the new Eurocode 3: BS EN 1993 and the associated national annexes. The frames are designed to carry lightweight roof and wall claddings of the same loading conditions as those for precast concrete portal frames (see Chapter 4.2). These cladding systems provide a structural diaphragm to the frame to resist lateral loadings, such as wind. To ensure that this occurs, it is common to provide diagonal wire ties to brace both the roof purlins and the wall cladding rails. Where additional wind bracing is required, this is usually in the form of hot-rolled steel tubes or flat plates, arranged as diagonal braces aligned with both the external portal columns and the portal rafters. You can find details of the different cladding systems, function and construction in Chapter 6 of this book.

Single-span portal frames in steel are available up to 40 m. Where a larger area needs to be housed, portal frames can be doubled or linked up in multi-span portal frame systems. In multi-span portals it is common practice to use valley beams to eliminate some internal columns, to ensure flexibility of the available space. Generally alternate columns are omitted and the valley of the frame is supported on a valley-beam spanning between the columns of adjacent frames (see Fig. 4.3.2). This arrangement is often referred to as 'hit and miss' frames, the frames with columns being the 'hit' frames.

FOUNDATIONS AND FIXINGS

The foundation is usually a reinforced concrete isolated base or pad foundation designed to suit loading and ground-bearing conditions. The connection of the frame to the foundation can be by one of three basic methods.

- **Pocket connection** The foot of the supporting member is inserted and grouted into a pocket formed in the concrete foundation, as for precast concrete portal frames. To facilitate levelling, some designs have gussets welded to the flanges of the columns, as shown in Fig. 4.3.2. This is generally considered to provide a fixed base connection: it does not allow any rotation at the connection and so transfers the bending moment into the base foundations. This enables the use of relatively slender steelwork for the portal, but requires substantial pad or inclined pile foundations to resist the bending.

- **Baseplate connection** This is a traditional structural steelwork column-to-foundation connection, using a steel baseplate fixed to a reinforced concrete foundation with cast-in holding-down bolts (see Fig. 4.3.2). Depending on the number of holding-down bolts, these connections are usually considered to be partially pinned, so they allow some rotation at the base connection.

- **Pin or hinge connection** Special bearing plates designed to accommodate true pin or rocker devices are fixed by holding-down bolts to the concrete foundation to give the required low degree of rigidity at the connection, as shown in Fig. 4.3.3.

ADVANTAGES OF STEEL FRAMED PORTALS

- Factory-controlled production gives a standard range of manufacturer's systems.

- Steel framed portals offer good wind resistance.

- As with precast concrete portal frames, site assembly is easy, and can use semi-skilled operatives.

- Generally, the overall dead load of a steel portal frame is less than that of a comparable precast concrete portal frame.

However, steel has the disadvantage of being a corrosive material, which will require a long-life protection of a patent coating or regular protective maintenance, usually through coats of paint.

Steel has a lower fire resistance than precast concrete, but if the frame is for a single-storey building, structural fire protection may not be required under the Building Regulations (see Approved Document B); it depends very much on the purpose group or function of the building and the amount of space it occupies. However, the building's insurers usually require more effective fire protection than that determined by current building legislation, and may apply the Loss Prevention Certification Board's *Loss Prevention Standards (Fire and Security)*. Where fire-resistance of steel is required this can be achieved by using intumescent paints, sprayed concrete grout or plasterboard claddings.

Typical details of steel portal frames, cladding fixings, splicing and hinges are shown in Figs 4.3.1 to 4.3.3.

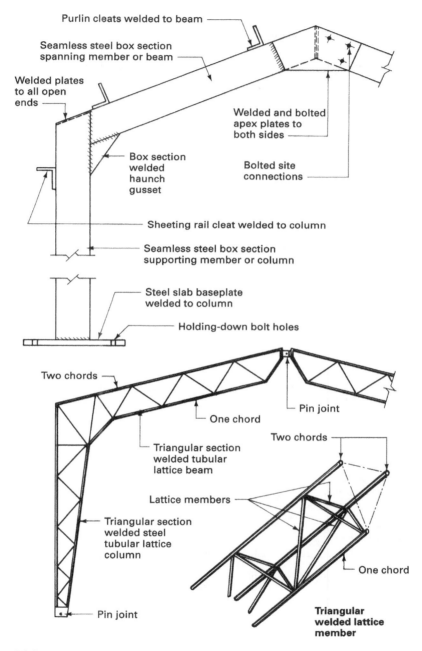

Purlin cleats welded to beam

Seamless steel box section spanning member or beam

Welded plates to all open ends

Welded and bolted apex plates to both sides

Box section welded haunch gusset

Bolted site connections

Sheeting rail cleat welded to column

Seamless steel box section supporting member or column

Steel slab baseplate welded to column

Holding-down bolt holes

Two chords

One chord

Pin joint

Triangular section welded tubular lattice beam

Two chords

Lattice members

Triangular section welded steel tubular lattice column

One chord

Pin joint

Triangular welded lattice member

Figure 4.3.1   Typical steel portal frames

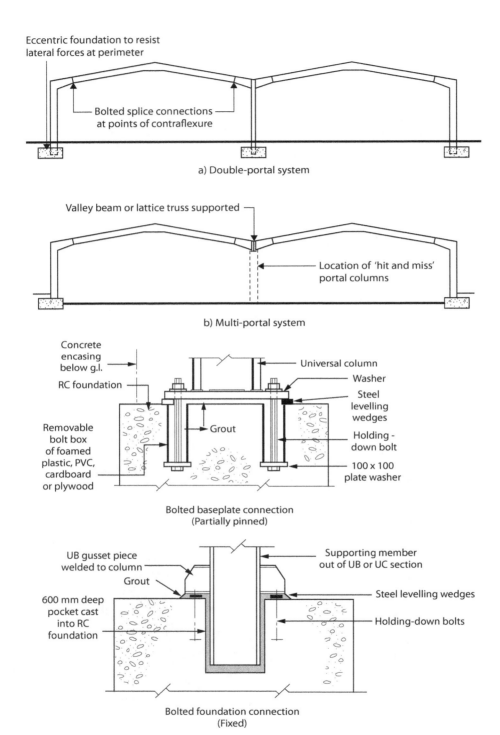

Eccentric foundation to resist
lateral forces at perimeter

Bolted splice connections
at points of contraflexure

a) Double-portal system

Valley beam or lattice truss supported

Location of 'hit and miss'
portal columns

b) Multi-portal system

Concrete
encasing
below g.l.

RC foundation

Universal column

Washer

Steel
levelling
wedges

Grout

Holding -
down bolt

Removable
bolt box
of foamed
plastic, PVC,
cardboard
or plywood

100 x 100
plate washer

Bolted baseplate connection
(Partially pinned)

UB gusset piece
welded to column

Supporting member
out of UB or UC section

Grout

Steel levelling wedges

600 mm deep
pocket cast
into RC
foundation

Holding-down bolts

Bolted foundation connection
(Fixed)

Figure 4.3.2   Steel portal frame details: 1

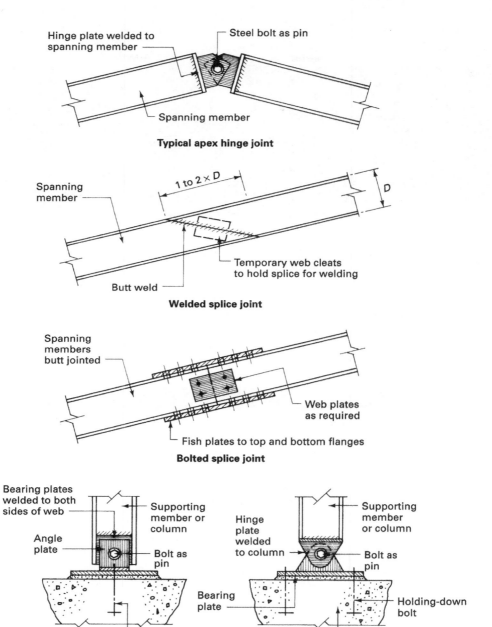

Typical apex hinge joint

Hinge plate welded to spanning member

Steel bolt as pin

Spanning member

Welded splice joint

Spanning member

1 to 2 × D

D

Temporary web cleats to hold splice for welding

Butt weld

Bolted splice joint

Spanning members butt jointed

Web plates as required

Fish plates to top and bottom flanges

Typical base hinges

Bearing plates welded to both sides of web

Supporting member or column

Angle plate

Bolt as pin

Holding-down bolt

Hinge plate welded to column

Supporting member or column

Bolt as pin

Bearing plate

Holding-down bolt

RC foundation

Figure 4.3.3   Steel portal frame details: 2

Timber portal frames

4.4

Timber portal frames can be manufactured using a range of methods, producing a light, strong frame of pleasing appearance suitable for buildings such as churches, halls and gymnasiums, where clear space and architectural appearance are important. The most common type used today is glued laminated portal frames, where the method of fabrication is known as 'glulam' construction. The other timber systems are traditional timber latticework with plywood stressed skins, solid timber and modern laminated veneer timber (lumber).

SOLID TIMBER PORTAL FRAMES

Solid timber portal frames were traditionally developed to provide a simple and economic timber portal frame for clear-span buildings of up to 9 m, using ordinary tools and basic skills. They are seldom specified today in the UK mainly due to the advent of various types of engineered joist (see Chapter 5.3 of *Construction Technology*). Nevertheless, the general concept of this form of frame has its applications in developing countries.

These frames are spaced close together (at 600, 900 and 1,200 mm centres), and clad with an unglued plywood sheath; the finished structure acts as a shell, giving a lightweight building that is very rigid and strong. The frames are often supplied in two halves and assembled by fixing the plywood apex gussets on site before erection; alternatively, they can be supplied as a complete frame, ready for site erection.

The foundation for this form of timber portal frame either consists of a ground beam or the frames can be fixed to the edge of a raft slab. A timber spreader or sole plate is used along the entire length of the building at dpc level to receive and distribute the thrust loads of the frames. Connection to this spreader plate is made using standard galvanised steel joist hangers or galvanised steel angle cleats. Standard timber windows and doors can be inserted into the side walls by trimming in the conventional way and infilling where necessary with studs, noggins and rafters.

Typical details are shown in Fig. 4.4.1.

STRESSED-SKIN PORTAL FRAMES

Stressed-skin portal frames are also suitable for short spans in the region of 9 m and again rely on labour-intensive craft skills rather than low-skill pre-manufactured components. The portal frames are in essence boxed beams, consisting of a skeleton core of softwood members faced on both sides with plywood, which takes the bending stresses. The hollow form of the construction enables electrical and other small services to be accommodated within the frame members. Design concepts, fixing and finishes are as for glued laminated portal frames. Typical details are shown in Fig. 4.4.2. Again these sorts of frame have their limitations due to the availability of stronger and relatively more cost-effective engineered timber frame solutions, such as those that are described next.

GLUED LAMINATED (GLULAM) PORTAL FRAMES

The use of timber in an engineered 'glulam' form is becoming far more common in the UK and is just one example of the development and use of structural timber composites (STC). The main objective of forming a laminated member consisting of glued layers of thin-section timber members is to obtain an overall increase in strength of the complete component compared with that which could be expected from a similar-sized solid section of a particular species of timber.

This type of portal frame is usually manufactured by a specialist firm, because the jigs required would be too costly for small outputs. The selection of suitable-quality softwoods of the right moisture content is also important for a successful design. Glulam elements of rectangular cross-section are normally made from 45 mm or 33 mm laminations, in widths corresponding to sawmills' standard ranges. After completion of most of the manufacturing, the side faces of the glulam are planed. In common with other timber portal frames, these can be fully rigid, two- or three-pin structures. If pinned, typical spans possible are between 12 and 25 m; if fully fixed, spans of up to 40 m are possible.

Site work is simple, consisting of connecting the foot of the supporting member to the metal shoe fixing or to a pivot housing bolted to the concrete foundation, and connecting the joint at the apex or crown with a bolt fixing or a hinge device. Most glued laminated timber portal frames are fabricated in two halves, which eases transportation problems and gives maximum usage

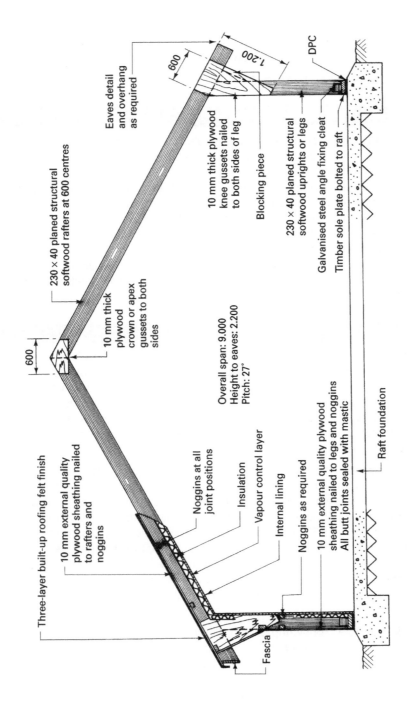

Three-layer built-up roofing felt finish

10 mm external quality plywood sheathing nailed to rafters and noggins

Eaves detail and overhang as required

600

1:200

DPC

600

230 × 40 planed structural softwood rafters at 600 centres

10 mm thick plywood knee gussets nailed to both sides of leg

Blocking piece

230 × 40 planed structural softwood uprights or legs

Galvanised steel angle fixing cleat

Timber sole plate bolted to raft

10 mm thick plywood crown or apex gussets to both sides

Overall span: 9.000
Height to eaves: 2.200
Pitch: 27°

Noggins at all joint positions

Insulation

Vapour control layer

Internal lining

Noggins as required

10 mm external quality plywood sheathing nailed to legs and noggins
All butt joints sealed with mastic

Raft foundation

Fascia

Figure 4.4.1 Typical solid timber and plywood gusset portal frame

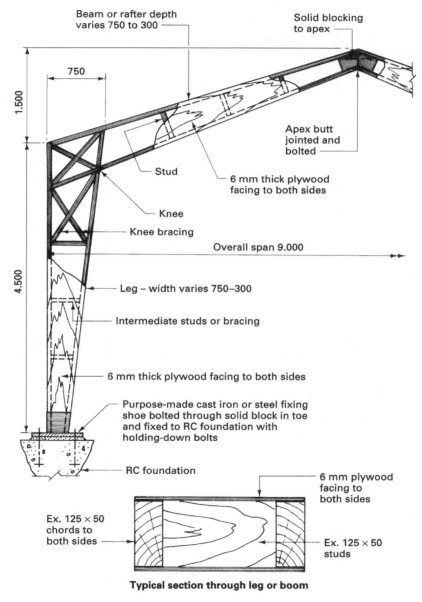

Beam or rafter depth
varies 750 to 300

Solid blocking
to apex

750

1.500

Apex butt
jointed and
bolted

Stud

6 mm thick plywood
facing to both sides

Knee

Knee bracing

Overall span 9.000

4.500

Leg – width varies 750–300

Intermediate studs or bracing

6 mm thick plywood facing to both sides

Purpose-made cast iron or steel fixing
shoe bolted through solid block in toe
and fixed to RC foundation with
holding-down bolts

RC foundation

6 mm plywood
facing to
both sides

Ex. 125 × 50
chords to
both sides

Ex. 125 × 50
studs

Typical section through leg or boom

Figure 4.4.2   Stressed-skin portal frame

of the assembly jigs. The frames can be linked together at roof level with timber purlins and clad with a lightweight sheeting or decking; alternatively, they may be finished with traditional roof coverings. Any form of walling can be used in conjunction with these frames, provided such walling forms comply with any of the applicable Building Regulations.

Typical details are shown in Fig. 4.4.3.

LAMINATED VENEER LUMBER PORTAL FRAMES

Laminated veneer lumber (LVL) is a modern engineered structural timber composite product, manufactured from rotary-peeled veneers glued vertically together to form continuous elements. LVL is recognised in Eurocode 5: BS EN 1995 *Design of timber structures*. These components look similar to traditional solid timber joists, but they can be fabricated up to depths of 900 mm and breadths of 215 mm. They have excellent bending resistance, tension and compression properties, and greater load-bearing capacity than traditional solid timber. They have a reduced tendency to twist, warp or split, and so are ideal for fabricating portal frames. If the portals are pinned, typical spans are similar to glulam beams (12 to 25 m); if fully fixed, 40 m spans are possible.

LVL portal rafters and columns are brought to site in separate pieces then fixed together with pre-marked plywood gusset plates, which identify the position of all the nails required. The knee and ridge gusset may also be strengthened with additional timber stiffeners where required, to improve the moment transfer at the joint.

Typical details are shown in Fig. 4.4.4.

ADVANTAGES OF TIMBER PORTAL FRAME TYPES

- Timber is a sustainable, natural material.
- The frames are relatively inexpensive when compared with other materials.
- They are light in weight and easy to transport and erect.
- Solid timber portals can be trimmed and easily adjusted on site.
- With engineered solutions, metal connections are often largely concealed within the element and designated connection nodes are pre-planned for site assembly.
- Protection against fungal and/or insect attack can be by impregnation or surface application.
- Their condition is predictable in terms of long-term performance, stiffness and serviceability.
- They have a pleasing architectural appearance, either as a natural timber finish or painted.

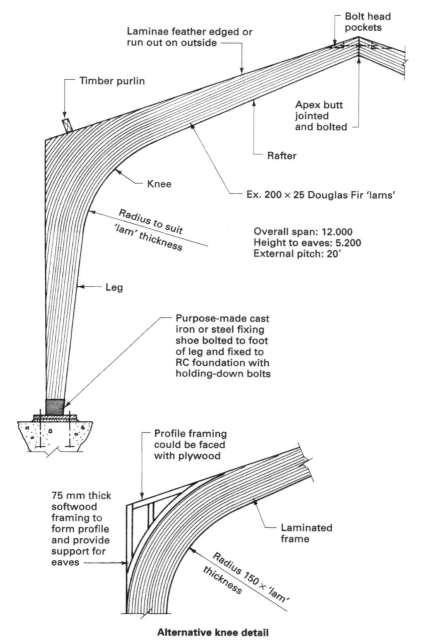

Laminae feather edged or
run out on outside

Bolt head
pockets

Timber purlin

Apex butt
jointed
and bolted

Rafter

Knee

Ex. 200 × 25 Douglas Fir 'lams'

Radius to suit
'lam' thickness

Overall span: 12.000
Height to eaves: 5.200
External pitch: 20°

Leg

Purpose-made cast
iron or steel fixing
shoe bolted to foot
of leg and fixed to
RC foundation with
holding-down bolts

Profile framing
could be faced
with plywood

75 mm thick
softwood
framing to
form profile
and provide
support for
eaves

Laminated
frame

Radius 150 × 'lam'
thickness

Alternative knee detail

Figure 4.4.3   Typical glued laminated portal frame

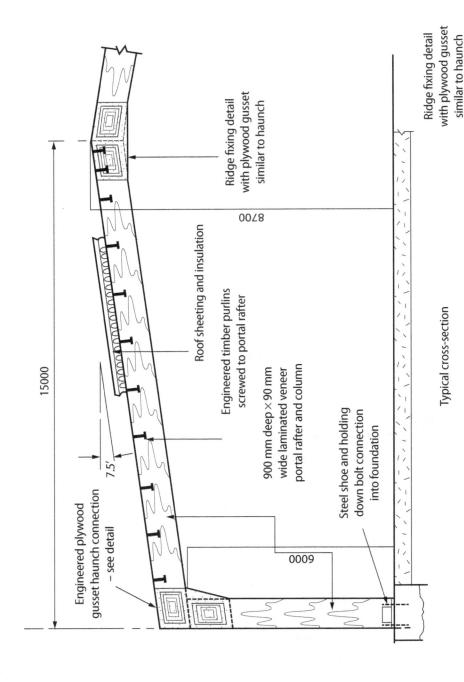

Ridge fixing detail with plywood gusset similar to haunch

Ridge fixing detail with plywood gusset similar to haunch

8700

Roof sheeting and insulation

Engineered timber purlins screwed to portal rafter

900 mm deep × 90 mm wide laminated veneer portal rafter and column

Steel shoe and holding down bolt connection into foundation

Typical cross-section

15000

7.5'

Engineered plywood gusset haunch connection – see detail

6000

Figure 4.4.4   Typical laminated veneer lumber portal frame

5

Fire protection

The problem of fire 5.1

Fire has always been an essential element of our technological advancement, providing heat, energy and light. In modern construction, it is used with fuels to heat water and provide space heating of radiators. The use of fires within living rooms of homes is controlled through the use of a gas fire or solid-fuel open or stove fireplaces.

A heat-producing appliance can cause a hazard when it is covered by a combustible object. If a combustible object falls or is placed near to a heat-producing appliance, over time it can start to combust, giving off smoke and flames that can quickly engulf a home. Occupants' smoking is often a dangerous way to start a fire, and old non-fire-retardant furniture can create a lethal cocktail of combustion gases. It is the smoke that is the main cause of occupant deaths, due to inhalation of the gases that poison the human respiratory system.

Fashionable wood-burning stoves now provide a controlled environment in which to burn solid fuel without the risk of a spark flying out of an open fire and setting light to a room. They allow the slow combustion of coal or wood within an environment where airflow can be regulated. An ash pan collects the results of combustion, which can be disposed of safely when the appliance has cooled.

Early scientists discovered that air consists of one-fifth oxygen and that the other main gas, nitrogen, accounted for the bulk of the remaining four-fifths. They showed that oxygen played an important part in the process of combustion, and that nitrogen does not support combustion. This discovery of the true nature of fire led to the conclusion that fire is a chemical reaction whereby atoms of oxygen combine with other atoms, such as carbon and hydrogen, releasing water, carbon dioxide and energy in the form of heat. The chemical reaction will only start at a suitable temperature, which varies according to the substance or fuel involved. During combustion, gases will be given off, some of which are more inflammable than the fuel itself and therefore ignite and appear as flames, giving light due to tiny particles being heated to a point at which they glow. Smoke is an indication of incomplete combustion and can give rise to deposits of solid carbon, commonly known as soot.

From the discovery of the true nature of fire and processes of combustion it can be concluded that there are three essentials to all fires:

- **fuel** – generally any organic material is suitable;
- **heat** at the correct temperature to promote combustion of a particular fuel. Heat can be generated deliberately, which is termed **ignition**, or can be **spontaneous**, when the fuel itself ignites;
- **oxygen** – air is necessary to sustain and support the combustion process.

The need for these three essentials is often called the **triangle of fire**; remove any one of the three essentials and combustion cannot take place. This fact provides the whole basis for fire prevention, fire protection and firefighting.

If non-combustible materials were used in the construction and furnishing of buildings, fires would not develop as no fuel would ignite. However, this would be far too restrictive on the designer and contractor. Instead, combustible materials are used and protected with layers or coverings of non-combustible materials: for example, plasterboard linings to wooden frames and combustible insulants.

Guidance on minimum periods of fire resistance to structural elements is provided within Tables A1 and A2 in Approved Document B of the Building Regulations. There is also provision for materials of limited combustibility, such as applications to stair construction and roof decking, provided the extent of exposure is restricted. See Table A7 in Approved Document B to the Building Regulations and associated references to BS 476-11: *Method for assessing the heat emission from building products*. Firefighters try to remove one side of the fire triangle; to remove the fuel is not generally practicable, but by using a cooling agent, such as water, the heat can be reduced to a safe level; alternatively by using a blanketing agent, such as foam, the supply of oxygen can be cut off and the fire extinguished and cooled.

Annual statistics for occurrence of fires and types of building affected are produced by the Office for National Statistics (ONS) and the Department for Communities and Local Government (DCLG). Further data are also produced by the fire service and insurance companies. The overall number of annual fatalities from fire (not just in buildings) in the UK averages 350 persons in recent years, and has been dropping steadily thanks to campaigns encouraging the fitting of smoke detectors in homes. Fatalities from fires in domestic dwellings exceed 300 a year on average, hence the importance of employment of fire protection measures within domestic homes.

Residential buildings are the most vulnerable, accounting for over 50,000 incidents annually or about 50 per cent of all fires. In excess of 3,000 fires in dwellings are attributed to children playing with fire, in the form of matches and other easily ignitable home gadgets. Another high-risk area is commercial distribution, where large quantities of inflammable goods are held in store. One of the main causes of fires in non-domestic buildings is faulty electrical equipment and wiring. This is not an indication that the plant or installation is poor, but that regular maintenance, renewal and routine checks must be carried out and be documented by qualified and experienced personnel. In residential buildings, the same provision should be made for electrical and gas services.

Given the impressive fire suppression record for sprinklers in commercial and industrial premises, building insurers and the fire prevention industry and professions have been considering and promoting the installation of sprinkler systems in residential accommodation. The Government has responded with inclusion of these in Part B (Fire safety) of the Building Regulations. Initially, this guidance is limited to dwelling houses, residential care homes and in flats over 30 m in height. However, there have also been several Government initiatives in supplying and fitting smoke detector alarms into vulnerable people's homes. The Building Regulations now make provision for the inclusion of linked mains-wired smoke and heat detectors within domestic dwellings. This has seen a remarkable and steady decline in the number of fatalities due to fire over the past ten years.

The annual cost of fires in buildings is difficult to assess exactly, but probably runs into figures expressed in billions of pounds; the personal loss is impossible to value. The cost to manufacturers and employers in terms of loss of goodwill, loss of production, effect of fatalities and injuries to employees and the delay in returning to full production or working are almost incalculable. The seriousness of fires within buildings cannot be overstated.

Obviously designers and builders alike cannot be held responsible for the actions or non-actions of the occupants of the buildings they create, but they can ensure that these structures are designed and constructed in such a manner that they give the best possible resistance to the action of fire should it occur, allowing sufficient time for the structure to resist fire while the occupants escape from the building.

Several fire precautions can be taken within buildings to prevent a fire occurring – or, if it should occur, to contain it within the region of the outbreak, provide a means of escape for people in the immediate vicinity and fight the fire. In the following sections we will examine methods of:

- structural fire protection;
- means of escape from a building.

Structural fire protection is generally known as a **passive** measure of fire protection, and is incorporated within the design specification of each element of construction. Firefighting and fire control mechanisms such as sprinklers, fire alarms, fusible link-operated shutters and doors are very much **active**.

Structural fire protection 5.2

The purpose of structural fire protection is to ensure that during a fire the temperature of structural members that are supporting the dead and live loads of a structure does not increase to a point at which their strength would be adversely affected.

There are two main methods available to prevent the spread of a fire throughout a structure:

- increasing the resistance of a covering material to prevent combustion or reduction in strength of the structural member;
- forming compartments that have clearly defined zones of fire resistance measured in hours.

It is not economical to give all of the elements of a building complete fire resistance as it would make such a feature not economically feasible for the project. Therefore compartmented elements are given a fire resistance for a certain period of time that it is anticipated will give:

- sufficient delay to the spread of fire enabling escape;
- delay on the ultimate collapse of the structure; and
- time for firefighting to be commenced safely.

These periods vary depending on the function or purpose group and occupancy of the building, and the size/height of the top floor above ground and the depth of basement. See the relevant tables in Appendix A to Approved Document B of the Building Regulations for specific requirements.

Before a fire-resistance period can be determined it is necessary to consider certain factors:

- fire-load intensity of the building (amount of combustible material per m² of floor area);
- behaviour of materials under fire conditions;
- behaviour of combinations of materials under fire conditions;
- Building Regulation requirements as laid down in Part B.

FIRE RESISTANCE

The resistance of elements of a structure, such as walls, floors and roof, is graded as to the amount of overall fire resistance in minutes required by taking into account the following factors:

- height of top floor above ground level;
- building occupancy purpose group – for example, industrial or office;
- load-bearing capacity, integrity and amount of fire insulation.

The above factors have been incorporated into the tables within the Appendix of Part B of the Building Regulations. These provide a comprehensive list of requirements for elements and the performance of materials, products and structures.

In order for materials to resist for the time specified within Part B of the Building Regulations, they must be tested under certified conditions. BS 476: *Fire tests on building materials and structures* is the standard that must be met for the material to comply with the fire rating required. The standard is split down into a number of separate parts that include the measurement of the effects of temperature, spread of flames and methods for determining the fire resistance of load-bearing elements of the construction. Components can also be studied under this standard as separate entities, with regard to:

- their behaviour when subjected to the intense heat encountered during a fire;
- their ability to support fire spread over their exposed surfaces.

Structural steel does not behave well under conditions of extreme heat, although its surface fire spread is negligible. As the fire takes hold and progresses, the temperature of steel increases and there is an actual gain in the ultimate strength of mild steel. This gain in strength decreases back to normal over the temperature range of 250 to 400 °C. The decrease in strength continues, and by the time the steel temperature has reached 550 °C it will have lost most of its useful strength and therefore its ability to support loads. As the rise in temperature during the initial stages of a fire is rapid, this figure of 550 °C can be reached very quickly. The loss of the strength of the steel at this temperature can result in the collapse of a supporting steel member, causing the redistribution of its working loads, subsequently causing other members to be overstressed, and the resultant progressive collapse of the whole building.

Reinforced concrete structural members have excellent fire-resistance properties and, being non-combustible, do not contribute to the spread of flame over their surfaces. However, it is possible, under the intense and prolonged heat of a fire, that the bond between the steel reinforcement and the concrete will be broken. This is mainly due to the thermal expansion of the steel reinforcement within the concrete. This generally results in spalling of the concrete, where small pieces are blown off the face, which decreases both the protective cover of the concrete over the steel and the cross-sectional area that is supporting the loads. As for structural steel members, this can result in a redistribution of stresses leading to overloading of certain members, culminating in progressive collapse.

Timber, strange as it may seem, behaves very well structurally under the action of fire. This is due to its slow combustion rate, the strength of its core failure remaining fairly constant. The ignition temperature of timber is low (250–300 °C), but during combustion the timber chars at about 0.5–1.0 mm per minute, depending on the species and extent of heat and flame. The layer of charcoal so formed slows down the combustion rate of the core. Although its structural properties during a fire are good, timber, being an organic material and therefore combustible, will spread fire over its surface, which makes it unsuitable in most structural situations. Intumescent paints will provide a limited resistance to fire, but more successful protection is achieved by use of fire-rated plasterboard.

From the above brief considerations it is obvious that designers and builders need to have data on the performance, under the conditions of fire, of materials and especially combinations of materials forming elements. Such information is available in BS 476: *Fire tests on building materials and structures*. The BS is divided into parts that relate to the various fire tests applied to building materials and structures.

FIRE TESTS ON BUILDING MATERIALS AND STRUCTURES – BS 476

BS 476 consists of 17 parts numbered intermittently between 3 and 33. Omissions account for withdrawals such as Part 1, which was replaced by Part 8, which, in turn, was replaced by Parts 20 to 23. The following is a summary of the more significant parts of the standard.

PART 3 *CLASSIFICATION AND METHOD OF TEST FOR EXTERNAL FIRE EXPOSURE TO ROOFS*

This part covers a series of tests for grading roof construction in terms of the time it can resist external penetration by fire, and the amount of flame spread across the external surface of the roof. It can include the roof supporting joist and the lining below.

Three tests are applied to a sample of roof structure 1.50 m × 1.20 m and consist of a:

■ preliminary ignition test;

■ fire penetration test;

■ roof surface spread of flame test.

After testing, the specimen or form of roof construction can be graded by a double letter designation in the range A–D. An AA designation is the highest fire rated and DD the lowest. The initial letter relates to the time of fire penetration; the second letter is a measure of the surface spread of flame:

First letter (penetration)

A. No penetration within 1 hour.

B. Specimen penetrated in not less than ½ hour.

C. Specimen penetrated in less than ½ hour.

D. Specimen penetrated in the preliminary flame test.

Second letter (spread of flame)

A. No spread of flame.

B. Not more than 533 mm spread.

C. More than 533 mm spread.

D. Specimens that continue to burn for 5 minutes after withdrawal of the test flame or spread more than 381 mm across the region of burning in the preliminary flame test.

Specimens can be tested from a 45° inclination down to 1° (effectively flat), such that this matches the in-situ pitch of the actual roof, and may be prefixed EXT.S or EXT.F accordingly. If during the test any dripping from the underside of the specimen, any mechanical failure and/or development of holes is observed, a suffix 'X' is added to the designation thus:

EXT.S.AB–EXT.S.ABX

The Building Regulations, Approved Document B *Fire safety*, Sections 10 (Volume 1) and 14 (Volume 2): Roof coverings, has adopted the use of the double letter classification. It is shown in Table 5.2.1 compared with European Standard BS EN 13501: *Fire classification of construction products and building elements.*

Table 5.2.1 Building Regulations classification compared with BS classification

Building Regulations classification	BS EN 13501-5 classification
AA, AB, AC	$B_{ROOF}(t4)$
BA, BB, BC	$C_{ROOF}(t4)$
CA, CB, CC	$D_{ROOF}(t4)$
AD, BD, CD	$E_{ROOF}(t4)$
DA, DB, DC, DD	$F_{ROOF}(t4)$

Notes:
The suffix (t4) indicates test number 4 in the Standard.
Classes A1 and A2 also occur in BS EN 13501. Products tested to these classifications satisfy the same criteria as Class B, and have no significant contribution to the fire load or fire growth.

PART 4 *NON-COMBUSTIBILITY TEST FOR MATERIALS*

This part deals with the classification of a material as to whether it is combustible or non-combustible according to their behaviour in the non-combustibility test.

The material is prepared to a specified dimension and placed within a furnace under the required test conditions and monitoring. A material is classified by this test as non-combustible if none of the three specimens tested:

■ causes the temperature reading from either of the two thermocouples within the furnace to rise by 50 °C or more above the initial furnace temperature;

■ is observed to flame continuously for 10 seconds or more inside the furnace.

Otherwise the material shall be deemed combustible.

PART 7 *METHOD OF TEST TO DETERMINE THE CLASSIFICATION OF THE SURFACE SPREAD OF FLAME OF PRODUCTS*

This test is used to measure the lateral spread of flame for essentially flat materials, including composites that may be applied as exposed wall or ceiling finishes.

The sample for testing must have any surfacings or coatings applied in the usual manner, such as emulsion paint on plasterboard. The 270 mm × 885 mm sample is fixed in the holder of the apparatus and subjected to radiant heat from the test furnace at the base of the sample. During the first minute of the test a luminous gas flame is applied to the furnace end of the specimen. After one minute it is extinguished and at intervals of 1.5 minutes and 10 minutes the extent of flaming in mm along the material specimen is measured and recorded.

Approved Document B2: Sections 3 (Volume 1) and 6 (Volume 2) specifies a higher class than class 1, called class 0, for wall and ceiling linings. This is defined as a non-combustible material throughout and is used in locations such as school noticeboards, and hospital corridors where any ignition of paper products needs to limit flame spread.

Table 5.2.2 BS 476-7, Table 1: Classification of flame spread against measured lengths of samples

Classification	Spread of flame at 1½ minutes		Final spread of flame	
	Limit	Limit for one specimen in sample	Limit	Limit for one specimen in sample
	(mm)	(mm)	(mm)	(mm)
Class 1	165	165 +25	165	165 +25
Class 2	215	215 +25	455	455 +45
Class 3	265	265 +25	710	710 +75
Class 4		Exceeding the limits for class 3		

PART 10 *GUIDE TO THE PRINCIPLES AND APPLICATION OF FIRE TESTING*

This is the guide that relates to the general principles and methods of applying fire tests to building products, components and elements of construction for other parts of BS 476. It provides an introduction and an overview to the various detailed parts of BS 476.

PART 12 *METHOD OF TEST FOR IGNITABILITY OF PRODUCTS BY DIRECT FLAME IMPINGEMENT*

This is used to establish the response of materials when subjected to impingement of flames having varying size and heat intensity. Several specimen sizes are used, ranging from 100 mm × 150 mm up to 500 mm × 750 mm, with a range of flame applications to the sample surfaces and edges. The material reactions are measured over exposures of 1 second to 180 seconds. Results are letter designated:

I   = Ignition occurred

T   = Transient ignition

N   = No ignition

W = Flaming or glowing to material edge or within 10 seconds of ignition source removal

PARTS 20 TO 24

These include provision for determining the fire resistance of various parts, elements and components of construction. The term 'fire resistance' relates to a complete element of building construction and not to the individual materials of which that element is composed. The tests enable elements of construction to be assessed according to their ability to retain their composition, to resist the passage of flame and hot gases, and to provide the necessary resistance to heat transmission in a defined pressure environment. This is measured against retention of fire separating and/or loadbearing facilities over a specified time.

PART 20 *METHOD FOR DETERMINATION OF THE FIRE RESISTANCE OF ELEMENTS OF CONSTRUCTION (GENERAL PRINCIPLES)*

The general principles outlined in this part of BS 476 include details for preparing specimens – that is, the number and size for specific tests – and descriptions of the apparatus for specimen support and exposure. Procedures for effecting the various tests are also provided. Details of criteria for performance are shown measured in terms of **integrity** (loadbearing capacity and fire containment) and **insulation** (thermal transmittance) with reference to test report format and guidance on application of method. Test results are given in minutes from the start of the test until failure occurs.

PART 21 *METHOD FOR DETERMINATION OF THE FIRE RESISTANCE OF LOADBEARING ELEMENTS OF CONSTRUCTION*

This document defines the application to beams, columns, floors, flat roofs and walls.

BEAMS

The specimen is to be full size or have a minimum span of 4.000 m and be located to simulate actual site conditions. If the beam is exposed to fire on three faces, associated construction as in practice shall be included in the specimen.

Fire resistance

The test specimen is deemed to have failed when it can no longer support its design loading. This occurs if either of the following is exceeded first:

- the deflection exceeds $L/20$, where L is the clear span of the specimen in mm;
- the rate of deflection (mm/min) calculated at one-minute intervals commencing one minute from the heat application exceeds the limit set by $L^2/9,000d$, where d is the distance from the top of the structural section to the bottom of the design tension zone in mm.

COLUMNS

The specimen is to be full-size or to have a minimum length of 3.000 m and be loaded to simulate actual site conditions. The specimen is heated on all exposed faces.

Fire resistance

This is the time taken for the specimen to no longer support the test load – that is, to show a noticeable change in the rate of deformation.

FLOORS AND FLAT ROOFS

The specimen is to be full-size or a minimum of 3.000 m wide × 4.000 m span and be loaded to simulate actual site conditions. If a ceiling is intended to add to the fire resistance, it must be included with the test specimen. This also applies to any construction or expansion joints, light diffusers or any other components integral to the actual installation. The specimen is heated from one side by a furnace to produce a positive pressure at standard heating conditions until failure occurs or the test is terminated.

Fire resistance

Loading as for columns, plus criteria to assess integrity and insulation.

- **Integrity** A 100 mm × 100 mm × 20 mm-thick cotton wool pad of mass 3 to 4 g is held over the centre of any crack through which flames and gases can pass. The pad is held 25 mm from and parallel to the crack for a period of 10 seconds to determine whether hot gases can cause ignition.

- **Insulation** The unexposed face of elements having a separating function is observed at intervals of not more than five minutes. Failure is deemed to occur if:

  - mean temperature rises more than 140 °C above initial temperature;
  - point temperature rises more than 180 °C above initial temperature.

WALLS

The specimens must include provision for any mechanical joints and be tested from both sides. They must be full size or a minimum of 3.000 m × 3.000 m and loaded to simulated actual site conditions.

Fire resistance

Time taken to failure by any one of three observations:

- noticeable change in the rate of deformation;
- loss of integrity (as for floors and flat roofs);
- loss of insulation (as for floors and flat roofs).

PART 22 *METHODS FOR DETERMINATION OF FIRE RESISTANCE OF NON-LOADBEARING ELEMENTS OF CONSTRUCTION*

This is applicable to partitions, door sets and shutters (fully insulated, partially insulated and uninsulated), ceiling membranes and glazing.

PARTITIONS

Specimens to be at least 3.000 m × 3.000 m and tested from one or either side. They must include any provision for mechanical joints such as built-in facilities for expansion and contraction.

Fire resistance

Failure time for integrity and insulation.

DOOR SETS AND SHUTTERS

Specimens are the same size as for partitions and are tested from both sides.

Fire resistance (fully insulated)

Integrity and insulation as previously described and/or sustained flaming – that is, a visible flame occurring for at least 10 seconds.

CEILINGS

The specimen to be at least 4.000 m × 3.000 m and exposed to test from the underside only.

Fire resistance

For integrity and insulation as previously described.

GLAZED ELEMENTS

The specimen to be at least 3.000 m × 3.000 m and exposed to test from both sides.

Fire resistance

For integrity and insulation as previously described.

PART 23 *METHODS FOR DETERMINATION OF THE CONTRIBUTION OF COMPONENTS TO THE FIRE RESISTANCE OF A STRUCTURE*

This part is in two sections:

1. the contribution made by components such as suspended ceilings when protecting steel beams;
2. the use of intumescent materials/seals with single-acting latched timber fire-resisting door assemblies.

SUSPENDED CEILINGS PROTECTING STEEL BEAMS

The specimen to be full size or a minimum of 4.000 m × 3.000 m, with the underside only exposed to the furnace. The steel beams used to support the specimen have the top flange covered with a lightweight concrete floor at least 130 mm thick. Provision for any light fittings or similar outlets must be made within the specimen.

Fire resistance

The limit of effective protection is deemed to have been reached when either of the following are exceeded:

■ with an unloaded specimen, the beam temperature at any point attains 400 °C;

■ for a loaded specimen, see reference to Part 21: Beams.

INTUMESCENT SEALS

This test is suitable for evaluating the effect of intumescent materials or seals between door-to-frame assemblies for doors with fire resistance specified at up to one hour. Full-size specimens are exposed to the furnace on one side only, as described in Part 22.

Fire resistance

As previously described for integrity, or failure of the seal due to the presence of continuous flaming on the unexposed door face.

PARTS 31 TO 33

PART 31 *METHODS FOR MEASURING SMOKE PENETRATION THROUGH DOOR SETS AND SHUTTER ASSEMBLIES*

Section 31.1: *Methods of measurement under ambient temperature conditions* provides guidance on smoke penetration potential through door sets by applying pressurised air tests and taking measurements of the air leakage rate. Its purpose is for smoke control only, and it has no reference to fire resistance. Full-size specimens are subjected to fan-delivered variable air pressures between the two door faces, in increments up to 100 Pa. Air penetration and leakage is measured in m^3/h. See also Building Regulations, Approved Document B: Appendix B, Table B1, where fire doors for certain situations (suffixed S) are required to have an air leakage rate not exceeding 3 m^3/h at 25 Pa.

PART 32 *GUIDE TO FULL-SCALE FIRE TESTS WITHIN BUILDINGS*

PART 33 *FULL-SCALE ROOM TEST FOR SURFACE PRODUCTS*

These procedures are intended to simulate fires in buildings from ignition through to full maturity. Tests can represent specific fire situations and materials to determine potential for flashover conditions, development of fire growth, and behaviour of various products in a continuous fire environment. In addition to visual interpretations of flame spread both vertically and horizontally, measurements can also include temperature, air flow and gas emissions.

BUILDING REGULATIONS: PART B

One of the major aims of this part of the Building Regulations is to limit the spread of fire. This is achieved by considering:

- the use of a building;

- the fire resistance of structural elements and surface finishes;

- the size of the building or parts of a building;

- the degree of isolation between buildings or parts of buildings.

Part B to the Building Regulations contains five regulations that are concerned with fire under the headings of:

- Means of warning and escape – Regulation B1;

- Internal fire spread (linings) – Regulation B2;

- Internal fire spread (structure) – Regulation B3;

- External fire spread – Regulation B4;

- Access and facilities for the fire service – Regulation B5.

Approved Document B provides comprehensive guidance and gives recommendations for meeting the performance requirements set out in the actual regulations.

A full understanding of the terminology used is important to comprehend the recommendations being made in the Approved Document, and this is given in Appendix E of Volume 2, under the heading of definitions to assist in the understanding of the regulations. Figure 5.2.1 illustrates some of the general definitions that are given in Appendix E. Other interpretations that must be clearly understood are elements of structure, fire stops, and boundaries. These are illustrated in Figs 5.2.2, 5.2.3 and 5.2.4 respectively.

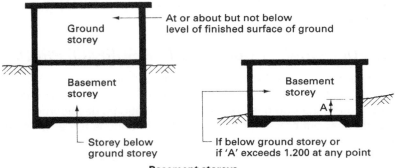

At or about but not below
level of finished surface of ground

Ground
storey

Basement
storey

Storey below
ground storey

Basement
storey

A

If below ground storey or
if 'A' exceeds 1.200 at any point

Basement storeys

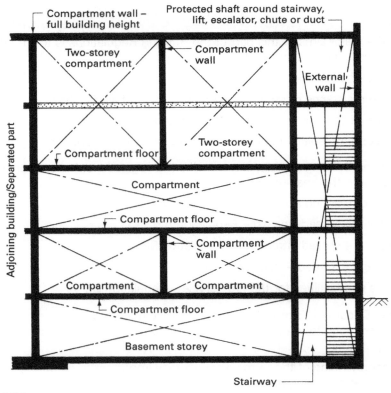

Compartment wall –
full building height

Protected shaft around stairway,
lift, escalator, chute or duct

Two-storey
compartment

Compartment
wall

External
wall

Two-storey
compartment

Compartment floor

Compartment

Compartment floor

Compartment
wall

Compartment

Compartment

Compartment floor

Basement storey

Adjoining building/Separated part

Stairway

Figure 5.2.1   Approved Document B: general definitions

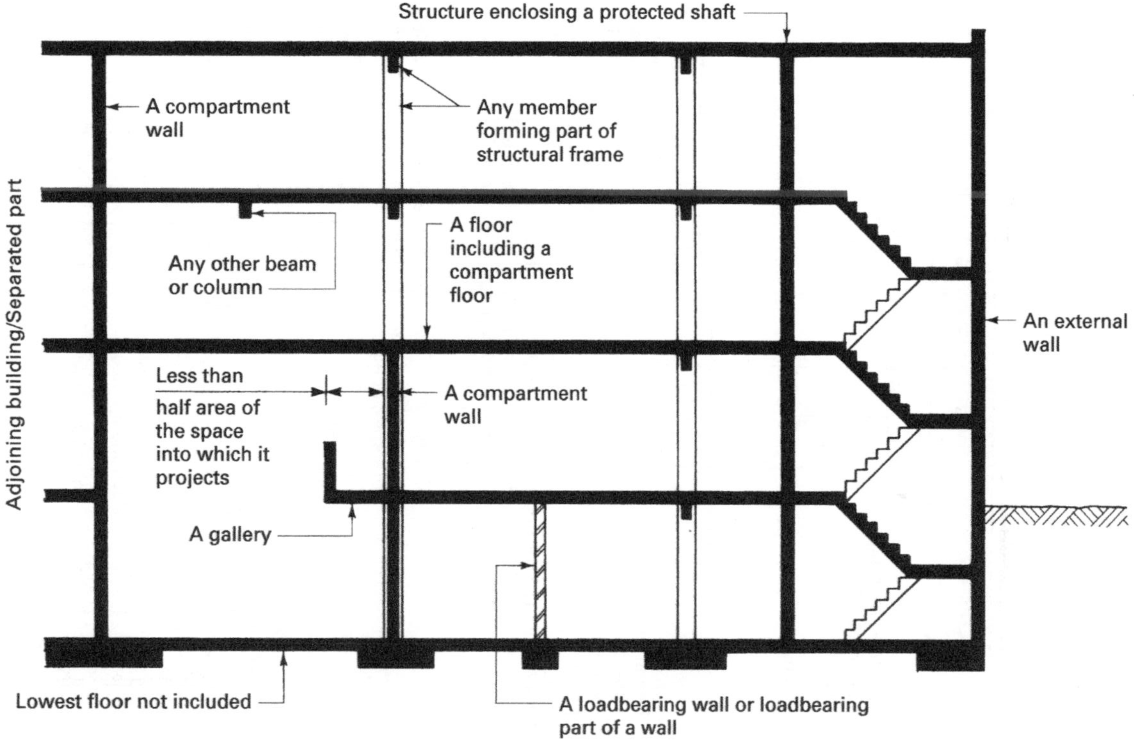
Structure enclosing a protected shaft
A compartment wall
Any member forming part of structural frame
Adjoining building/Separated part
Any other beam or column
A floor including a compartment floor
An external wall
Less than half area of the space into which it projects
A compartment wall
A gallery
Lowest floor not included
A loadbearing wall or loadbearing part of a wall

Figure 5.2.2 **Approved Document B: elements of structure**

Fire-stop – a seal which closes an imperfection of fit or design tolerance between elements or components, to prevent the passage of smoke or flame.

Cavity barrier – construction which closes a concealed space against smoke or flame penetration.

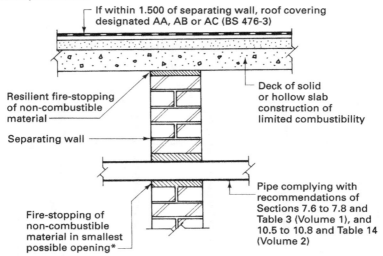

If within 1.500 of separating wall, roof covering designated AA, AB or AC (BS 476-3)

Resilient fire-stopping of non-combustible material

Deck of solid or hollow slab construction of limited combustibility

Separating wall

Pipe complying with recommendations of Sections 7.6 to 7.8 and Table 3 (Volume 1), and 10.5 to 10.8 and Table 14 (Volume 2)

Fire-stopping of non-combustible material in smallest possible opening*

* *Alternatively*: 1. Proprietary sealing system or collar.
2. Sleeving for pipes of low melting point, e.g. uPVC, extending ≮1 m beyond separating element. Sleeve stopped as shown.

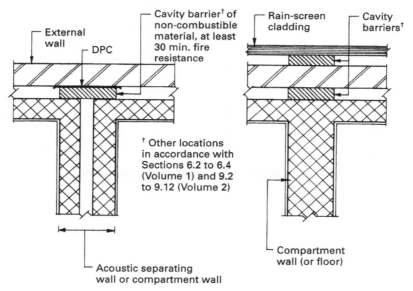

External wall

DPC

Cavity barrier† of non-combustible material, at least 30 min. fire resistance

Rain-screen cladding

Cavity barriers†

† Other locations in accordance with Sections 6.2 to 6.4 (Volume 1) and 9.2 to 9.12 (Volume 2)

Acoustic separating wall or compartment wall

Compartment wall (or floor)

Figure 5.2.3 Approved Document B: fire-stopping and cavity barriers

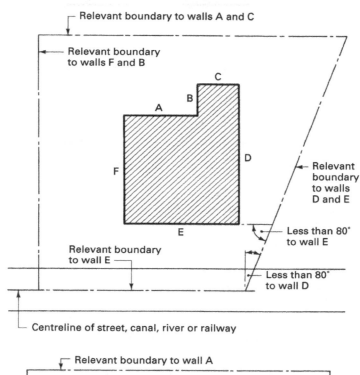

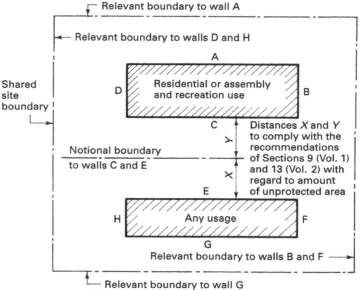

Figure 5.2.4   Approved Document B: boundaries and space separation

BUILDING USE

The use of a building enables it to be classified into one of the seven purpose groups given in Approved Document B Volume 2, Appendix D, Table D1. These purpose group classifications are used in the further application of the regulation of elements of their construction, compartmentation and surrounding structure. The purpose groups can apply to a whole building, or to a separated part or compartment of a building. All buildings covered by Part B of the Building Regulations should be included in one of these purpose groups. Each purpose group can be summarised as follows.

1. **Residential (dwellings)** Subdivides into three groups:

   a. flat;

   b. dwelling house with habitable accommodation containing a floor level over 4.500 m above the ground;

   c. dwelling house without a habitable storey, containing floor levels more than 4.500 m above ground.

Note: Where a flat also functions as a workplace for its occupants and others, provisions for escape in event of a fire are given in Approved Document B1 (Vol. 2), *Means of escape from flats*, Section 2.52, Live/work units.

2. a. **Residential (institutional)** Hospital, care home, school or other similar establishment where persons sleep on the premises;

   b. **Other** Hotel, boarding house, hostel, hall of residence and any other residential purpose not described above.

3. **Office** Any premises used for office and administration work.

4. **Shop and commercial** Includes business premises used for any form of retail trade, such as a restaurant or hairdressers, or where members of the public can enter to deliver goods for repair or treatment.

5. **Assembly and recreation** A public building or place of assembly of people for functions, which could be social, recreational, entertainment, exhibition, education, etc.

6. **Industrial** Premises used for manufacture, production, processing, repairs, etc.

7. **Storage and other non-residential** Subdivides into two categories:

   a. place for storage, deposit or placement of goods and materials (vehicles excluded), and any other non-residential purpose not described in 1–6, except for detached garages and carports not exceeding 40 m², which are included in the dwelling-house purpose group;

   b. car parks solely for the accommodation of cars, motorcycles, passenger and light goods vehicles of less than 2,500 kg gross weight.

The use of fire-resistant cells or compartments within a building (Fig. 5.2.1) is a means of confining an outbreak of fire to the site of origin for a reasonable time, to allow the occupants a chance to escape and the firefighters time to tackle, control and extinguish the fire. The Approved Document B3, Sections 5 (Volume 1) and 8 (Volume 2), provides guidance on maximum recommended dimensions for buildings or compartments. These are expressed in terms relating to height of building, floor area and cubic capacity for the various purpose groups. Table A2 (Vol 2) in Appendix A gives the recommended minimum periods of fire resistance for all elements of structure (see Fig. 5.2.5). These minimum periods of fire resistance are stated in minutes according to:

- purpose group;
- ground or upper storey (height of top floor above ground or of separated part of building);
- basement storey including floor over (depth of lowest basement).

The tables give recommended fire resistances in minutes but do not give specification details on how these are to be achieved in the construction of each element. The Building Research Establishment (BRE) Report BR 128: *Guidelines for the construction of fire-resisting structural elements* gives appropriate and common methods of construction for the various notional periods of fire resistance for walls, beams, columns and floors. The guidelines are written and presented in a tabulated format, which needs to be translated into working details. Figures 5.2.6 to 5.2.12 show typical examples taken from this document and other sources of reference, such as manufacturers' tested data. Another useful resource is *Loss Prevention Standards* produced by the Loss Prevention Certification Board. These provide details of minimum standards to satisfy building insurers, which are often higher than those specified in the Building Regulations. It is most important for building designers and detailers to ensure that their specifications satisfy all authorities, as post-construction changes are very expensive!

The examples in Figs 5.2.7 and 5.2.8 for fire protection to structural sections assume a **section factor** greater than 140 (section factor = H_p/A (m^{-1})). Where this number is specified as less than 140, a reduced amount of fire insulation can apply and less fire resistance can be expected. However, this will be adequate for many situations. Calculation of the section factor is based on:

H_p = Perimeter of section exposed to fire (m) [The heated perimeter].

A = Cross-sectional area of the steel section (m^2) – data available in BS 4: *Structural steel sections* and other tables of standard steel sections.

Note: The section factor is defined as the ratio of heated perimeter to section area. Heated perimeter is the inside perimeter of fire protection. For protection of box sections, take the perimeter of the smallest rectangle enclosing the section.

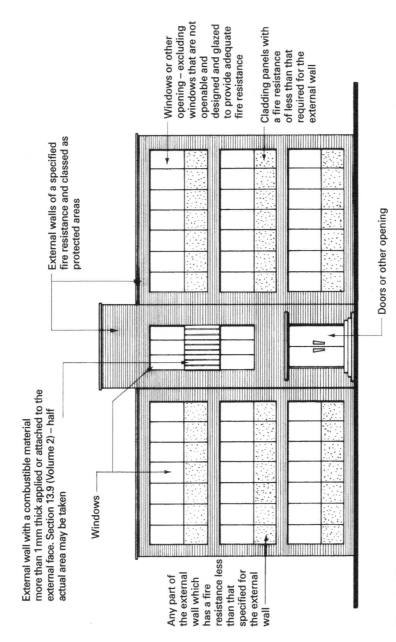

Windows or other opening – excluding windows that are not openable and designed and glazed to provide adequate fire resistance

Cladding panels with a fire resistance of less than that required for the external wall

External walls of a specified fire resistance and classed as protected areas

External wall with a combustible material more than 1 mm thick applied or attached to the external face. Section 13.9 (Volume 2) – half actual area may be taken

Windows

Any part of the external wall which has a fire resistance less than that specified for the external wall

Doors or other opening

Figure 5.2.5   Approved Document B: unprotected areas – fire resistance less than that given in Appendix A, Table A2

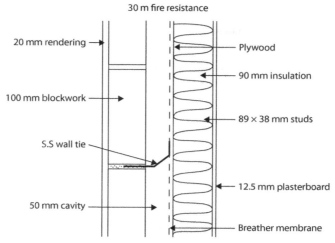

30 m fire resistance

20 mm rendering →

Plywood

90 mm insulation

100 mm blockwork →

89 × 38 mm studs

S.S wall tie

12.5 mm plasterboard

50 mm cavity

Breather membrane

Typical section through external timber framed wall

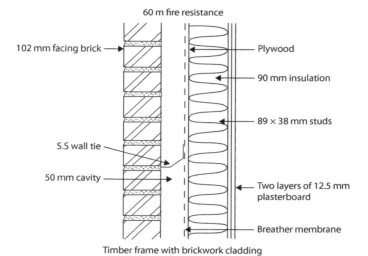

60 m fire resistance

102 mm facing brick →

Plywood

90 mm insulation

89 × 38 mm studs

S.S wall tie

Two layers of 12.5 mm plasterboard

50 mm cavity

Breather membrane

Timber frame with brickwork cladding

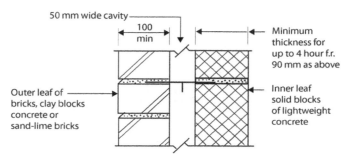

50 mm wide cavity

100 min

Minimum thickness for up to 4 hour f.r. 90 mm as above

Outer leaf of bricks, clay blocks concrete or sand-lime bricks →

Inner leaf solid blocks of lightweight concrete

Typical traditional cavity wall fire-resistant construction

Figure 5.2.6   Fire resistance: walls of masonry construction

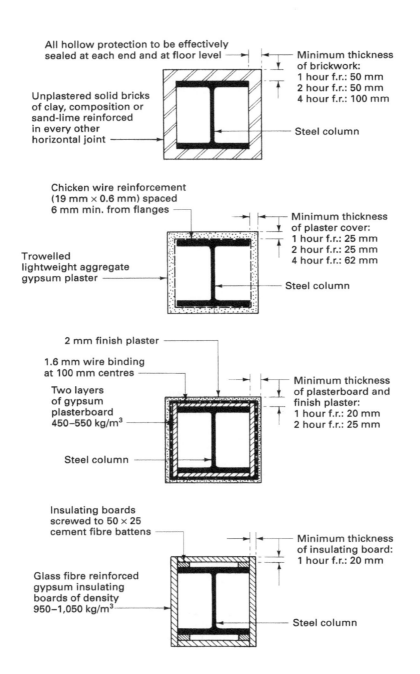

All hollow protection to be effectively sealed at each end and at floor level

Minimum thickness of brickwork:
1 hour f.r.: 50 mm
2 hour f.r.: 50 mm
4 hour f.r.: 100 mm

Unplastered solid bricks of clay, composition or sand-lime reinforced in every other horizontal joint

Steel column

Chicken wire reinforcement (19 mm × 0.6 mm) spaced 6 mm min. from flanges

Minimum thickness of plaster cover:
1 hour f.r.: 25 mm
2 hour f.r.: 25 mm
4 hour f.r.: 62 mm

Trowelled lightweight aggregate gypsum plaster

Steel column

2 mm finish plaster

1.6 mm wire binding at 100 mm centres

Two layers of gypsum plasterboard 450–550 kg/m³

Minimum thickness of plasterboard and finish plaster:
1 hour f.r.: 20 mm
2 hour f.r.: 25 mm

Steel column

Insulating boards screwed to 50 × 25 cement fibre battens

Minimum thickness of insulating board:
1 hour f.r.: 20 mm

Glass fibre reinforced gypsum insulating boards of density 950–1,050 kg/m³

Steel column

Figure 5.2.7   Fire resistance: hollow protection to steel columns (section factor > 140)

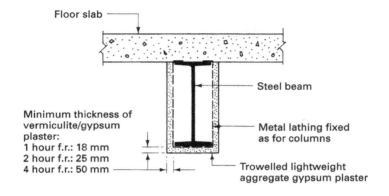

Floor slab

Steel beam

Minimum thickness of vermiculite/gypsum plaster:
1 hour f.r.: 18 mm
2 hour f.r.: 25 mm
4 hour f.r.: 50 mm

Metal lathing fixed as for columns

Trowelled lightweight aggregate gypsum plaster

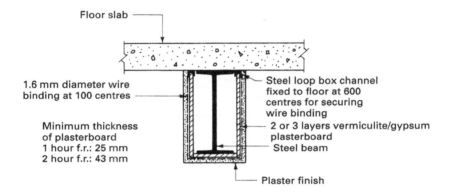

Floor slab

1.6 mm diameter wire binding at 100 centres

Steel loop box channel fixed to floor at 600 centres for securing wire binding

Minimum thickness of plasterboard
1 hour f.r.: 25 mm
2 hour f.r.: 43 mm

2 or 3 layers vermiculite/gypsum plasterboard

Steel beam

Plaster finish

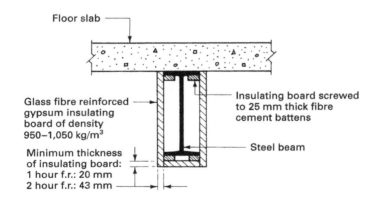

Floor slab

Glass fibre reinforced gypsum insulating board of density 950–1,050 kg/m³

Insulating board screwed to 25 mm thick fibre cement battens

Minimum thickness of insulating board:
1 hour f.r.: 20 mm
2 hour f.r.: 43 mm

Steel beam

Figure 5.2.8   Fire resistance: hollow protection to steel beams (section factor > 140)

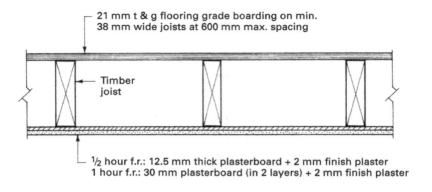

21 mm t & g flooring grade boarding on min.
38 mm wide joists at 600 mm max. spacing

Timber
joist

½ hour f.r.: 12.5 mm thick plasterboard + 2 mm finish plaster
1 hour f.r.: 30 mm plasterboard (in 2 layers) + 2 mm finish plaster

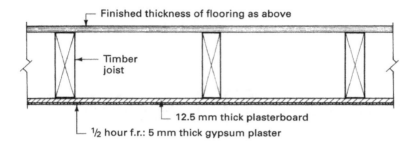

Finished thickness of flooring as above

Timber
joist

12.5 mm thick plasterboard

½ hour f.r.: 5 mm thick gypsum plaster

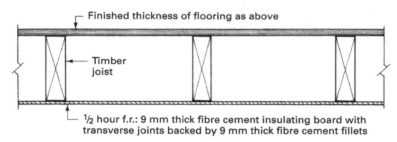

Finished thickness of flooring as above

Timber
joist

½ hour f.r.: 9 mm thick fibre cement insulating board with
transverse joints backed by 9 mm thick fibre cement fillets

Note: Plasterboard nailed at max. 150 mm spacing, 20 mm min. nail length
beyond board.

Figure 5.2.9   Fire resistance: timber floors

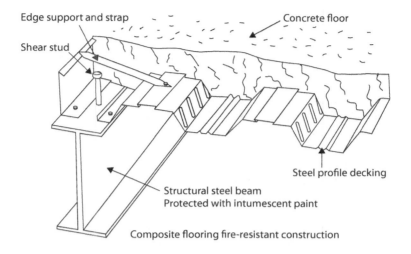

Edge support and strap

Shear stud

Concrete floor

Steel profile decking

Structural steel beam
Protected with intumescent paint

Composite flooring fire-resistant construction

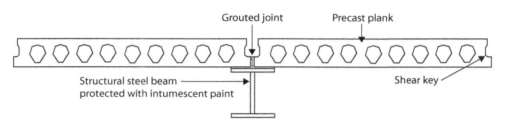

Grouted joint

Precast plank

Structural steel beam
protected with intumescent paint

Shear key

Hollow-core precast concrete plank on steel beam support

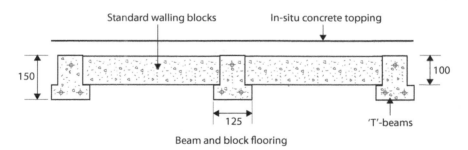

Standard walling blocks

In-situ concrete topping

150

100

125

'T'-beams

Beam and block flooring

Figure 5.2.10 **Fire resistance of modern floor construction**

Source: Kingspan Limited – contact www.kingspan.com for latest details.

Readers are also encouraged to study manufacturers' literature on the many patent ready-cut, easy-to-fix fire protection systems for standard structural members, and other methods such as the use of intumescent paints and materials, which expand to form a thick insulating coating or strip on being heated by fire.

FIRE RESISTANCE: SPACE SEPARATION

The degree of space separation needed between buildings to limit the spread of fire by radiation is covered in Part B4: Sections 8 to 10 (Volume 1) and 12 to 14 (Volume 2) of Approved Document B. Radiation spread of fire is by the warming up of other construction elements by absorbing heat until they start to combust. Part B4 deals with external walls and roofs and, in particular, unprotected areas permitted in relationship to the distance of the building or compartment within the building from the relevant boundary. It differentiates between buildings with external walls occurring within or over one metre from the relevant boundary. External wall surfaces must satisfy the fire-resistance requirements given in Tables A1 and A2 of Appendix A (Vol. 2), or be constructed with a limited amount of combustible material (see Diagram 40, Approved Document B4, Section 12 (Vol. 2)).

The Approved Document B4 (External Fire Spread) provides four methods to determine the acceptable amount of unprotected external wall area. An unprotected wall area is defined as the part of an external wall which has less resistance to fire than an appropriate amount given in Table A2 Appendix A.

METHOD 1

The first is concerned with small residential buildings not included in the Institutional group, over one metre from the relevant boundary. A simple table and diagram given in the Approved Document shows the relationship between boundary distance, maximum length of building side and the maximum total of unprotected area permitted (Sections 9.16, Diagram 22 (Volume 1) and 13.19, Diagram 46 (Volume 2)).

METHOD 2

This method can be used for other buildings or compartments that are not less than one metre from the relevant boundary and not over 10 metres high (excludes open-sided car parks). Data are interpolated from Approved Document B4, Sections 9, Table 4 (Volume 1) and 13, Table 15 (Volume 2), to obtain an acceptable percentage of unprotected area relative to the various purpose grouping.

Alternatively, other methods described in the BRE Report BR 187: *External fire spread: building separation and boundary distances* may be used. These are known as the **enclosing rectangle method** and the **aggregate notional area method**.

METHOD 3

The enclosing rectangle method can be used to ascertain the maximum unprotected area for a given boundary position or to find the nearest position of the boundary for a given building design. The method consists of placing an enclosing rectangle around the unprotected areas, noting the width and height of the enclosing rectangle, and calculating the unprotected percentage in terms of the enclosing rectangle. This information will enable the distance from the boundary to be read direct from the tables given in the BRE report. Note that in compartmented buildings the enclosing rectangle is taken for each compartment and not for the whole façade. Typical examples are shown in Fig. 5.2.11.

METHOD 4

In the aggregate notional area method, the principle of separation is still followed, but by a more precise method, which will involve more investigation than reference to an enclosing rectangle. Reference is made to an aggregate notional area, which is calculated by taking the sum of each relevant unprotected area and multiplying by a factor given in the BRE report. Which unprotected areas are relevant is given in the Appendix. The method entails dividing the relevant boundary into a series of 3.000 m spacings called vertical data and projecting from each point a datum line to the nearest point on the building. A baseline is drawn through each vertical datum at right angles to the datum line and a series of semicircles of various radii drawn from this point represent the distance and hence the factors for calculating the aggregate notional area (see Fig. 5.2.12). For buildings or compartments with a residential, assembly or office use, the result should not exceed 210 m^2; for other purpose groups the result should not exceed 90 m^2. In practical terms this method would not normally be used unless the situation was critical or the outline of the building was irregular in shape.

The specific recommendations for compartment walls, openings in compartment walls, compartment floors, protected shafts, doors, stairways and protecting structure are set out in Approved Document B. Each purpose group is considered separately, although many of the recommendations are similar for a number of purpose groups.

FIRE RESISTANCE: FIRE DOORS

Doors have to be used throughout a building for accessibility and movement of occupants both vertically and horizontally. Where doors are fitted into the compartmental elements with regard to fire resistance, they must be constructed and installed in accordance with Approved Document Part B Appendix B and Table B1.

Note: Enclosing rectangle figures used are the nearest figures given in tables above actual dimensional figures.

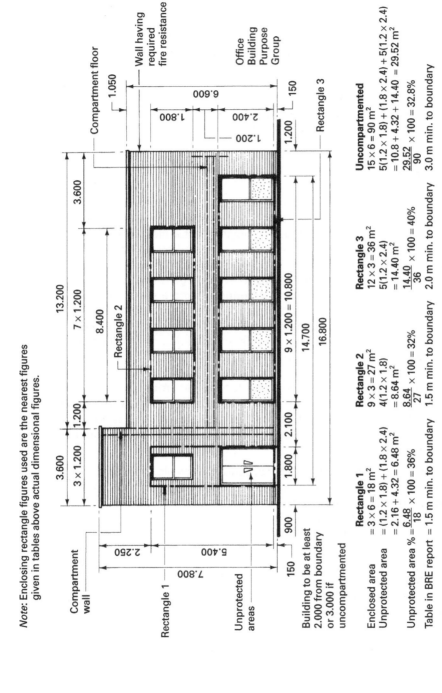

Rectangle 1
Enclosed area = 3 × 6 = 18 m²
Unprotected area = (1.2 × 1.8) + (1.8 × 2.4)
= 2.16 + 4.32 = 6.48 m²
Unprotected area % = $\frac{6.48}{18}$ × 100 = 36%
Table in BRE report = 1.5 m min. to boundary

Rectangle 2
9 × 3 = 27 m²
4(1.2 × 1.8)
= 8.64 m²
$\frac{8.64}{27}$ × 100 = 32%
1.5 m min. to boundary

Rectangle 3
12 × 3 = 36 m²
5(1.2 × 2.4)
= 14.40 m²
$\frac{14.40}{36}$ × 100 = 40%
2.0 m min. to boundary

Uncompartmented
15 × 6 = 90 m²
5(1.2 × 1.8) + (1.8 × 2.4) + 5(1.2 × 2.4)
= 10.8 + 4.32 + 14.40 = 29.52 m²
$\frac{29.52}{90}$ × 100 = 32.8%
3.0 m min. to boundary

Figure 5.2.11 Unprotected areas: enclosing rectangle method (reference: BRE Report BR 187)

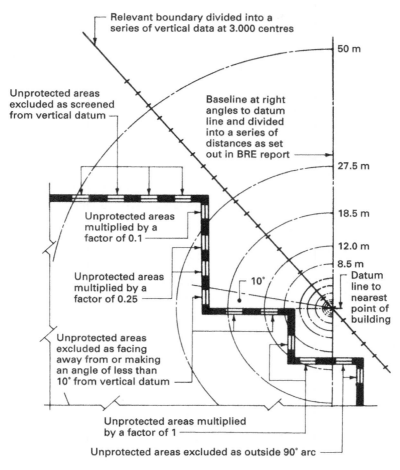

Relevant boundary divided into a series of vertical data at 3.000 centres

Unprotected areas excluded as screened from vertical datum

Baseline at right angles to datum line and divided into a series of distances as set out in BRE report

50 m

27.5 m

18.5 m

12.0 m

8.5 m

Datum line to nearest point of building

Unprotected areas multiplied by a factor of 0.1

Unprotected areas multiplied by a factor of 0.25

10°

Unprotected areas excluded as facing away from or making an angle of less than 10° from vertical datum

Unprotected areas multiplied by a factor of 1

Unprotected areas excluded as outside 90° arc

Notes: 1. Procedure repeated for each vertical datum position.
2. Calculations carried out for any side of a building or compartment.
3. For each position aggregate notional area of the unprotected areas must not exceed:
    $210 \ m^2$ for residential, assembly or office use.
    $90 \ m^2$ for all other purpose groups.

Figure 5.2.12 Unprotected areas: aggregate notional area method (reference: BRE Report BR 187)

The classification for doors follows the procedures laid down within Part 22 of BC 476. Doors are classified as:

- fire door (FD);
- rating in minutes;
- smoke control door.

These three requirements are incorporated into Table B1 – for example, a FD30S is a 30-minute smoke-seal fitted door. Under the European standard the 'FD' is replaced with an 'E'. All fire doors should be fitted with a door closer which automatically shuts the door when it is used. This maintains the integrity of the fire compartment or element in which the door is fitted. Except for doors within flats, bedroom doors and lift entrance doors, all fire doors must be fitted with appropriate fire safety signs reading 'Fire door keep shut', 'Keep locked' or 'Keep clear'.

FIRE RESISTANCE: ROOF COVERINGS

For roof coverings the Building Regulations are concerned mainly with the performance of roofs when exposed externally to fire, which questions the application and use of plastics and other combustibles, particularly in rooflights. Thermoplastic materials divide broadly into two categories: TP(a) (rigid) or TP(b). TP(a) considers rigid rooflight products having a class 1 rating to BS 476-7, or satisfying the test requirements of BS 2782-5, *Methods of testing plastics*. TP(b) applies to rigid polycarbonates less than 3 mm thick and other products satisfying tests to BS 2782-5. Diagram 23 and Tables 6 and 7 in Section 10 (Volume 1) and Diagram 47 and Tables 17 and 18 in Section 14 (Volume 2) of Approved Document B4 indicate minimum acceptable spacings for rooflights, distances from boundaries and maximum areas applicable to these categories of thermoplastics.

Means of escape in case of fire   5.3

Means of escape from within a building is concerned with making sure people can escape if an outbreak of fire should occur. It is designed to provide the occupants with the opportunity to reach an area of designated safety, which is normally an external allocated point where numbers of occupants can be established. This takes into account factors such as the risks to human life, disabled occupants, lack of familiarity with the building's internal layout, problems of smoke and the short space of time available to evacuate the premises before the fire devastates the building.

Fear is a natural human response when confronted with uncontrolled fire and in particular fear of **smoke**, which is justified by the fact that more deaths are caused by smoke and heated gases than by burns. Statistics show that approximately 54 per cent of deaths in fires are caused by smoke, 40 per cent by burns and scalds, and 6 per cent by other causes. Smoke may be defined as the visible suspension in the atmosphere of solid and/or liquid particles resulting from combustion or pyrolysis. The main danger with the combustion of fuel within the fire is the carbon dioxide and carbon monoxide gases, and other poisonous gases produced from furniture and furnishings, that are released and cause rapid asphyxiation of any unfortunate trapped occupants.

The presence of these gases does not always cause the greatest hazard in human terms, because the density of smoke is more likely to create fear than the undetectable gases. Buoyant and mobile dense smoke will spread rapidly within a building or compartment during a fire, masking or even obliterating exit signs and directions. Gases other than those mentioned previously are generally irritants that can affect the eyes, causing watering, which further impairs the vision, and can also affect the respiratory organs, causing the slowing of reactions and a loss of directional sense. Smoke rises with the fire's heat towards a ceiling, often leaving a gap towards the floor that is the only possible escape and source of breathable air.

Carbon dioxide is odourless and invisible and is always present in the atmosphere; however, because it is a by-product of combustion, its volume increases at the expense of oxygen during a fire. The gas is not poisonous but can cause death by asphyxia as oxygen is depleted from the air. The normal amount of oxygen present in the air is approximately 21 per cent; if this is reduced to 12 per cent, abnormal fatigue can be experienced; down to about 6 per cent it can cause nausea, vomiting and loss of consciousness; below 6 per cent respiration is difficult, which can result in death. Carbon dioxide will not support combustion, and can cause a fire to be extinguished if the content by volume exceeds 14 per cent, a fact used by firefighters in their efforts to deal with an outbreak of fire.

Carbon monoxide, like carbon dioxide, is odourless, but it is extremely poisonous and, having approximately the same density as air, will spread rapidly. A very small concentration (0.2 per cent by volume) of this colourless gas can cause death in about 40 minutes. The first effects are dizziness and headaches, followed in five to ten minutes by loss of consciousness, leading to death. As the concentration increases, so the time lapse from the initial dizziness to death decreases, so that by the time the concentration has reached about 1.3 per cent by volume, death can take place within a minute or two. Carbon monoxide is also a by-product of the combustion of natural gas, and any leak within the flue discharge system can cause the death of any of the occupants. The smoke from a fire acts as an initial indicator, and often moves occupants away from the potential of carbon monoxide to act on their circulation systems.

The **heat** that is associated with fire and smoke can also be injurious and even fatal. Temperatures in excess of 100 °C can cause damage to the windpipe and lungs, resulting in death within 30 minutes or sooner as the temperature rises. Injuries caused by heat are generally in the form of burns, often followed by shock, which can be fatal in many cases.

The importance of a satisfactory means of escape cannot be overestimated within our domestic and commercial buildings, especially in high occupancy categories such as hospitals, student accommodation, flats and high-rise accommodation. To this end, a substantial amount of legislation and advisory documentation exists to guide the designer in planning escape routes without being too restrictive on the overall design concept. In the UK this has included regional by-laws, the Building Regulations, the Health and Safety at Work etc. Act 1974, the Fire Safety and Safety of Places of Sports Act 1987, the Management of Health and Safety at Work Regulations 1999, and references to BS 9999: *Code of practice for fire safety in the design, management and use of buildings*. The Regulatory Reform (Fire Safety) Order 2005 has repealed a number of pieces of fire safety legislation and consolidated fire safety regulations under one regulatory framework.

Building Regulations Approved Document Part B requires that, in case of fire, a means of escape leading from the building to a place of safety outside the building must be capable of being safely and effectively used at all times. The Regulation covers all building types with the exception of prisons. Approved Document B is divided into two volumes as follows:

- Volume 1: domestic dwelling houses,
- Volume 2: buildings other than dwellings:
  - flats;
  - general provisions for the common parts of flats;
  - design for horizontal escape;
  - design for vertical escape;
  - general provisions.

APPROVED DOCUMENT B, VOLUME 1: MEANS OF ESCAPE FROM DWELLING HOUSES

PLANNING ESCAPE ROUTES

When escape routes are being planned, the type of occupancy must be considered: for example, occupants of flats will be familiar with the layout of the premises, whereas customers in a shop may be completely unfamiliar with their surroundings. In schools the fundamental principle is the provision of an alternative means of escape, and in hospitals the main concern is with the adequacy of the means of escape from all parts of the building, as many occupants will be in poor health.

Regarding the means of escape in case of a fire, the building and its contents are of secondary importance; the primary importance is getting the occupants out of the building safely. The provision of a safe escape route should, however, allow at the same time an easy access for the fire service using the same routes, and because these routes are protected the risk of fire spread is minimised. In practice the provision of an adequate means of escape and structural fire protection of the building and its contents are virtually inseparable. Each building has to be considered as an individual exercise, but certain common factors prevail in all cases.

- An outbreak of fire does not necessarily imply the evacuation of the entire building.
- Rescue facilities of the local fire service should not be considered as part of the planning of means of escape.

- Persons should be able to reach safety without assistance when using the protected escape routes.

- All possible sources of an outbreak and the course the fire is likely to take should be examined and the escape routes planned accordingly.

Dwelling houses

Provision for escape from fire in dwellings can be categorised under three headings:

- **Category 1**: Dwelling houses of one or two storeys with floors not more than 4.500 m above ground level;

- **Category 2**: Dwelling houses with one floor more than 4.500 m above ground level;

- **Category 3**: Dwelling houses of two or more storeys with more than one floor over 4.500 m above ground level.

Category 1

With the exception of kitchens, each habitable room should open directly onto a hallway or protected stairway, which in turn connects to the property entrance door or a window through which escape can be made. Such windows should have minimum opening dimensions of 0.450 m high × 0.450 m wide and an opening area of at least 0.330 m². The lowest level of opening in the window is 0.800 m and the highest is 1.100 m above floor level. With skylights and dormers the minimum opening height above floor level can reduce to 0.600 m (see Building Regulations, Approved Document K: *Protection from falling, collision and impact*).

Inner rooms (those that have to be reached through another room) are discouraged unless they are a kitchen, laundry or utility room, dressing room or bathroom, or they have their own openable window that could be used as a means of escape.

Balconies and flat roofs may be considered a means of escape if the roof is part of the same building and the route leads to a storey exit or external escape route. The escape route should be constructed to at least a 30-minute fire-resisting standard, including any opening within 3.000 m of that escape route.

A basement stairway could be blocked by fire or smoke. Therefore, if a basement contains habitable accommodation, an alternative means for egress should be provided. This may be:

- an external door; or

- a window of dimensions specified above; or

- a protected stairway to a final exit.

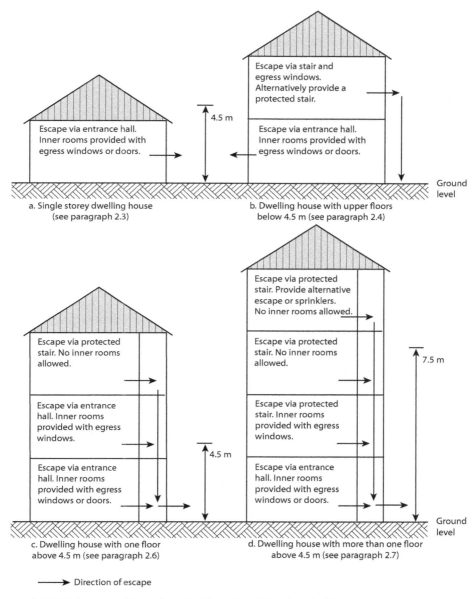

a. Single storey dwelling house
(see paragraph 2.3)

4.5 m

Escape via entrance hall.
Inner rooms provided with
egress windows or doors.

b. Dwelling house with upper floors
below 4.5 m (see paragraph 2.4)

Escape via stair and
egress windows.
Alternatively provide a
protected stair.

Escape via entrance hall.
Inner rooms provided with
egress windows or doors.

Ground
level

c. Dwelling house with one floor
above 4.5 m (see paragraph 2.6)

Escape via protected
stair. No inner rooms
allowed.

Escape via entrance
hall. Inner rooms
provided with egress
windows.

Escape via entrance
hall. Inner rooms
provided with egress
windows or doors.

4.5 m

d. Dwelling house with more than one floor
above 4.5 m (see paragraph 2.7)

Escape via protected
stair. Provide alternative
escape or sprinklers.
No inner rooms allowed.

Escape via protected
stair. No inner rooms
allowed.

Escape via protected
stair. Inner rooms
provided with egress
windows.

Escape via entrance
hall. Inner rooms
provided with egress
windows or doors.

7.5 m

Ground
level

⟶ Direction of escape

Note: This diagram must be read in conjunction with all of the relevant guidance.

Figure 5.3.1 Categories of dwelling house for fire escape purposes

Loft conversions from a two-storey house currently attract certain provisions as defined in Section 2 of Approved Document B to the Building Regulations. These are generally as described in the next category for houses with one floor more than 4.500 m above ground level. New access stairs should be in an enclosure continuing from the existing stairs or accessing the existing stairs. Depending on the age and type of construction of the house, the new floor joists will probably need structurally upgrading from their former function as ceiling ties, and the floor construction will need to achieve a 30-minute fire resistance. Escape windows from within the roof space are as previously described.

Category 2

Dwelling houses with a floor more than 4.500 m above ground level should have at least two internal stairways. Each of these stairways is to provide an effective means of escape. If it is impractical to install two stairways in a dwelling, all upper storeys should be served by a protected stairway enclosed by 30-minute fire-resisting construction. This stairway should extend directly to a final exit. Alternatively, the stairway can give access to at least two escape routes to final exits at ground level, each route separated from the other by 30-minute fire-resisting construction and fire doors.

A different approach is to separate the upper storey from the lower storeys with 30-minute fire-resisting construction, and provide the upper storey with a designated fire escape route to its own final exit.

Fire-resisting construction includes the need for fire doors. Traditionally these have self-closing devices, but current thinking indicates that, where used in dwellings, these devices can pose a hazard to children. Also, many occupants remove door closers as they regard them as an inconvenience. Therefore, guidance promotes the benefits of manually closing these doors, particularly at night.

Category 3

Dwelling houses of two or more storeys with more than one floor over 4.500 m above ground level have the same requirements as the previous category. The following additional provisions apply.

- Each storey over 7.500 m above ground level should have an alternative escape route with access via the protected stairway to an upper storey, or from a landing within the protected stairway to an alternative escape route on the same storey. At or about 7.500 m above ground level the protected stairway should be separated from the lower storeys by fire-resisting construction.

- Alternatively, the whole house should be fitted with a domestic sprinkler system. See BS 9251: *Sprinkler systems for residential and domestic occupancies. Code of practice.*

COMPARTMENTATION AND SEPARATION

Semi-detached dwellings

Every wall separating semi-detached and terraced houses should be constructed as a fire-resisting compartment wall, with each house considered as a separate building. The fire-resistance period is usually 60 minutes: see Table A2 in the Approved Document to Part B of the Building Regulations and in particular note 2 which states that 'separating walls are increased to 60 minutes fire resistance for compartment walls separating buildings'.

Attached garages

An attached or integral garage should be separated, where in contact with the rest of the house, by 30-minute fire-resisting construction. If the floor of the garage cannot be sloped towards the outside to allow for fuel spills to flow away from any internal door, then alternatively the door opening into the garage should be at least 100 mm above the garage floor level and be fitted with a self-closing FD30S fire door.

APPROVED DOCUMENT B, VOLUME 2: MEANS OF ESCAPE BUILDINGS OTHER THAN DWELLINGS

MEANS OF ESCAPE FROM FLATS

It does appear a contradiction in terms in that a flat would be considered not to be a dwelling house, but the guidance contained within Volume 2 covers multi-storey provisions, which tend to be specialised in the methods of planning escape routes.

At one time, only accommodation over 24.000 m above ground level would have been considered, this being the height over which external rescue by the fire service was impracticable. It has become apparent that, even with dwellings within reach of firefighters' ladders, external rescue is not always possible. This is because present-day traffic conditions and congestion may prevent the appliance from approaching close to the building or may delay the arrival of the fire service. Section 2 of Approved Document B1, Volume 2 to the Building Regulations specifically relates to this class of buildings and has special provision for accommodation with floors over 4.500 m above ground level.

As with other forms of buildings, the only sound basis for planning means of escape from apartments is to identify the positions of all possible sources of any outbreak of fire and to predict the likely course the fire, smoke and gases would follow. The planning should be considered in three stages:

- **Stage 1**: risk to occupants of the dwelling in which the fire originates;
- **Stage 2**: risk to occupants of adjoining dwellings should the fire or smoke penetrate the horizontal escape route or common corridor;
- **Stage 3**: risk to occupants above the level of the outbreak, particularly on the floor immediately above the source of the fire.

Stage 1

Most of the serious accidents and deaths occur in the room in which the outbreak originates. Fires in bedrooms have increased, mainly because of the increased use of electrical appliances, such as heated blankets, that have been improperly maintained or wired. Social habits, such as watching television, have enabled fires starting in other rooms to develop to a greater extent before detection. Fires occurring in the circulation spaces, such as halls and corridors, are a serious hazard, and the use of naked-flame heaters and smoking in these areas should be discouraged. Risks to occupants in maisonettes are higher than those incurred in flats because fire and smoke will spread more rapidly in the vertical direction than in the horizontal direction.

Stage 2

This is concerned mainly with the safety of occupants using the horizontal escape route, where the major aim is to ensure that, should a fire start in any one dwelling, it will not adversely affect or obstruct the escape of occupants of any other dwelling on the same or adjoining floor.

Stage 3

This is concerned with occupants using a vertical escape route, which in fact means a stairway, for in this context lifts are not considered as a means of escape because of:

■ time delay in lift answering call;

■ limited capacity of lift;

■ possible failure of the electricity supply in the event of a fire.

The main objective is to remove the risk of fire or smoke entering a stairway and rendering it impassable above that point. This objective can be achieved by taking a number of protective measures.

■ Where there is more than one stairway serving a ventilated common corridor, there should be a smoke-stop door across the corridor between the doors in the enclosing walls of the stairways to ensure that both stairs are not put at risk in the event of a fire (see Fig. 5.3.2).

■ Where there is only one stairway serving a ventilated common corridor, there must be a smoke-stop door between the door in the enclosing walls of the stairway and any door covering a potential source of fire, and the lobby so formed should be permanently ventilated through an adjoining external wall (see Fig. 5.3.3).

■ Means for ventilating common corridors are necessary because some smoke will inevitably penetrate the corridor as occupants escape through the entrance door to an affected apartment. Ventilation may be by natural or mechanical means.

Here is a list of the basic means of escape requirements for flats.

- Every flat should have a protected entrance hall of 30-minute fire resistance.
- Every living room should have an exit into the protected entrance hall.
- Bedrooms should be nearer to the entrance door than the living rooms or kitchen, unless an alternative exit is available from the bedroom areas.
- Doors opening onto a protected entrance hall should be self-closing FD30 fire doors.
- The maximum travel distance from any habitable room exit to the entrance door should be 9.000 m; if exceeded, an alternative route should be provided.
- Cooking facilities should be remote from entrance doors.

A typical flat example is shown in Fig. 5.3.3.

The basic means of escape requirements for maisonettes can be similarly listed.

- All maisonettes should have a protected entrance hall and stairway.
- All habitable rooms should have direct access to this hall or stairway.
- Doors opening on to hall or stairway should be self-closing FD30.
- The hall and stairway should have 30-minute fire-resistant walls.
- Alternatively, means of escape from each habitable room must be provided except that at entrance level: for example, exit through windows (dimensions as described for dwelling houses).

A typical maisonette example is shown in Fig. 5.3.4.

Readers are encouraged to study the many examples of means of escape planning for flats given in Section 2 of Approved Document B1, Volume 2 to the Building Regulations and BS 5588-1: *Fire precautions in the design, construction and use of buildings. Code of practice for residential buildings*.

DESIGN FOR HORIZONTAL ESCAPE – OTHER BUILDINGS

Planning for escape

Volume 2 of the Approved Document B plans for the escape of the occupants through the consideration of a number of variables:

- number of escape routes available;
- number of occupants;
- entry/exit controls;
- travel distances;
- alternative escape routes;
- provision of inner rooms.

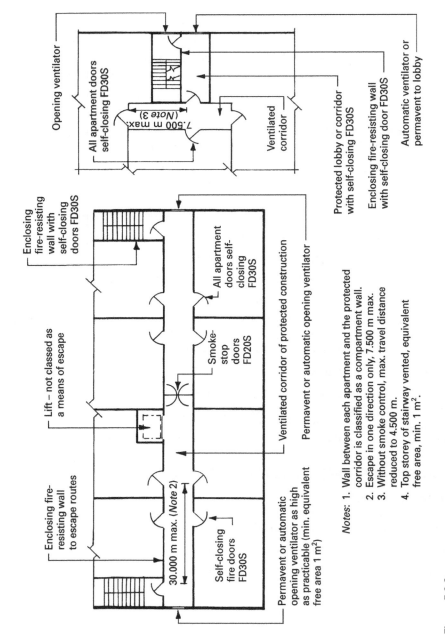

Opening ventilator

All apartment doors self-closing FD30S

7.500 m max. (Note 3)

Ventilated corridor

Protected lobby or corridor with self-closing FD30S

Enclosing fire-resisting wall with self-closing door FD30S

Automatic ventilator or permavent to lobby

Enclosing fire-resisting wall with self-closing doors FD30S

Lift – not classed as a means of escape

All apartment doors self-closing FD30S

Smoke-stop doors FD20S

Ventilated corridor of protected construction

Permavent or automatic opening ventilator

Enclosing fire-resisting wall to escape routes

30.000 m max. (Note 2)

Self-closing fire doors FD30S

Permavent or automatic opening ventilator as high as practicable (min. equivalent free area 1 m²)

Notes: 1. Wall between each apartment and the protected corridor is classified as a compartment wall.
2. Escape in one direction only, 7.500 m max.
3. Without smoke control, max. travel distance reduced to 4.500 m.
4. Top storey of stairway vented, equivalent free area, min. 1 m².

Figure 5.3.2 Main stairway protection: flats

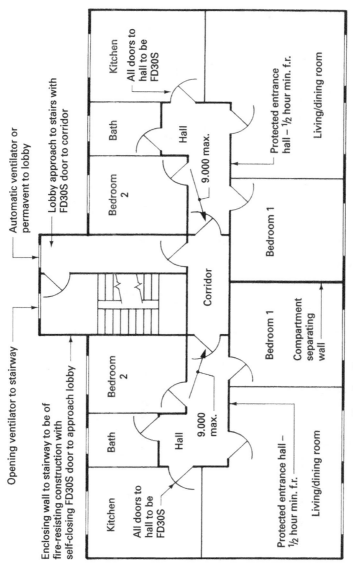

Opening ventilator to stairway

Automatic ventilator or permavent to lobby

Lobby approach to stairs with FD30S door to corridor

Enclosing wall to stairway to be of fire-resisting construction with self-closing FD30S door to approach lobby

Kitchen
All doors to hall to be FD30S

Bath

Bedroom 2

Corridor

Hall

9.000 max.

Protected entrance hall – ½ hour min. f.r.

Living/dining room

Bedroom 1

Kitchen
All doors to hall to be FD30S

Bath

Bedroom 2

Hall

9.000 max.

Bedroom 1

Compartment separating wall

Protected entrance hall – ½ hour min. f.r.

Living/dining room

Note: Dwelling entrance to lobby/corridor door 7.500 max.

Figure 5.3.3 Means of escape: example of flats with only one stairway

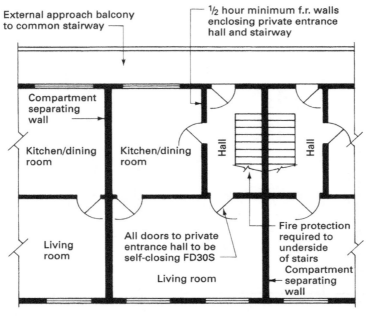

External approach balcony to common stairway

½ hour minimum f.r. walls enclosing private entrance hall and stairway

Compartment separating wall

Kitchen/dining room

Kitchen/dining room

Hall

Hall

Living room

All doors to private entrance hall to be self-closing FD30S

Fire protection required to underside of stairs

Compartment separating wall

Living room

Lower floor plan

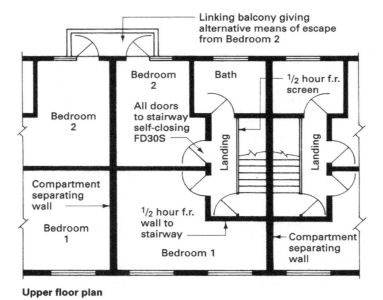

Linking balcony giving alternative means of escape from Bedroom 2

Bedroom 2

Bath

½ hour f.r. screen

Bedroom 2

All doors to stairway self-closing FD30S

Landing

Landing

Compartment separating wall

Bedroom 1

½ hour f.r. wall to stairway

Compartment separating wall

Bedroom 1

Upper floor plan

Figure 5.3.4   Means of escape: maisonette example

Number of escape routes

Best practice would always be to design for an alternative escape route from a complex building. However, Approved Document Part B makes provision for circumstances where a single escape route may exist providing that it meets certain constraints on the number of occupants and the travel distances in one direction. Part B makes a direct correlation between the number of occupants and the number of escape routes that have been provided. Reference to Table 3 gives the number of exits as one for a maximum of 60 occupants, two for a maximum of 600 occupants, and three for occupancy of over 600 people.

Access controls

Many buildings have access controls that are activated by a swipe card or key fob for occupants to gain entry into the building. These must be able to safely evacuate the occupant at times when they are in use: for example, out of hours working where fewer security staff may be on hand in case of evacuation of the building.

Travel distances

Part B places limitations on the distance that an occupant has to travel to the escape exit from the floor that they are on. This information is produced within Table 2, which contains the classification element of the building purpose or type: for example, industrial, office or institutional.

Against each of these purpose groups, Table 2 provides information on distances within two maximums. The first is where travel is in one direction only (for example, an office with only one entrance door) and the second is for travel in two directions (for example, an office that has two doors at opposite or diagonally opposing ends for entry and exit). Travel distances are therefore reduced where there is only travel in one direction. Typical travel distances for an office are 18 metres for one direction and 45 metres in two directions. Both the industrial and storage classifications are further subdivided into normal hazard and higher hazard travel distances. The higher the hazard the shorter the design travel distances.

Alternative escape routes

Great care should be made with the design of alternative escape routes. They need to be as far apart as possible to prevent the dual disablement of both fire exits in the event of a fire. To this end the regulations make provision for the formation of 45° angles between escape routes. The angle between two escape routes from the point of starting to escape should be more than 45°. If this angle is less than 45°, the routes must be separated by a fire-resistant construction.

Inner room provision

An inner room is one that has the only means of escape through another room, defined as the access room. Such inner room arrangements are only acceptable if the following conditions apply:

- the capacity of the occupants within the inner room does not exceed 60, or 30 with an institutional classification;

- the room is not a bedroom;
- the inner room has to be directly off the access room, with no corridor;
- travel distances from Table 3 must not be exceeded;
- a vision panel should be sited within the door to enable the occupants to be alerted in case of a fire outside the room or the walls left 0.50 m below ceiling level or an automatic fire-detection system operates in the access room to warn occupants within the inner room.

THE DESIGN OF ESCAPE ROUTES

The Regulations place several constraints on the design of the escape route. These ensure the safe exit of the occupants from a building which may be on fire when there may be some measure of fear and panic with occupants who want to exit as fast as possible. The major constraints that a designer must ensure are met are:

- the height of the escape route which must be not less than 2 m, including the height from the toe on a tread to the sloping ceiling within stairways;
- the width of the escape route depends upon the number of occupants that are likely to be using them (see Table 5.3.1);
- the horizontal exit capacity onto the vertical escape route has to be calculated should one of the final exits off the horizontal storey become blocked.

Table 5.3.1   **Width of escape routes**

Max. number of persons	Max. width (mm)
60	750
110	860
220	1,060
+220	+5 per person

Calculation for final horizontal exit

The following formula is used to calculate the width of the final exit in metres:

$$W = \frac{((N/2.5) + (60 \times S))}{80}$$

where W = width of final exit
N = number of people

DESIGN FOR VERTICAL ESCAPE – BUILDINGS OTHER THAN FLATS

Vertical escape can cover several different methods such as escape lifts, external and internal stairways, and refuge areas. Normally it is the use of internal stairways which is the principle escape provision within a multi-storey structure. The design for vertical escape routes is covered within the Regulations under several headings:

- the protection of the escape stairway;
- number of escape stairs;
- width of escape stairs;
- use of refuges;
- escape from complex buildings.

NUMBER OF ESCAPE STAIRS

The number of escape stairs that are required will be affected by a number of factors:

- the total number of occupants in the building;
- the final horizontal exit design;
- a requirement for access for different occupants within an institution;
- provision within the design of a single stair;
- sufficient design width availability;
- provision for access by firefighting services.

The above factors can be considered in the following example. A retail shop development at ground and first-floor level has flat accommodation within the mansard roof space. Separate stairs for the flats and first-floor shop access must be provided to remove the use of a common stairway, which could become blocked in case of a fire.

Single escape stairs are allowed under the Regulations in the following situations:

- from a basement that has a protected stairway and a window provision for means of escape and meets the minimum travel distances;
- where the storeys above the basement do not exceed 11 metres at floor level and they each have separate escape routes (except small premises);
- where the building is single occupancy and no more than three storeys including the basement, with no floor greater than 280 m^2;
- in a five-storey office building complying with the travel distances and additional escape routes where the floor level exceeds 11 metres;
- where the use is for a factory no more than two storeys high;
- in a process plant building with occupant levels at no more than 10 people.

USE OF REFUGES

A refuge is a safe waiting area for a short period of time. A refuge can be an extended landing on a stairwell, a purpose-built fire compartment or an area that is open to the air, such as a roof top or balcony. The Building Regulations make use of compartmentation of one area into two where, by default, the area unaffected by fire becomes a refuge area. A refuge area should be clearly labelled by a safety sign. Each refuge area should be provided with a means of communication to those that will instigate the evacuation of a building or the firefighting services.

WIDTH OF ESCAPE STAIRS

In the design of the vertical escape stairway there are several considerations that must be taken into account.

1. The width of any horizontal exit entering the stairway must be matched by the vertical element and not reduce in width.

2. If the height is more than 30 m, then they must not exceed 1,400 mm in width.

3. They must not reduce in width at any point towards the final exit.

4. They must conform to the requirements of Table 6 Part B, namely:

   a. institutional building – max 150 people – width 1,000 mm;

   b. assembly building – max 220 people – width 1,100 mm;

   c. other buildings occupancy over 50 or over 220 people – Tables 7 and 8;

   d. any other stair – max 50 people – width 800 mm.

The Building Regulations Part B considers whether the evacuation of a building during a fire would be undertaken either as a phased event or simultaneously. This obviously has a direct bearing on the number of occupants using the escape stairway at any one time.

Table 7 of Part B Volume 2 gives an indication of the total number of persons that can be simultaneously evacuated for different stair widths and storeys. As an alternative to the table, the Regulations allow the calculation of the stair widths over 1,100 mm using the following formula:

$$P = 200w + 50 (w-0.3) (n-1), \text{ or}$$

$$w = \frac{P + 15n - 15}{150 + 50n}$$

where P = number of people using the stair
w = the width of the stair
n = number of storeys.

For example, in a 15-storey building with 1,000 people, $P = 1,000$ and $n = 15$. Therefore, using the formula:

$$
\begin{aligned}
1,000 &= 200w + 50(w - 0.3)(15 - 1) \\
&= 200w + (50w - 15)(14) \\
&= 200w + 700w - 210 \\
&= 900w - 210 \\
\therefore w &= 1,210/900 = 1.345 \text{ m}
\end{aligned}
$$

Table 8 of Part B Volume 2 gives an indication of the minimum width of the stairs designed for phased evacuation and is based on the maximum number of people in any storey.

A typical example of means of escape from an office building is shown in Fig. 5.3.5.

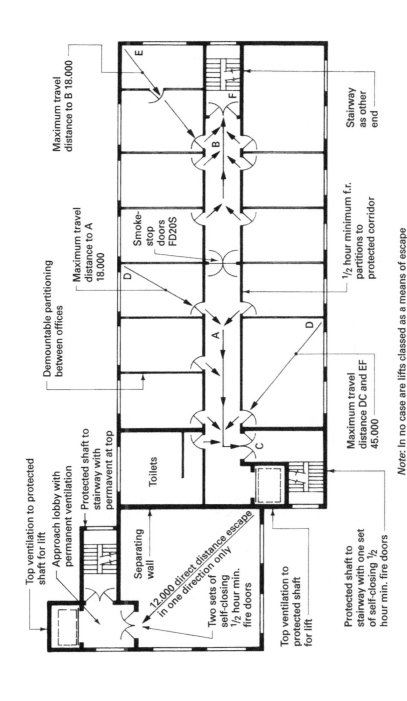

Maximum travel distance to B 18.000

Maximum travel distance to A 18.000

Demountable partitioning between offices

Smoke-stop doors FD20S

½ hour minimum f.r. partitions to protected corridor

Stairway as other end

E

D

B

D

A

F

C

Note: In no case are lifts classed as a means of escape

Maximum travel distance DC and EF 45.000

Top ventilation to protected shaft for lift

Approach lobby with permanent ventilation

Protected shaft to stairway with permavent at top

Toilets

Separating wall

12.000 direct distance escape in one direction only

Two sets of self-closing ½ hour min. fire doors

Top ventilation to protected shaft for lift

Protected shaft to stairway with one set of self-closing ½ hour min. fire doors

Figure 5.3.5 Means of escape: office buildings – upper floor level < 5.000 m above ground

ESCAPE FROM COMPLEX BUILDINGS

BS 9999: *Code of practice for fire safety in the design, management and use of buildings* covers shops, sales areas and other similar types of premises, such as restaurants and bars. The main objectives of the code's requirements are to provide safety from fire by means of risk planning and providing protection of both horizontal and vertical escape routes for any area threatened by fire, thus enabling any person confronted by an outbreak of fire to make an unassisted escape.

BS 9999 covers the approach of the designer on three levels: general, advanced and fire safety engineering. It complements Building Regulations Approved Document B at the general level. This standard examines management of fire safety, means of escape, escape protection and the provision of fire-fighting facilities in buildings such as theatres, offices, shopping complexes and bars.

The standard uses risk profiling to consider the rate of fire spread, the type of occupant, whether they are familiar with the building and finally if they are asleep within the building. This produces a risk rating, which is then used in several guidance tables.

The standard considers the minimum package of fire protection measures that will aid and enhance escape from a complex building, including:

- the installation of a fire alarm system with heat and smoke detection that may incorporate automatic closing measures for doors and compartmentation of zones;
- the use of emergency lighting and signage;
- the operation, opening and fastenings of doors;
- the final exit design from the building.

The horizontal means of escape from a complex building is considered under two areas: an assessment of the maximum numbers of occupants likely to be in the building, and the minimum number of escape routes (for example, 600-plus occupants should have three escape routes available to them).

Small shops with floor areas of not more than 280 m² in one occupancy and of not more than two storeys plus a basement are treated separately. The maximum travel or direct distance is governed by the floor's position within the building. Basement and upper floor distances are 18.000 m and DD 12.000 m for single exits, with ground-floor distances being 27.000 m and DD 18.000 m respectively (see Fig. 5.3.6). Where more than one exit exists, the distances are 45.000 m and DD 30.000 m respectively. Note direct distances (DD) are 2/3 of the maximum travel values within Table 5 of Part B.

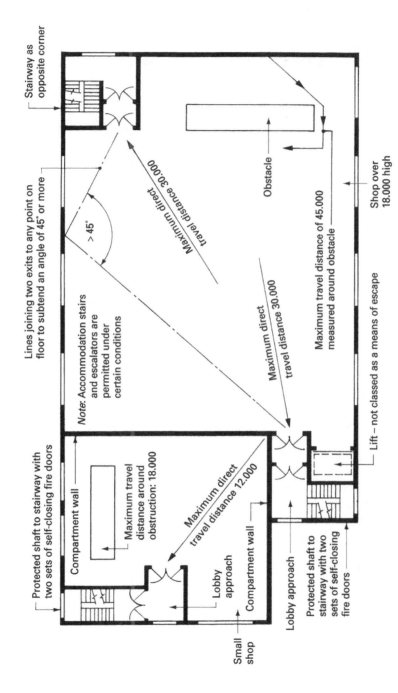

Figure 5.3.6 Means of escape: small and large shops – upper floor example

Labels within figure:

- Stairway as opposite corner
- Lines joining two exits to any point on floor to subtend an angle of 45° or more
- > 45°
- Maximum direct travel distance 30.000
- Obstacle
- Maximum direct travel distance 30.000
- Maximum travel distance of 45.000 measured around obstacle
- Shop over 18.000 high
- Lift – not classed as a means of escape
- Note: Accommodation stairs and escalators are permitted under certain conditions
- Protected shaft to stairway with two sets of self-closing fire doors
- Compartment wall
- Maximum travel distance around obstruction: 18.000
- Maximum direct travel distance 12.000
- Lobby approach
- Compartment wall
- Lobby approach
- Protected shaft to stairway with two sets of self-closing fire doors
- Small shop

These should be of non-combustible materials and continuous, leading ultimately to the final exit door to the place of safety. The recommended dimensions of going, rise, handrails and maximum number of risers per flight are shown in Fig. 5.3.7. Fire-resistant glazing is permitted but should be restricted to the portion of the wall above the handrails and preferably designed in accordance with the recommendations of BS 476-22.

DOORS

Factors relating to the integrity, insulation and stability of doors in the presence of fire are considered in some detail in Chapter 8.1 of the accompanying volume, *Construction Technology*, and in Chapter 5.2 of this volume. Specific reference is also made to BS 476-22: *Methods for determination of fire resistance of non-loadbearing elements of construction*. See also Fig. 5.3.8 for a typical example of a fire door with variations to achieve FD30 and FD60 ratings.

SMOKE-STOP DOORS

The function of this form of door is obvious from its title, and no special requirements are recommended as to door thickness or area of glazing when using 6-mm wired glass. The most vulnerable point for the passage of smoke is around the edges, and therefore a maximum gap of 3 mm is usually specified, together with a brush draught-excluder seal.

AUTOMATIC FIRE DOORS

In compartmented industrial buildings it is not always convenient to keep the fire doors in the closed position as recommended. A closed door in such a situation may impede the flow and circulation of people and materials, thus slowing down or interrupting the production process. To seal the openings in the compartment walls in the event of a fire, automatic fire doors or shutters can be used. These can be held in the open position in normal circumstances by counterbalance weights or electromagnetic devices, and should a fire occur they will close automatically, usually by gravitational forces. The controlling device can be a simple, fusible link, with a smoke or heat detector, which could be linked to the fire alarm system. A typical detail of an automatic sliding fire door is shown in Fig. 5.3.9.

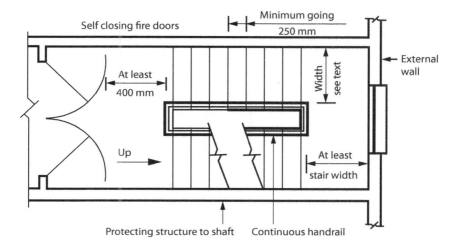

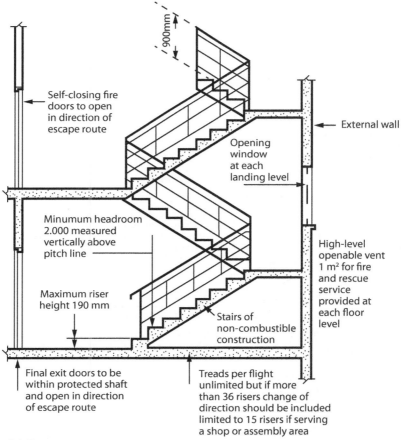

Figure 5.3.7   Typical escape stairway details for single stairs

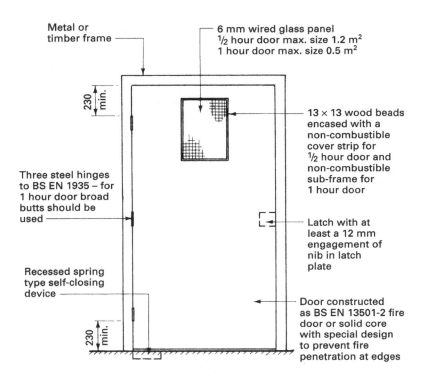

Metal or timber frame

6 mm wired glass panel
½ hour door max. size 1.2 m²
1 hour door max. size 0.5 m²

230 min.

13 × 13 wood beads encased with a non-combustible cover strip for ½ hour door and non-combustible sub-frame for 1 hour door

Three steel hinges to BS EN 1935 – for 1 hour door broad butts should be used

Latch with at least a 12 mm engagement of nib in latch plate

Recessed spring type self-closing device

230 min.

Door constructed as BS EN 13501-2 fire door or solid core with special design to prevent fire penetration at edges

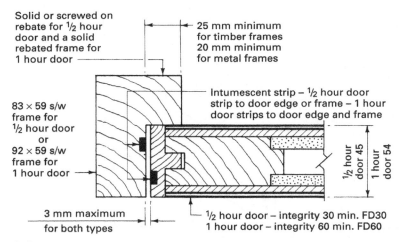

Solid or screwed on rebate for ½ hour door and a solid rebated frame for 1 hour door

25 mm minimum for timber frames
20 mm minimum for metal frames

83 × 59 s/w frame for ½ hour door or 92 × 59 s/w frame for 1 hour door

Intumescent strip – ½ hour door strip to door edge or frame – 1 hour door strips to door edge and frame

½ hour door 45   1 hour door 54

3 mm maximum for both types

½ hour door – integrity 30 min. FD30
1 hour door – integrity 60 min. FD60

References:
BS EN 1935: *Building hardware. Single axis hinges.*
BS EN 13501-2: *Fire classification of construction products and building elements.*

Figure 5.3.8   Fire doors

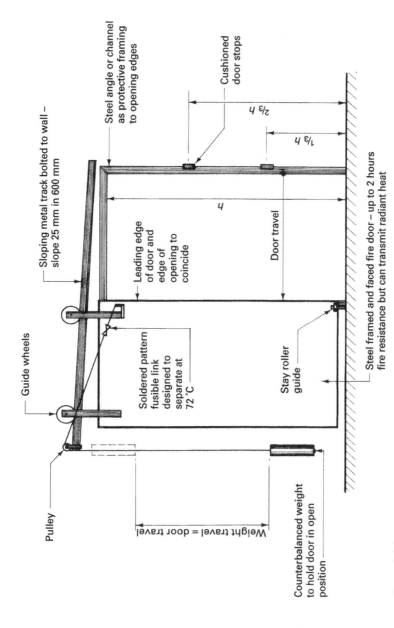

Figure 5.3.9 Automatic sliding fire door details

Steel angle or channel as protective framing to opening edges

Sloping metal track bolted to wall – slope 25 mm in 600 mm

Cushioned door stops

$^2/_3$ h

$^1/_3$ h

h

Door travel

Leading edge of door and edge of opening to coincide

Guide wheels

Soldered pattern fusible link designed to separate at 72 °C

Stay roller guide

Steel framed and faced fire door – up to 2 hours fire resistance but can transmit radiant heat

Pulley

Weight travel = door travel

Counterbalanced weight to hold door in open position

The traditional means of keeping escape stairways clear of smoke during a fire by having access lobbies and/or natural ventilations are not always acceptable in modern designs where, for example, the stairways are situated in the core of the building or where the building has a full air conditioning and ventilation design. Pressurisation is a method devised to prevent smoke logging in an unfenestrated stairway by a system of continuous pressurisation, which will keep the stairway clear of smoke if the doors remain closed. Only a small fan installation is required to maintain a nominal pressure of not less than 50 Pa. In addition to controlling smoke, pressurisation also increases the fire resistance of doors opening into the protected area. It is essential with these installations that the doors are well sealed to prevent pressure loss.

EXTERNAL ESCAPE STAIRS

THE REGULATORY REFORM (FIRE SAFETY) ORDER 2005

This relatively new piece of legislation was brought in to reform some of the vast amount of fire safety legislation and bring it under one umbrella. It has removed the issuing of fire certificates for certain buildings and has placed the responsibility on the building owner, who has to assess the risks associated with operating their property and act accordingly within the regulations. The Reform Order covers several important areas, which we will examine in detail for each relevant regulation.

Regulation 8 *Duty to take general fire precautions*
This regulation uses the term 'responsible person'. This must be a person who could be the employer or owner, or a person who has control over the premises, business, trade or other undertaking. This makes the responsible person ensure that general fire precautions are undertaken for the safety of all employees and the general public.

Regulation 9 *Risk assessment*
This is the key element of the regulations concerning assessing the risk of fire to the building or premises, or the presence of dangerous substances. The responsible person must make a 'suitable and sufficient' assessment of the risks to which persons are exposed, to identify the fire precautions that will be required to control the hazard identified. Any dangerous substances must be controlled or the risk from them must be eliminated. Conditions to this effect are covered with Regulation 11.

Regulation 10 *Principles of prevention*
The methods of prevention that should be followed in relation to the risk are avoidance, evaluation, combatting at source, substitution, policy, collective protection and instructions and training to employees.

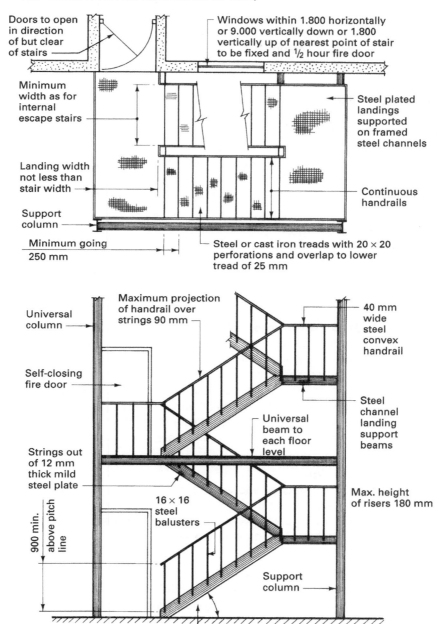

Note: Protective roof and sides omitted for clarity

Doors to open in direction of but clear of stairs

Windows within 1.800 horizontally or 9.000 vertically down or 1.800 vertically up of nearest point of stair to be fixed and ½ hour fire door

Minimum width as for internal escape stairs

Steel plated landings supported on framed steel channels

Landing width not less than stair width

Continuous handrails

Support column

Minimum going 250 mm

Steel or cast iron treads with 20 × 20 perforations and overlap to lower tread of 25 mm

Universal column

Maximum projection of handrail over strings 90 mm

40 mm wide steel convex handrail

Self-closing fire door

Steel channel landing support beams

Universal beam to each floor level

Strings out of 12 mm thick mild steel plate

Max. height of risers 180 mm

900 min. above pitch line

16 × 16 steel balusters

Support column

(See also Figs 12.2.1 and 12.2.2)

Maximum pitch limited by twice rise + going total between 550 and 700 mm

Figure 5.3.10 Typical external steel escape stairway details

Regulation 13 *Firefighting and detection*

Under this regulation the responsible person has to ensure that adequate fire-fighting equipment, alarms and fire detectors are provided in order to protect the safety of persons on the premises in the event of a fire. This may involve the training of nominated and competent employees on the use of firefighting equipment, and developing lines of communication with the emergency services.

Regulation 14 *Escape routes and exits*

The responsible person must ensure that the building's escape routes lead to a place of safety and are kept clear and free from obstructions at all times. The number of escape exits, their position and size must be balanced against the number of occupants using the building. Any door on the escape route must open outwards towards the final exit. No emergency door must be kept locked and they must be clearly indicated with a sign: 'Emergency exit keep clear'.

Regulation 15 *Procedures*

This regulation covers such aspects as fire and safety drills in the building or premises, appointment of competent persons to assist in this requirement, and the procedures to deal with the evacuation of persons from an area of serious and imminent danger.

Regulations 19/20 *Provision of information*

This places a duty to inform all employees of the risks that have been identified, the preventative and protective measures and the procedures for a fire evacuation. The competent persons must be made known to the employees. Where a dangerous substance is present in the place of work, this must be made known to the employees along with the procedures for dealing with the hazards associated with the substance. This information must also be conveyed to the self-employed and any employers from outside undertakings.

Regulation 21 *Training*

Training is essential at induction and when the risk changes, along with the introduction of new equipment or technology. This regulation ensures that adequate training is given to all new and existing employees and is repeated where required and takes place during working hours.

Regulation 23 *Employees' duties*

These duties are very similar to those of the HASAWA 1974 in ensuring that employees take care of themselves and others who might be affected, to cooperate with an employer and inform if any equipment or provision is found to be defective.

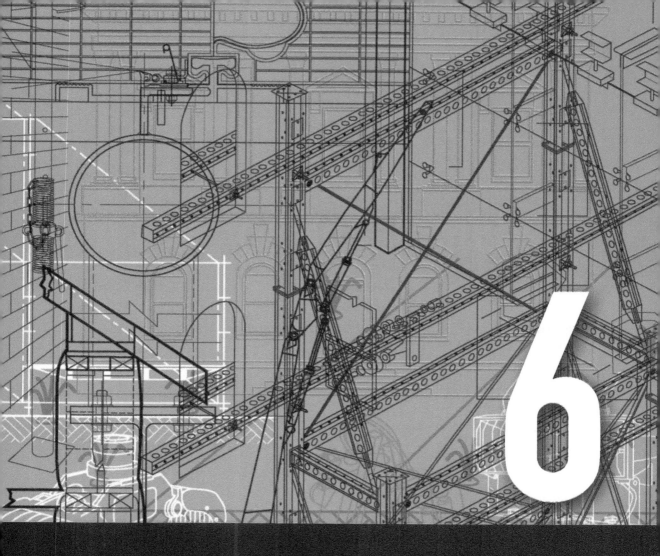

6

Claddings to framed structures

Cladding panels   6.1

Claddings are a form of masking or infilling a structural frame, and can be considered under the following headings:

- façade system cladding;
- precast concrete cladding panels;
- light infill panels;
- glazed curtain walling systems.

All forms of cladding should:

- be able to self-support between the framing members;
- resist rain penetration;
- resist both positive and negative wind pressures;
- resist wind penetration;
- provide adequate thermal insulation;
- resist sound transmission;
- provide fire resistance;
- enable entry of natural daylight and ventilation;
- be constructed to a suitable size for handling and placing;
- provide durability with limited maintenance;
- be aesthetically pleasing.

Over the following subsections we will examine the different types of cladding and the variety of materials that are used to produce the modern commercial building. The new façade systems will be detailed that include the extensive use of composite insulated panels and tiled systems fixed to insulated panels that mimic brickwork, or provide support to the tile on a rail finish fixed to the outside of the insulated panel.

The variety of systems to complete the building envelope is now so extensive they cannot be readily categorised. The following subsections try to place the described system within the most appropriate classification, as it would be generally known and illustrated through manufacturers' details and literature.

With the advances in materials technology the variety of materials that can be used is endless, from glass tiles, to aluminium, stone, steel mesh and sustainable timber claddings; the combinations available to the designer are limitless.

PRECAST CONCRETE CLADDING PANELS

These are factory-produced quality units with a finished face to the client's requirements, in a storey-height or undersill panel format. The storey-height panel is designed to span vertically from beam to beam and, if constructed to a narrow module, will give the illusion of a tall building. Undersill panels span horizontally from column to column and are used where a high wall to window ratio is required. Combinations of both formats are also possible.

Concrete cladding panels should be constructed of a dense concrete mix and suitably reinforced with bar reinforcement or steel-welded fabric. The reinforcement should provide the necessary tensile resistance to the stresses induced when hung in the final position, and to the stresses set up during transportation and the final hoisting into position. A method should be incorporated into the design to ensure that the panels are hoisted in the correct manner, so that unwanted stresses are not induced. This may require the use of a specialist lifting beam attached to the crane. The usual specification for cover of concrete over reinforcement is 25 mm minimum. If thin panels are being used, the use of galvanised or stainless steel reinforcement should be considered to reduce the risk of corrosion and the eventual spalling of the finished face.

When designing or selecting a panel, the following must be taken into account:

- column or beam spacing, as this will determine the span of the panels;
- the weight of the finished panel, as this will determine the type of cranage required;
- the method of jointing panel to panel;
- the degree of exposure conditions;
- any special requirements as to finish, texture or bonded facing elements.

The greatest problem facing the designer and installer of concrete panels is how to allow for structural and thermal movements and at the same time provide an adequate long-term joint (see the following sections on jointing, sealing and gaskets). Typical examples of storey-height and undersill panels are shown in Figs 6.1.1 and 6.1.2.

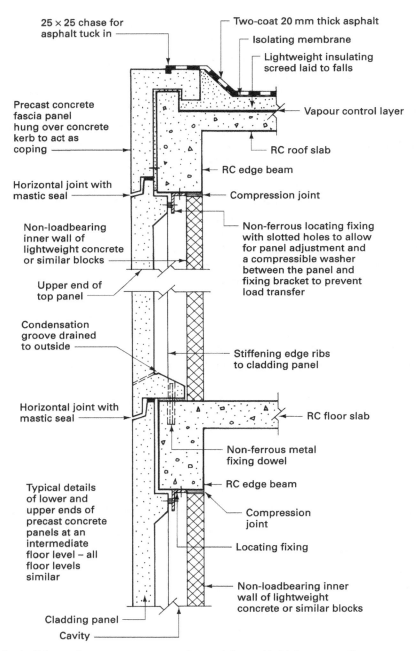

25 × 25 chase for asphalt tuck in

Two-coat 20 mm thick asphalt

Isolating membrane

Lightweight insulating screed laid to falls

Precast concrete fascia panel hung over concrete kerb to act as coping

Vapour control layer

RC roof slab

RC edge beam

Horizontal joint with mastic seal

Compression joint

Non-ferrous locating fixing with slotted holes to allow for panel adjustment and a compressible washer between the panel and fixing bracket to prevent load transfer

Non-loadbearing inner wall of lightweight concrete or similar blocks

Upper end of top panel

Condensation groove drained to outside

Stiffening edge ribs to cladding panel

Horizontal joint with mastic seal

RC floor slab

Non-ferrous metal fixing dowel

RC edge beam

Compression joint

Locating fixing

Typical details of lower and upper ends of precast concrete panels at an intermediate floor level – all floor levels similar

Non-loadbearing inner wall of lightweight concrete or similar blocks

Cladding panel

Cavity

Note: For buildings where energy conservation and thermal bridging prevention measures are required (UK – see Building Regulation L), a supplementary thickness of insulation and inner walling can extend over the edge beam. Additional insulation is applied to the soffit of the RC roof/floor slabs. Cavity insulation may also be considered.

Figure 6.1.1 **Typical storey-height concrete cladding panel**

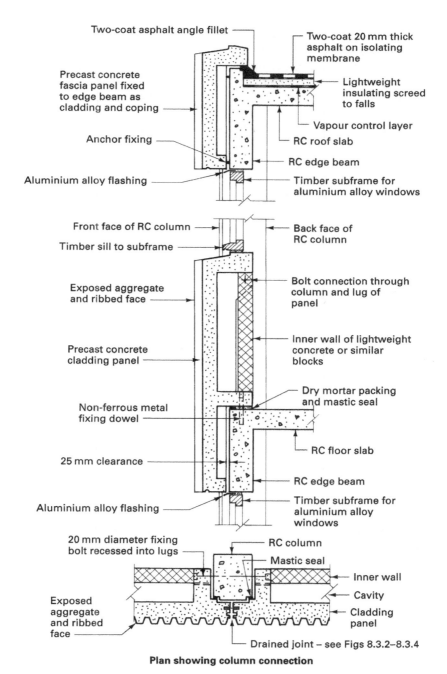

Two-coat asphalt angle fillet

Two-coat 20 mm thick asphalt on isolating membrane

Precast concrete fascia panel fixed to edge beam as cladding and coping

Lightweight insulating screed to falls

Anchor fixing

Vapour control layer

RC roof slab

Aluminium alloy flashing

RC edge beam

Timber subframe for aluminium alloy windows

Front face of RC column

Back face of RC column

Timber sill to subframe

Exposed aggregate and ribbed face

Bolt connection through column and lug of panel

Precast concrete cladding panel

Inner wall of lightweight concrete or similar blocks

Non-ferrous metal fixing dowel

Dry mortar packing and mastic seal

25 mm clearance

RC floor slab

Aluminium alloy flashing

RC edge beam

Timber subframe for aluminium alloy windows

20 mm diameter fixing bolt recessed into lugs

RC column

Mastic seal

Inner wall

Cavity

Exposed aggregate and ribbed face

Cladding panel

Drained joint – see Figs 8.3.2–8.3.4

Plan showing column connection

Figure 6.1.2 Typical undersill concrete cladding panel (see note to Fig. 6.1.1)

Precast concrete panels can be finished in a variety of secondary finishes from stone or brick slips, to a textured, exposed aggregate finish. As such they are a composite panel that has to bond the wet concrete process to the second external finish. One example is the use of a reconstituted stone-veneer composite panel. These panels have the strength and reliability of precast concrete panel design and manufacture, but the appearance of traditional stonework. This is achieved by casting a concrete backing to a suitably keyed natural or reconstituted stone facing and fixing it to the frame by traditional masonry fixing cramps or by conventional fixings (see Fig. 6.1.3).

Thermal insulation can be achieved when using precast concrete panels by creating a cavity, as shown in Figs 6.1.1 and 6.1.2. However, continuity of insulation will require an insulative material lining with supplementary blockwork extending over the face of the edge beams to coincide with mineral wool (or similar) insulation slabs to the roof/floor slab soffit. These slabs can be faced with plasterboard containing an integral vapour-control layer. It may also be possible to incorporate insulation in a sandwich cladding panel.

Concrete cladding panels can be large and consequently heavy. To reduce the weight they are often designed to be relatively thin (50 to 75 mm) across the centre portion and stiffened around the edges with suitably reinforced ribs, which usually occur on the back face but can be positioned on the front face as a feature, which can also limit the amount of water that can enter the joint.

Further relevant reading includes the Building Research Establishment's (BRE) Cladding Pack (2005). This is a collection taken from their BRE Digests, Information Papers and Good Building Guides, which include:

- *Insulated external cladding systems* (GG 31);
- *Metal cladding: assessing the thermal performance of built-up systems which use 'Z' spacers* (IP 10/02);
- *Reinforced plastics cladding panels* (DG 161);
- *Wall cladding defects and their diagnosis* (DG 217);
- *Wall cladding: Designing to minimise defects due to inaccuracies and movements* (DG 223).

The BRE also publishes *A cladding bibliography of selected publications*: BRE Report 194.

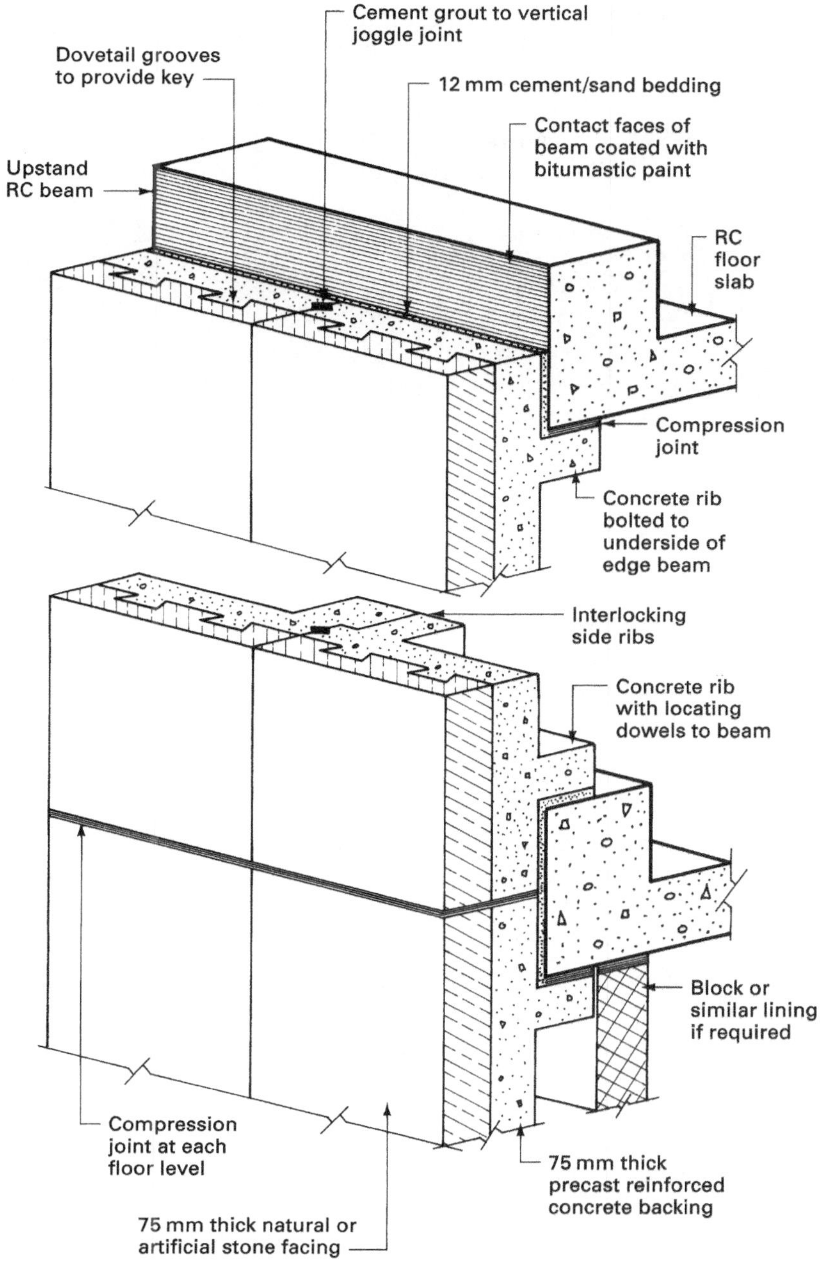
Cement grout to vertical joggle joint
Dovetail grooves to provide key
12 mm cement/sand bedding
Contact faces of beam coated with bitumastic paint
Upstand RC beam
RC floor slab
Compression joint
Concrete rib bolted to underside of edge beam
Interlocking side ribs
Concrete rib with locating dowels to beam
Block or similar lining if required
Compression joint at each floor level
75 mm thick precast reinforced concrete backing
75 mm thick natural or artificial stone facing

Figure 6.1.3 Typical storey-height composite panel

Infill panels 6.2

As its name suggests, an infill panel (sometimes referred to as a 'spandrel panel') sits between the supporting structural elements of a building and as such does not carry any of the structural loads from the building. Infill panels are lightweight and usually glazed to provide natural daylighting into the internal space of the building. The layout of the infill panels can be configured to expose some or all of the structural members, creating various aesthetical impressions for the viewer. For example, if horizontal panels are used, leaving only the beams exposed, an illusion of extra length and/or reduced height can be created (see Fig. 6.2.1). These practices are typical of many existing structures and continue to be appropriate where thermal insulation and thermal bridging are not a design priority, such as in car parks. Current energy conservation requirements under Part L of the Building Regulations provide an opportunity to over-clad exposed structural members with a superficial cladding when refurbishing and updating these buildings and increasing the level of insulation (see Chapter 6.7 on rainscreen cladding).

A wide variety of materials or combinations of materials can be employed, such as timber, steel, aluminium or plastic. Double-glazing techniques can be used to achieve the desired sound or thermal insulation. The glazing module should be such that a reasonable thickness of glass can be specified: for example, toughened if the windowsill is below 800 mm.

The design of the 'solid' panel is of great importance because this panel must provide the necessary resistance to fire, heat loss, sound penetration and interstitial condensation, in order to meet the requirements of the Building Regulations for a commercial building. Most of these panels are of composite or sandwich construction, as shown in Figs 6.2.2 and 6.2.3. The extent and type of glazing, plus the thickness of insulating quilt in the lower panels, will depend on local climatic conditions and associated legislation (see note to Fig. 6.2.2).

The main issue with infill lightweight panelling is the junction with the structural frame, and the allowance for moisture or thermal movement, which is usually accommodated by using a suitable mastic or sealant with a backing medium (see Chapter 6.5). Masonry infill panels of contemporary cavity construction can be used to preserve traditional features and to attain high standards of sound and thermal insulation and fire protection. Masonry walls are tied to the structural frame with wall ties cast into the concrete columns or with a purpose-made cladding support system. However, the method of fixing must accommodate any thermal movement between the different materials that make up the structural frame and the masonry infill panel. Figures 6.2.4 and 6.2.5 show the principles of application. Where infill panels consist of one skin of brickwork coupled with a drylined system, then vertical movement ties will be required between the brickwork and any intermediate supporting structure for the wall to accommodate thermal expansion and contraction.

Most infill panels are supplied as a manufacturer's modular system, because purpose-made panels can be uneconomic. However, whichever method is chosen the design aims remain constant: that is, to provide a panel that fulfils all the required functions and has a low long-term maintenance factor. Note that many of the essentially curtain-walling systems are adaptable as infill panels, which gives the designer a wide range of systems from which to select the most suitable.

One of the maintenance problems encountered with infill panels, and probably to a lesser extent with the concrete claddings, is the cleaning of the façade and in particular the glazing. All buildings collect dirt, the effects of which can vary with the material: concrete and masonry tend to accept dirt and weather naturally, whereas impervious materials such as metals and glass do not accept dirt and can corrode or become less efficient. The design of the panels with regard to the method in which water is thrown off the building face will have an effect on staining and areas of dirt collection as rainwater washes down the face of the building, carrying with it collected dirt particles.

If glass is allowed to become coated with dirt its visual appearance is less acceptable, its optical performance lessens because clarity of vision is reduced and the useful penetration of natural daylight diminishes. There are two solutions to this problem. The first is traditional window cleaning, which, on a multi-storey structure, may require a specialist cleaning contractor. The second method is the use of 'self-cleaning glass'. The technological advances in glass manufacture now enable organic particles to react with a special coating, which breaks these down. Further advances in the surface finish of the glass break the surface tension on droplets of water, spreading these out so they wash away any dirt particles.

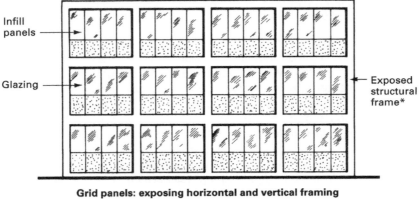

Grid panels: exposing horizontal and vertical framing

Infill panels

Glazing

Exposed structural frame*

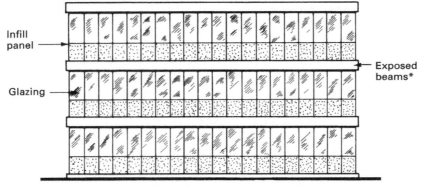

Horizontal panels: beams exposed to create illusion of length

Infill panel

Glazing

Exposed beams*

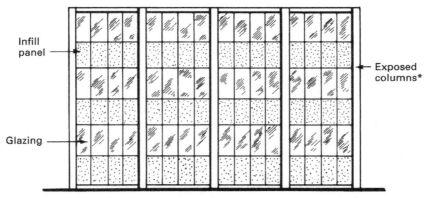

Vertical panels: columns exposed to create illusion of height

Infill panel

Glazing

Exposed columns*

*See introductory text on previous page.

Figure 6.2.1   Typical infill panel arrangements

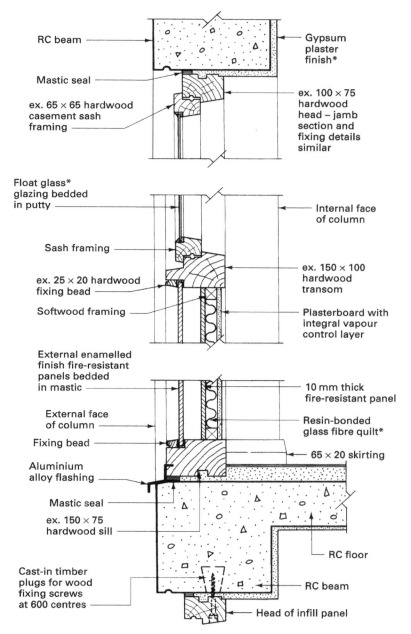

RC beam

Mastic seal

ex. 65 × 65 hardwood casement sash framing

Gypsum plaster finish*

ex. 100 × 75 hardwood head – jamb section and fixing details similar

Float glass* glazing bedded in putty

Sash framing

ex. 25 × 20 hardwood fixing bead

Softwood framing

Internal face of column

ex. 150 × 100 hardwood transom

Plasterboard with integral vapour control layer

External enamelled finish fire-resistant panels bedded in mastic

External face of column

Fixing bead

Aluminium alloy flashing

Mastic seal

ex. 150 × 75 hardwood sill

Cast-in timber plugs for wood fixing screws at 600 centres

10 mm thick fire-resistant panel

Resin-bonded glass fibre quilt*

65 × 20 skirting

RC floor

RC beam

Head of infill panel

*Note: Construction shown can be upgraded to improve thermal insulation standards by applying thermal laminate gypsum plasterboard to interior of exposed RC beams and underside of the RC floor, installing double glazing (see Construction Technology) and an insulated screed, and increasing the thickness of the lower panel to contain more glass-fibre insulant.

Figure 6.2.2   Typical timber infill panel details

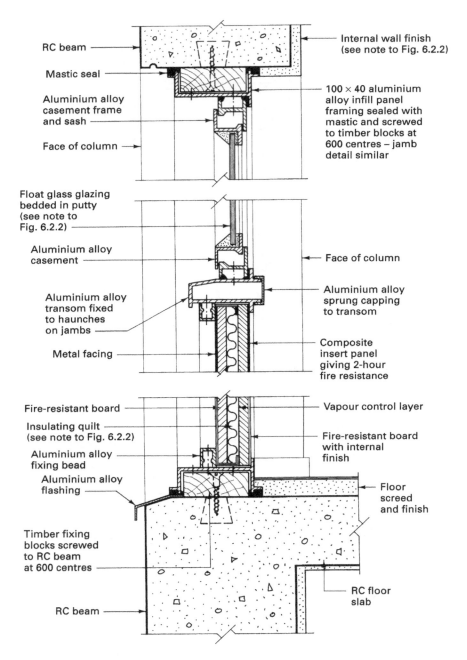

RC beam

Mastic seal

Aluminium alloy casement frame and sash

Face of column

Internal wall finish (see note to Fig. 6.2.2)

100 × 40 aluminium alloy infill panel framing sealed with mastic and screwed to timber blocks at 600 centres – jamb detail similar

Float glass glazing bedded in putty (see note to Fig. 6.2.2)

Aluminium alloy casement

Aluminium alloy transom fixed to haunches on jambs

Metal facing

Face of column

Aluminium alloy sprung capping to transom

Composite insert panel giving 2-hour fire resistance

Fire-resistant board

Insulating quilt (see note to Fig. 6.2.2)

Aluminium alloy fixing bead

Aluminium alloy flashing

Vapour control layer

Fire-resistant board with internal finish

Floor screed and finish

Timber fixing blocks screwed to RC beam at 600 centres

RC beam

RC floor slab

Figure 6.2.3   Typical metal infill panel details

The traditional method of cleaning windows will require access, which can be external or internal. Windows at ground level present no access problems and present only the question of choice of method such as hand cloths or telescopic poles with squeegee heads. By using a pole, first-floor two-storey building windows can be adequately cleaned. Low- and medium-rise structures can be reached by ladders or a mobile scaffold tower and usually present very few problems. However, high-rise structures need careful consideration. External access to windows is gained by using a cradle suspended from roof level. The usual method is to employ permanent support systems, which are incorporated as part of the building design, and are more efficient. These consist of a track on which a mobile trolley is mounted and from which davit arms can be projected beyond the roof edge to support the cradle; on completion of the cleaning work, the whole set-up can be stored away on the roof and is not visible. A single track fixed in front of the roof edge could also be considered; this is simple and reasonably efficient, but the rail is always visible and can therefore mar the building's appearance.

Internal access for cleaning the external glass face can be achieved by using windows such as reversible, horizontal and vertical sliding sashes, but designers are restricted in their choice to the reach possible by the average person. It cannot be over-emphasised that such windows can be a very dangerous hazard, unless they are carefully designed so that all parts of the glazed area can be reached by the person cleaning the windows while standing firmly on the floor. Specific anchorage points may have to be provided internally if the operative has to use ladders to reach the higher parts of a window, or lean externally to clean while the window is open, so that they can utilise a protective harness. With any method of cleaning the windows, the method of rescue in the event of mechanical or human failure must be carefully considered.

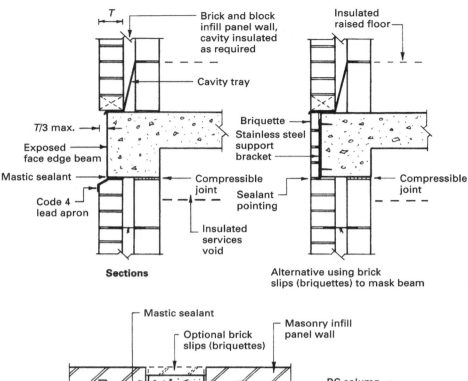

T	Brick and block infill panel wall, cavity insulated as required
	Cavity tray
T/3 max.	Briquette
Exposed face edge beam	Stainless steel support bracket
Mastic sealant	Compressible joint
Code 4 lead apron	Sealant pointing
Insulated services void	Compressible joint

Insulated raised floor

Sections

Alternative using brick slips (briquettes) to mask beam

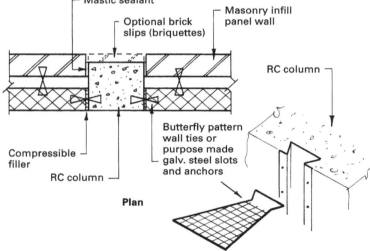

Mastic sealant

Optional brick slips (briquettes)

Masonry infill panel wall

RC column

Compressible filler

RC column

Butterfly pattern wall ties or purpose made galv. steel slots and anchors

Plan

Figure 6.2.4   **Non-loadbearing masonry infill panel**

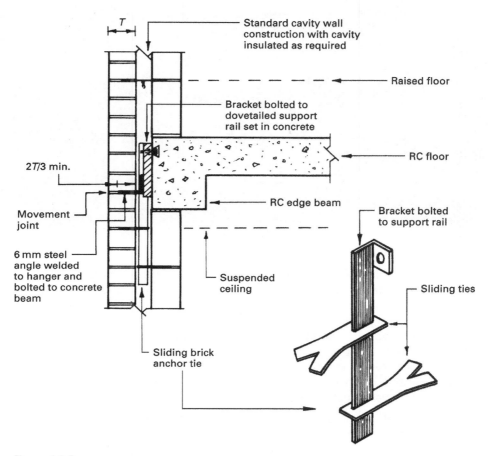

T

2T/3 min.

Standard cavity wall
construction with cavity
insulated as required

Raised floor

Bracket bolted to
dovetailed support
rail set in concrete

RC floor

Movement
joint

RC edge beam

6 mm steel
angle welded
to hanger and
bolted to concrete
beam

Bracket bolted
to support rail

Suspended
ceiling

Sliding ties

Sliding brick
anchor tie

Figure 6.2.5   Non-loadbearing masonry cladding panel

Composite panel façade systems 6.3

Composite panel façade systems consist of a core of insulation to which a metal skin in steel or aluminium is wrapped and bonded around the insulation core. This forms a composite panel that can be used in two functional ways: as aesthetic wall panels and as a substrate to fix a façade system to. The panel has a cleverly designed tongue-and-groove system that locks each panel into place, and they can be used horizontally and vertically. Different manufacturers have alternative methods of fixing and jointing panels both horizontally and vertically. The joint provides movement for thermal expansion and seals the wall structure against the external elements.

The inner sheet that acts as a liner panel has a standard finish applied to its face, whereas the external panel can have a number of different pressed applications made to it to form aesthetic/architectural features, which can enhance a commercial building's appeal.

Figure 6.3.1 illustrates the detail at the edge of each panel. In the figure you can see the method of detailing of the panels such that the base of one panel fits into the top of the next to form a sealed joint. The fixing screws for the panel are therefore hidden from view, to provide a clean look to the wall.

ARCHITECTURAL PANELS

The use of large modular insulated panels has several advantages for the building to which it is applied.

- Because the panel has an insulated core, the U-values stated under Part L of the Building Regulations can be easily met, as thermal resistance is an integral part of the design.

- Panels can be fire-rated to provide fire resistance under BS 476.

- Modular panel widths provide efficiency of construction and reduce wastage.

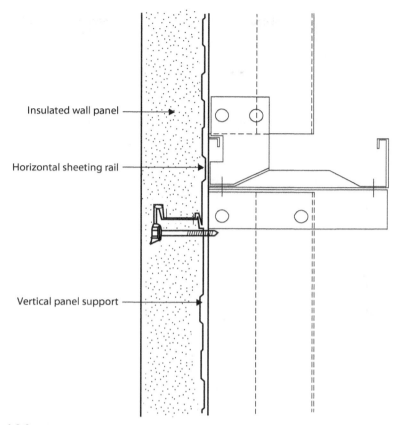

Insulated wall panel

Horizontal sheeting rail

Vertical panel support

Figure 6.3.1 **Section through typical horizontal joint**

Source: Kingspan Limited – contact www.kingspan.com for latest details.

- Joints within panels meet the air leakage requirements of the Building Regulations.
- Doors and openings for windows can be integrated into the wall on a modular basis to suit the panel width.
- A range of applied powder-coated colours is available.
- Fixing screws are concealed in the panel design.
- The method of construction is faster.
- Metal panels can be recycled after lifecycle use.
- They can be retro-fitted to an existing structure to refurbish the external envelope.

The architectural merits of using such panels can be enhanced by the use of several accessories to the system. Joints can be featured using curtain-walling components such that an aluminium cover cap is snapped into position over a pre-formed aluminium joint fixing plate. Figure 6.3.2 illustrates such a feature joint that can be used both vertically and horizontally to create a grid effect. This enables a colour to be applied to the joints. A second method is to manufacture the panels with a groove within them that mimics large tiles. The groove can be enhanced with different-coloured inserts. This gives any designer a vast number of options to finish the external envelope of a commercial building, while maintaining a high degree of control on the aesthetics for a client.

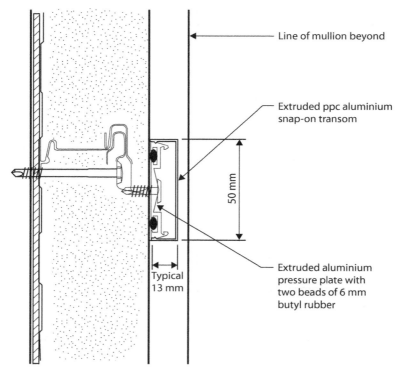

Line of mullion beyond

Extruded ppc aluminium snap-on transom

50 mm

Typical 13 mm

Extruded aluminium pressure plate with two beads of 6 mm butyl rubber

Figure 6.3.2   **Section through enhanced mullion horizontal joint**

Source: Kingspan Limited – contact www.kingspan.com for latest details.

INSULATED FAÇADE SYSTEMS

Insulated façade systems use the functional properties of the composite panel to full effect. They use a standard panel with a budget face fixed to the structure as before, then a secondary finish can be applied to the face. This is done by the use of aluminium extruded rails to which fixing clips attach in order to secure a variety of façade finishes. The choice in the final finish can be extensive, from terracotta tiles, aluminium tiles, stone and sustainable timber boards to rendered cementatious panels.

There are several advantages to using such a system.

- It can be retro-fitted to an existing building to increase its thermal performance and extend the life of the building, giving it a modern feel and aesthetics.
- A more economical composite panel can be used.
- There is a variety of finishes in many different colours.
- The system uses modular design dimensions.
- It has good fire resistance, and moderate sound resistance.
- Designs can mimic external brickwork to blend in with planning constraints.
- Dry trade installation is faster, shortening contract programmes.
- Interchangeable components can add a variety of banding and colour schemes to a design.

Figure 6.3.3 illustrates the use of a manufactured modular brick component. The bricks are in fact manufactured from high-strength concrete that contains a colouring agent, and the face is textured to simulate brickwork. The front face design fits together to form a typical 75 mm-joint spacing, as would traditional brickwork bonded in cement mortar. Various detailing components are available for the installation of external corners, plinth details and soldier courses. The base plinth will need careful detailing and design as it has to carry the additional weight of the concrete bricks.

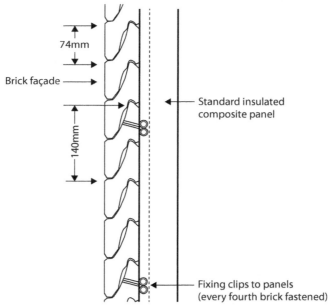

Figure 6.3.3   **Vertical section through detail of brick façade**

Source: Kingspan Limited – contact www.kingspan.com for latest details.

Figure 6.3.4 illustrates an alternative façade system in which a secondary rail system supports a tile finish, which is secured to aluminium rails fastened to the insulated composite panel. The flexibility in the system allows a number of different materials to be made into tiles, such as reconstituted stone, aluminium, steel, ceramics and plastics.

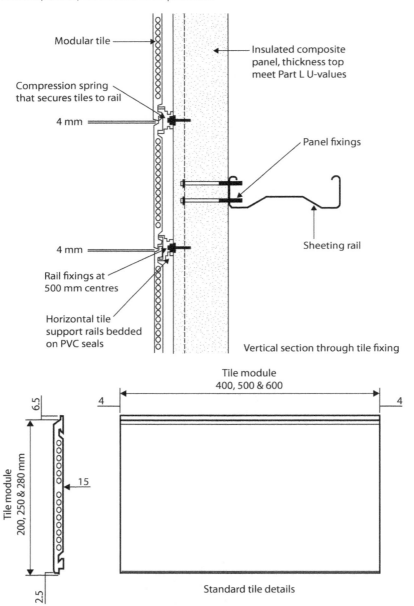

Figure 6.3.4   **Vertical section through tile fixing**

Source: Kingspan Limited – contact www.kingspan.com for latest details.

The advantages of such a system are that:

- different tiles can be used within a wall to provide aesthetic and architectural colour schemes;
- damaged tiles can easily be replaced;
- window systems can be modular, and can be coordinated into the design;
- interlocking of the tiles enables a completely weatherproof system;
- the construction is frost-resistant;
- economical composite panels can be used for the insulation element.

A sustainable material, such as timber, can still be accommodated within the façade system. The aluminium rails are replaced with vertical treated timber bearers, to which sustainable timber planks can be fixed to provide an attractive feature wall.

The advantages of such a system are that:

- sustainable timber products can be used;
- timber does not require any treatment and will weather naturally;
- it offers a green solution to covering external walls;
- the finish is recyclable;
- it can be used with other products;
- it gives a lightweight external finish;
- a void is created behind the timber face.

Figure 6.3.5 illustrates such a system. This uses traditional timber battens which are fixed to the insulated composite panel using fasteners and the timber planks are then secured using screwed fixing to the battens.

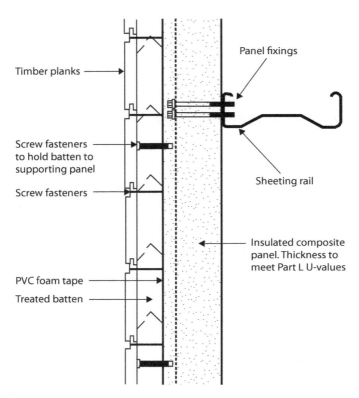

Figure 6.3.5 **Vertical section through a timber façade fixing**

Source: Kingspan Limited – contact www.kingspan.com for latest details.

The following labels appear in the figure:

- Timber planks
- Screw fasteners to hold batten to supporting panel
- Screw fasteners
- PVC foam tape
- Treated batten
- Panel fixings
- Sheeting rail
- Insulated composite panel. Thickness to meet Part L U-values

Jointing 6.4

When incorporating any cladding panels into a framed structure, making the joints waterproof is of paramount importance as this stops moisture getting into the structure, which can have a detrimental effect on the internal environment. Various manufacturers employ different systems to prevent the ingress of weather through the joint. In this section you will look at precast concrete, composite panel and infill panel joints.

Joints should be designed so that they:

- exclude wind, rain and snow;
- allow for structural, thermal and moisture movement;
- have a durability for the lifespan of the building;
- are easily maintained;
- maintain the thermal and sound insulation properties of the surrounding cladding;
- are easily made or assembled on site.

Bad design, poor workmanship and a lack of understanding of the function of a joint can lead to water penetration through the joints between cladding panels. To overcome this, it is essential that both the designer and the site operative fully appreciate the design principles behind the joint construction, the need for accurate installation, and the production of a quality seal.

PRECAST CONCRETE PANEL JOINTS

Suitable joints can be classified under two headings:

- filled joints;
- drained joints.

Filled joints are generally satisfactory if the cladding panel module is small; if incorporated in large module panels, filled joints can crack and allow water to penetrate, as there is a bigger thermal expansion with a longer panel. This failure is due either to the filling materials being incapable of accommodating movement or to a breakdown of adhesion between the filling material and the panel sides. Research has shown that, if the above failures are possible, the most effective alternative is the drained joint.

FILLED JOINTS

Filled joints are not easy to construct and rely mainly on mortars, sealants, mastics or preformed gaskets to provide the barrier against the infiltration of wind and rain. They are limited in their performance by the amount to which the sealing material(s) can accommodate movement and, to a certain extent, by their weathering properties, such as their resistance to ultraviolet rays. The disadvantages of filled joints are:

- difficulty in making and placing the joints accurately, particularly with combinations of materials;
- providing for structural, thermal and moisture movement;
- suitable only for small module claddings.

For typical details of a precast concrete panel joint, see Fig. 6.4.1.

DRAINED JOINTS

Drained joints have been designed and developed to overcome the disadvantages of the filled joint by:

- designing the joint to have a drainage zone within its construction;
- providing an airtight seal at the rear of the joint.

Drained joints have two components: the vertical joint and the horizontal joint.

VERTICAL JOINTS

These consist basically of a deep, narrow gap between adjacent panels, where the rear of the joint is adequately sealed to prevent the passage of air and moisture. The width of the joint does not significantly affect the amount of water reaching the rear seal, for two main reasons.

- Most of the water entering the joint (approximately 80 per cent) will do so by following over the face of the panel; the remainder (approximately 20 per cent) will enter the joint directly, and most of this water entering the joint will drain down within the first 50 mm of the joint depth. Usually the deciding factor for determining the joint width is the type of mastic or sealant being used and its ability to accommodate movement.

■ There are checks on the amount of water entering the drainage zone, such as ribs to joint edges, exposed aggregate external surfaces and the use of baffles.

Baffles are loose strips of material such as neoprene, butyl rubber or plasticised PVC, which are unaffected by direct sunlight and act as a first line of defence to water penetration. The baffles are inserted after the panels have been positioned and fixed, either by pulling them through prepared grooves or by direct insertion into the locating grooves from the face or back of the panel, according to the joint design. They work by acting as a barrier: moisture clings to the baffle and, under the action of gravity, moves downwards, where it is directed by a suitable flashing out of the bottom of the joint. You must take care when inserting baffles by the pulling method because they invariably stretch during insertion; you must allow them to return to their original length before trimming off the surplus, to ensure adequate cover at the intersection of the vertical and horizontal joint.

Adequate sealing at the back of the joint is of utmost importance, because some water will usually penetrate past the open drainage zone or the baffle. The back of the joint is the final point of resistance to moisture reaching the inside environment, and any air movement through the joint seal will also assist the passage of water or moisture. Drained joints that have only a back seal or a baffle and seal can have a cold-bridge effect on the internal face, giving rise to local condensation. You must consider how to maintain the continuity of the thermal insulation value of the cladding (see typical details in Figs 6.4.1 and 6.4.2).

HORIZONTAL JOINTS

Horizontal joints are usually in the form of a rebated lap joint, the upper panel being lapped over the top edge of the lower panel. As with vertical joints, the provision of an adequate back seal to prevent air movement through the joint is of paramount importance. The seal must also perform the function of a compression joint, so the sealing strip is of a compressible material, such as bituminised foamed polyurethane or a preformed cellular rubber strip.

The profile of the joint is such that any water entering the gap by flowing from the panel face or by being blown in by the wind is encouraged to drain back onto the face of the lower panel. The depth of joint overlap is usually determined by the degree of exposure, and ranges from 50 mm for normal exposure to 100 mm for severe exposure (see Fig. 6.4.3 for typical details). Note that the effective overlap of a horizontal joint is measured from the bottom edge of the baffle in the vertical joint to the seal, and not from the rebated edge of the lower panel (see Fig. 6.4.4).

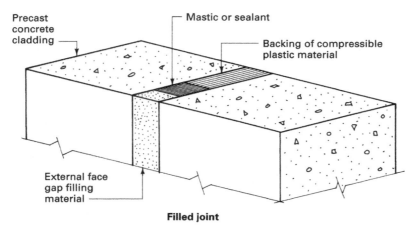

Precast concrete cladding

Mastic or sealant

Backing of compressible plastic material

External face gap filling material

Filled joint

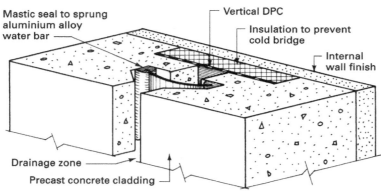

Mastic seal to sprung aluminium alloy water bar

Vertical DPC

Insulation to prevent cold bridge

Internal wall finish

Drainage zone

Precast concrete cladding

Drained joint with water bar

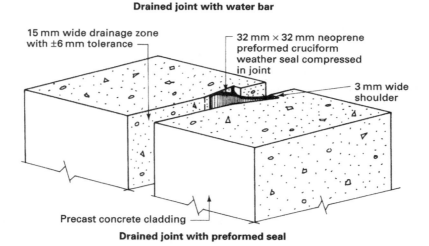

15 mm wide drainage zone with ±6 mm tolerance

32 mm × 32 mm neoprene preformed cruciform weather seal compressed in joint

3 mm wide shoulder

Precast concrete cladding

Drained joint with preformed seal

Figure 6.4.1   Typical filled and drained joints

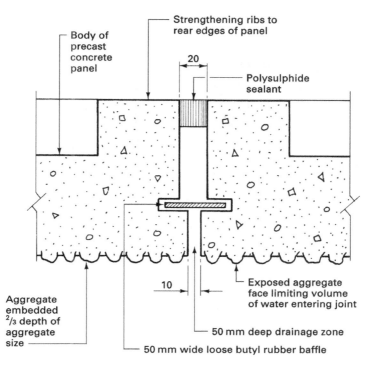

Body of precast concrete panel

Strengthening ribs to rear edges of panel

20

Polysulphide sealant

Aggregate embedded $^2/_3$ depth of aggregate size

10

Exposed aggregate face limiting volume of water entering joint

50 mm deep drainage zone

50 mm wide loose butyl rubber baffle

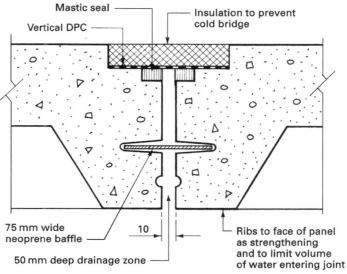

Mastic seal

Vertical DPC

Insulation to prevent cold bridge

75 mm wide neoprene baffle

10

50 mm deep drainage zone

Ribs to face of panel as strengthening and to limit volume of water entering joint

Figure 6.4.2   Typical drained joints using baffles

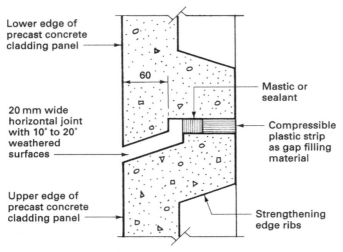

Joint for low to moderate exposure

Lower edge of precast concrete cladding panel

60

20 mm wide horizontal joint with 10° to 20° weathered surfaces

Upper edge of precast concrete cladding panel

Mastic or sealant

Compressible plastic strip as gap filling material

Strengthening edge ribs

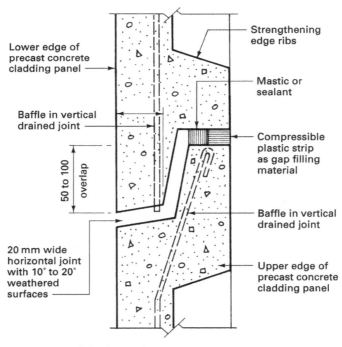

Joint for moderate to severe exposure

Strengthening edge ribs

Lower edge of precast concrete cladding panel

Baffle in vertical drained joint

50 to 100 overlap

Mastic or sealant

Compressible plastic strip as gap filling material

Baffle in vertical drained joint

20 mm wide horizontal joint with 10° to 20° weathered surfaces

Upper edge of precast concrete cladding panel

Figure 6.4.3   Typical horizontal joints

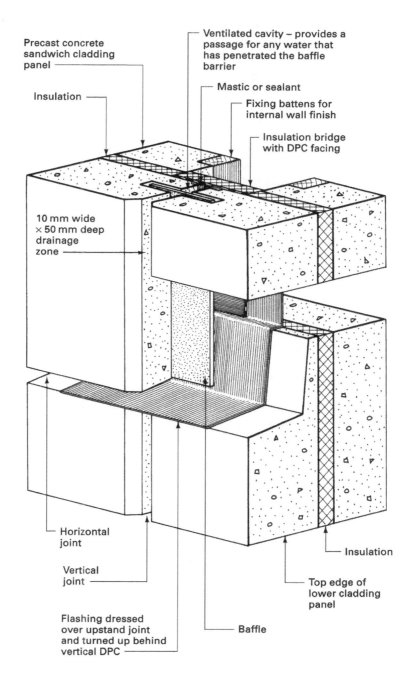

Precast concrete sandwich cladding panel

Insulation

Ventilated cavity – provides a passage for any water that has penetrated the baffle barrier

Mastic or sealant

Fixing battens for internal wall finish

Insulation bridge with DPC facing

10 mm wide × 50 mm deep drainage zone

Horizontal joint

Vertical joint

Insulation

Top edge of lower cladding panel

Flashing dressed over upstand joint and turned up behind vertical DPC

Baffle

Figure 6.4.4 Typical drained joint intersection detail

The intersection of the joints is an important feature of drained joint design and detail. It is necessary to shed any water draining down the vertical joint on to the face of the lower panels where the vertical and horizontal joints intersect, as the joints are designed to cater only for the entry of water from any one panel connection at a time. The usual method is to use a flashing, starting at the back of the panel, dressed over and stuck to the upper edge of the lower panel, as shown in Fig. 6.4.4. The material for the flashing must be chosen carefully, because it must accept the load of the upper panel and any movements made while the panel is positioned and secured. Also, it should be a material that is durable but will not give rise to staining of the panel surface. Suitable materials are bitumen-coated woven glasscloth and synthetic rubber sheet.

COMPOSITE PANEL JOINTS

HORIZONTAL

With the advent of composite panelling systems, the filled joint has developed into one incorporated into the panel design. A horizontal joint is secured by a locking weathering key, with dual factory-applied seals at the horizontal joint between panels, enabling a very effective seal to be maintained. Figure 6.4.5 illustrates such a key, where the micro-rib panel has a front projecting lip on the adjacent panel and an elongated joint that slopes towards the outside, draining any moisture away. A factory-applied seal between the two panels provides a watertight and airtight joint. Because of this profile design, the fixing, which could be subjected to weathering, is hidden from the elements. This also gives a flat, unobstructed design to the panel finish.

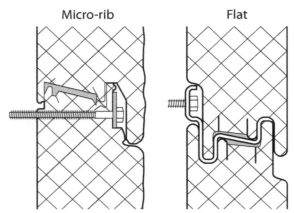

Micro-rib   Flat

Figure 6.4.5 **Different horizontal composite panel joints**

VERTICAL

Where panels are fixed vertically, this increases the complexity of working with moisture, as it will not simply run off the joint under the force of gravity; instead, capillary action can now take place. As you can see from Fig. 6.4.6, the vertical joint is of complex construction, so it takes a great deal of care on site to achieve a product of the quality to outlive the lifespan of the project. The joint illustrated has an extruded aluminum section, with inserts that can be coloured to enhance the aesthetics of the joint. Various enhanced joint details are available to give designers a variety of projections and colour options. With this joint, the thermal properties of the panels are continuous and any cold bridging at the connection is avoided.

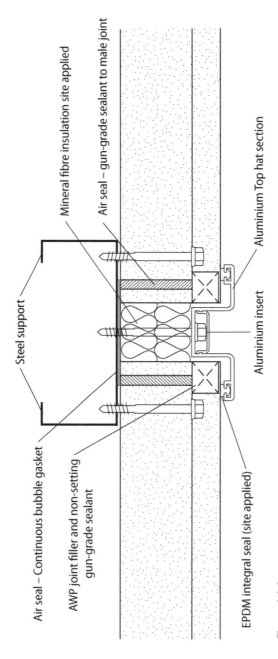

Steel support

Air seal – Continuous bubble gasket

AWP joint filler and non-setting
gun-grade sealant

Mineral fibre insulation site applied

Air seal – gun-grade sealant to male joint

Aluminium Top hat section

Aluminium insert

EPDM integral seal (site applied)

Figure 6.4.6   Vertical enhanced joint in composite panelling

Mastics, sealants and gaskets 6.5

The provision of an effective seal between panels or the supporting structure is essential to maintain the weather- and airtightness of the internal structure. Materials that are to be used for sealing joints, whether in the context of claddings or for sealing the gap between a simple frame and the reveals, have to fulfil the following requirements:

- provide a weathertight seal;
- accommodate movement due to thermal expansion, wind loadings, structural movement and/or moisture movement;
- accommodate and mask tolerance variations;
- be sufficiently self-supporting and remain in position;
- should not deteriorate and stain adjacent materials.

Mastics and sealants have a limited life ranging from 10 to 25 years, which in most cases is less than the initial design life of the structure. Therefore all joints must be designed in such a way that the seals can be renewed with reasonable ease, efficiency and economy. The common form of sealing material, called **putty**, is unsuitable for many applications because it hardens soon after application and cannot therefore accept the movements that can be expected in claddings and similar situations, but is the common method of retaining glazing into casement frames.

A wide variety of mastics and sealants is available to the designer and builder, giving a variety of properties and applications. We are not going to examine and conduct a complete analysis of all the types available, but will compare some of the most frequently used.

Mastics and sealants can be applied in a variety of ways, such as gun or knife applications. The most common method is by hand-held gun, using disposable cartridges fitted into the body of the applicator. Most guns can be fitted with various disposable nozzles that can be shaped to produce a neat bead of the required size. Application using this method is the most common as it is neater, causes less waste and enables a controlled amount of sealant to be spread.

The joint should be carefully prepared to receive the mastic or sealant by ensuring that the contact surfaces are free from all dirt, grease and oil. The contact surfaces must be perfectly dry, and in some instances they may have to be primed before applying the jointing compound.

MASTICS

These are materials that are applied in a plastic state and form a surface skin over the core, which remains pliable for a number of years.

BUTYL MASTICS

Butyl mastics are basic mastics with the addition of butyl rubber or related polymers, making them suitable for glazing. They have a durability of up to 10 years and can accommodate both negative and positive movements up to 5 per cent, using a maximum joint width of 20 mm and a minimum joint depth of 6–10 mm.

OIL-BOUND MASTICS

Oil-bound mastics are the most widely used mastics, made with non-drying oils. They have similar properties to the butyl mastics but in some cases a better durability of up to 15 years. They have a joint width of 25 mm maximum, although some special grades can be up to 50 mm, but in all cases the minimum joint depth is 12 mm.

SEALANTS

Sealants are capable of accommodating greater movement than mastics; they are more durable, but dearer in both material and installation costs. They are applied in a plastic state and are converted by chemical reactions into an elastomer or synthetic rubber.

TWO-PART POLYSULPHIDE AND POLYURETHANE SEALANTS

Two-part polysulphide and polyurethane sealants are well-established elastomer sealants with an excellent durability of 25 years or more, and a movement accommodation of approximately 15 per cent. They are supplied as two components: a polysulphide or polyurethane base and a curing or vulcanising agent. Dense, impervious surfaces, such as glass, may need priming, particularly before polyurethane is applied. The two parts are mixed shortly before use, the mixture having a pot life of approximately four hours. The two parts chemically react to form a synthetic rubber, and should be used with a maximum joint width of 25 mm and a minimum joint depth of 6 mm when used in conjunction with metal or glass, or a 10 mm minimum joint depth when used in connection with concrete. Two-part polysulphide sealants should conform to BS EN ISO 11600: *Building Construction. Jointing products. Classification and requirements for sealants*.

ONE-PART POLYSULPHIDE, SILICONE OR URETHANE SEALANTS

One-part polysulphide, silicone or urethane sealants do not require premixing on site before application. They convert to a synthetic rubber or an elastomer by chemical reaction with absorbed moisture from the atmosphere. The joint sizing is similar to that described for a two-part polysulphide, but the final movement accommodation is less at approximately plus or minus 12.5 per cent. Silicones and urethanes cure quite quickly, whereas polysulphides can take up to two months, and during this period the movement accommodation is very low. One-part polysulphides are generally specified for pointing and where early or excessive movements are unlikely to occur. They should conform to BS EN ISO 11600: *Building Construction. Jointing products. Classification and requirements for sealants*.

SILICONE RUBBER SEALANT

Silicone rubber sealant is a one-part sealing compound that converts to an elastomer by absorbing moisture from the atmosphere, and has similar properties to the one-part polysulphide. It is available as pure white, translucent or in a wide range of colours, which makes it suitable for a variety of internal and external applications, such as sealing tiles to bath fittings and window/door frames to masonry. It should conform to BS EN ISO 11600: *Building Construction. Jointing products. Classification and requirements for sealants*.

See also BS EN 26927: *Building construction. Jointing products. Sealants. Vocabulary*.

GASKETS

As an alternative to using mastics and sealants for jointing, preformed gaskets of various designs and shapes are available. They are classified as **structural** or **non-structural**. Structural gaskets are produced from a compound of vulcanised polychloroprene rubber and specified to BS 4255-1: *Specification for non-cellular gaskets*. Non-structural gaskets can be made from synthetic rubber and various plastics in solid or tubular profiles. They too should conform to BS 4255-1. These can have better durability than mastics and sealants, but the design and manufacture of the joint profile requires a high degree of accuracy if a successful joint is to be obtained.

Curtain walling   6.6

Curtain walling is a form of external lightweight cladding attached to a supporting framed structure forming a complete envelope around the structural frame, apart from the roof. These are essentially non-loadbearing claddings, which have to support not only their own dead weight but also any imposed wind loadings that are transferred to the structural frame through connectors, which are usually positioned at intermediate floor levels. For clarification, BS EN 13830: *Curtain walling – product standard* defines curtain walling as 'an external vertical building enclosure produced by elements mainly of metal, timber or plastic'. The basic conception of most curtain walls is a series of vertical mullions spanning from floor to floor interconnected by horizontal transoms forming openings into which can be fixed panels of glass or infill panels of opaque materials. Most curtain walls are constructed using a patent or proprietary system produced by profile-metal fabricators.

The primary objectives of using curtain-walling systems are to:

- provide an enclosure to the structure that will give the necessary protection against the elements;
- satisfy the requirements of regulations on external wall thermal properties of refurbishments;
- upgrade and retro-fit to enhance an existing structure;
- make use of dry construction methods;
- impose onto the structural frame the minimum load in the form of claddings;
- provide an aesthetic architectural feature.

To fulfil its primary functions a curtain wall must meet certain requirements.

- **Resistance to the elements** The materials used in curtain walls are usually impervious, such as glazing, but, by virtue of the way in which they are fabricated in modular formats, a large number of joints occur. These joints must be made as impervious as the surrounding materials or designed as a drained joint. The aesthetics of the curtain façade must also be kept in mind with the design of the joint, in order to give a uniform appearance to the finished building envelope. The jointing materials must also allow for any local thermal, structural or moisture movement, and generally consist of mastics, sealants and/or preformed gaskets of synthetic rubber or PVC.

- **Assistance in maintaining the designed internal temperatures** As curtain walls usually include a large percentage of glass, the overall resistance to the transfer of heat is low, and therefore preventive measures have to be incorporated into the design to avoid excessive solar heat gain and associated cooling costs. Heat gain is where short-wave radiation from the sun passes through the glazing and warms up the air and surfaces of the internal walls, equipment and furniture. These surfaces will in turn radiate this acquired heat in the form of long-wave radiation that cannot pass back through the glazing, creating an internal heat build-up. There are a number of solutions to this problem that can be used in the design of the exterior. The first is the use of louvres fixed within a curtain-walling system that will reduce some of the sun's glare. A system of non-transparent external louvres or sunshades will reduce the heat gain slightly by absorbing heat and radiating it back to the external air. Other methods employed to solve the problem of internal heat gain are:

  - deep recessed windows;

  - balanced internal heating and ventilation systems;

  - use of special solar-control glass, such as reflective glasses that are modified during manufacture by depositing a metallic or dielectric reflective layer on the surface of the glass. The energy from the sun's rays is reflected off the glass, reducing the amount of short-wave energy that passes through to the inside.

- **Adequate strength** Although curtain walls are classified as non-loadbearing, they must be able to carry their own self-weight and resist both positive and negative wind pressure. The magnitude of this latter loading will depend on three basic factors:

  - height of the building;

  - degree of exposure to prevailing winds;

  - geographical location of building.

The strength of curtain walling relies mainly on the stiffness of the vertical component or mullion, together with its anchorage or fixing to the structural frame. Glazing beads and the use of compressible materials also add to the resistance of possible wind damage of the glazed and infill panel areas, by enabling these units to move independently of the curtain-wall framing.

- **Required degree of fire resistance** This is probably one of the greatest restrictions encountered when using curtain-walling techniques because of the large proportion of unprotected areas as defined in the Building Regulations, Approved Document B4: Section 12.7. See also Chapter 5 of this volume for matters relating to external fire spread. By using suitable fire-resistant glass and infill materials, the curtain wall can normally achieve the required fire resistance to allow continuity of fire compartmentation.

- **Ease of assembly and fixing** The principal member of a curtain-walling system is usually the mullion, which can be a solid or box section that is fixed to the structural frame at floor levels by means of adjustable anchorages or connectors. The infill framing and panels may be obtained as a series of individual components or as a single, prefabricated unit. Prefabrication provides a faster way to site-fix curtain-walling units.

- **Required degree of sound insulation** Sound originating from within the structure may be transmitted vertically through the curtain-walling members. The chief source of this form of structure-borne sound is machinery, and this may be reduced by isolating the offending machines, mounting them on resilient pads and/or using resilient connectors in the joints between mullion lengths. Airborne sound can be troublesome with curtain-walling systems because the lightweight cladding has little mass to offer in the form of a sound barrier, the weakest point being the glazed areas. A reduction in the amount of sound transmitted can be achieved by:

  - reducing the areas of glazing;
  - using sealed windows of increased glass thickness;
  - double glazing in the form of inner and outer panes of glass with an air space of 150 to 200 mm between them.

- **Provision for thermal and structural movements** As curtain walling is situated on an external face of the structure, it will be more exposed than the structural frame and will therefore be subject to greater amounts of temperature change, resulting in high thermal movement. This movement will travel around the structure with the rotation of the sun across the sky, with one part of a building subject to heating and the opposing face subject to cooling. These differential movements mean that the curtain-walling systems should be designed, fabricated and fixed so that the attached cladding can move independently of the structure. The usual methods of providing for this required movement are to have slotted bolt connections and, to allow for movement within the curtain walling itself, to have spigot connections and/or mastic-sealed joints. Figures 6.6.1 and 6.6.2 show typical established curtain-walling examples to illustrate these principles. To reduce the risk of condensation on the interior of support framing and thermal bridging, a thermal break is incorporated as shown in Fig. 6.6.3.

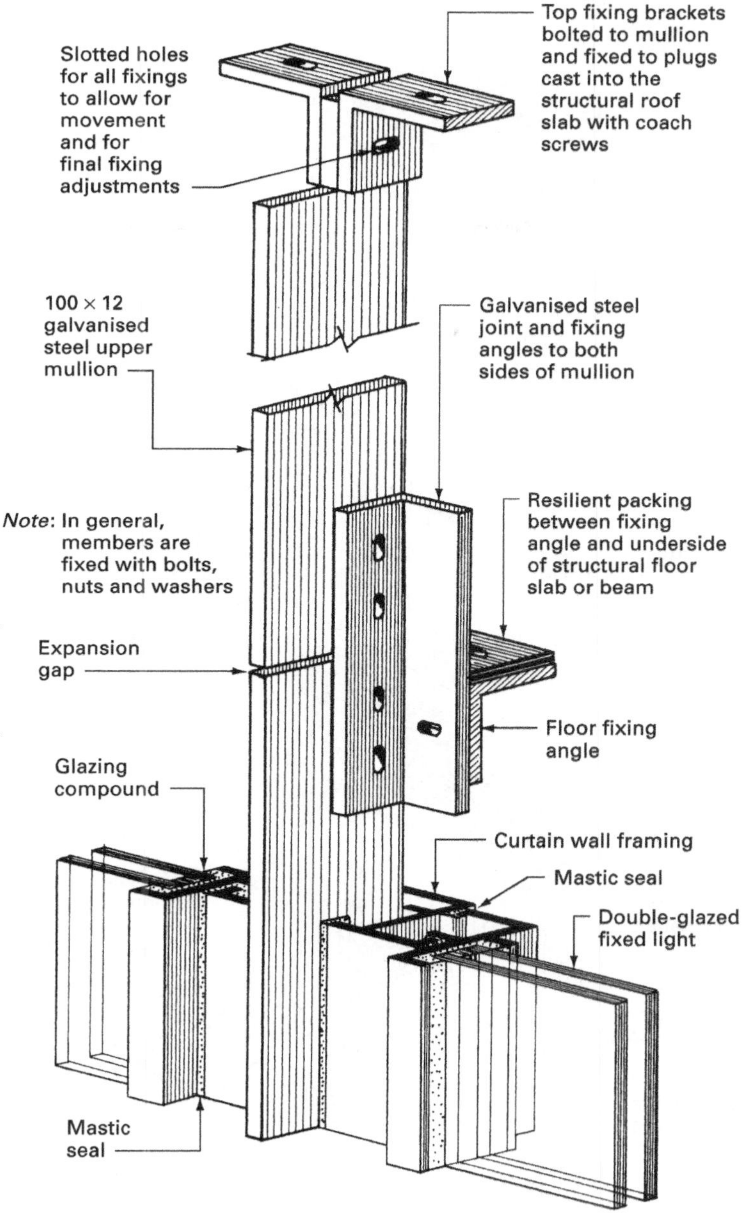

Slotted holes for all fixings to allow for movement and for final fixing adjustments

Top fixing brackets bolted to mullion and fixed to plugs cast into the structural roof slab with coach screws

100 × 12 galvanised steel upper mullion

Galvanised steel joint and fixing angles to both sides of mullion

Note: In general, members are fixed with bolts, nuts and washers

Resilient packing between fixing angle and underside of structural floor slab or beam

Expansion gap

Floor fixing angle

Glazing compound

Curtain wall framing

Mastic seal

Double-glazed fixed light

Mastic seal

Figure 6.6.1   Typical curtain-walling details: 1

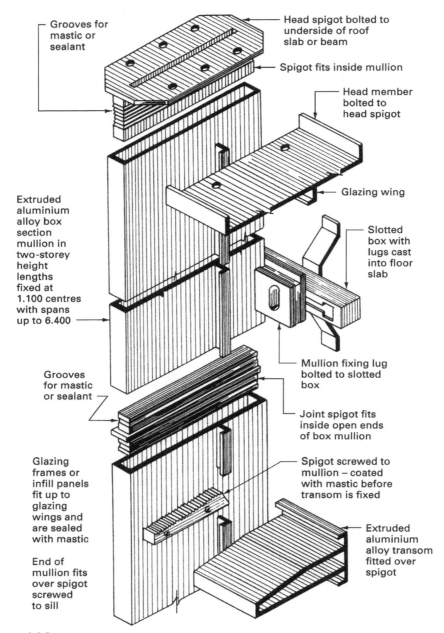

Grooves for mastic or sealant

Head spigot bolted to underside of roof slab or beam

Spigot fits inside mullion

Head member bolted to head spigot

Extruded aluminium alloy box section mullion in two-storey height lengths fixed at 1.100 centres with spans up to 6.400

Glazing wing

Slotted box with lugs cast into floor slab

Grooves for mastic or sealant

Mullion fixing lug bolted to slotted box

Joint spigot fits inside open ends of box mullion

Glazing frames or infill panels fit up to glazing wings and are sealed with mastic

Spigot screwed to mullion – coated with mastic before transom is fixed

Extruded aluminium alloy transom fitted over spigot

End of mullion fits over spigot screwed to sill

Figure 6.6.2 Typical curtain-walling details: 2

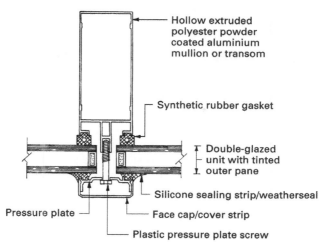

Labels on figure:
- Hollow extruded polyester powder coated aluminium mullion or transom
- Synthetic rubber gasket
- Double-glazed unit with tinted outer pane
- Silicone sealing strip/weatherseal
- Pressure plate
- Face cap/cover strip
- Plastic pressure plate screw

Figure 6.6.3 **Extruded aluminium mullion or transom incorporating thermal breaks**

INFILL PANELS

To overcome the visibility of the frame from the outside, opaque panels are used to hide the floor and internal partition junctions. The panels used to form the opaque areas in a curtain-walling system should have the following properties:

- lightweight;
- rigidity;
- impermeability;
- suitable fire resistance;
- suitable resistance to heat transfer;
- good durability, requiring little or no maintenance.

No one material has all the properties listed above, so infill panels are usually manufactured in the form of a sandwich or composite panel. One of the main problems encountered with any form of external sandwich panel is interstitial condensation, which is usually overcome by including a vapour control layer of suitable material situated near the inner face of the panel. A vapour control layer can be defined as a membrane with a vapour resistance greater than 200 MN/g. Suitable materials include adequately lapped sheeting, such as aluminium foil, waterproof building papers and polythene sheet. Modern insulation has a vapour control layer already bonded to its outer surfaces. Care must be taken when positioning vapour control layers to ensure that an interaction is not set up between adjacent materials, such as the alkali attack of aluminium if placed next to concrete or fibre cement.

The choice of external facing for these infill cladding materials is very important because of their direct exposure to the elements. Plastic and plastic-coated materials are obvious choices provided they meet the minimum fire regulations as set out in Approved Document B. One of the most popular materials for the external facing of infill panels is vitreous-enamelled steel or aluminium sheets of 0.7 and 0.8 mm thickness respectively. In the preparation process a thin coating of glass is fused onto the metal surface at a temperature of between 800 and 860 °C, resulting in an extremely hard, impervious, acid- and corrosion-resistant panel that will withstand severe abrasive action; also, the finish will not be subject to crazing or cracking, resulting in an attractive finish with the strength of the base metal. Aluminium sheeting can be finished with a polyester powder coating in a variety of colours. When used in combination with materials such as Styrofoam®, mineral fibre and polyurethane insulants, a lightweight infill panel can be achieved giving U-values that comply with Part L. Typical infill panel examples are shown in Fig. 6.6.4. Fire performance for both internal and external surfaces must satisfy the Building Regulations for class 0, tested in accordance with BS 476-6 and 7. The insulating core should not ignite when tested to BS 476-12. Panels can be manufactured up to 3 m long by 1 m wide.

GLAZING

The primary function of any material fixed into an opening in the external façade of a building is to provide illumination. Glass allows daylight illumination of the interior of the building for the carrying out of specific tasks; at the same time it can provide a view out or visual contact with the outside world. Sustainable buildings require making full use of the amount of natural daylight that enters through the glazing to reduce the reliance on energy usage for artificial lighting. Glazing also provides the psychological attachment to the outside world, the weather and the time of the day.

The nature of work to be carried out and the position of the working plane will largely determine whether daylight from glazed areas alone can provide sufficient illumination for specific work tasks. Due to the depth of buildings and the different heights of the sun during the calendar year, the amount of useful daylight cannot be entirely relied upon. Many of these tasks are carried out on a horizontal surface, which is best illuminated by vertical light: therefore window size, window position and possible daylight factors need to be carefully assessed by the designer if glazed areas are to provide the main source of task illumination. There is often a need for a view from tall buildings that carries a marketing premium. Large areas of glazing will provide this view but will require careful planning, bearing in mind the size, orientation and view obtained (for example, the positioning of mullions so they do not block the view).

The major problem remaining when using glass in the façades of high-rise structures, and in particular for curtain walling, is providing a means of access

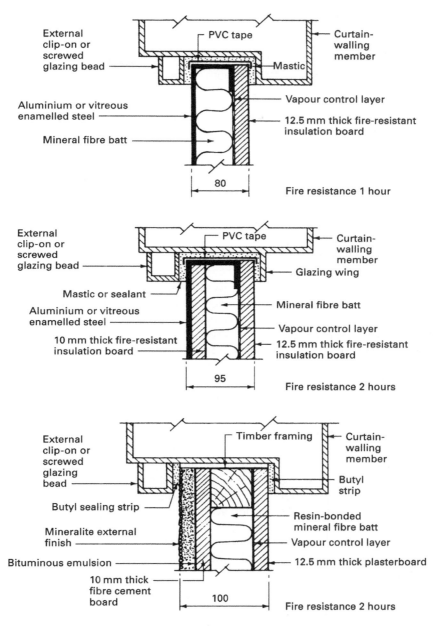

External
clip-on or
screwed
glazing bead

PVC tape

Curtain-
walling
member

Mastic

Aluminium or vitreous
enamelled steel

Vapour control layer

12.5 mm thick fire-resistant
insulation board

Mineral fibre batt

80

Fire resistance 1 hour

External
clip-on or
screwed
glazing bead

PVC tape

Curtain-
walling
member

Glazing wing

Mastic or sealant

Aluminium or vitreous
enamelled steel

Mineral fibre batt

Vapour control layer

10 mm thick fire-resistant
insulation board

12.5 mm thick fire-resistant
insulation board

95

Fire resistance 2 hours

External
clip-on or
screwed
glazing
bead

Timber framing

Curtain-
walling
member

Butyl
strip

Butyl sealing strip

Mineralite external
finish

Resin-bonded
mineral fibre batt

Vapour control layer

Bituminous emulsion

12.5 mm thick plasterboard

10 mm thick
fibre cement
board

100

Fire resistance 2 hours

Figure 6.6.4   Typical curtain-walling infill panel details

for cleaning and maintenance. The cleaning of windows on multi-storey buildings can be a dangerous and costly process, but such glazed areas do need cleaning, for the following reasons:

■ prevention of dirt accumulation resulting in obscurity of the view;

■ maintaining the designed daylight illumination levels;

■ maintaining the clarity of vision out;

■ prevention of deterioration of the glazing materials and seals due to chemical and/or dirt attack.

The usual method of cleaning windows is by washing with water using swabs, chamois leather, scrims and squeegee cleaners, all of which are hand-held, requiring close access to the glass to be cleaned. Cleaning the internal surfaces does not normally present a problem but, unless pivot or tilt-and-turn windows are used, the cleaning of the outside surface will require a means of external access. In low- to medium-rise structures, access can be by means of window-cleaning poles and mobile elevated platforms.

Access for external cleaning of curtain-wall façades in high-rise structures is generally by the use of suspended cradles, which can be of a temporary nature as shown in Fig. 2.6.2, or of a permanent system designed and constructed as an integral part of the structure. The simplest form is to install a universal beam section at roof level positioned about 450 mm in front of the general façade line and continuous around the perimeter of the roof. A conventional cradle is attached by means of castors located on the bottom flange of the ring or edge beam. Control is by means of ropes from the cradle, which must be lowered to ground level for access purposes.

A better form is where the transversing track is concealed by placing a pair of rails to a 750 mm gauge, to which a trolley is fixed. The trolley can have projecting beams or davits that can be retracted and/or luffed, which is particularly useful for negotiating projections in the general façade. The trolley may be manually operated from the cradle for transversing, or it may be an electric-powered trolley, giving vertical movement of between 5 and 15 m per minute or horizontal movement of 5 to 12 m per minute. This form of trolley is usually considered essential for heights over 45 m. In all cases the roof must have been structurally designed to accept the load of the apparatus.

Some form of cable restraint must be incorporated in the design to overcome the problem of unacceptable cradle movement due to the action of wind around the structure. Winds moving up the face of the building will cause the cradle to swing at right angles to the face, giving rise to possible impact damage to the face of the structure, as well as placing the operatives in a hazardous situation.

Crosswinds can cause the cradle to move horizontally along the face of the building, with equally disastrous results. Methods of cradle and cable restraint available are:

■ the use of suction grips;

■ eyebolts fixed to the façade through which the suspension ropes can be threaded;

■ suspension ropes suitably tensioned at ground level;

■ electrical cutouts at intervals of 15 m, which will prevent further cradle movement until a special plug through which the hoist line passes has been inserted;

■ using the mullions as a guide for rollers that are either in contact with the mullion face to prevent lateral movement, or using castors located behind a mullion flange to prevent outward movement.

Rainscreen cladding 6.7

Rainscreen cladding is a facing or façade applied as part of the construction of new buildings or as a protective or decorative overcladding to existing structures that are being refurbished or upgraded. The latter application provides an opportunity to modernise the exterior appearance of a dated building and improve its sound and thermal insulation properties, bringing them up to the revised standards contained in the Building Regulations Approved Documents.

As the name suggests, rainscreen cladding provides an outer layer that screens the rain down its impervious surface; however, the joints may often be open and allow any excess moisture to pass through, which is then caught by a drained cavity and dropped out at ground level by a drip. The 'loose-fit' nature of rainscreen cladding with screw, rivet or clip-on fixings allows for any future changes, such as variations in corporate image rebranding or changing owner requirements. A large number of finishes can be used with this system including timber, ceramic tiles, terracotta, render, metal, laminates and aluminium composites.

The key features of this type of exterior treatment are the provision of:

- an outer weather-resistant layer;
- an outer decorative layer;
- a vertical support frame secured to the structural wall;
- a drained and ventilated cavity between cladding and wall;
- a breathable vapour control layer;
- a facility for integral insulation.

The concept is shown in Fig. 6.7.1.

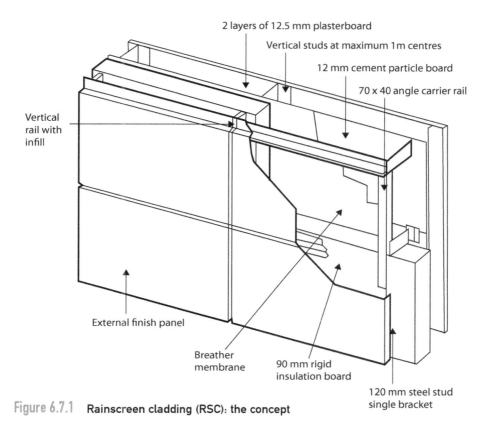

2 layers of 12.5 mm plasterboard

Vertical studs at maximum 1m centres

12 mm cement particle board

70 x 40 angle carrier rail

Vertical
rail with
infill

External finish panel

Breather
membrane

90 mm rigid
insulation board

120 mm steel stud
single bracket

Figure 6.7.1   **Rainscreen cladding (RSC): the concept**

NEW CONSTRUCTION

Rainscreen cladding allows the designer to use an inexpensive single skin of lightweight concrete blockwork as infilling between the structural frame. This provides the background support for superimposed water-resistant insulation, such as resin-bonded rock-fibre batts in thicknesses up to 150 mm. Traditional details such as cavity trays and damp-proof membranes/courses are not required. Also, the continuity of insulation considerably reduces any possibility for cold bridging or the occurrence of interstitial condensation.

This type of rainscreen cladding is also suited to a drylining system that is fixed to steel 'C'-section studs that are fixed to the ceilings and floors of a composite concrete and steel structure. This eliminates the wet trade of infill blockwork and enables faster construction. The system consists of:

■   internal 'C'-section stud steel channels that form the support for the two layers of internal plasterboard dry lining;

■   particle cement board that provides a weatherproof screen;

- a vapour control layer consisting of a breather membrane;
- rigid insulation boards;
- fire barriers at compartmentation zones;
- fixing brackets to hold vertical or horizontal finish carrier rails;
- horizontal or vertical carrier rails;
- external finish module.

Insulation can be used internally or externally to provide the requirements of Part L of the Building Regulations. The void created by the use of 'C'-section studs and internal plasterboard drylining allows a service corridor to be created for mechanical and electrical installations. Thicknesses of external walls using this system are in the region of 650 mm wide.

EXISTING CONSTRUCTION

Weather-damaged, dated and generally unattractive building façades can be considerably improved and protected by supplementary cladding. However, local planning restrictions with particular regard to conservation areas must be consulted before proceeding with the obliteration of an existing façade. Significant change of appearance, modernisation and change of use will necessitate a formal planning application. Nevertheless, the alternative interior approach to updating to comply with current sound and thermal regulations and periods of fire resistance may prove difficult and expensive. The practicalities of application to the interior of walls can be more difficult to effect and more disruptive for the building's occupants, and can impose on valuable floor area.

APPLICATION TO EXISTING WALLS

Vertical battens of timber or metal-profiled framing are secured to the background wall at spacings to suit the cladding. Treated timber battens of 50 mm × 38 mm and 100 mm × 38 mm at panel junctions are normally secured at between 400 and 600 mm vertical spacing, to allow continuity of ventilation and drainage. There should be no horizontal battens, noggins or struts. Figure 6.7.2 shows the principles of application, which must be supplemented with anti-bird mesh at upper and lower ventilation openings to prevent entry by birds, vermin or insects. Ventilation openings must also be provided at interruptions, such as windows and projecting ring beams. The size of ventilation openings can vary with the height of the building:

- up to 5 storeys – 10 mm continuous or equivalent;
- 5 to 15 storeys – 15 mm continuous or equivalent;
- over 15 storeys – 20 mm continuous or equivalent.

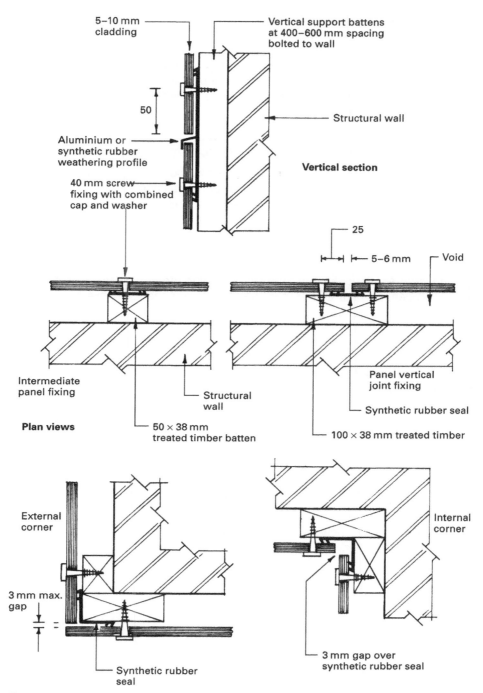

5–10 mm cladding

Vertical support battens at 400–600 mm spacing bolted to wall

50

Aluminium or synthetic rubber weathering profile

Structural wall

Vertical section

40 mm screw fixing with combined cap and washer

25

5–6 mm

Void

Intermediate panel fixing

Structural wall

Panel vertical joint fixing

Synthetic rubber seal

Plan views

50 × 38 mm treated timber batten

100 × 38 mm treated timber

External corner

Internal corner

3 mm max. gap

Synthetic rubber seal

3 mm gap over synthetic rubber seal

Figure 6.7.2   RSC with timber support battens

Incorporating insulation behind timber-supported cladding is unlikely to be economical or practical when considering the space required. Several manufacturers produce purpose-made aluminium sections with extendable brackets that can suspend the cladding up to 200 mm from the wall, as shown in Fig. 6.7.3. This enables sufficient depths to be used for the increased thicknesses of insulation required to meet the U-values in Part L of the Building Regulations.

FIRE SPREAD

As for any other void or cavity, the cavity created behind the cladding is treated by the Building Regulations as a potential passage for conveyance of fire. Therefore fire barriers must be located in accordance with the spacings specified in Approved Document B3, Section 9 (Volume 2). The vertical support battens or framing are normally adequate, but these will need to be supplemented horizontally. In order to preserve the continuity of ventilation and drainage, horizontal barriers are manufactured from glass-fibre rigid batts with an intumescent coating that expands at high temperatures, closing the ventilation gap. The fire barrier must be securely mechanically fixed using steel fixings so it cannot move (see Fig. 6.7.4).

CLADDING MATERIALS

Rainscreen claddings are fire-tested to BS 476-6 and 7 for fire propagation and surface spread of flame respectively, and are also classified for fire resistance depending on purpose grouping. Standards apply to both sides of the cladding, with criteria for assessment depending on building height, purpose group and distance from the boundary. These are determined in the Building Regulations, Approved Document B4, Section 11.6, with reference to class 0 materials of limited combustibility and an Index (I) < 20 (see reference to BS 476-6 in Chapter 7.2).

Sheet claddings are available in thicknesses ranging from 5 to 10 mm, and in a wide variety of colours. Popular sheet material applications include fibre cement, zinc, copper, polyester-coated aluminium and stainless steel.

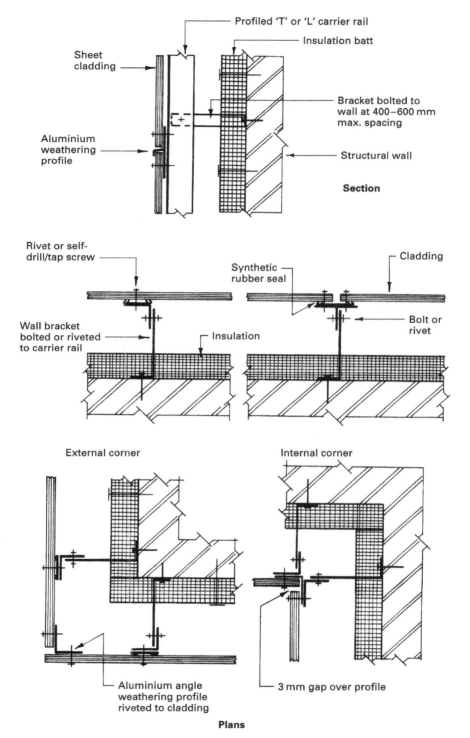

Profiled 'T' or 'L' carrier rail

Insulation batt

Sheet cladding

Bracket bolted to wall at 400–600 mm max. spacing

Aluminium weathering profile

Structural wall

Section

Rivet or self-drill/tap screw

Synthetic rubber seal

Cladding

Wall bracket bolted or riveted to carrier rail

Insulation

Bolt or rivet

External corner

Internal corner

Aluminium angle weathering profile riveted to cladding

3 mm gap over profile

Plans

Figure 6.7.3   RSC with profiled aluminium carriers and brackets

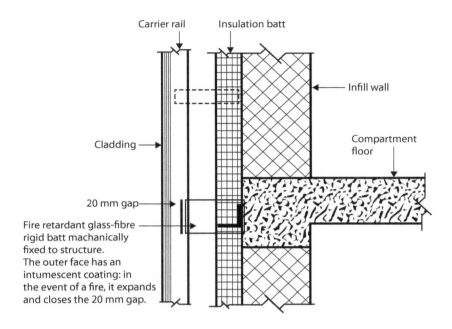

Carrier rail

Insulation batt

Infill wall

Cladding

Compartment floor

20 mm gap

Fire retardant glass-fibre
rigid batt machanically
fixed to structure.
The outer face has an
intumescent coating: in
the event of a fire, it expands
and closes the 20 mm gap.

Figure 6.7.4 **Cavity fire barrier**

Structural glass cladding 6.8

Glass is a material that can be bent, patterned or etched, and now can be electrically charged to change its translucency. As a building material it does have weaknesses that have to be engineered out before use. The brittleness of glass has to be countered with a toughening process, undertaken after the glass has been cut to size and any holes or openings cut within it. This annealing process straightens out the stresses within the glass, making it more durable to compressive and tensile forces that will be exerted by the rain and wind loads.

Toughened glass sheeting of 10 or 12 mm thickness can be considered under two headings in its use as a façade system:

- structural sealant (silicone) glazing;
- mechanically fixed glazing.

STRUCTURAL SEALANT (SILICONE) GLAZING

This system relies on the bonding characteristics of silicone or other elastomeric sealants between the glass and the support frame. Sealants are also selected with regard to their long-term resilience when exposed to ultraviolet light, heat, rain, varying temperatures, thermal movement, atmospheric pollutants and so on.

Two types of glazing are used: two-edge and four-edge. The two-edge system uses conventional head and sill rebates with glass retention beads, and the sides are silicone-bonded to mullions or styles. The four-edge system uses a structural adhesive to bond all four sides of the glass pane to a frame, with the sealant functioning as an integral structural link between glazing and frame. The lower edge of each glass pane is supported on resilient setting blocks or spacers of precured silicone to ensure adequate provision of sealant under the lower edge. The advantage of both systems is that a double-glazed unit can be installed into the rebated frames that enables the requirements of Part L in thermal U-values to be met. Total control of production quality (particularly for four-edge systems) will require factory assembly of glazing modules and close supervision of the bonding process. Standard application is to factory-bond the glass to an aluminium frame with in-situ silicone finishing and pointing provided between the pane and frame, as shown in Fig. 6.8.1. Some coated glass, and absorbent backgrounds such as timber or masonry, must be carefully primed before application of sealant.

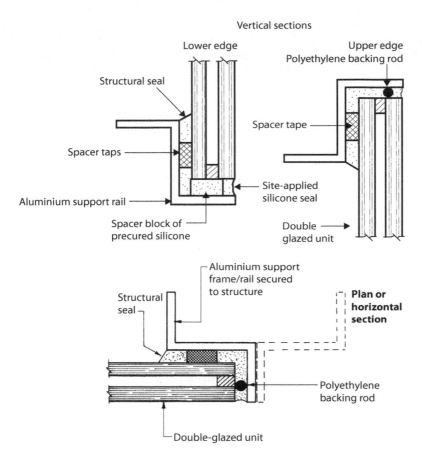

Figure 6.8.1   **Structural sealant glazing**

The advances in the fixing technology associated with utilising glass as a wall and ceiling enclosure have enabled large areas of glass to be used as frameless walls that make maximum use of its dramatic aesthetic effect and daylight light levels.

Mechanical fixing of façade glass can be subdivided into the systems that are used to provide the support system for the weight and loading from the glass. These are:

- bolted glazing – utilises the corner fixing of glass back to a supporting steel structural frame;
- cable-supported glazing – utilises tension wires and bolted connections;
- truss-supported glazing – utilises a steel lightweight truss and bolted connections;
- fin-supported glazing – utilises an opposing glass fin to add lateral support;
- suspended glazing – hung from the structural framework.

BOLTED SYSTEM

The bolted system utilises an aluminium or stainless steel specialist fitting that bolts to the glass through pre-prepared holes and then is bolted back to the supporting structure. An example of the bolted system is the Pilkington Planar System. This variation of frameless glazing reduces the visual impact of fixing plates. It has been developed by Pilkington UK Ltd on the principle of direct bearing of the glass via nylatron polyamide bushes onto the fixing bolts. Each pane is separately fixed independent of the pane below, with bolts having minimal clamping effect owing to attachment to a spring plate by predetermined torque settings (see Fig. 6.8.3) The top left drawing illustrates a single glass fixing element for single glazing. The system can utilise a spider bracket, which accommodates all four corner fixings in one cast-aluminium unit, allowing variable movement about the fitting. The spring plate system creates flexibility by reducing the concentration of stresses around the 'planar' fitting. A further advantage is application to either single or double glazing, the latter providing opportunity for variations in the selection of glass. A typical specification for a double-glazed system could be 10 or 12 mm toughened glass outer layer, tinted to deflect solar gains, with a 6 mm clear toughened glass inner layer and a 16 mm air space, as shown in Fig. 6.8.2. Also shown in this figure is the steel hollow-section frame support that the loads from the spider fitting are transferred over to relieve any stresses on the glazing; the bolted connection can easily be taken directly onto the structure of the building.

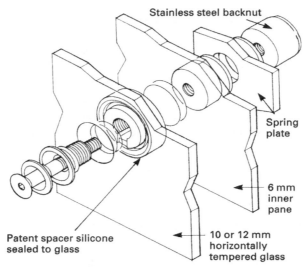

Stainless steel backnut

Spring plate

6 mm inner pane

10 or 12 mm horizontally tempered glass

Patent spacer silicone sealed to glass

Planar spring plate attachment for double glazing

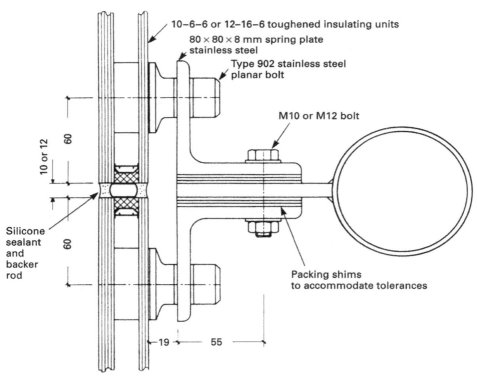

10–6–6 or 12–16–6 toughened insulating units

80 × 80 × 8 mm spring plate stainless steel

Type 902 stainless steel planar bolt

M10 or M12 bolt

10 or 12

60

60

Silicone sealant and backer rod

Packing shims to accommodate tolerances

19   55

Planar type 902 bolt assembly for attaching double-glazed panels to internal structure

Figure 6.8.2   'Planar' double glazing (courtesy of Pilkington Glass)

CABLE SYSTEM

The cable system uses tension rods or cables to provide a supporting network on the inside of the glass. The system works as a uniform balance: when wind pressure pushes against one side of the glass, it dissipates it along the cables and wires into the structure. By tensioning the cables the system can be made rigid to resist the lateral forces placed upon it. Cable frameworks are placed horizontally at each corner junction of the glazing panels, and vertically; in this way, forces are resisted in two directions (see Fig. 6.8.3. The cable system by design provides a very clear and uninterrupted appearance to the façade and can be considered as an architectural design feature.

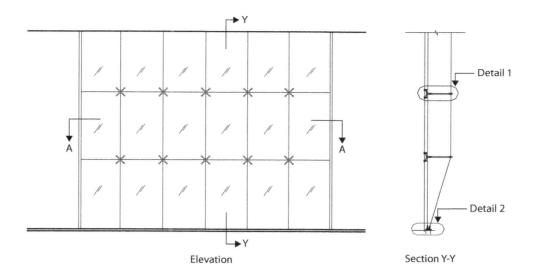

Elevation   Section Y-Y

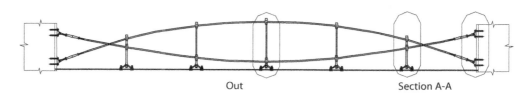

Out   Section A-A

Figure 6.8.3 **Fixing details of cable support system**

Source: Linox Technology

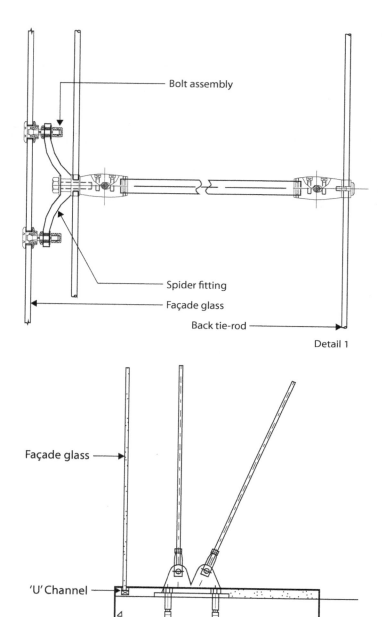

Bolt assembly

Spider fitting

Façade glass

Back tie-rod

Detail 1

Façade glass

'U' Channel

Detail 2

Figure 6.8.3 *(Continued)*

Source: Linox Technology

TRUSS SYSTEM

The truss is a more solid structure that is able to resist higher wind loads from the glazing. These architectural, designed structures are fixed at the top and bottom to secure them against twisting forces. The truss transfers the loads from the glazing bolted connections down the floor slab and foundation. Trusses are normally fabricated from circular hollow-section steel and are fully welded. The number of vertical trusses will depend on the load and the area of glass that is presented to the wind loadings. In extreme climate areas this may require a truss at every vertical junction of the glass. Figure 6.8.4 illustrates a typical steel truss support.

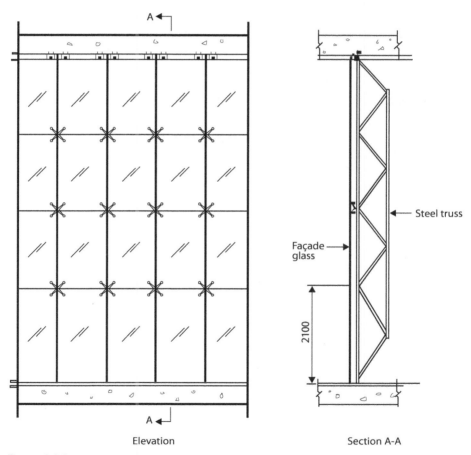

Elevation

Section A-A

Figure 6.8.4   **Steel truss support system to spider glazing fixings**

FIN SYSTEM

This system makes full use of a vertical 'T' shape, which incorporates a 90° form in glass. Forming a T-shape provides rigidity to the vertical sheets of glass by the use of small 'fins', which are bolted together at their tops and bottoms to form a continuous support. As with the steel trusses, these are bolted to the supporting structural frame of the building. Figure 6.8.5 illustrates the method employed along with a horizontal section through the joint.

The bottom row of structural glass is supported in a floor channel that uses structural silicone to secure the base of the glass. The finished floor screed hides the base of the channel, giving a seamless appearance to the finished structure. Stainless steel fittings are used throughout to provide protection against corrosion and extend the lifecycle required for maintenance. The vertical joint between the fin and the two opposing sheets of structural glass does however rely on a structural silicone sealant in order to maintain a waterproof internal environment and to assist with the distribution of stresses when under wind-loading conditions.

SUSPENDED SYSTEM

As the name suggests this system relies partly on the structure at the top of the glazed opening. By incorporating an attachment system the glass can be 'hung' from the frame at the top. Figure 6.8.6 illustrates in diagrammatic form the suspension system that is employed to hold the glass in place at the top. This allows only lighter fittings and fins to be used to support the rest of the structure.

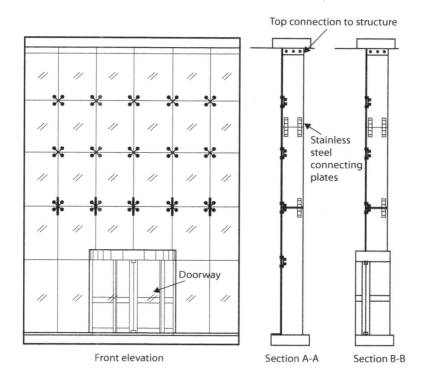

Top connection to structure

Stainless
steel
connecting
plates

Doorway

Front elevation

Section A-A

Section B-B

Cross-section through glass fin and external glazing

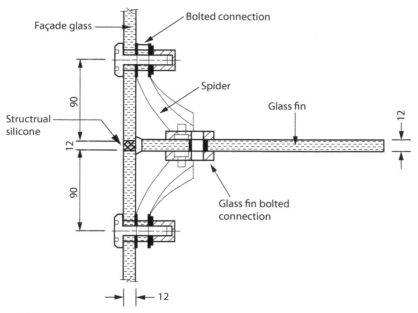

Façade glass

Bolted connection

Spider

Glass fin

Structrual
silicone

Glass fin bolted
connection

90

12

90

12

12

Figure 6.8.5   Glass fin supports to structural glazing

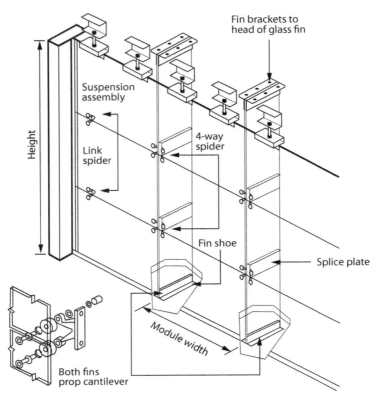

Figure 6.8.6 **A suspended system**
Source: Metro GlassTech

Sustainable construction 6.9

Existing buildings originally designed specifically for commercial, industrial or assembly purposes are too often demolished, as it is not financially or practically viable to adapt them for alternative use. Modern premises require vast areas for cabling and other services, but much of our building heritage was designed solely for its initial function, without consideration for the high-technology facilities needed today. The result is a relatively short-term cycle of demolition and new build. When designing and building framed structures and their cladding, the economic and environmental benefits of creating sustainable, versatile and adaptable construction are now recognised. The concept evolved gradually over the latter part of the last century. In 1968, the designer Robert Propst published *The office: A facility based on change*. This reference book presented the idea of buildings being adaptable to differing uses and functions as a strategic lifecycle design plan. Later, in the early 1970s, the architect Alex Gordon developed a design formula that became known as 'long life, loose fit, low energy'. These concepts, although radical 40 years ago, have proved now to be correct, given the effect that climate change and global warming are having and the dwindling supply of fossil fuels that we consume on earth.

LONG LIFE

The idea of 'long life' is to ensure that the design of a building is not just based on the limited life of its initial function. Planned obsolescence may be appropriate for inexpensive commodities, which can include temporary buildings and site huts. However, commercial buildings represent a huge financial investment, considerable construction time and significant use of diminishing material resources. Designers have a responsibility to society to ensure maximum benefit from a facility, and a responsibility to their clients to provide maximum financial return. Long life takes into consideration that a building is not only for tomorrow; it should have an in-built design to adapt and upgrade to future changes of use and occupation. The structure should be a shell or skeletal frame with a number of solutions for division of space and provision for services.

LOOSE FIT

'Loose fit' develops the concept further, with provision for accommodating numerous uses over the design life of a building.

The accommodation must be flexible in layout, so that successive occupants can configure and reconfigure the interior to suit their changing needs, working layouts and processes. This requires demountable partitioning, generally of modular format, suspended ceilings and raised flooring to facilitate state-of-the-art service installations.

The outward appearance or aesthetic perspective can be considered with regard to architectural fashion changes and the corporate image of the building's occupants. Loose fit in the form of replacement curtain-walling and over-cladding systems permits a complete exterior change to an existing frame, thereby considerably enhancing a structure without the cost of demolition and rebuilding. Such changes can run parallel with upgrading to current legislative standards, particularly thermal insulation and airtightness.

LOW ENERGY

'Low energy' refers to the last consideration under loose fit, which ensures that refurbishment associated with ongoing changes of use is subject to building control and is in accordance with current energy conservation measures, such as high-efficiency fuel-burning appliances, solar controls and double glazing. Low energy also refers to the time, labour and material cost or wastage that would otherwise be associated with the repetitive cycle of building and demolition that has been needed to meet ongoing architectural design trends and client requirements.

All structural investments do have an economic life: that is, because of deterioration and limited adaptability they reach a point at which the value of the site and building becomes no more than the value of the land on which the building stands. The objective of sustainable construction is to maximise the investment value in a building by ensuring that it has a long and useful life.

SUSTAINABLE TECHNIQUES

This section gives an outline of the techniques that are now available to construct a sustainable building. Each could be the subject of a full chapter in itself, but here we offer a short appraisal of a few of the more common techniques that are proving their worth in the long term in saving energy and finite resources.

THERMAL MASS

Increasing density by adding thermal mass to a building is a modern concept that is slowly being introduced. It utilises the capacity of some building materials to absorb heat and release it slowly, avoiding the extremes of heat that sometimes occur with modern homes. Dense concrete blocks and concrete can be used, but cement production does produce a high CO_2 emission.

ETFE ENVELOPES

ETFE envelopes are manufactured from ethyltetrafluroethylene (ETFE) and form an envelope roof structure to commercial buildings. They are manufactured into bags that can be single, double or triple layered. The ETFE is resistant to UV light and so does not degrade during exposure to sunlight. Low-pressure air is pumped into the bag to keep it constantly inflated. A perimeter retaining structure prevents any movement of the bags. The transmission of light through the material is excellent and the multi-layer construction gives an excellent U-value. Large spans are possible and printing onto the surface of the material reduces the amount of solar heat passing through the material. Under the action of rainwater, the surface self-cleanses, washing away any dirt. The material reacts well under fire and self-extinguishes. ETFE is a recyclable product, reduces costs in the supporting framework as it is lightweight, and reduces artificial lighting costs. Figure 6.9.1 shows the typical fixing arrangements for securing the EFTE bags.

THERMAL EFFICIENCY

STRAW BALE CONSTRUCTION

This sustainable method of construction is normally used as a single-storey application. Straw is not a loadbearing material that can carry excessive intermediate floor loads, so a supporting timber structure may be required as the straw is subjected to settlement. Straw's usefulness is as a thermal insulating material that is relatively cheap to obtain and can be rendered externally and plastered internally. The bales must be packed to a good density for use in constructing walls and are laid in a stretcher bond coursing. The foundation must be carefully designed to allow for drainage away from the base of the wall. Figure 6.9.2 illustrates details of straw bale construction.

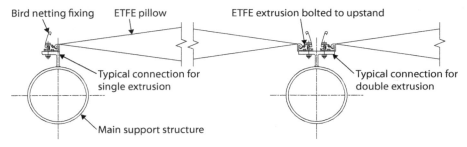

Typical section through an ETFE cushion

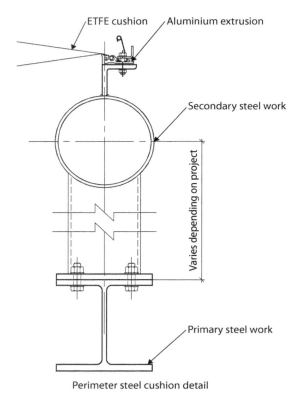

Perimeter steel cushion detail

Figure 6.9.1 **ETFE cushion roof structure**

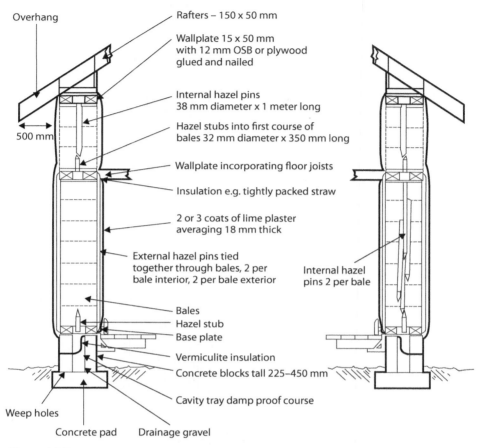

Section through a lightweight frame wall

Section through a loadbearing wall

Overhang

Rafters – 150 x 50 mm

Wallplate 15 x 50 mm
with 12 mm OSB or plywood
glued and nailed

500 mm

Internal hazel pins
38 mm diameter x 1 meter long

Hazel stubs into first course of
bales 32 mm diameter x 350 mm long

Wallplate incorporating floor joists

Insulation e.g. tightly packed straw

2 or 3 coats of lime plaster
averaging 18 mm thick

External hazel pins tied
together through bales, 2 per
bale interior, 2 per bale exterior

Internal hazel
pins 2 per bale

Bales

Hazel stub

Base plate

Vermiculite insulation

Concrete blocks tall 225–450 mm

Cavity tray damp proof course

Weep holes

Concrete pad   Drainage gravel

Figure 6.9.2   **Section through straw bale wall**

ENERGY EFFICIENCY

RAINWATER HARVESTING

Rainwater harvesting recycles rainwater and surface water, which is collected within an underground tank, where it is filtered and pumped to a service tank within the loft. This water is then used in non-potable applications, saving valuable resources on drainage for disposal, chemicals for water treatment and pollution. Figure 6.9.3 shows the complexity involved with the storage and use of rainwater.

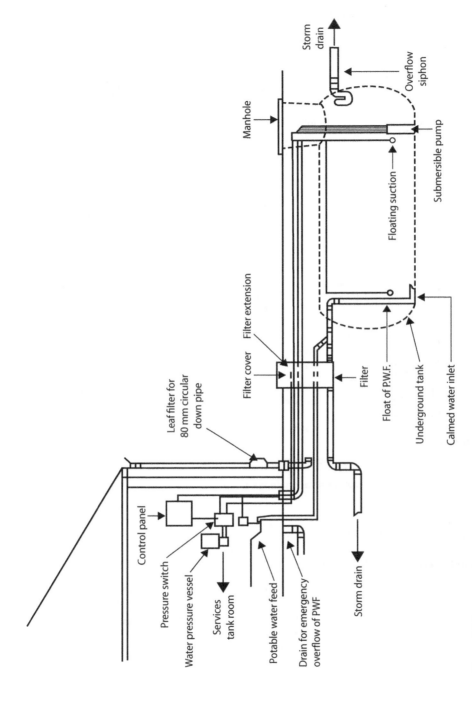

Control panel

Pressure switch

Water pressure vessel

Services
tank room

Potable water feed

Drain for emergency
overflow of PWF

Storm drain

Leaf filter for
80 mm circular
down pipe

Filter cover

Filter extension

Filter

Float of P.W.f.

Underground tank

Calmed water inlet

Manhole

Storm
drain

Overflow
siphon

Floating suction

Submersible pump

Figure 6.9.3   Section through rainwater harvesting system

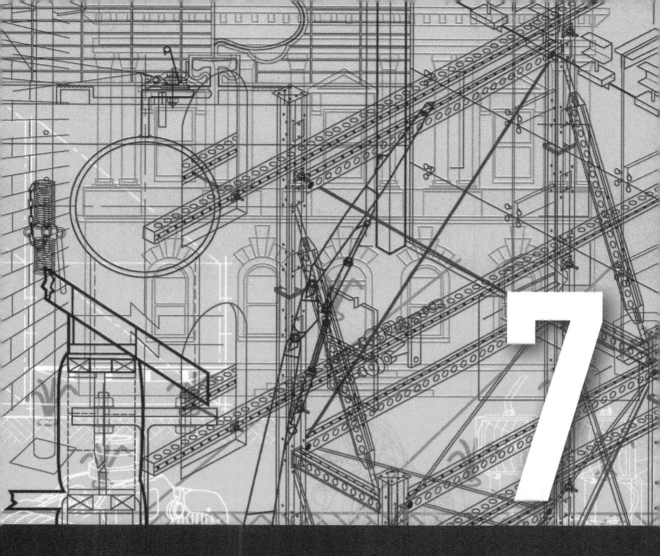

7

Formwork systems

Formwork   7.1

An introduction to the simple forms of formwork was provided in Chapter 10.3 of *Construction Technology*. Before considering different applications, it is worth defining some aspects of this work. Formwork is used generally to describe the temporary work associated with the formation of concrete structures and components. However, it can be subdivided into several areas.

■ **Falsework** This is timber, scaffolding or other temporary structural framing to support the forms, shutters or moulds for concrete work or masonry arches. Falsework can be of sawn timber but is more likely to be steel sections, scaffold tube and couplings, or a proprietary aluminum or steel system. The structure must have sufficient strength to prevent deformation, deflection and collapse under the full load of wet concrete, operatives and equipment. Falsework must remain in position until the permanent structure is sufficiently strong to support itself and ongoing work. In multi-storey structures, often lower-level props only are left in place as the upper floors are being constructed. Known as 'back-propping' or 're-propping', this is common procedure to speed up the rate of construction while the lower floors are still gaining their full strength.

■ **Formwork shuttering** This includes all site-erected systems that form the physical shape and texture of structural in-situ concrete. Formwork is the material in contact with the wet concrete during the curing process. The forms need to be constructed so as to attain the required surface specification of the concrete. Formwork shuttering materials comprise planed timber, plywood sheeting, steel and other materials used for supporting and shaping the finished concrete on site. Timber products will require mould oil treatment to the surface exposed to concrete, to stop cement adhering to the timber and subsequently spalling when the shuttering is struck. Timber also needs to be sourced from legal sources as recognised by either the Forest Stewardship Council (FSC) or Programme for the Endorsement of Forest Certification (PEFC) scheme.

- **Formwork moulds** These are generally associated with the off-site manufacture of precast concrete units, such as columns, panels, blocks and lintels, and feature work to mimic stonework, such as sills and coping stones. Moulds can be of planed timber, steel or plastics. Plastic materials are often used as a smooth lining or facing to timber.

The key standards and referencing guides for the use of concrete formwork and formwork systems are:

- BS 5975 *Code of practice for temporary works procedures and the permissible stress design of falsework*;
- CIRIA Report 136 *Formwork striking times: Criteria, prediction and method of assessment*.

Formwork provides the facility for wet concrete and reinforcement to form a particular shape with a predetermined strength. Depending on the complexity of the form, the relative cost of formwork to concrete can be as high as 75 per cent of the total cost to produce the required member. A typical breakdown of percentage costs could be:

- **concrete**   (materials 28%; labour 12%) = 40%;
- **reinforcement** (materials 18%; labour 7%) = 25%;
- **formwork**   (materials 15%; labour 20%) = 35%.

This breakdown shows that a building contractor drawing up a competitive tender will have to use an economic method of providing the necessary formwork, as this is the factor over which the company has most control.

Here are the economic essentials of formwork.

- **Low cost** Only spend the minimum amount necessary to produce the required form.
- **Strength** Select formwork materials and falsework carefully to get the most economic balance in terms of quantity used and continuing site activity around the assembled formwork.
- **Finish** Select method, materials and, if necessary, linings to produce the desired result direct from the formwork. Applied finishes are usually specified, so the method is the only real factor over which the builder has any economic control.
- **Transportation and assembly** Consider the use of patent systems and mechanical handling plant, and how to store, maintain and transport them safely.
- **Material** Consider the advantages of using timber, steel or glass-reinforced plastics. Generally, timber's light weight and adaptability make it the best material; steel and glass-reinforced plastics have more uses than timber, but cannot be repaired as easily.

- **Design** Within the confines of the architectural and/or structural design, formwork should be as repetitive and adaptable as possible. Timber forms have limited reuse – five or six is usual – with exposed surfaces suffering the most damage. Steel and plastics can be reused indefinitely if cared for. These are pressed or moulded into fixed shapes, with specific application to repetitive uses, such as columns and beams in multi-storey buildings.

- **Joints** These should be tight enough to prevent grout leakage. Seal joints with compressible plastic tape between components or with mastic or silicone sealant applied to junctions of adjacent pieces of formwork.

You will need to achieve a balance of these essentials, preferably at pre-tender stage, to calculate an economic and competitive cost.

Before reading on, you should re-read the basic principles of formwork in Chapter 10.3 of *Construction Technology*.

WALL FORMWORK

In principle the design, fabrication and erection of wall formwork are similar to those for column formwork (see *Construction Technology*). Several basic methods are available that will enable a wall to be cast in large quantities, defined lifts or continuous from start to finish.

TRADITIONAL WALL FORMWORK

Traditional wall formwork usually consists of standard framed panels tied together over their backs with horizontal members called **walings**. The walings fulfil the same function as yokes or column clamps, providing the resistance to the horizontal force of wet concrete. A 75 mm-high concrete kicker is formed at the base of the proposed wall to enable the forms to be accurately positioned and to help prevent the loss of grout by seepage at the base.

The usual assembly is to erect one side of the wall formwork, ensuring that it is correctly aligned, plumbed and strutted. The steel reinforcement cage is inserted and positioned before the formwork for the other side is erected and fixed. Keeping the forms parallel and at the correct distance from one another is most important; this can be achieved using precast concrete spacer blocks that are cast in, or steel spacer tubes that are removed after casting and curing, the voids created being made good, or by using one of the many proprietary wall-tie spacers available (see Fig. 7.1.1).

To keep the number of ties required within acceptable limits, horizontal members or walings are used; these also add to the overall rigidity of the formwork panels and help with alignment. Walings are best if composed of two members, with a space between to accommodate the shank of the wall-tie bolt: this will give complete flexibility in positioning the ties and leave the waling timbers undamaged for eventual reuse. To ensure that the loads are evenly distributed over the pair of walings, plate washers should be specified.

Plywood sheet is the common material used for wall formwork, but is vulnerable to edge and corner damage. The usual format is to make up wall forms as framed panels on a timber studwork principle, with a plywood facing sheet screwed to the studs so that it can be easily removed and reversed, for the maximum number of uses.

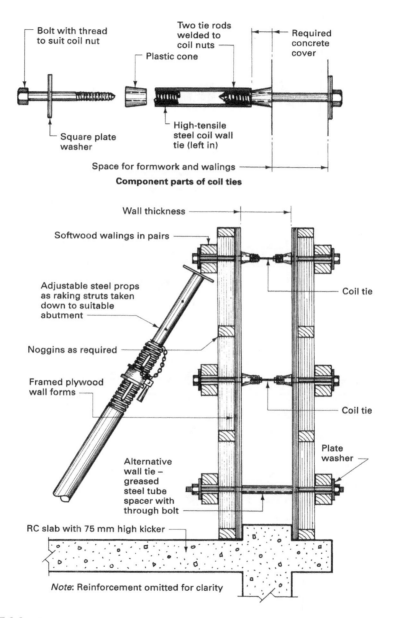

Figure 7.1.1 Traditional wall formwork details: 1

Corners and attached piers need special consideration, because the increased pressure at these points could cause the abutments between panels to open up, giving rise to unacceptable grout escape and a poor finish to the cast wall. The walings can be strengthened by including a loose tongue at the abutment position, and extra bracing could be added to internal corners (see Fig. 7.1.2). When considering formwork for attached piers, it is usually necessary to have special panels to form the reveals.

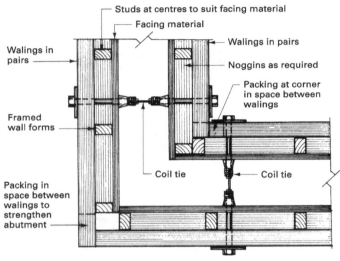

Plan on corner formwork

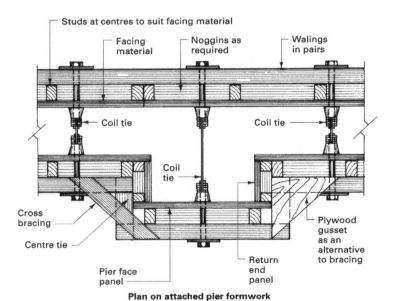

Plan on attached pier formwork

Figure 7.1.2   Traditional wall formwork details: 2

JUMP FORMWORK

Jump formwork systems are often described as climbing forms, although they do not 'self-climb'. They are suitable for construction of multi-storey vertical concrete elements in high-rise structures, such as:

- loadbearing walls and shear walls;
- lift, service and stair shafts;
- various tall civil engineering structures, such as bridge piers and pylons.

Jump forms combine the formwork and working platforms for the purposes of cleaning and setting up the formwork, fixing the steel reinforcement and carrying out the in-situ concreting. The formwork supports itself on the concrete cast earlier, usually on steel gallows brackets, so does not rely on independent falsework, such as dead shores or props. In general, there are two types of jump form that require the assistance of a crane for repositioning:

1. standard jump form – where units are individually lifted off the structure and relocated at the next construction level using a crane;
2. guided jump form – which also uses a crane, but offers greater safety and control during lifting as units remain anchored to the structure.

Jump formwork is a highly productive system, designed to increase speed and efficiency while minimising labour and crane time. Systems are normally modular, so can be joined to form long lengths to suit varying rectilinear or curved construction geometries. Note that modern jump forms tend to be patented systems, usually made from steel or laminated waterproof timberboard materials.

SLIP-FORM FORMWORK

Slip-form formwork is a system that slides continuously up the face of the wall being cast, by climbing up and being supported by a series of hydraulic jacks operating on jacking rods. The whole wall is cast as a monolithic and jointless structure, making the method suitable for tall structures such as water towers, chimneys and the cores of multi-storey buildings with repetitive floors. This technique is often referred to as 'slipforming'.

Because the system is a continuous operation, good site planning and organisation are essential, and will involve:

- round-the-clock working, with shift working and artificial lighting to enable work to carry on outside normal daylight hours;
- careful control of concrete supply to avoid stoppages of the lifting operation, and potentially having standby plant as an insurance against mechanical breakdowns;
- suitably trained staff used to this method of constructing in-situ concrete walls.

The architectural and structural design must be suitable for the application of a slipform system. Generally the main requirements are a wall of uniform thickness with a minimum number of openings and a height of at least 20 m, to make the cost of equipment, labour and planning economic.

Here are the basic components of slip formwork.

- **Side forms** Of timber and/or steel construction, these need to be strongly braced, and are loadbearing. Steel forms are heavier than timber, and more difficult to assemble and repair, but have lower frictional loading, are easier to clean and have better durability. Timber forms are lighter, have better flexibility and are easier to repair. A typical timber form would consist of a series of 100 mm × 25 mm planed straight-grained staves assembled with a 2 mm-wide gap between consecutive boards to allow for swelling, which could give rise to unacceptable friction as the forms rise. The forms are usually made to a height of 1.2 m, with an overall sliding clearance of 6 mm by keeping the external panel plumb and the internal panel tapered so that it is 3 mm in at the top and 3 mm out at the bottom, giving the true wall thickness in the centre position of the form. The side forms must be adequately stiffened with horizontal walings and vertical puncheons to resist the lateral pressure of concrete and transfer the loads of working platforms to the supporting yokes.

- **Yokes** These assist in supporting the suspended working platforms and transfer the platform and side-form loads to the jacking rods. Yokes are usually made of framed steelwork suitably braced and designed to provide the necessary bearings for the working platforms.

- **Working platforms** Three working levels are usually provided. The first is situated above the yokes at a height of about 2 m above the top of the wall forms for the use of the steel fixers. The second level is a platform over the entire inner floor area at a level coinciding with the top of the wall forms, and is used by the concrete gang for storage of materials and to carry levelling instruments and jacking control equipment. This decking could ultimately be used as the soffit formwork to the roof slab, if required. The third platform is in the form of a hanging or suspended scaffold, usually to both sides of the wall. This gives access to the exposed freshly cast concrete below the slip formwork, for finishing operations.

- **Hydraulic jacks** The jacks used are usually specified by their loadbearing capacities, such as 3 tonnes or 6 tonnes, and consist of two clamps operated by a piston. The clamps operate on a jacking rod of 25 to 50 mm diameter according to the design load, and are installed in banks operated from a central control to give an all-round consistent rate of climb. The upper clamp grips the jacking rod, and the lower clamp, being free, rises, pulling the yoke and platforms with it until the jack extension has been closed. The lower clamp now grips the climbing rod while the upper clamp is released and raised to a higher position, when the lifting cycle is recommended. Factors such as temperature and concrete quality affect the rate of climb, but typical speeds are between 150 and 450 mm per hour.

The upper end of the jacking rod is usually encased in a tube or sleeve to overcome the problem of adhesion between the rod and the concrete.

The jacking rod remains loose in the cast wall and can be recovered at the end of the jacking operation. The 2.5 to 4 m lengths of rod are usually joined together with a screw joint arranged so that all such joints do not occur at the same level.

A typical arrangement of sliding formwork is shown diagrammatically in Fig. 7.1.3.

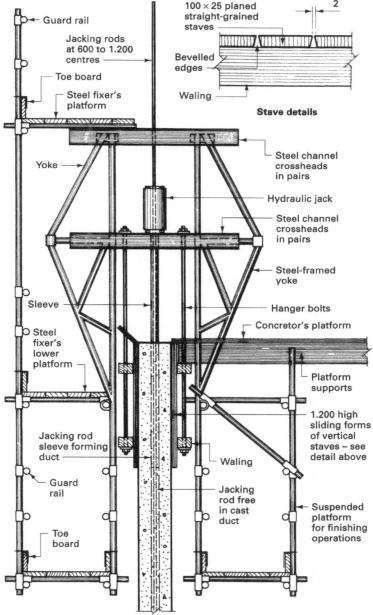

Figure 7.1.3 **Diagrammatic arrangement of sliding formwork**

The site operations start with the formation of a substantial kicker 300 mm high incorporating the wall and jacking rod starter bars. The wall forms are assembled and fixed together with the yokes, upper working platforms and jacking arrangement, after which the initial concrete lift is poured. The initial rate of climb must be slow, to allow time for the first batch of concrete to reach a suitable condition before emerging from beneath the sliding formwork. A standard or planned rate of lift is usually reached within about 16 hours after starting the lifting operation.

Openings can be formed in the wall using framed formwork with splayed edges, to reduce friction, tied to the reinforcement. Small openings can be formed using blocks of expanded polystyrene, which should be 75 mm less in width than the wall thickness so that a layer of concrete is always in contact with the sliding forms, to eliminate friction. The concrete cover is later broken out and the blocks removed. Chases for floor slabs can be formed with horizontal boxes drilled to allow the continuity reinforcement to be passed through and to be bent back within the thickness of the wall; when the floor slabs are eventually cast, the reinforcement can be pulled out into its correct position.

PERMANENT WALL FORMWORK

In certain circumstances formwork is left permanently in place because of the difficulty and/or cost of removing it once the concrete has been cast. It can also be a means of using the facing material as both formwork and thermal insulation, especially in the construction of in-situ reinforced concrete walls. This system is known as insulating concrete formwork (ICF), and is based on hollow, lightweight expanded polystyrene block components that lock together without the need for bedding materials, such as mortar or adhesives into which an in-situ concrete core is poured. It has been used elsewhere in Europe for some years, and is now beginning to be used in the UK, particularly for basement wall construction, although it is equally useful for constructing superstructural walls.

This unit provides a permanent formwork unit which can be constructed with relatively low-tech equipment and skills: for example, the blocks can be cut and shaped with a simple hand saw. Wall heights of up to 3 m can be cast in one pour, without the use of temporary bracing. This is normally achieved with a pump-grade concrete mix using 10 mm maximum aggregate size.

A number of different ICF systems are available. Some arrive on site in flat packs, which can be clipped or slotted together; others come as preformed box units. A typical expanded polystyrene (EPS) box unit is shown in Fig. 7.1.4. These EPS units have good thermal properties, with U-values as low as 0.11 W/m^2 K possible. The ICF wall can be rendered with a standard waterproof render or clad in a variety of modern and traditional materials, such as brickwork or rain-screen cladding. Some systems replace the EPS with a 24 mm-thick cement-bound particle board.

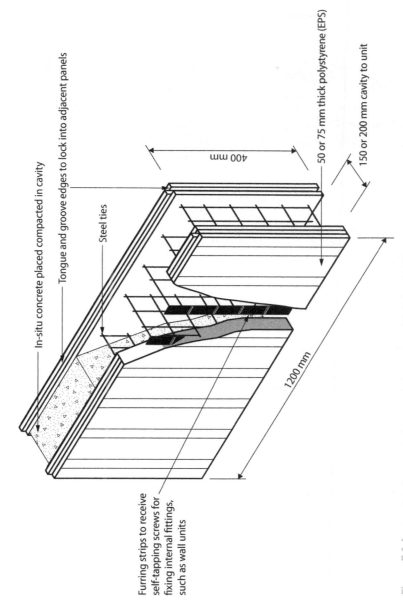

In-situ concrete placed compacted in cavity

Tongue and groove edges to lock into adjacent panels

Steel ties

400 mm

50 or 75 mm thick polystyrene (EPS)

150 or 200 mm cavity to unit

1200 mm

Furring strips to receive self-tapping screws for fixing internal fittings, such as wall units

Figure 7.1.4 Insulated concrete formwork (courtesy of PolySteel UK)

Patent formwork 7.2

Patent formwork is sometimes called system formwork, and is usually identified by the manufacturer's name. All proprietary systems have the same common aim, and most are similar in their general approach to solving some of the problems encountered with formwork for modular or repetitive structures. As explained in the previous section, formwork is one area where the contractor has most control over the method and materials to be used.

In trying to design or formulate the ideal system for formwork, the following must be considered:

- **strength** to carry the concrete and working loads;
- **lightness without strength reduction** to enable maximum-size units to be employed;
- **durability without prohibitive costs** to gain maximum usage of materials;
- **good and accurate finish straight from the formwork** to reduce the costly labour element of making good and patching, which in itself is difficult to do without it being obvious that this kind of treatment was necessary;
- **erection and dismantling times**;
- **ability to employ unskilled or semi-skilled labour**.

Patent or formwork systems have been devised to satisfy most of these requirements by standardising forms and using easy methods of securing and bracing the positioned formwork.

The major components of any formwork system are unit panels, which should be:

- available in a wide variety of sizes based on a standard module (usually multiples and submultiples of 300 mm);

- manufactured from durable materials;

- covered with a facing material that is durable and capable of producing the desired finish;

- interchangeable so that they can be used for beams, columns and slabs;

- formed so that they can be easily connected together to form large unit panels;

- lightweight so that individual unit panels can be handled without mechanical aid;

- designed so that the whole formwork can be assembled and dismantled easily by unskilled or semi-skilled labour, with fixing mechanisms as shown in Fig. 7.2.1;

- capable of being adapted so that non-standard width inserts of traditional formwork materials can be included where lengths or widths are not exact multiples of the unit panels;

- equipped with integral safe working platforms, ladders and edge protection to reduce construction hazards, particularly falls from height when fixing reinforcement and placing or compacting the wet concrete.

Examples of some patent formwork manufacturers and systems available in the UK and Europe are shown in Table 7.2.1. For more detailed information, readers can study the various systems from the manufacturers' websites. Essentially, each manufacturer provides a kit of interchangeable components comprising form lining, panel or grid supports, raking or vertical props and safe access platforms. The main configurations used in patented formwork are:

- framed panel systems;

- lightweight grid systems;

- braced beam systems.

Most unit panels and grids consist of a framed tray made from angle or channel sections of galvanised light metal or aluminum, stiffened across the width as necessary. The edge framing is usually perforated with slots or holes to take the fixing connectors and waling clips or clamps. The facing is of sheet metal or plywood; some manufacturers offer a choice of sealed or plastic-coated finishes for improved reuse. Longitudinal stiffening and support are given by clamping, over the backs of the assembled panels, special walings of hollow section or, in many systems, standard scaffold tubes. Where required, vertical support can be given by raking standard adjustable steel 'push-pull' props with pivoted end plates. Where heavy loadings are encountered, most systems have special vertical stiffening and support arrangements based on designed girder principles, which also provide support for an access and working platform. Spacing between opposite forms is maintained using wall ties or similar devices, as described for traditional formwork.

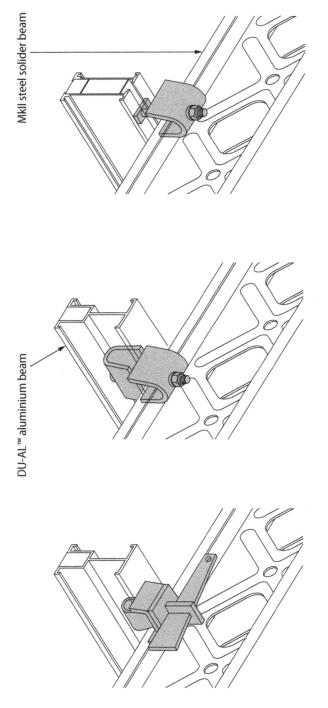

MkII steel solider beam

DU-AL™ aluminium beam

Wedge clamp
A quick-action fixing method suitable for connecting any combination of DU-AL™ and MkII Soldier beams.

Universal clamp
Connects secondary beams to primary beams. Recommended where forms are to be handled repeatedly or are likely to be subject to considerable vibration.

Universal anchor clamp
A longer version of the T-bolt combined with a universal clamp arm and captive nut.

Figure 7.2.1 Typical fixing mechanisms for modern formwork systems

Table 7.2.1 Types of proprietary wall and slab form systems currently available

| Proprietary name | Examples of typical formwork product systems available | | Website link |
	Walls	Slabs	
PERI	*Maximo* – steel frame/coated plywood panels and walls ties *Vario* – vertical timber lattice girders and plywood panels braced with push-pull props *Rundflex* – vertical steel box-girder frame with adjustable curved horizontal wallers, allowing curved plywood forms to be fixed	*Skydeck* – aluminum props and panels with plastic-coated plywood forms – see *Construction Technology*, Chapter 10.3 *Gridflex* – lightweight aluminium props and flat grids fixed to coated plywood forms	www.peri.ltd.uk
Harsco-SGB Formwork	*Manto* – hot-dip galvanised steel frames with high-grade plastic-coated form linings *Multiform A-frames* – used to support single-sided formwork support, as in basement wall construction. Uses modular sizes with a sliding foot feature to tighten from against kicker	*Topmax* – table forms made from hot-dip galvanised and powder-coated steel frame, covered with plastic-coated form linings *Topec* – large, lightweight steel grid-based timber ply panel slabs forms: see Fig. 7.2.2 (b) *MkII & DU-AL Beams* – primary and secondary aluminum beams with timber inserts for nailing plywood forms: can also be used for walls with support	www.harsco-i.co.uk
Doka	*Framex Xlife* – steel-framed system for large-scale wall casts *Dokaset* – steel panel frame with plastic-coated plywood screed from the rear to leave no imprints on the concrete	*Dokaflex 30 Tec* – versatile lightweight hand-set timber beams system supported on steel props *Dokamatic table form system* – prefabricated formwork tables for large-scale flat slab construction	www.doka.com
RMD Kwikform	*Minima* – lightweight, modular formwork panel system designed to take up to 60 kN/m^2 *Trapeze* – variable circular formwork system that can form radii of 2.5+ m and withstand concrete pressures up to 60 kN/m^2	*Airodek* – lightweight soffit formwork system, capable of supporting concrete up to 450 mm thick	www.rmdkwikform.com

Walls that are curved in plan can be formed in a similar manner to the straight wall techniques described above, using a modified tray that has no transverse stiffening members, making it flexible. Climbing formwork can also be carried out using system formwork components: instead of reversing the forms as for traditional formwork, climbing shoes are bolted to the cast section of the wall to act as bearing corbels to support the soldiers for each lift.

When forming beams and columns, the unit panels are used as side or soffit forms held together with steel column clamps or, in the case of beams, conventional strutting, or using wall ties through the beam thickness. Some manufacturers produce special beam box clamping devices that consist of a cross-member surmounted by attached and adjustable triangulated struts, to support the side forms.

Many patent systems for the construction of floor slabs that require propping during casting use basic components of unit panels, narrow width (150 mm) filler panels, special drop-head adjustable steel props, joists and standard scaffold tubes for bracing. The steel, aluminium or composite timber lattice joists are lightweight and purpose-made, and are supported on the secondary head of the prop in the opposite direction to the filler panels, which are also supported by the secondary or drop head of the prop. The unit panels are used to infill between the filler panels and pass over the support joist; the upper head of the prop is at the same level as the slab formwork, and is part of the slab soffit formwork.

After casting the slab and allowing it to gain sufficient strength, the whole of the slab formwork can be lowered and removed, leaving the undisturbed prop head to give the partially cured slab a degree of support. This enables the formwork to be removed at a very early stage, releasing the unit panels, filler pieces and joists for reuse, and at the same time accelerating the drying out of the concrete by allowing a free air flow on both sides of the slab.

Most lightweight grid systems can be erected by one person from the floor below, by hooking the grid onto the tops of the vertical props and pushing them up with erection poles. This avoids operatives working from height, keeping them safe from falls, as can be seen in the SGB Topec system shown in Fig. 7.2.2.

Slabs of moderate spans (up to 7.5 m) that are to be cast between loadbearing walls or beams without the use of internal props can be formed using unit panels supported by steel telescopic floor centres. These centres are made in a simple lattice form, extending in one or two directions according to span, and are light enough to be handled by one operative. They are pre-cambered to compensate for any deflections when loaded. Typical examples of system formwork are shown in Figs 7.2.3 to 7.2.5.

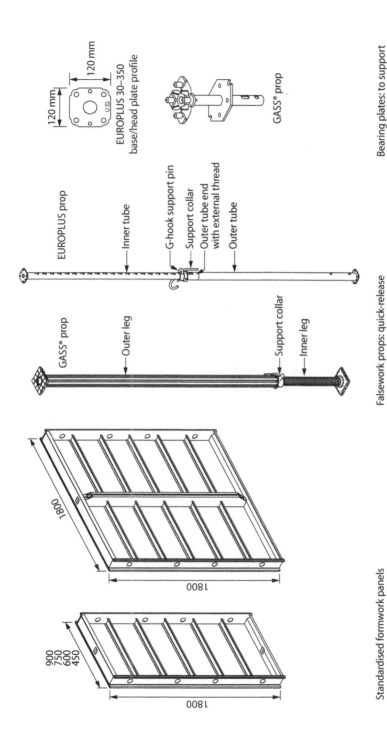

120 mm

120 mm

EUROPLUS 30–350
base/head plate profile

GASS® prop

Bearing plates: to support
four floor panels and for
quick release

EUROPLUS prop

Inner tube

G-hook support pin

Support collar

Outer tube end
with external thread

Outer tube

Outer tube

Falsework props: quick-release
lightweight steel standard
EUROPLUS prop and an alternative
heavier-duty GASS prop for greater
flexibility in layout of props

GASS® prop

Outer leg

Support collar

Inner leg

Standardised formwork panels
to fit grid with integral 10 mm
thick 7-ply timber board coated
with plastic

1800

1800

900
750
600
450

1800

a) Harsco Infrastructure 'Topec' falsework and formwork components

Figure 7.2.2 'Harsco SGB' slab formwork

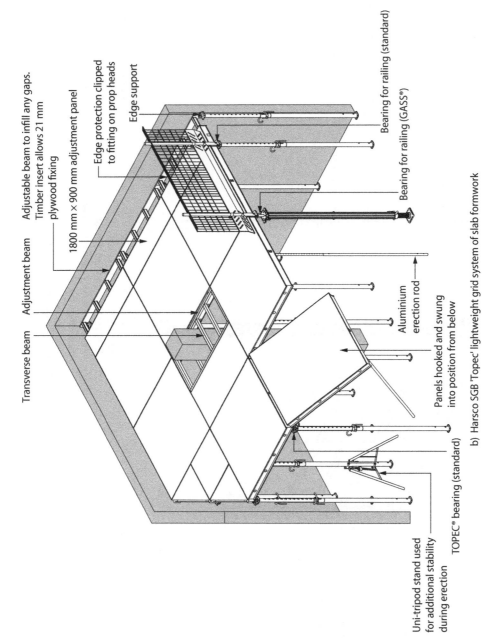

Adjustable beam to infill any gaps.
Timber insert allows 21 mm
plywood fixing

1800 mm × 900 mm adjustment panel

Adjustment beam

Transverse beam

Edge protection clipped
to fitting on prop heads

Edge support

Bearing for railing (standard)

Bearing for railing (GASS®)

Bearing for railing

Aluminium
erection rod

Panels hooked and swung
into position from below

TOPEC® bearing (standard)

Uni-tripod stand used
for additional stability
during erection

b) Harsco SGB 'Topec' lightweight grid system of slab formwork

Figure 7.2.2 *Continued*

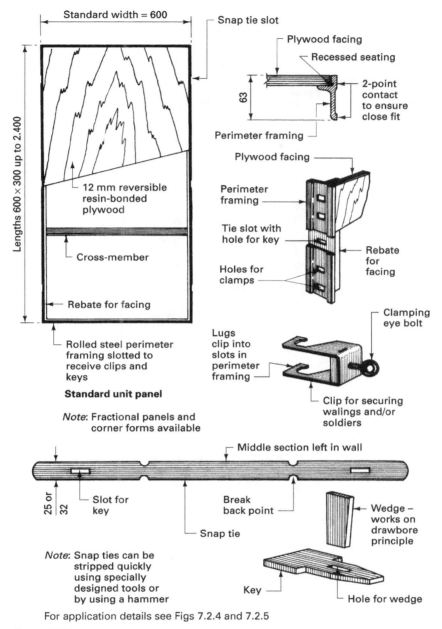

Standard width = 600

Snap tie slot

Plywood facing

Recessed seating

2-point contact to ensure close fit

63

Perimeter framing

Lengths 600 × 300 up to 2.400

12 mm reversible resin-bonded plywood

Plywood facing

Perimeter framing

Tie slot with hole for key

Rebate for facing

Cross-member

Holes for clamps

Rebate for facing

Clamping eye bolt

Rolled steel perimeter framing slotted to receive clips and keys

Lugs clip into slots in perimeter framing

Standard unit panel

Note: Fractional panels and corner forms available

Clip for securing walings and/or soldiers

Middle section left in wall

25 or 32

Slot for key

Break back point

Wedge – works on drawbore principle

Snap tie

Note: Snap ties can be stripped quickly using specially designed tools or by using a hammer

Key

Hole for wedge

For application details see Figs 7.2.4 and 7.2.5

Figure 7.2.3 System formwork: typical components

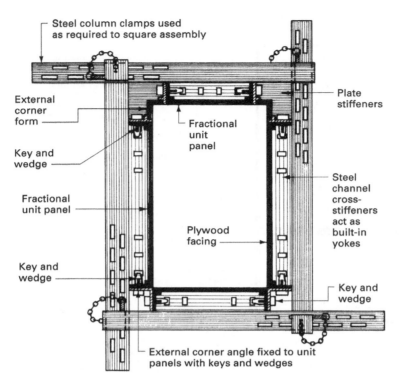

Steel column clamps used
as required to square assembly

Plate
stiffeners

External
corner
form

Fractional
unit
panel

Key and
wedge

Steel
channel
cross-
stiffeners
act as
built-in
yokes

Fractional
unit panel

Plywood
facing

Key and
wedge

Key and
wedge

External corner angle fixed to unit
panels with keys and wedges

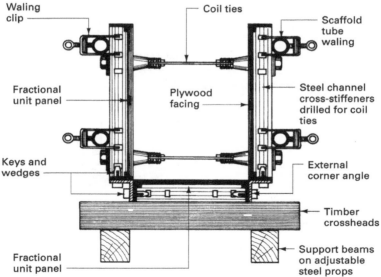

Waling
clip

Coil ties

Scaffold
tube
waling

Fractional
unit panel

Plywood
facing

Steel channel
cross-stiffeners
drilled for coil
ties

Keys and
wedges

External
corner angle

Timber
crossheads

Fractional
unit panel

Support beams
on adjustable
steel props

Figure 7.2.4 System formwork: columns and beams

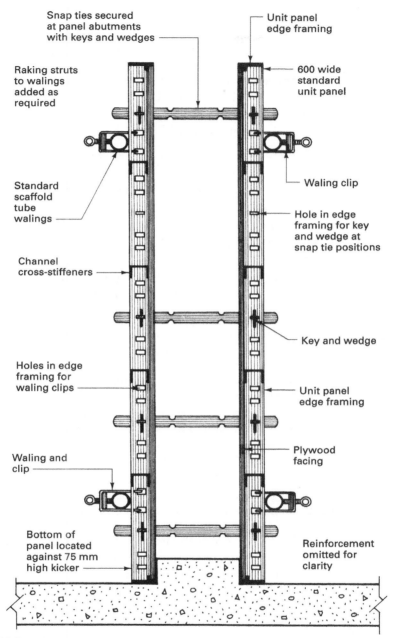

Snap ties secured at panel abutments with keys and wedges

Unit panel edge framing

Raking struts to walings added as required

600 wide standard unit panel

Standard scaffold tube walings

Waling clip

Channel cross-stiffeners

Hole in edge framing for key and wedge at snap tie positions

Key and wedge

Holes in edge framing for waling clips

Unit panel edge framing

Waling and clip

Plywood facing

Bottom of panel located against 75 mm high kicker

Reinforcement omitted for clarity

Figure 7.2.5   System formwork: walls

This special class of formwork has been devised for use when casting large, repetitive floor slabs in medium- to high-rise structures. The main objective is to reduce the time factor in erecting, striking and re-erecting slab formwork by creating a system of formwork that can be struck as an entire unit, removed, hoisted and repositioned without any dismantling. These table forms are largely formed from panel or gridded component systems, as above.

The basic requirements for a system of table formwork are:

- means of adjustment for aligning and levelling the forms;
- adequate means of lowering the forms so that they can be dropped clear of the newly cast slab; generally the provision for lowering the forms can also be used for final levelling purposes;
- means of manoeuvring the forms clear of the structure to a point where they can be attached to the crane for final extraction, lifting and repositioning, ready to receive the next concrete-pour operation;
- means of providing a working platform at the external edge of the slab to eliminate the need for an independent scaffold, which would be obstructive to the system.

The basic support members are usually a modified version of inverted adjustable steel props. These props, suitably braced and strutted, carry a framed decking that acts as the soffit formwork. To manoeuvre the forms into a position for attachment to the crane, a framed wheeled arrangement can be fixed to the rear end of the tableform, so that the whole unit can be moved forward with ease. The tableform is picked up by the crane at its centre of gravity by removing a loose centreboard to expose the framework. The unit is then extracted clear of the structure, hoisted in the balanced horizontal position and lowered onto the recently cast slab for repositioning.

Another method, devised by RMD Kwikform Ltd, uses a special lifting beam suspended from the crane at predetermined sling points, which are lettered so that the correct balance for any particular assembly can be quickly identified, each table having been marked with the letter point required. The lifting beam is connected to the working platform attachment of the tableform so that, when the unit formwork is lowered and then extracted by the crane, the tableform and the lifting beam are in perfect balance. This method of system formwork is illustrated in Fig. 7.2.6.

The selection of formwork for multi-storey and complex structures is not a simple matter; it requires knowledge and experience in design appreciation, suitability of materials and site operations. For major works, main contractors will generally prefer to limit their direct involvement to administration only, and seek to subcontract this area of work to a specialist company that provides a combined design and install service. These subcontractors will have

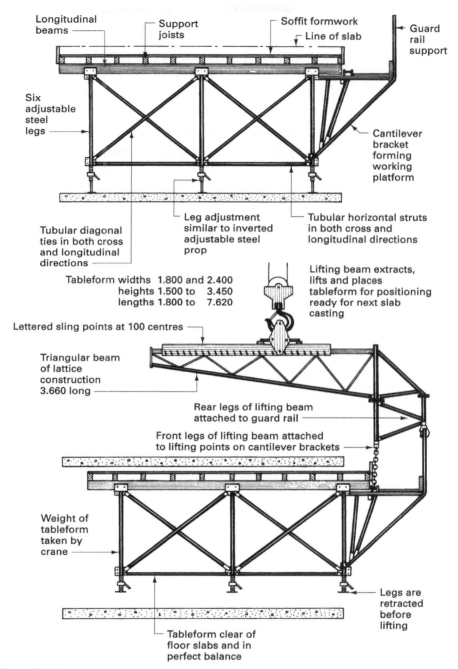

Longitudinal beams

Support joists

Soffit formwork

Line of slab

Guard rail support

Six adjustable steel legs

Cantilever bracket forming working platform

Tubular diagonal ties in both cross and longitudinal directions

Leg adjustment similar to inverted adjustable steel prop

Tubular horizontal struts in both cross and longitudinal directions

Tableform widths 1.800 and 2.400
heights 1.500 to 3.450
lengths 1.800 to 7.620

Lifting beam extracts, lifts and places tableform for positioning ready for next slab casting

Lettered sling points at 100 centres

Triangular beam of lattice construction 3.660 long

Rear legs of lifting beam attached to guard rail

Front legs of lifting beam attached to lifting points on cantilever brackets

Weight of tableform taken by crane

Legs are retracted before lifting

Tableform clear of floor slabs and in perfect balance

Figure 7.2.6 Tableforms: 'Kwikform' system

experienced and specialist site staff to fabricate the formwork or, if using a patent system, to supervise the semi-skilled or unskilled labour being used. Often the formwork manufacturers provide a package of design and logistical services to both main and specialist r.c. subcontractors, rather than just hiring out formwork kit and equipment. This package includes:

- assisting in the project development phase;
- re-engineering the temporary work for better buildability;
- providing consultation during tendering and programming phases;
- offering practical guidance and consultancy in the logistics of transportation and erection of the forms;
- advising on final dismantling and return of the equipment.

Although formwork is essentially a temporary operation, it still needs all the care and thought with regards to quality as if it were a permanent part of the works. Badly designed or erected formwork can result in dangerous working conditions for operatives and failure of the structure during the construction period, or inaccurate and unacceptable members being cast, both of which can be financially disastrous for a company and ruinous to the firm's reputation.

Concrete surface finishes 7.3

The appearance of concrete members is governed by their surface finish, which is influenced by colour and texture. The method used to produce the concrete member will have some degree of influence over the finish; greater control of quality is usually possible with precast concrete techniques under factory-controlled conditions. Most precast concrete products can be cast in the horizontal position, which again promotes better control over the resultant casting, whereas in-situ casting of concrete walls and columns must be carried out using vertical-casting techniques, which not only have a lesser degree of control but also limit the types of surface treatment possible direct from the mould or formwork.

COLOUR

The colour of concrete as produced direct from the mould or formwork depends on the colour of the cement being used and, to a lesser extent, the colour of the fine aggregate. The usual methods for varying the colour of finished concrete are:

- **using a coloured cement** – the range of colours available is limited, and most are pastel shades; if a pigment is used, the cement content of the mix will need to be increased by approximately 10 per cent to counteract the loss of strength due to the colouring additive;

- **using the colour of the coarse aggregate** – the outer matrix or cement paste is removed to expose the aggregate, which imparts not only colour but also texture to the concrete surface.

Concrete can become stained during the construction period, resulting in a mottled appearance. Here are some of the causes of this form of staining.

- **Using different-quality timber within the same form** Generally timber that has a high absorption factor will give a darker concrete than timber with a low absorption factor. The same disfigurement can result from using old and new timber in the same form, because older timber tends to give darker concrete than new timber with its probable higher moisture content and hence lower absorption.

- **Formwork detaching itself from the concrete** This allows dirt and dust to enter the space and attach itself to the 'green' concrete surface.
- **Type of release agent used** Generally the thinner the release agent used, the better the result; even coating over the entire contact surface of the formwork or mould is also of great importance.

Staining occurring on mature concrete, usually after completion and occupation of the building, is often due to bad design, poor selection of materials or poor workmanship. Large overhangs to parapets and sills without adequate throating can create damp areas that are vulnerable to algae or similar growths and pollution attack. Poor detailing of damp-proof courses can create unsightly stains because they fail to fulfil their primary function of providing a barrier to dampness infiltration. Efflorescence can occur on concrete surfaces, although it is not as common as on brick walls. The major causes of efflorescence on concrete are allowing water to be trapped for long periods between the cast concrete and the formwork, and poorly formed construction or similar joints allowing water to enter the concrete structure. Removal of efflorescence is not easy or always successful, so the emphasis should always be on good design and workmanship in the first instance. Methods of efflorescence removal range from wire brushing through various chemical applications to mechanical methods, such as grit-blasting.

TEXTURE

When concrete first became acceptable as a substitute for natural stone in major building works, the tendency was to try to recreate the smooth surface and uniform colour possible with natural stones. This kind of finish is difficult to achieve using the concrete because:

- natural shrinkage of concrete can cause hairline cracks on the surface;
- texture and colour can be affected by the colour of the cement, water content, degree of compaction and the quality of the formwork;
- pinholes on the surface can be caused by air being trapped between the concrete and the formwork;
- rough patches can result from the formwork adhering to the concrete face;
- grout leakage from the formwork can cause fins or honeycombing on the cast concrete.

Under site conditions, it is not easy to keep sufficient control over the casting of concrete members to guarantee that these faults will not occur. However, greater control is possible with the factory-type conditions prevailing in a well-organised precast concrete works. Note that attempts to patch, mask or make good the above defects are nearly always visible.

Methods that can be used to improve the appearance of a concrete surface include:

- finishes obtained direct from the formwork or mould intended to conceal the natural defects by attracting the eye to a more obvious and visual point;
- special linings placed within the formwork to produce a smooth or profiled surface;
- removal of the surface matrix to expose the aggregate;
- applied surface finishes, such as ceramic tiles, renderings and paints.

Formwork can be designed and constructed to highlight certain features, such as the joints between form members or between concrete pours or lifts, by adding to the inside of the form small fillets to form recessed joints, or conversely by recessing the formwork to form raised joints. The axiom here is, if you cannot hide or mask a joint, make a feature of it. Using sawn boards to imprint the grain pattern can give a visually pleasing effect to concrete surfaces, particularly if the boards used are narrow.

A wide variety of textured, patterned and profiled surfaces can be obtained by using different linings within the formwork. Typical materials used are thermoplastics, glass fibre mouldings, moulded rubber and PVC sheets; all of these materials can be obtained to form various patterns, giving many uses, and are easily removed from the concrete surface. Glass fibre and thermoplastic linings have the advantage of being mouldable to any reasonable shape or profile. Ribbed and similar profiles can be produced by fixing materials such as troughed or corrugated steel into the form, as a complete or partial lining, to give a panelled effect. Coarse fabrics, such as hessian, can also be used to produce various textures, provided the fabric is first soaked with a light mineral oil and squeezed to remove the excess oil before fixing, as a precaution against the fabric becoming permanently embedded on the concrete surface.

The most common method of imparting colour and texture to a concrete surface is to expose the coarse aggregate, which can be carried out by a variety of methods depending generally on whether the concrete is 'green', mature, or yet to be cast. Brushing and washing is a common method employed on 'green' concrete, and is carried out using stiff wire or bristle brushes to loosen the surface matrix and clean water to remove the matrix and clean the exposed aggregate. The process should be carried out as soon as practicable after casting, which is usually within two to six hours of the initial pour, but certainly not more than 18 hours after casting. When treating horizontal surfaces, brushing should start at the edges, whereas brushing to vertical surfaces should start at the base and be finished by washing from the top downwards. Skill is required with this method of exposing aggregates to ensure that only the correct amount of matrix is removed, the depth being dependent on the size of aggregate used. Retarding agents can be employed, but it is essential that these are used only in strict accordance with the manufacturer's instructions.

Treatments that can be given to mature concrete to expose the aggregate include tooling, bush hammering and blasting techniques. These methods are generally employed when the depth of the matrix to be removed is greater than that usually associated with the brushing method described above. The most suitable age for treating the mature concrete surface depends on such factors as the type of cement used and the conditions under which the concrete is cured, the usual period being from two to eight weeks after casting.

The usual hand-tooling methods that can be applied to natural stone, such as point tooling and chiselling, can be applied to a mature concrete surface but can be expensive if used for large areas. Tooling methods should not be used on concrete if gravel aggregates have been specified, because these tend to shatter and leave unwanted pits on the surface. Hard point tooling near to sharp arrises should also be avoided because of the tendency to spalling at these edges; to overcome this problem a small plain margin at least 10 mm wide should be specified.

Except for small areas, or where special treatment is required, bush hammering has largely replaced the hand tooling techniques to expose the aggregate. Bush hammers are power-operated, hand-held tools to which two types of head can be used to give a series of light but rapid blows (such as 1,750 blows per minute) to remove about 3 mm of matrix for each passing. Circular heads with 21 cutting points are easier to control if working on confined or small areas than the more common and quicker roller head, with its 90 cutting points.

Blasting techniques using sand, shot or grit are becoming increasingly popular because these methods do not cause spalling at the edges and generally result in a very uniform textured surface. Blasting can be carried out at almost any time after casting but preferably within 16 hours to three days of casting, which gives a certain degree of flexibility to a programme of work. This is a specialised method, and is normally carried out by a subcontractor who can usually expose between 4 and 6 m² of surface per hour. Sand is usually dispersed in a jet of water, whereas grit is directed onto the concrete surface from a range of about 300 mm in a jet of compressed air, and is the usual method employed. Shot is dispersed from a special tool with an enclosed head that collects and recirculates the shot.

Any method of removing the surface skin or matrix will reduce the amount of concrete cover over the reinforcement, so if this type of surface treatment is to be used, an adequate cover of concrete should be specified, such as 45 mm minimum, before surface treatment.

When exposing aggregates by the above methods the finished result cannot be assured, because the distribution of the aggregates throughout the concrete is determined by the way in which the fine and coarse aggregates disperse themselves during the mixing, placing and compacting operations. If a particular colour or type of exposed aggregate is required, it will be necessary to use the chosen aggregate throughout the mix, which may be both uneconomic and undesirable.

The aggregate transfer method can be used for casting in-situ concrete, and will ensure the required distribution of aggregate over the surface while enabling a selected aggregate to be used on the surface. This method entails sticking the aggregate to the rough side of pegboard sheets with a mixture of water-soluble cellulose compounds and sand fillers. The resultant mixture has a cream-like texture and is spread evenly over the pegboard surface to a depth of one-third the aggregate size. The aggregate may be sprinkled over the surface and lightly tamped; alternatively it may be placed by hand, after which the prepared board is allowed to set and dry for about 36 hours. This prepared pegboard is used as a liner to the formwork, with a loose baffle of hardboard or plywood immediately in front of the aggregate face as protection during the pouring of the cement. The baffle is removed as work proceeds. This is not an easy method of obtaining a specific exposed aggregate finish to in-situ work, and it is generally recommended that, where possible, you should consider simpler precast methods that are just as effective.

Precast concrete casting can be carried out in horizontal or vertical moulds. Horizontal casting can have surface treatments carried out on the upper surface or on the lower surface that is in contact with the mould base. Plain, smooth surfaces can be produced by hand-trowelling the upper surface, but this is expensive in labour costs and requires a high degree of skill to get a level surface of uniform texture and colour, whereas a good, plain surface can usually be achieved direct from the mould using the lower-face method. Various surface textures can be obtained from upper-face casting using patterned rollers, profiled tamping boards, a straight-edged tamping board drawn over the surface with a sawing action, or scoring the surface with brooms or rakes. Profiled finishes from lower-face casting can be obtained without any difficulty using a profiled mould base or a suitable base liner.

Exposed aggregate finishes can be obtained from either upper- or lower-face horizontal casting by spraying and brushing or by the tooling methods previously described. Alternatives for upper-face casting are to trowel selected aggregates into the freshly cast surface or to trowel in a fine (10 mm) aggregate dry mortar mix and use a float with a felt pad to pick up the fine particles and cement, leaving a clean, exposed aggregate finish.

The sand bed method involves covering the bottom of the mould with a layer of sand, placing the selected aggregate into the sand bed, and immediately casting the concrete over the layer of aggregate. This is the usual method employed with lower-face casting. Cast-on finishes, such as bricks, tiles and mosaic, can be fixed in a similar manner to those described above for exposed aggregates. For all practical purposes, vertical casting of precast concrete members needs the same considerations as those described for in-situ casting.

Another surface treatment that could be considered for concrete is grinding and polishing. However, this treatment is expensive and time-consuming, so is usually applied to surfaces such as slate, marble and terrazzo. The finish can be obtained using a carborundum rotary grinder incorporating a piped water supply to the centre of the stone. The grinding operation should be carried out as soon as the mould or formwork has been removed, by first wetting the surface and then grinding, allowing the stone dust and water to mix, forming an abrasive paste. The operation is completed by washing and brushing to remove the paste residue. Dry grinding is possible, but this creates a great deal of dust, so operatives must wear protective masks to shield their eyes and lungs.

It is often wrongly assumed that adding texture or shape to a concrete surface will mask any deficiency in design, formwork fabrication or workmanship. This is not so; a textured surface, like the plain surface, will only be as good as the design and workmanship involved in producing the finished component, and any defects will be evident with all finishes.

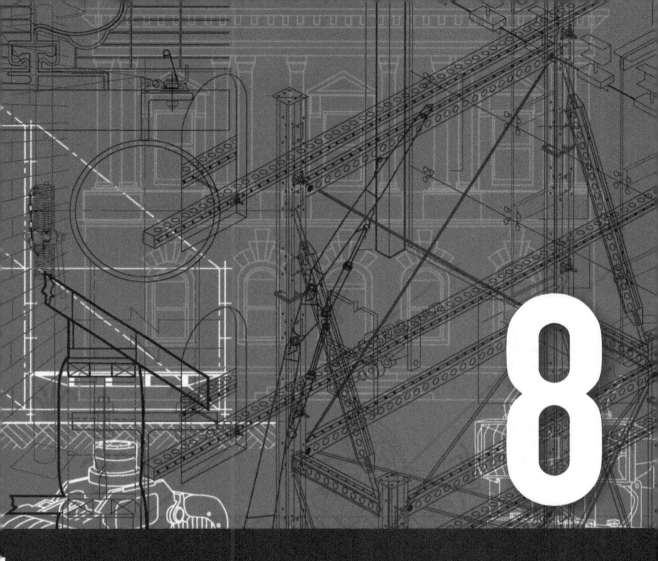

8

Prestressed concrete

Prestressed concrete: principles and applications 8.1

The basic principle of prestressing concrete is very simple. If a material has little tensile strength, it will fracture as soon as its own tensile strength is exceeded; however, if such a material is given an initial compression, when load-creating tension is applied the material will be able to withstand the force of this load as long as the initial compression is not exceeded. At this stage in the study of construction technology, readers will already be familiar with the properties of concrete, which result in a material of high compressive strength with low tensile strength. By combining concrete with steel reinforcing bars of the correct area and pattern, ordinary concrete can be given an acceptable amount of tensile strength. Prestressing techniques are applied to concrete to make full use of the material's high compressive strength.

TERMS USED FOR PRESTRESSING ELEMENTS

- **Bars** Reinforcement of solid section complying with the requirements of BS 4486. Generally only supplied in straight lengths, but often threaded for prestressing applications. Diameter 12–75 mm.

- **Wire** Reinforcement of solid section complying with the requirements of BS 5896 *High tensile steel wire and strand for the prestressing of concrete.* Generally supplied in coil form. Diameter 2–8 mm. In general use, the smallest diameter should be 4 mm.

- **Strands** A group of wires spun in helical form around a common longitudinal axis complying with the requirements of BS 5896. Two basic types available, having 7 or 19 wires. The 7-wire strand comprises six wires helically wound to form a single layer about a straight inner core; once formed the wire is subjected to heat treatment.

- **Tendon** A stretched element used in concrete member to impart prestress to the concrete. May consist of individual hard-drawn bars, wires or strands.
- **Cable** A group of tendons, typically specified for suspension structures and lifting equipment.

PRESTRESSING MATERIALS

In normal reinforced concrete the designer cannot make full use of the high tensile strength of steel or the high compressive strength of the concrete. When loaded above a certain limit, tension cracks will occur in a reinforced concrete member, which should not generally be greater than 0.3 mm in width, as recommended in BS 8110: *Structural use of concrete*. This stage of cracking will normally be reached before the full strength potential of either steel or concrete has been obtained. In prestressed concrete, the steel is stretched and securely anchored within its mould box before the concrete is placed. After concreting, the steel will try to regain its original length but, because it is fully restricted, it will be subjecting the concrete to a compressive force throughout its life. Figure 8.1.1 shows a comparison of methods.

Concrete will shrink while curing; it will also suffer losses in cross-section due to creep when subjected to pressure. Shrinkage and creep in concrete can normally be reduced to an acceptable level by using a material of high strength with a low workability. Mild steel will also suffer from relaxation losses, where the stresses in steel under load decrease towards a minimum value after a period of time. This can be counteracted by increasing the initial stress in the steel. If mild steel is used to induce a compressive force into a concrete member, the amount of shrinkage, creep and relaxation that occurs will cancel out any induced stress. However, the special alloy steels used in prestressing have different properties, enabling the designer to induce extra stress into the concrete member, counteracting any losses due to shrinkage and creep, and at the same time maintaining the induced compressive stress in the concrete component.

The high-quality strength concrete specified for prestress work should take into account the method of stressing. For pre-tensioned work, a minimum 28-day cube strength of 40 N/mm^2 is required; for post-tensioned work, a minimum 28-day cube strength of 30 N/mm^2 is required. Steel in the form of wire or bars used for prestressing should conform to the recommendations of BS 5896: *Specification for high tensile steel wire and strand for the prestressing of concrete*, which covers steel wire manufactured from cold-drawn plain carbon steel. The wire can be plain round, crimped or indented, with a diameter range of 2 to 7 mm. Crimped and indented bars will develop a greater bond strength than plain round bars, and are available in 4, 5 and 7 mm diameters. Another form of stressing wire or tendon is strand, which consists of a straight core wire around which further wires are helically wound to form a 6 over 1 or 7-wire strand, or a 9 over 9 over 1 giving a 19-wire strand tendon.

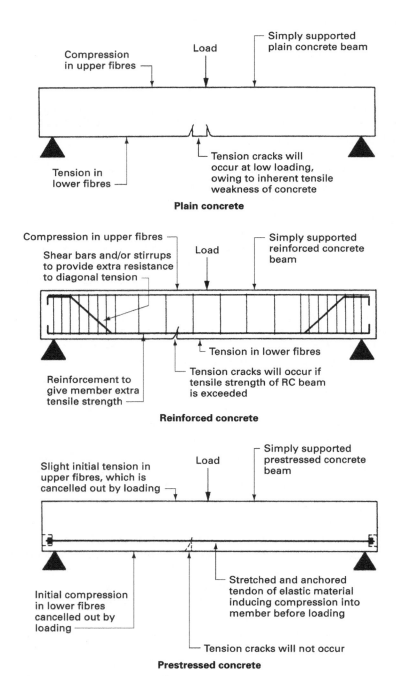

Figure 8.1.1   Structural concrete: comparison of methods

Seven-wire strand is the easiest to manufacture and is in general use for tendon diameters up to 15 mm. The wire used to form the strand is cold-drawn from plain carbon steel, as recommended in BS 5896. To ensure close contact of the individual wires in the tendon, the straight core wire is usually 2 per cent larger in diameter than the outer wires, which are helically wound around it at a pitch of 12 to 16 times the nominal diameter of the strand.

Tendons of strand can be used singly or in groups to form a multi-strand cable. The two major advantages of using strand are that:

- a large prestressing force can be provided in a restricted area;
- it can be produced in long, flexible lengths so can be stored on drums, saving site space and reducing site labour requirements.

BASIC PRINCIPLES

Prestressing works in the same way that compressing several bricks together stops the middle ones falling out. A prestressing force inducing precompression into a concrete member can be achieved by anchoring a suitable tendon at one end of the member, and applying an extension force at the other end; this end can be anchored when the desired extension has been reached. Once released, the anchored tendon will try to regain its original length, inducing a compressive force into the member. Figure 8.1.2 shows a typical arrangement in which the tendon inducing the compressive force is acting about the neutral axis, and is stressed so that it will cancel out the tension induced by the imposed load W. The stress diagrams show that the combined or final stress will result in a compressive stress in the upper fibres equal to twice the imposed load. The final stress must not exceed the characteristic strength of the concrete as recommended in BS 8110. If the arrangement given in Fig. 8.1.2 is adopted, the stress induced by the imposed load will be only half its maximum.

To get a better economic balance, the arrangement shown in Fig. 8.1.3 is normally adopted, where the stressing tendon is placed within the lower third of the section. The basic aim is to select a stress that, when combined with the dead load, will result in a compressive stress in the lower fibres equal to the maximum stresses induced by any live loads. This will result in a final stress arrangement with, in the upper fibres, a compressive stress equal to the characteristic strength of the concrete and, in the bottom fibres, a zero stress. This is the pure theoretical case, and is almost impossible to achieve in practice; however, provided any induced tension occurring in the lower fibres is not in excess of the tensile strength of the concrete used, an acceptable prestressed condition will exist.

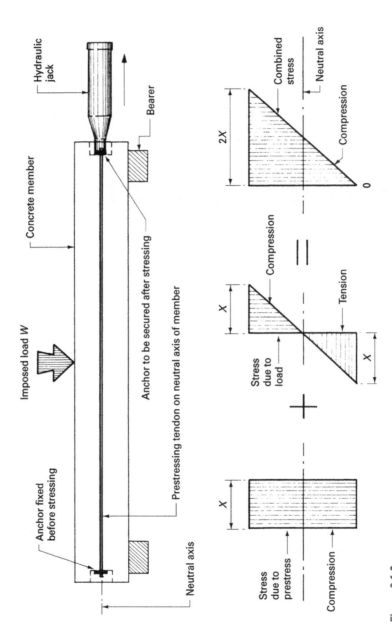

Figure 8.1.2 Prestressing principles: 1

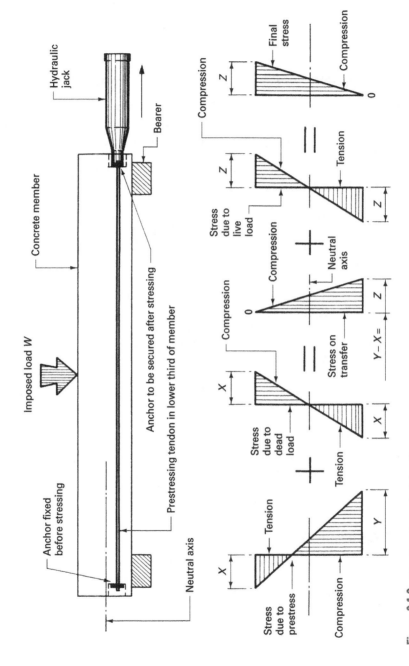

Figure 8.1.3 Prestressing principles: 2

There are two methods of producing prestressed concrete:

1. pre-tensioning;
2. post-tensioning.

PRE-TENSIONING

In pre-tensioning, the wires or cables are stressed before the concrete is cast around them. The stressing wires are anchored at one end of the mould and are stressed by hydraulic jacks from the other end, until the required stress has been obtained. It is common practice to overstress the wires by about 10 per cent to counteract the anticipated losses due to creep, shrinkage and relaxation. After stressing the wires, the side forms of the mould are positioned and the concrete is placed around the tensioned wires; the casting is then usually steam-cured for 24 hours to get the desired characteristic strength, a common specification being 28 N/mm^2 in 24 hours. The wires are cut or released, and the bond between the stressed wires and the concrete will prevent the tendons from regaining their original length, thus inducing the prestress.

At the extreme ends of the members, the bond between the stressed wires and concrete will not be fully developed. This is because low frictional resistance will result in a contraction and swelling at the ends of the wires, forming what is in effect a cone-shape anchor. The distance over which this contraction takes place is called the **transfer length** and is equal to 80 to 120 times the wire diameter. Usually, small-diameter wires (2 to 5 mm) are used so that, for any given total area of stressing wire, a greater surface contact area is obtained. The bond between the stressed wires and concrete can also be improved by using crimped or indented wires.

Pre-tensioning is the prestressing method used mainly by manufacturers of precast components such as floor units and slabs, employing the long-line method of casting, where precision metal moulds up to 120 m long can be used with spacers or dividing plates positioned along the length to create the various lengths required. A typical arrangement is shown in Fig. 8.1.4.

POST-TENSIONING

In post-tensioning, the concrete is cast around ducts in which the stressing tendons can be housed, and the stressing is carried out after the concrete has hardened. The tendons are stressed from one or both ends; when the stress required has been reached, the tendons are anchored at their ends to prevent them from returning to their original length, thus inducing the compressive force. The anchors used form part of the finished component. The ducts for housing the stressing tendons can be formed using flexible steel tubing or

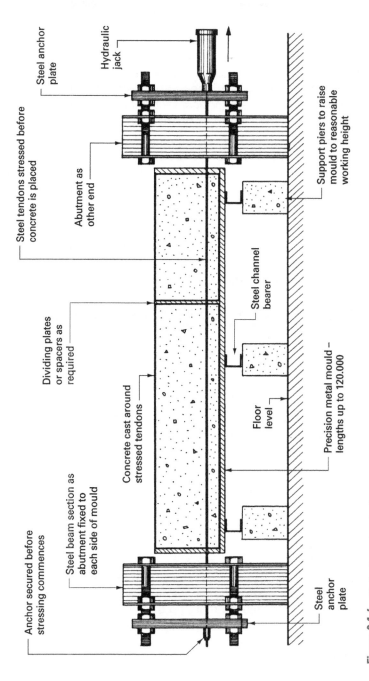

Hydraulic jack

Steel anchor plate

Steel tendons stressed before concrete is placed

Abutment as other end

Support piers to raise mould to reasonable working height

Dividing plates or spacers as required

Steel channel bearer

Concrete cast around stressed tendons

Floor level

Precision metal mould – lengths up to 120.000

Anchor secured before stressing commences

Steel beam section as abutment fixed to each side of mould

Steel anchor plate

Figure 8.1.4 Typical pre-tensioning arrangement

inflatable rubber tubes. The void created by the ducting will enable the stressing cables to be threaded before placing the concrete; alternatively, they can be positioned after the casting and curing of the concrete has been completed. In both cases, the remaining space within the duct should be filled with grout to stop any moisture present setting up a corrosive action, and to assist in stress distribution. A typical arrangement is shown in Fig. 8.1.5. For comparison of anchorages and their performance under test, see BS EN 13391: *Mechanical tests for post-tensioning systems*. System types and comparisons are shown in the next part.

Post-tensioning is the method usually employed where stressing is to be carried out on site. Its benefits include reducing the amount of complicated deck formwork needed and reducing the nett weight of steel reinforcement that needs to be fixed. Curved tendons can be used where the complete member is to be formed by joining a series of precast concrete units, or where negative bending moments are encountered. Figure 8.1.6 shows various methods of overcoming negative bending moments at fixed ends and for continuous spans. Figure 8.1.7 shows a typical example of the use of curved tendons in the cross-members of a girder bridge. Another application of post-tensioning is in the installation of ground anchors.

GROUND ANCHORS

A ground anchor is basically a prestressing tendon embedded and anchored into the soil to provide resistance to structural movement of a member by acting on a 'tying back' principle. Common applications include anchoring or tying back retaining walls and anchoring diaphragm walls, particularly in the context of deep excavations. This latter application also has the advantage of providing a working area entirely free of timbering members, such as struts and braces. Ground anchors can also be used in basements and similar constructions for anchoring the foundation slab, to resist uplift pressures and to prevent flotation, especially during the early stages of construction.

Ground anchors are known by their method of installation, such as grouted anchors, or by the nature of the subsoil into which they are embedded, such as rock anchors.

ROCK ANCHORS

Rock anchors have been used successfully for many years, and can be formed by inserting a prestressing bar into a pre-drilled hole. The leading end of the bar has expanding sleeves that grip the inside of the bored hole when the bar is rotated to a recommended torque, to obtain the desired grip. The anchor bar is usually grouted over the fixed or anchorage length before being stressed and anchored at the external face.

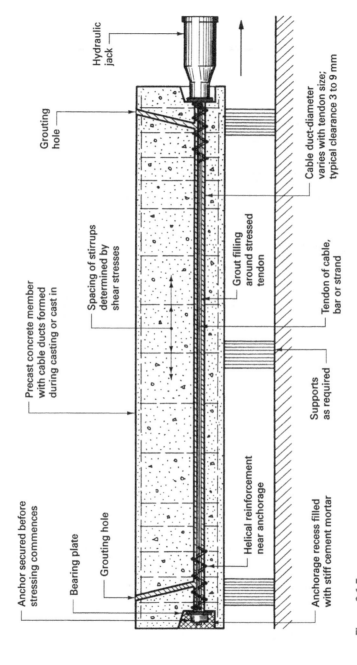

Anchor secured before stressing commences

Bearing plate

Grouting hole

Precast concrete member with cable ducts formed during casting or cast in

Grouting hole

Hydraulic jack

Spacing of stirrups determined by shear stresses

Helical reinforcement near anchorage

Grout filling around stressed tendon

Cable duct–diameter varies with tendon size; typical clearance 3 to 9 mm

Tendon of cable, bar or strand

Supports as required

Anchorage recess filled with stiff cement mortar

Figure 8.1.5 Typical post-tensioning arrangement

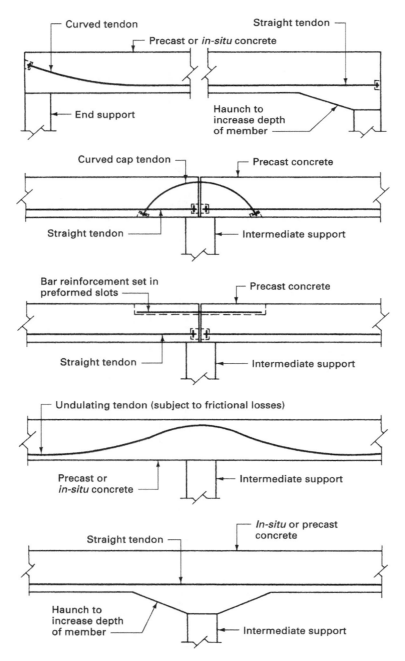

Curved tendon

Straight tendon

Precast or *in-situ* concrete

End support

Haunch to
increase depth
of member

Curved cap tendon

Precast concrete

Straight tendon

Intermediate support

Bar reinforcement set in
preformed slots

Precast concrete

Straight tendon

Intermediate support

Undulating tendon (subject to frictional losses)

Precast or
in-situ concrete

Intermediate support

Straight tendon

In-situ or precast
concrete

Haunch to
increase depth
of member

Intermediate support

Figure 8.1.6 Prestressing: overcoming negative bending moments

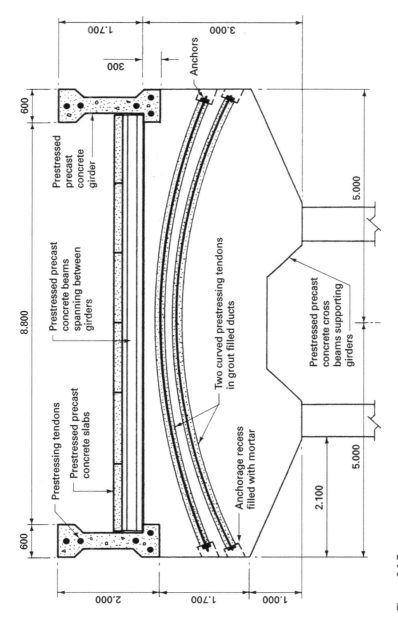

Figure 8.1.7 Typical example of the structural use of prestressed concrete

Alternatively, the anchorage of the leading end can be provided by grout injection, relying on the bond developed between a ribbed sleeve and the wall of the bored hole. A dense, high-strength grout is required over the fixed length to develop sufficient resistance to pull out when the tendon is stressed. The unbonded or elastic length will need protection against corrosion, which can be provided by protective coatings such as bitumen or rubberised paint, casings of PVC or wrappings of greased tape. Alternatively, a full-length protection can be given by filling the void with grout after completion of the stressing operation (see Fig. 8.1.8).

INJECTION ANCHORS

The knowledge and experience gained in the use of rock anchors has led to the development of suitable ground anchorage techniques for most subsoil conditions, except for highly compressible soils, such as alluvial clays and silts. Injection-type ground anchorages have proved to be suitable for most cohesive and non-cohesive soils. A hole is bored into the soil using a flight auger, with or without water-flushing assistance; casings or linings can be used where the borehole would not remain open if unlined. The prestressing tendon or bar is placed into the borehole and pressure-grouted over the anchorage length. For protection purposes, the unbonded or elastic length can be grouted under gravity for permanent ground anchors, or covered with an expanded polypropylene sheath for temporary anchors. Anchor boreholes in clay soils are usually multi-under-reamed to increase the bond, using a special expanding cutter or brush tools. Gravel placement ground anchors can also be used in clay and similar soils for lighter loadings. In this method, an irregular gravel is injected into the borehole over the anchorage length. A small casing with a non-recoverable point is driven into the gravel plug to force the aggregate to penetrate the soil around the borehole. The stressing tendon is inserted into the casing and pressure-grouted over the anchorage length as the casing is removed. Typical ground anchor examples are shown in Fig. 8.1.9.

ADVANTAGES AND DISADVANTAGES

As already explained, in pre-tensioning, it is the bond between the tendon and the concrete that prevents the prestressing wire from returning to its original length, while in post-tensioning, it is the anchorages that prevent the stressing tendon returning to its original length. Here is a summary of the advantages and disadvantages of prestressed concrete when compared with conventional reinforced concrete.

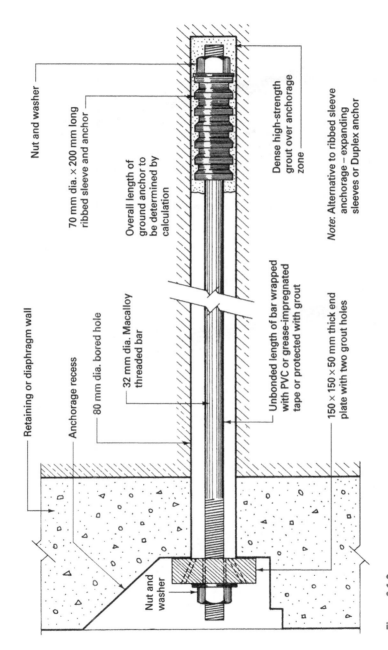

Nut and washer

70 mm dia. × 200 mm long
ribbed sleeve and anchor

Overall length of
ground anchor to
be determined by
calculation

Dense high-strength
grout over anchorage
zone

Note: Alternative to ribbed sleeve
anchorage – expanding
sleeves or Duplex anchor

Retaining or diaphragm wall

Anchorage recess

80 mm dia. bored hole

32 mm dia. Macalloy
threaded bar

Unbonded length of bar wrapped
with PVC or grease-impregnated
tape or protected with grout

150 × 150 × 50 mm thick end
plate with two grout holes

Nut and
washer

Figure 8.1.8 Typical rock ground anchor

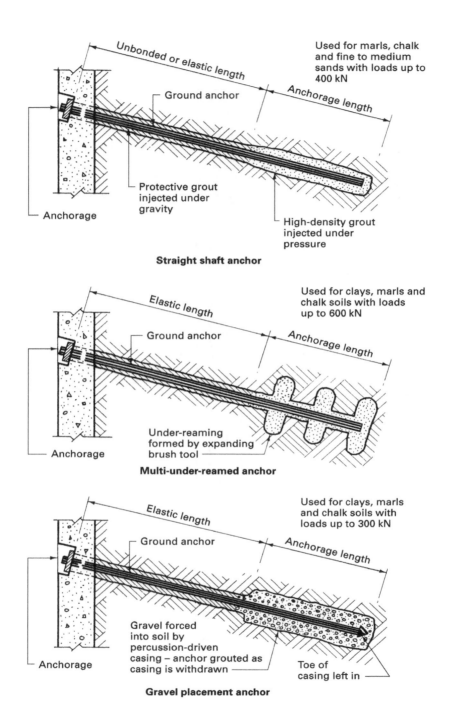

Figure 8.1.9 Typical injection ground anchors

ADVANTAGES

- Makes full use of the inherent compressive strength of concrete.
- Makes full use of the special alloy steels used to form prestressing tendons.
- Eliminates tension cracks, thus reducing the risk of corrosion of steel components.
- Reduces shear stresses.
- For any given span and loading condition a member with a smaller cross-section can be used, giving a reduction in weight and height.
- Individual units can be joined together to act as a single member.
- Avoids the need for intermediate downstand beams, allowing a flat floor soffit to the underside of large floor plates, which is easier and cheaper to form.
- Can reduce the nett weight of steel reinforcement required in an r.c. member or allow for a shallower depth of concrete for a given span.

DISADVANTAGES

- Requires a high degree of control of materials, design and workmanship.
- Special alloy steels are more expensive than mild steels.
- Special jacking and monitoring equipment is needed to carry out stressing activities.

As a general comparison between these two structural media, it is usually found that:

- for spans of up to 6 m, traditional reinforced concrete is the most economic method;
- for spans between 6 and 9 m, the two mediums are compatible;
- for spans of over 9 m, prestressed concrete is generally more economical than reinforced concrete.

Prestressed concrete systems 8.2

The prestressing of concrete is usually carried out by a specialist contractor, or alternatively by the main contractor using a particular system and equipment. The basic conception and principles of prestressing are common to all systems; only the type of tendon, type of anchorage and stressing equipment vary. The following established systems are typical and representative of the methods available.

BBRV (SIMON-CARVES LTD)

The BBRV system was one of the first prestressed systems to be developed for major civil engineering projects in the 1930s. This is a unique system of prestressing developed by four Swiss engineers – Birkenmaier, Brandestini, Ros and Vogt – whose initials are used to name the system. It differs from other systems in that multi-wire cables capable of providing prestressing forces from 300 to 7,800 kN are used, and each wire is anchored at each end by means of enlarged heads formed on the wire. The cables or tendons are purpose-made to suit individual requirements, and may comprise up to 121 wires. The high-tensile steel wire conforms to the recommendations of BS 5896: *Specification for high tensile steel wire and strand for the prestressing of concrete*. It is cut to the correct length, and any sheathing required is threaded on, together with the correct anchorage, before the button heads are formed using special equipment. The completed tendons can be coiled or left straight for delivery to site. Four types of stressing anchor are available, the choice depending on the prestressing force being induced, whereas the fixed anchors come in three forms. The finished tendon is fixed in the correct position to the formwork before concreting; when the concrete has hardened, the tendons are stressed, and grout is injected into the sheathing. If unsheathed tendons are to be drawn into preformed ducts, the anchor is omitted from one and is fixed after drawing through the tendon, the button heads being formed with a portable machine.

Tendons in this system are tensioned using a special hydraulic jack of the centre-hole type, with capacities ranging from 30 to 800 tonnes. A pull rod or pull sleeve (depending on anchor type) is coupled to the basic element carrying the wires. A locknut, stressing stool, hydraulic jack and dynamometer are then threaded on. The applied stressing force can be read from the dynamometer, while the actual extension achieved can be seen on the scale engraved on the jack.

After stressing, the locknut is tightened up and the jack released before grouting takes place. Losses due to friction, shrinkage and creep can be overcome by restressing at any time after the initial stressing operation, as long as the tendons have not been grouted.

In common with most other prestressing systems, tendons in long, continuous members can be stressed in stages without breaking the continuity of the design. The first tendon length is stressed and grouted before the second tendon length is connected to it using a coupling anchor, after which it is stressed and grouted, before repeating the procedure for any subsequent lengths. Alternatively, lengths of unstressed tendons can be coupled together and the complete tendon stressed from one or both ends. Figure 8.2.1 shows a typical BBRV stressing arrangement.

CCL SYSTEMS

Established in 1935, CCL has become a leading prestressed concrete engineering specialist.

The company established two basic systems: Cabco and Multiforce. These have since been further developed by the company, but are retained here to illustrate a principle of post-tensioning.

In the Cabco system, each strand or wire is stressed individually, the choice of hydraulic jack being governed by the size of strand being used. The system is fast and, being manually operated, eliminates the need for lifting equipment. To form the tendon, this post-tensioning system uses a number of strands, ranging from 4 to 31, according to the system being employed, giving a range of prestressing forces from 450 to 5,000 kN. By applying the total tendon force in stages, problems such as differential elastic shortening and out-of-balance forces are reduced. Curved tendons are possible using this system without the need for spacers, but spacers are recommended for tendons over 30 m long. The alternative Multiforce system uses the technique of simultaneously stressing all the strands forming the tendon.

In both systems, the basic anchorages are similar in design. The fixed or dead-end anchorage has a tube unit to distribute the load, a bearing plate and a compressing grip fitted to each strand. If the fixed anchorage is to be totally embedded in concrete, a compressible gasket and a bolted-on retaining plate are also used, a grout vent pipe being inserted into the grout

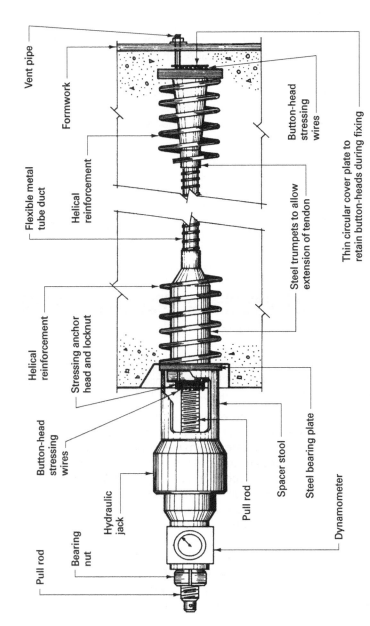

Pull rod

Bearing nut

Hydraulic jack

Button-head stressing wires

Helical reinforcement

Stressing anchor head and locknut

Flexible metal tube duct

Helical reinforcement

Vent pipe

Formwork

Button-head stressing wires

Thin circular cover plate to retain button-heads during fixing

Steel trumpets to allow extension of tendon

Pull rod

Spacer stool

Steel bearing plate

Dynamometer

Figure 8.2.1 Typical BBRV prestressing arrangement

hole of the tube unit. The compression grips consist of an outer sleeve with a machined and hardened insert, which can be fitted while making up the tendon or installed after positioning the tendon. The stressing anchor is a similar device, consisting of a tube unit, bearing plate and wedges working on a collet principle to secure the strands. The wedges are driven home to a 'set' by the hydraulic jack used for stressing the tendon before the jack is released and the stress is transferred. Figure 8.2.2 shows a typical CCL Cabco stressing arrangement.

DYWIDAG-SYSTEMS INTERNATIONAL

DYWIDAG-Systems International is based in the UK and was established in 1968. It currently concentrates its work in the geotechnical field and has become the leading specialist for prestressed ground anchors and associated fixings, such as soil nails and rock bolts. Their post-tensioning system uses single- or multiple-bar tendons with diameters ranging from 12 to 36 mm for single-bar applications and 16 mm diameter threaded bars for multiple-bar tendons, giving prestressing forces up to 950 kN for single-bar tendons and up to 2,000 kN for multiple-bar tendons. Both forms of tendon are placed inside a thin-wall corrugated sheathing, which is filled with grout after completion of the stressing operation. The single-bar tendon can be of a smooth bar with cold-rolled threads at each end to provide the connection for the anchorages, or it can be a threadbar that has a coarse thread along its entire length, providing full mechanical bond. A threadbar is used for all multiple-bar tendons.

Two forms of anchorage are available: the bell anchor and the square or rectangular plate anchor. During stressing, using a hydraulic jack acting against the bell or plate anchor, the tendon is stretched and at the same time the anchor nut is being continuously screwed down, to provide the transfer of stress when the specified stress has been reached and the jack is released. A counter shows the number of revolutions of the anchor nut and the amount of elongation of the tendon. As in other similar systems, the tendons can be restressed at any time before grouting. Figure 8.2.3 shows typical details of a single-bar tendon arrangement.

MACALLOY SYSTEMS

Macalloy, established in 1921, has concentrated on threaded bars systems and fittings. Its post tensioning bar system has a range of diameters from 20 mm to 75 mm threaded bars. Like the method previously described, the Macalloy system uses single- or multiple-bar tendons The bars used are of a cold-worked high-strength alloy steel threaded at each end to provide an anchorage connection, and are available in lengths up to 17.8 m for diameters between 20 mm and 50 mm and up to 8.4 m for 75 mm, giving prestressing forces ranging from 262 kN up to 3,495 kN for the 20 mm and 75 mm threaded bars respectively.

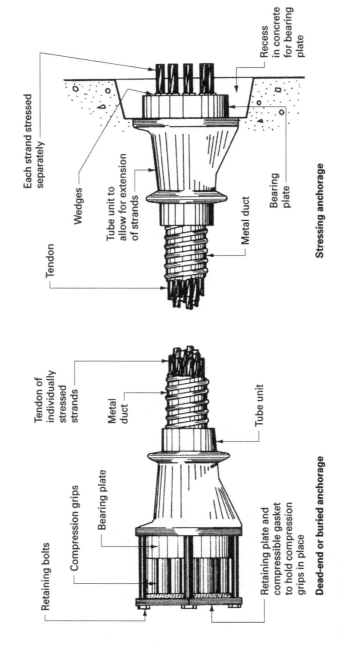

Each strand stressed separately

Wedges

Tendon

Tube unit to allow for extension of strands

Metal duct

Recess in concrete for bearing plate

Bearing plate

Stressing anchorage

Tendon of individually stressed strands

Metal duct

Tube unit

Retaining bolts

Compression grips

Bearing plate

Retaining plate and compressible gasket to hold compression grips in place

Dead-end or buried anchorage

Figure 8.2.2   Typical details of the CCL Cabco system of prestressing

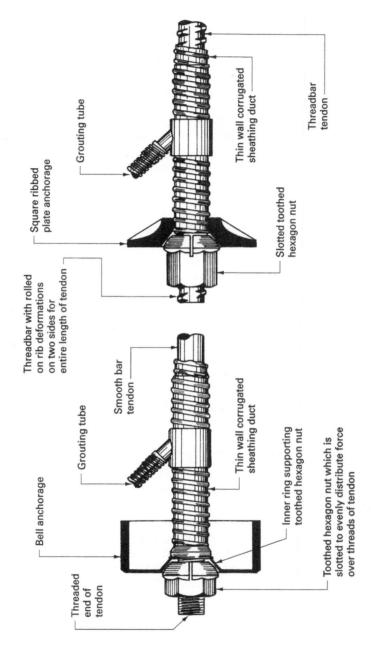

Figure 8.2.3 Typical stressing anchors for DYWIDAG post-tensioning system

Threadbar with rolled
on rib deformations
on two sides for
entire length of tendon

Grouting tube

Square ribbed
plate anchorage

Thin wall corrugated
sheathing duct

Threadbar
tendon

Slotted toothed
hexagon nut

Bell anchorage

Threaded
end of
tendon

Grouting tube

Smooth bar
tendon

Thin wall corrugated
sheathing duct

Inner ring supporting
toothed hexagon nut

Toothed hexagon nut which is
slotted to evenly distribute force
over threads of tendon

The fixed anchorage consists of an end plate drilled and tapped to receive the tendon, whereas the stressing anchor consists of a similar plate complete with a grouting flange and anchor nut. Stressing is carried out using a hydraulic jack operating on a drawbar attached to the tendon, the tightened anchor nut transferring the stress to the member upon completion of the stressing operation. The prestressing is completed by grouting in the tendon after all the stressing and any necessary restressing has been completed. Figure 8.2.4 shows typical details of a single-bar arrangement.

FREYSSINET LTD

This post-tension system uses strands to form the tendon and is available in three forms: the Freyssi-monogroup, Freyssinet multistrand and PSC monostrand. Monogroup tendons are composed of 7, 13, 15 or 19 wire strands stressed in a single pull by the correct model of hydraulic jack, giving prestressing forces of up to 5,000 kN. The anchorages consist of a cast-iron guide, shaped to permit the deviation of the strands to their positions in the steel anchor block, where they are secured by collet-type jaws. During stressing, 12 wires of the tendon are anchored to the body of the jack, the remaining wires passing through the jack body to an anchorage at the rear.

The multistrand system is based on the first system of post-tension ever devised (in 1928), and consists of a cable tendon made from 12 high-tensile steel wires laid parallel to one another and taped together, resulting in a tendon that is flexible and compact. Two standard cable diameters are produced, giving tendon sizes of 29 and 33 mm capable of taking prestressing forces up to 750 kN. The anchorages consist of two parts: the outer reinforced-concrete cylindrical body has a tapered hole to receive a conical wedge, which is grooved or fluted to receive the wires of the tendon. The 12 wires of the cable are wedged into tapered slots on the outside of the hydraulic jack body during stressing, and when this operation has been completed the jack drives home the conical wedge to complete the anchorage.

Monostrand uses a four- or seven-strand tendon for general prestressing or a three-strand tendon developed especially for prestressing floor and roof slabs. This system is intended for the small to medium range of prestressing work requiring a prestressing force not exceeding 2,000 kN, and as its title indicates each strand in the group is stressed separately, requiring only a light compact hydraulic jack. All system tendons are encased in a steel sheath and grouted after completion of the stressing operation. Typical details are shown in Fig. 8.2.5.

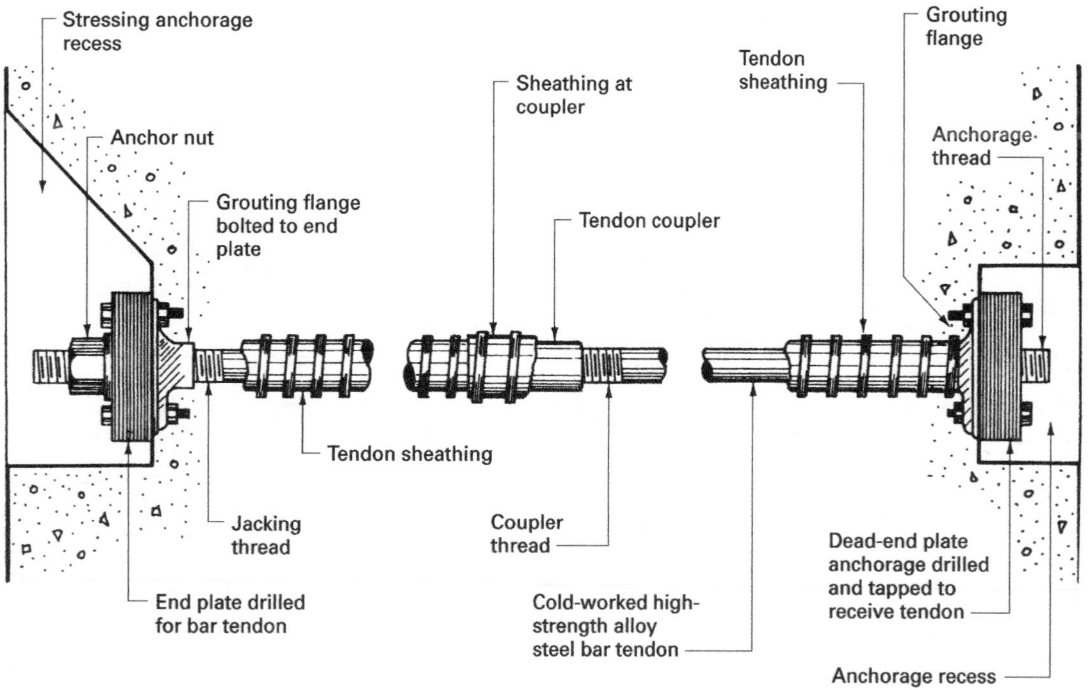
Stressing anchorage recess
Anchor nut
Grouting flange bolted to end plate
Sheathing at coupler
Tendon coupler
Tendon sheathing
Grouting flange
Anchorage thread
Tendon sheathing
Jacking thread
End plate drilled for bar tendon
Coupler thread
Cold-worked high-strength alloy steel bar tendon
Dead-end plate anchorage drilled and tapped to receive tendon
Anchorage recess

Figure 8.2.4 Typical Macalloy prestressing system details

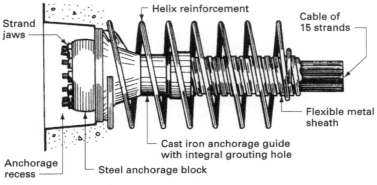

Strand jaws

Helix reinforcement

Cable of 15 strands

Flexible metal sheath

Cast iron anchorage guide with integral grouting hole

Steel anchorage block

Anchorage recess

Typical Freyssi-monogroup anchorage

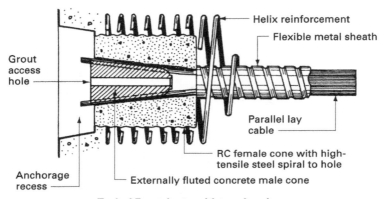

Helix reinforcement

Flexible metal sheath

Grout access hole

Parallel lay cable

RC female cone with high-tensile steel spiral to hole

Externally fluted concrete male cone

Anchorage recess

Typical Freyssinet multistrand anchorage

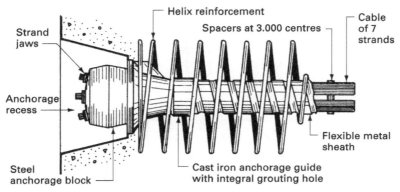

Helix reinforcement

Spacers at 3.000 centres

Cable of 7 strands

Strand jaws

Anchorage recess

Flexible metal sheath

Steel anchorage block

Cast iron anchorage guide with integral grouting hole

Typical PSC monostrand anchorage

Figure 8.2.5   Typical Freyssinet prestressing system

SCD (STRESSED CONCRETE DESIGN LTD)

This post-tensioning system offers three variations of tendon and/or stressing: multigrip circular, monogrip circular and monogrip rectangular. The multigrip circular system uses a tendon of 7, 12, 13 or 19 strands, forming a cable tendon capable of accepting prestressing forces up to 5,000 kN. The anchorages consist of a cast-iron guide plate, enabling the individual strands of the cable to fan out and pass through a bearing plate, where they are secured with steel collet-type wedges. The strands pass through the body of the hydraulic jack to a rear anchorage and are therefore stressed simultaneously.

The monogrip circular tendons are available as a single-, 4-, 7- or 12-strand cable in which each strand is stressed individually, and the whole tendon is capable of taking prestressing forces up to 3,200 kN. Each strand in the tendon is separated from adjacent strands by means of circular spacers at 2 m centres, or less if the tendon is curved. The anchorages are similar in principle to those already described for multigrip tendons.

The monogrip rectangular system, by virtue of using a rectangular tendon composed of 3 to 27 strands capable of taking a prestressing force of up to 3,900 kN, affords maximum eccentricity with a wide range of tendon sizes. The anchorage guide plate and bearing plates work on the same principle as described above for the multigrip method. In all methods the tendon is encased in a sheath unless preformed ducts have been cast; on completion of the stressing operation, all tendons are grouted in. See Fig. 8.2.6 for typical details.

Note: In all cases anchorage zone helix reinforcement would be used according to design

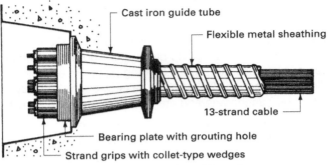

- Cast iron guide tube
- Flexible metal sheathing
- 13-strand cable
- Bearing plate with grouting hole
- Strand grips with collet-type wedges

Typical multigrip circular cable anchorage

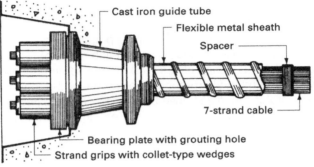

- Cast iron guide tube
- Flexible metal sheath
- Spacer
- 7-strand cable
- Bearing plate with grouting hole
- Strand grips with collet-type wedges

Typical monogrip circular cable anchorage

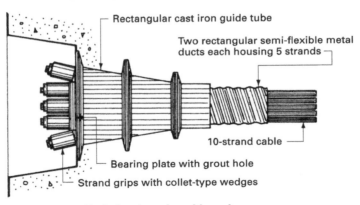

- Rectangular cast iron guide tube
- Two rectangular semi-flexible metal ducts each housing 5 strands
- 10-strand cable
- Bearing plate with grout hole
- Strand grips with collet-type wedges

Typical rectangular cable anchorage

Figure 8.2.6 Typical SCD prestressing system details

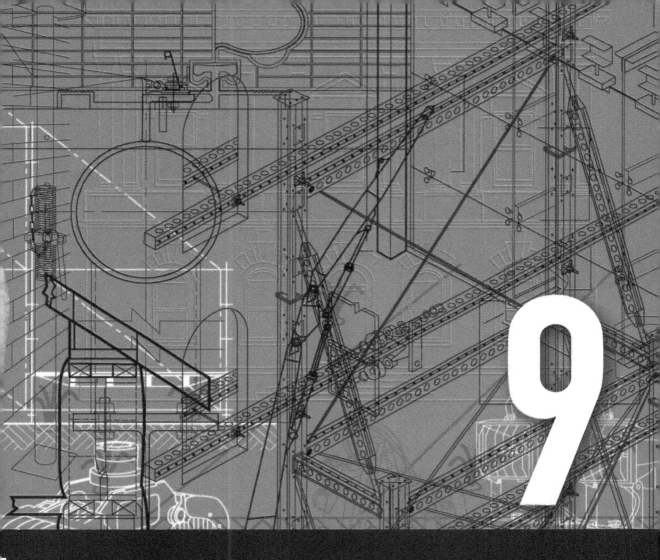

9

Buildings for industrial and storage use

Factory buildings: roofs 9.1

A roof's function, its construction technique and the methods of covering for non-industrial applications are covered in Parts 6 and 10 in the associated volume, *Construction Technology*. The reader is recommended to review these sections before continuing with this part.

The advanced level of study in this book is concerned mainly with particular building types, such as those suitable as small to large industrial or factory buildings. Buildings of this nature set the designer two main problems:

- production layout required at floor level and the consequent need for large unobstructed areas by omitting as far as practicable internal roof supports, such as loadbearing walls and columns;
- provision of natural daylight from the roof over the floor area below to reduce the buildings' energy costs.

In this part we will examine the function of a commercial roof in terms of the admittance of natural daylight, the thermal efficiency of a roof through energy conservation and the precautions that may be required for fire. The following parts examine in detail the construction details of more advanced roof construction.

PROVISION OF NATURAL DAYLIGHT

The amount of useful daylight that can penetrate into a building from openings in side and end walls is limited by the height of the end walls, the size of the window, the position of the window (low or high) and the orientation of the building (for example, south facing). Buildings with spans in excess of 18 m will generally need some form of overhead supplementary lighting at the working plane, as the penetration of the daylight will diminish towards the middle of the building. In single-storey buildings the supplementary lighting can take the form of glazed rooflights within the roof finish structure that are normally 10 per cent of the roof surface area. However, in multi-storey buildings the floors beneath the top floor will have to have the natural daylighting from the side and end walls supplemented by permanent artificial lighting.

The factors to be considered when designing or choosing a roof type or profile for a factory building in terms of rooflighting are:

■ amount of daylight required;

■ spread of daylight required over the working plane;

■ elimination of solar heat gain and solar glare.

The amount of daylight required within a building is usually based on the **daylight factor**, which can be defined as 'the illumination at a specific point indoors expressed as a percentage of the simultaneous horizontal illumination outdoors under an unobstructed overcast sky'. The minimum daylight factor for factories depends on usage and work function within the building: for example, detailed drawing work will require high daylight factors. Table 9.1.1 provides some comparison for different dwellings and room functions. A figure of 3–6 per cent may be acceptable in most factory assembly situations, depending on the intensity of work.

Table 9.1.1 **Daylight factors**

Building type	Daylight factor (%)	Recommendations
Dwellings:		
kitchen	2	Over at least 50% of floor area (min. approx. 4.5 m²)
living room	1	Over at least 50% of floor area (min. approx. 7.0 m²)
bedroom	0.5	Over at least 75% of floor area (min. approx. 5.5 m²)
Schools	2	Over all teaching areas and kitchens
Hospitals	1	Over all wards
Offices:		
general	1	With side lighting at approx. 3.75 m² penetration
general	2	With top light over whole area
drawing	6	On drawing boards
drawing	2	Over remainder of work areas
typing and computing	4	Over whole working area
Laboratories	3–6	Depending on dominance of side or top lighting
Factories	5	General recommendation
Art galleries	6	Max. on walls or screens where no special problems of fading
Churches	1	General, over whole area
Churches	1.5–2	In sanctuary areas
Public buildings	1	Depending on function, figure given can be taken as minimum

More accurate determination of the daylight factor can be achieved by plotting the particular roof profile and glass area using a **daylight protractor**. These are available from the Building Research Establishment Bookshop. The BRE Report BR 288: *Designing buildings for daylight* is also helpful, with worked examples and practical exercises relating to daylight design.

An estimation of glazed area can be made using a rule of thumb of one-fifth glass to floor area. For example, assuming a daylight factor of 3 per cent is required over a floor area of 500 m², the area of glass can be calculated as:

$$\text{Daylight factor} \times \text{floor area} \times 5 = \frac{3}{100 \times 500 \times 5} = 75 \text{ m}^2$$

It must be emphasised that this one-fifth rule of thumb is a preliminary design aid, and the result obtained should be checked by a more precise method through a building services engineer. It should also be compared with current thermal insulation requirements for maximum areas of glazing to industrial and storage buildings. This is considered later in this part and is defined in Building Regulations, Approved Document L2.

For an even spread of light over the working plane, the ratio between the spacing and height of the rooflights is particularly important. For typical ratios, see Fig. 9.1.1. If the ratios shown in Fig. 9.1.1 are not adopted, the result could be a marked difference in illumination values at the working plane, resulting in light and darker areas.

Another factor to be considered when planning rooflighting is the amount of obstruction to natural daylighting that could be caused by services and equipment housed in the roof void. Also, the effectiveness of the glazed areas will gradually deteriorate as dirt collects on the surfaces, and regular maintenance will be required to keep these levels.

The elimination of solar heat gain and solar glare can be achieved in various ways, such as fitting reflective louvres to the glazed areas and using solar-reflective film applied to the surface of the glass or using a tint in the glass units. Traditionally north light or monitor roofs were used for many factory units. With the advent of large-span portal frames and prefabricated lattice beams, this type of design along with monitor roofs has declined.

ROOFLIGHTS

The modern rooflight has evolved into a number of different shapes, applications and sizes that enable some aesthetics to be introduced into the industrial building as well as allowing the passage of daylight into the interior.

Rooflights can be subdivided into sub-classifications including:

- dome;
- barrel-vaulted;
- integrated;
- modular panel unit.

DOME ROOFLIGHTS

As the name suggests, dome rooflights are prefabricated polycarbonate domes that are fixed to an upstand kerb frame. Using multi-walled polycarbonate enables high levels of insulation to be achieved, while insulated kerbs enable Part L2 U-values to be maintained. The use of different tints within the glazed dome reduces the solar heat gain and glare on the working plane. Mechanical operating systems can be attached to a hinged dome to provide automatic ventilation with increased heat build-up in the roof space. Figure 9.1.1 illustrates a cross-section through a modern rooflight. You can see the triple-skin construction of the rooflight and the sloped insulated walls that reduce any cold bridges and the formation of condensation. Rooflights can be fitted with an auto-vent motor-driven system to open automatically or manually.

BARREL-VAULTED ROOFLIGHTS

A barrel rooflight consists of a modular unit that can be used in a continuous strip of vaults. This gives the system the ability to run a strip of rooflights up the slope of a roof in a vault that is attached to the existing roof sheeting structure. A special top-and-bottom unit is required for completion of the daylight vault. By vaulting the rooflight outwards, water will run off the light, washing any dirt with it over time. The rooflight drains as an integral part of the roof finishes. Internally a glass-reinforced plastic (GRP) liner sheet finishes the internal surfaces. Figure 9.1.2 illustrates a section through a vaulted rooflight, showing how it is integrated into the roof. On the right-hand side the finished upstand sloping kerb ensures maximum spread of any daylight entering the building, and provides a degree of aesthetics to the internal appearance.

INTEGRATED ROOFLIGHTS

Integrated rooflights are designed and built to fit a variety of standard composite or site-installed roofing sheet applications. Each is a sealed factory-manufactured unit that fully matches the profile of the main roof sheets, enabling it to fit into a standard sheet width and be fully waterproof. By producing these in a factory-assembled process, high thermal properties can be maintained to meet Part L of the Building Regulations Approved Documents. Integrated rooflights are fixed as roof sheets, provide high insulation values, do not UV-degrade and have a high impact safety factor for operatives working on the roof. They behave as a roof sheet, with lap joints and fixing methods as those of a roof sheet application. This type of rooflight can also be site-assembled, a process that involves fitting the outer and inner sheets separately as the roof sheeting is fitted. Integrated rooflights are not as thermally efficient as the factory-assembled system, but offer a more economical alternative in locations where thermal efficiency is not a high priority. Figure 9.1.3 illustrates the site fixing of a prefabricated rooflight.

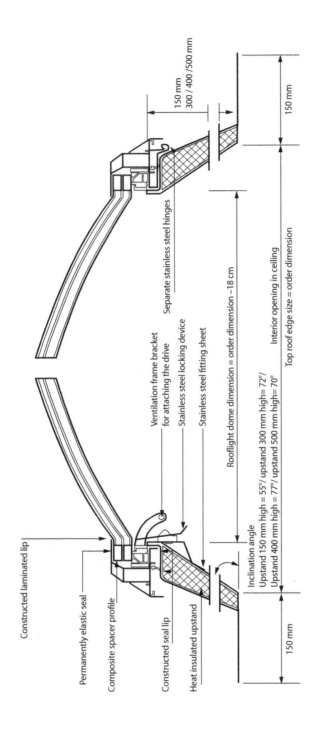

Constructed laminated lip

Permanently elastic seal

Composite spacer profile

Constructed seal lip

Heat insulated upstand

Ventilation frame bracket
for attaching the drive

Stainless steel locking device

Stainless steel fitting sheet

Separate stainless steel hinges

150 mm

300 / 400 /500 mm

150 mm

150 mm

Inclination angle
Upstand 150 mm high = 55°/ upstand 300 mm high= 72°/
Upstand 400 mm high = 77°/ upstand 500 mm high= 70°

Rooflight dome dimension = order dimension –18 cm

Interior opening in ceiling

Top roof edge size = order dimension

Figure 9.1.1   Cross-section through typical dome rooflight

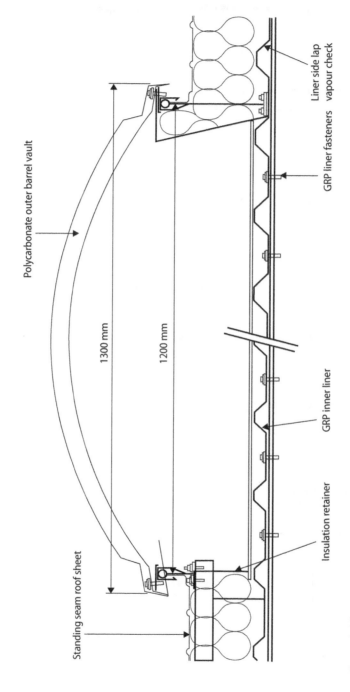

Polycarbonate outer barrel vault

Standing seam roof sheet

1300 mm

1200 mm

Insulation retainer

GRP inner liner

GRP liner fasteners

Liner side lap

vapour check

Figure 9.1.2 Cross-section through barrel-vaulted rooflight

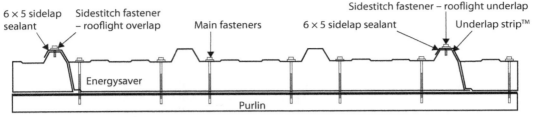

Cross-section through rooflight

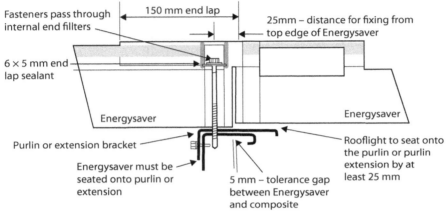

Joining rooflights

Figure 9.1.3 **Factory-assembled rooflights**

MODULAR PANEL UNITS

A modular panel unit system makes full use of the production of large polycarbonate twin wall units, which can be secured into an aluminium framing system. The polycarbonate sheets are cut into standard units that can be extended by the inclusion of the aluminium jointing system, enabling large areas of rooflights to be formed. The system is ideal for providing a flexible method of daylight entry into a building. They can be incorporated into any pitch of roof and utilise a kerb upstand system to waterproof the openings formed. Figure 9.1.4 illustrates the methods employed and the cross-section of the aluminum profile that retains the sheet in position, resisting wind and snow loadings.

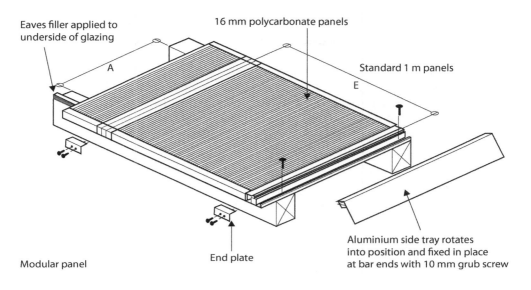

Eaves filler applied to underside of glazing

16 mm polycarbonate panels

Standard 1 m panels

A

E

Modular panel

End plate

Aluminium side tray rotates into position and fixed in place at bar ends with 10 mm grub screw

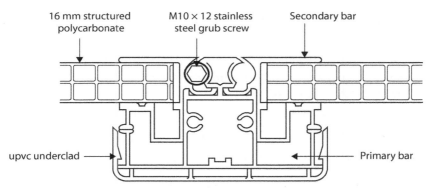

16 mm structured polycarbonate

M10 × 12 stainless steel grub screw

Secondary bar

upvc underclad

Primary bar

Cross-section through panel joint

Figure 9.1.4   **Modular panel system glazing**

The control of fuel energy consumption and measures to reduce the associated atmospheric contamination by CO_2 emissions from industrial buildings are included in the Approved Document to Building Regulation L2: *Conservation of fuel and power in buildings other than dwellings.* The Approved Document contains reference to other technical publications produced by the industry, research organisations and the professions. Among other provisions, the Approved Document and the additional references contain guidance for:

- limiting heat losses from hot water pipes, hot water storage vessels and warm air ducting;
- limiting heat gains by chilled water and refrigeration vessels;
- energy-efficient heating and hot water systems;
- limiting exposure to solar over-heating;
- controlling energy use in air conditioning and ventilation systems;
- energy-efficient artificial lighting and controls;
- documenting building and plant operating details to enable the enclosure to be maintained in an energy-efficient manner;
- avoidance of thermal bridges in construction and gaps in insulation layers;
- incorporating reasonably airtight construction with satisfactory sealing methods at construction interfaces;
- maximum acceptable elemental thermal transmittance coefficients (U-values);
- maximum acceptable window, door and other opening areas;
- calculation methodology for establishing carbon emissions.

Under the European Union's Energy Performance of Buildings Directive (EPBD), member states are required to have in place a methodology for calculating the energy performance of buildings and to display this as a heat rating chart. This calculation must incorporate every aspect of energy efficiency, not least heat losses through the fabric, daylight calculations, airtightness of the external envelope, heating and ventilating system efficiency, and internal lighting efficiency. In the UK the methodology can be satisfied by a standard carbon emission calculation. For dwellings (AD.L1), this is effected through the Government's recommended system for energy rating, known as **Standard Assessment Procedure** or SAP. For non-domestic buildings a National Calculation Method (NCM) has been devised by the Building Research Establishment to show compliance with Approved Document L2 in accordance with the EPBD. This is known as the **Simplified Building Energy Method** (SBEM). The SBEM incorporates all information relating to a building's design, principally its geometry, method of construction, function and energy-consuming services systems.

The final documentation is in the form of an asset-rating certificate. This may be used as part of the application for local authority building control approval and as an energy guide for building users and prospective purchasers.

The Approved Document L2 is not a package of prescriptive measures. It sets overall targets or benchmarks, with objectives to be satisfied under the NCM. Nevertheless, there are some parameters for construction practice that can be gauged. These include:

■ poorest acceptable fabric U-values;

■ air permeability, particularly at construction interfaces;

■ area allowances for windows, doors, roof windows and rooflights;

■ solar controls – maximum solar energy transmittance or window energy rating.

POOREST ACCEPTABLE FABRIC U-VALUES

In addition to limiting heat losses through the building fabric, establishment of maximum U-values will reduce the possibility of condensation. Table 9.1.2 shows the worst acceptable area-weighted average element U-values. Actual values can vary quite significantly, depending on the effectiveness of other energy conservation measures that may be employed within a building.

Table 9.1.2 **U-values**

Element of construction	Limiting area-weighted average (W/m² K)
Floor	0.25
Wall	0.35
Roof	0.25
Windows/rooflights	2.20
Personnel doors	2.20
Vehicle access and other large doors	1.50
High-use entrance doors	3.50
Roof and smoke vents	3.50

Notes: Table contains values taken from Part L2A, which is new non-domestic buildings

AIR PERMEABILITY

No infiltration of air should occur at intersections of elements of construction, such as wall/roof. Guidance for conformity is provided in Approved Document L2, with reference to the Chartered Institution of Building Services Engineers (CIBSE) publication TM23: *Testing buildings for air leakage*.

The approved procedure for pressure testing, published by ATTMA, is *Measuring air permeability of building envelopes*. A certificate must be produced by a person registered with the British Institute of Non-Destructive Testing.

All buildings that are not domestic are subject to pressure testing with the following exceptions:

■ floor areas less than 500 m² provided that an air permeability of 15 m³/(h. m²) at 50 Pa has been used in the building energy rating calculations;

■ a factory-manufactured modular building less than 500 m² with a use of less than two years;

■ large extensions where sealing off the extension from the remaining building is not possible and practical testing is deemed impossible;

■ large, complex buildings where pressure testing is impractical;

■ where a building has existing compartmentation with no connection between compartments.

Air permeability is determined by pressurising a building interior with portable fans. Smoke capsules are discharged to provide visual indication of any air leakage.

AREA ALLOWANCES FOR WINDOWS, DOORS, ROOF WINDOWS AND ROOFLIGHTS

The window and door opening area for industrial and storage buildings should not exceed 15 per cent of the internal area of exposed wall. Residential buildings, such as hotels and institutions, may have up to 30 per cent window and door opening area; places of assembly, offices and shops may have up to 40 per cent. Rooflight and dormer window area is a maximum of 20 per cent of the roof area. Purpose groups of buildings other than industrial (dwellings excluded) have the same roof openings limitations. These percentage values are taken from Part L2B.

SOLAR CONTROLS

Windows, roof windows and rooflights should be fitted with means for controlling solar gain to prevent overheating of the building interior. Solar control or solar heat block can be measured by solar energy transmittance or solar factor expressed in terms of a *g*-value, stated as a number between zero and one. For purposes of an approximate comparison, 0.48 equates to no curtains, 0.43 curtains open and 0.17 curtains closed. Table 9.1.3 shows the required standards for various building orientations.

Table 9.1.3 **Solar energy transmittance factors**

Windows/rooflight orientation	Maximum *g*-value
N	0.81
NE, NW and S	0.65
E, SE, SW and W	0.52
Horizontal	0.39

The heat loss from within a building is affected by the temperature difference between the internal and external environments.

To comply with various Acts of Parliament and to create good working conditions, it is desirable to maintain a minimum temperature for various types of activity. The ideal working temperature for any particular task is a subjective measure, but as a guide the following internal temperatures recommended by the Chartered Institution of Building Services Engineers are worth considering.

- Sedentary work: 18.4 °C minimum.
- Light work: 15.6 °C minimum.
- Heavy work: 12.8 °C minimum.

The advantages that can be gained by having a well-insulated roof (see Fig. 9.1.5) are:

- lower fuel bills;
- reduced capital outlay on heating equipment;
- better working conditions for employees and hence better working relationships.

The initial cost of a building will be higher if a good standard of thermal insulation is specified and installed but, in the long term, an overall saving is usually experienced, the increased capital outlay being recovered within the first five or so years by the savings made in running costs.

FIRE PROVISIONS

The inclusion of certain materials within a factory roof to comply with the thermal insulation requirements of this Act may introduce into the structure a fire hazard. To this end, the Act stipulates that the exposed surfaces of insulation materials used, even if within a cavity, must be at least Class 1 spread of flame, as defined in BS 476-7. Furthermore, where structural steel sections are used to make up the roof frame and provide lateral stability to external walls, they will also need resistance to fire by encasement in plaster profiles, sprayed vermiculite cement or other acceptable material (see Part B to the Building Regulations and the Loss Prevention Certification Board's Standards). By using suitable materials or combinations of materials and compartmentation, the risk of fire and fire spread in factories can be reduced considerably. The main objective, in common with all fires in buildings, is to confine the fire to the vicinity of the outbreak.

Factories can have compartment-type walls with automatic-closing fire-resistant doors as previously described, but if open planning is required, other precautions will be necessary. The roof volume can be divided into cells by

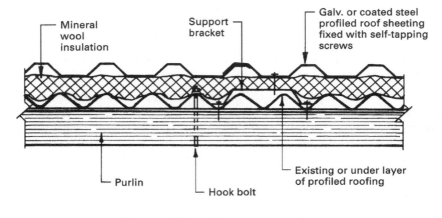

Mineral wool insulation

Support bracket

Galv. or coated steel profiled roof sheeting fixed with self-tapping screws

Purlin

Hook bolt

Existing or under layer of profiled roofing

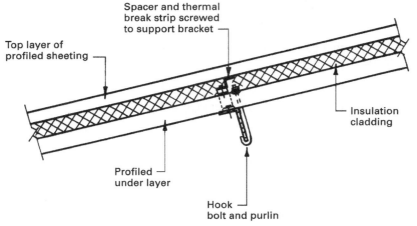

Spacer and thermal break strip screwed to support bracket

Top layer of profiled sheeting

Insulation cladding

Profiled under layer

Hook bolt and purlin

Upgraded or double-skin roof

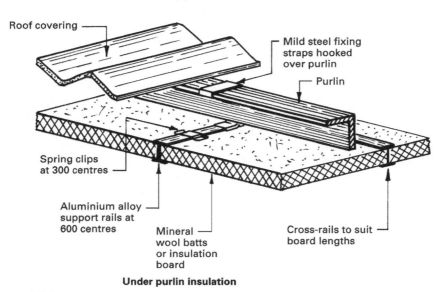

Roof covering

Mild steel fixing straps hooked over purlin

Purlin

Spring clips at 300 centres

Aluminium alloy support rails at 600 centres

Mineral wool batts or insulation board

Cross-rails to suit board lengths

Under purlin insulation

Figure 9.1.5   Industrial premises: typical insulation details

fitting permanent fire barriers within the triangulated profile of the roof structure, with non-combustible materials (such as fibre-cement sheet or galvanised wire mesh reinforced mineral wool) suitably fire-stopped within the profile of the roof covering. Beneath these fire barriers can be rolled curtains of fire-fibre cloth, controlled by a fusible link or similar fire-detection device, which will allow the curtain to fall forming a fire screen from roof to floor in the event of a fire. A similar curtain could also be positioned in the longitudinal direction under a valley beam.

Using the above method a fire can be contained for a reasonable period within the confined area, but the method can also create another problem or hazard, that of smoke logging. The smoke generated by a fire will rise to the roof level and then start to circulate within the screened compartment, completely filling the volume of the confined section within a short space of time. Apart from the hazard to people trying to escape, this can make it difficult for firefighters to:

- breathe;
- see the source of the fire;
- detect the nature of the outbreak;
- assess the extent of the outbreak.

One way to overcome this problem is to have automatic high-level ventilators that will allow the smoke to escape rapidly, giving the firefighters a chance to see clearly and enabling them to deal with the outbreak. The use of ventilators to overcome the problems of smoke logging will of course introduce more air, which aids combustion; however, this does not have the same negative effect as smoke logging, because the volume of air in this type of building is usually so vast that the introduction of more air will have very little effect on the intensity of the outbreak.

The design and position of automatic roof ventilators (also widely known as Automatic Opening Ventilators or AOVs) are normally the prerogative of a specialist designer, but as a guide the total area of opening ventilators should be between 0.5 and 5 per cent of the floor area, depending on the likely area of fire. The essential requirements for an automatic ventilator are that it must:

- open in the event of a fire, a common specification being when the heat around the ventilator reaches a temperature of 68 °C;
- be weatherproof in normal circumstances;
- be easy to fix;
- blend with the chosen roof covering material and profile.

Many automatic fire ventilators are designed to act as manually controlled ventilators under normal conditions. Typical examples are shown in Fig. 9.1.6.

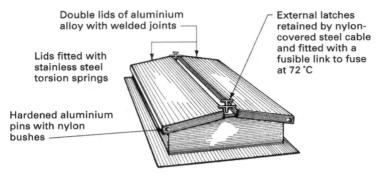

Double lids of aluminium alloy with welded joints

External latches retained by nylon-covered steel cable and fitted with a fusible link to fuse at 72 °C

Lids fitted with stainless steel torsion springs

Hardened aluminium pins with nylon bushes

Double flap automatic fire ventilator

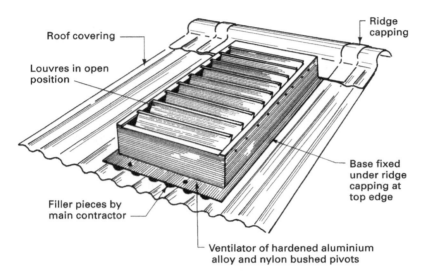

Roof covering

Ridge capping

Louvres in open position

Base fixed under ridge capping at top edge

Filler pieces by main contractor

Ventilator of hardened aluminium alloy and nylon bushed pivots

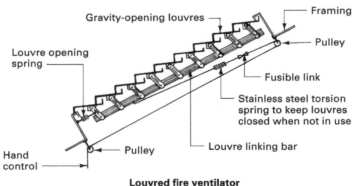

Gravity-opening louvres

Framing

Louvre opening spring

Pulley

Fusible link

Stainless steel torsion spring to keep louvres closed when not in use

Hand control

Pulley

Louvre linking bar

Louvred fire ventilator

Figure 9.1.6   Automatic roof fire ventilators

Factory buildings: walls 9.2

The walls of factory buildings have to fulfil the same functions as any enclosing wall to a building. They need to provide:

- protection from the elements;
- the required sound and thermal insulation values;
- the required degree of fire resistance;
- access to and exit from the interior;
- natural daylighting to the interior;
- reasonable security protection to the premises;
- resistance to anticipated wind pressures;
- reasonable durability to keep long-term maintenance costs down to an acceptable level;
- support for corporate signage.

Most contemporary factory buildings are constructed as framed structures using a three-dimensional space frame/columns or a system of portal frames linked with eaves beams, which means that the enclosing walls can be considered as non-loadbearing claddings supporting their own dead weight plus any wind loading. The wall can be designed as a complete envelope masking the structural framework entirely using brick or block walling, precast concrete panels, curtain-walling techniques or lightweight profiled steel wall claddings. Alternatively, an infill panel technique could be used, making a feature of the structural members, which can be exposed and painted.

The choice will depend on such factors as appearance, local planning requirements, short- and long-term costs and personal preference. If the factory is small and contains both production and offices within the same building, presentation of the company's image to business associates and clients is an important design consideration.

Many framed buildings use a lower portion of the outer wall built in traditional cavity wall and then clad the upper portion in lightweight sheeting. The brickwork and blockwork is harder-wearing and less prone to impact damage. The use of precast concrete cladding panels and lightweight infill panels is considered in Chapters 6.1 and 6.2 of this book. Many manufacturers of portal frame buildings provide a service that includes design, fabrication, supply and assembly of the complete structure, including the roof and wall coverings. Readers are recommended to study the publicity and data sheets issued by these companies for a comprehensive analysis of current practice. This section provides details of lightweight external wall claddings only, as applied to framed buildings in popular application to factory buildings.

LIGHTWEIGHT EXTERNAL WALL CLADDINGS

In common with other cladding methods for framed buildings, lightweight external wall claddings do not require high compressive strength because they only have to support their own dead load and any imposed wind loading, which will become more critical as the height or exposure increases. The subject of wind pressures is dealt with in greater detail in Chapter 9.3. Lightweight claddings are usually manufactured from impervious materials, which means that the run-off of rainwater can be high, particularly under storm conditions, when the discharge per minute could reach 2 litres per square metre of wall area exposed to the rain. Normally run-off is by the action of gravity down the face of the sheeting, which is then directed using drips onto the ground level. Detailing of the adjacent landscaping will be required to prevent splashback onto the base of the cladding.

A wide variety of materials can be used as a cladding medium, most being profiled to a corrugated or trough form because the shaping will increase the strength of the material over its flat-sheet form. Flat-sheet materials are available but are rarely applied to large buildings because of the higher strength obtained from a profiled sheet of similar thickness. Special contoured sheets have been devised by many manufacturers to give the designer a wide range of choice in terms of aesthetic appeal. Claddings in two layers or sandwich construction are also available to provide reasonable degrees of thermal and sound insulation, and to combat the condensation hazard that can occur with lightweight claddings of any nature. These were examined in detail in Chapter 6.3. Alternatively, an inner liner panel can be used to incorporate the required thermal insulation value into the wall as a site-assembled product, providing an attractive aesthetic inner finish. This may be more economical than using composite panelling.

The sheets are fixed in a similar manner to that for sheet roof coverings, except that support purlins are replaced in walls by similar steel angles called **sheeting rails**, fixed by angle cleats to the vertical structural frame. Alternatively, sheeting rails can be of zed or sigma profile for direct fixing to the structural frame. A significant difference occurs with the position of the sheet fixings, which in wall claddings are usually specified as being positioned in the trough of the profile as opposed to the crest when fixing roof coverings. This change in fixing detail is to ensure that the wall cladding is pulled tightly up to the sheeting rail or lining tray. Sheeting rails must therefore be level and vertical to prevent any distortion of the outer sheet.

Plastic protective caps for the heads of fixings are available, generally of a colour and texture that will blend with the wall cladding. A full range of fittings and trims is obtainable for most materials and profiles to accommodate openings, returns, top-edge and bottom-edge closing. Typical cladding details are shown in Figs 9.2.1 and 9.2.2.

Here are some of the materials commonly used for lightweight wall claddings.

■ **Fibre-cement sheets** Fibre cement, a non-combustible material, is shaped in corrugated and troughed sheets, which are generally satisfactory when exposed to the weather but are susceptible to impact damage. Fibre-cement sheets are a safer replacement for asbestos cement sheets, which are now steadily being removed and safely disposed of in licensed tips. They will match the existing profiles and can be used as replacement sheets for an existing roof. The average lifespan is about 20 years, which can be increased considerably by paint protection. Unpainted sheets lose their surface finish at the exposed surface by carbonation, and become ingrained with dirt. To achieve adequate thermal insulation standards, a lining material of mineral wool or similar will be required. This is normally sandwiched between the outer sheet cladding and an inner lining panel. This type of sheeting is ideal for an agricultural or industrial application, where aesthetics are not a great concern.

■ **Coated-steel sheets** Coated steel, a non-combustible material with a wide range of profiles, is produced in sheets by various manufacturers. The steel sheet forms the core of the cladding, providing its strength, and is covered with various forms of coatings to give weather protection, texture and colour. A typical specification would be a galvanised 0.9 mm-thick core, covered on both sides with silicone polyester for improved appearance and durability. Coated-steel sheets can now be incorporated and formed around a foam insulation core as a composite panel. The surface of the sheet can be powder-coated in a range of different colours, providing the designer with an aesthetic range.

■ **Aluminium alloy sheets** Aluminium alloy, a non-combustible material, is shaped in corrugated and troughed profiles, which are usually made

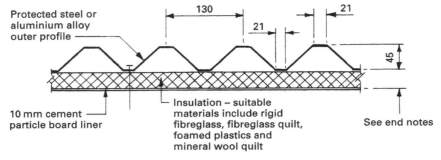

Protected steel or
aluminium alloy
outer profile

130

21

21

45

10 mm cement
particle board liner

Insulation – suitable
materials include rigid
fibreglass, fibreglass quilt,
foamed plastics and
mineral wool quilt

See end notes

Typical cladding profile

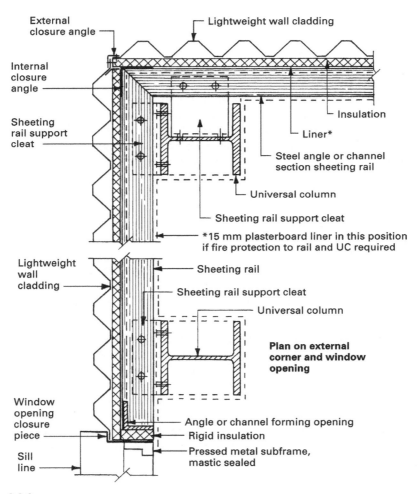

External
closure angle

Lightweight wall cladding

Internal
closure
angle

Sheeting
rail support
cleat

Insulation

Liner*

Steel angle or channel
section sheeting rail

Universal column

Sheeting rail support cleat

*15 mm plasterboard liner in this position
if fire protection to rail and UC required

Lightweight
wall
cladding

Sheeting rail

Sheeting rail support cleat

Universal column

**Plan on external
corner and window
opening**

Window
opening
closure
piece

Sill
line

Angle or channel forming opening

Rigid insulation

Pressed metal subframe,
mastic sealed

Figure 9.2.1   Lightweight wall cladding: typical details 1

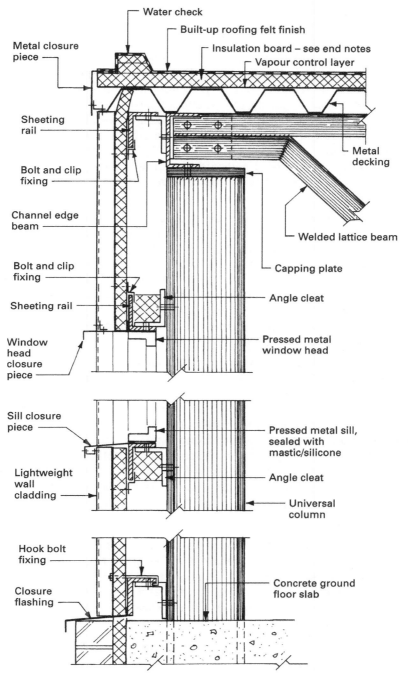

Note: See Fig. 9.2.1 re location of fire protection lining

Figure 9.2.2   Lightweight wall cladding: typical details 2

to the recommendations of BS 4868. As the sheets are manufactured from aluminium, there is an element of recycling in the manufacturing process and at the end of the product's life. Other profiles are also available as manufacturers' standards, so replacement sheets can be sourced for maintenance purposes. Durability will depend on the alloy specification, but can be increased by specialist paint applications or plastisol coatings as for steel. The sheets can be used as the manufactured finish but may need to be cleaned regularly. Aluminium alloys are unsuited to coastal locations, where the salty atmosphere will react with and corrode the metal surface. Fixings, fittings and the availability of linings are as indicated for other cladding materials. Again, the aluminium panel can be formed into a composite panel with an insulation core.

■ **Polyvinylchloride sheets** This type of sheet is manufactured from PVC, which is a hygienic surface, enabling the manufactured panels to be used in locations such as hospitals, chemical plants and food manufacturing units. Polyvinylchloride sheet tends to be manufactured in a strip form with interlocking profiles, so that it can mimic the use of timber. PVC cladding is hard-wearing, colour-fast, easily fixed and now available in a range of colours, from white to a natural wood finish. Large sheet sizes are impractical for commercial applications, so it is normally used in a shiplap configuration as a feature.

■ **Polycarbonate sheets** Advances in technology have enabled multi-walled polycarbonate sheets to be manufactured in long lengths that can be used as an external cladding. Figure 9.2.3 illustrates one such sheet and its fixing details. As can be seen, the sheet has incorporated into its construction a tongue-and-groove method of fitting one sheet to another and a hidden aluminium fixing plate. This enables large areas of external cladding to function as a clear, uninterrupted window where the whole wall allows the passage of daylight. The multi-walled construction of the sheets enables U-values as low as 0.7 to be obtained, as the units are sealed at their ends and trap air within their construction.

The importance of adequate design, detail and fixing of all forms of lightweight cladding cannot be overstressed, because the primary objective of these claddings is to provide a lightweight envelope to the building, giving basic weather protection and internal comfort at a reasonable cost. Claddings that will fulfil these objectives are very susceptible to wind damage unless properly secured to the structural frame.

Notes: Figs 9.2.1 and 9.2.2

1. Wall insulation thickness as required by national Building Regulations and building purpose. Approximate values using mineral fibre insulants:

Thickness (mm)	U-value (W/m² K)
None	2.80
60	0.55
80	0.45
100	0.35
150	0.25

2. Roof insulation thickness as required by national Building Regulations and building purpose. Approximate values using mineral fibre insulants:

Thickness (mm)	U-value (W/m² K)
None	2.60
60	0.54
80	0.40
100	0.33
130	0.25
150	0.22

Readers will need to refer to the insulation manufacturer's website for the data sheet and undertake a calculation to ensure the correct thickness is applied to meet the Part L Approved Document.

Typical cross section through wall panel   Building width 500 mm + 1 / – 1%

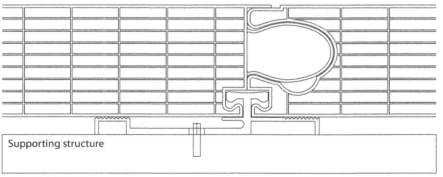

Supporting structure

Vertical joint detail

Figure 9.2.3   Typical 60 mm thick polycarbonate multi-walled panel

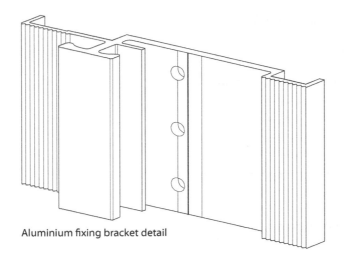

Aluminium fixing bracket detail

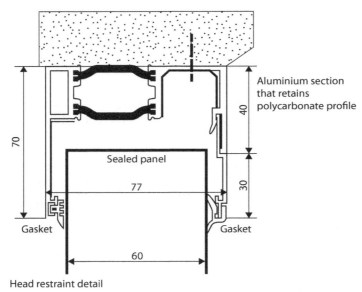

Aluminium section
that retains
polycarbonate profile

70

40

30

Sealed panel

77

Gasket

Gasket

60

Head restraint detail

Figure 9.2.3 (*continued*)

Wind pressures 9.3

Wind can be defined as a movement of air. The full nature of wind is not fully understood, but two major contributory factors that can be given are:

- convection currents caused by air being warmed at the Earth's surface, becoming less dense, rising and being replaced by colder air;
- transference of air between high- and low-pressure areas.

The speed with which the air moves in replacement or transfer is termed its **velocity**, and can be from 0 to 1.5 metres per second, when it is hardly noticeable, to speeds in excess of 24 metres per second, when considerable damage to property and discomfort to persons could be the result.

The physical nature of the ground or topography over which the wind passes will have an effect on local wind speeds, because obstructions, such as trees and buildings, can set up local disturbances by forcing the wind to move around the sides of the obstruction or funnel between adjacent obstacles. Where funnelling occurs, the velocity and therefore the pressure can be increased considerably. This pressure also works in reverse, by the action of the wind passing horizontally over a surface creating localised negative pressure that the construction has to resist. Experience and research have shown that the major damage to buildings is caused not by a wind blowing at a constant velocity but by short-duration bursts or gusts of wind of greater intensity than the prevailing mean wind speed. The durations of these gusts are usually measured in 3-, 5- and 15-second periods. Information on the likely maximum gust speeds for specific long-term time durations of 50 years or more is available from the Met Office.

When the wind encounters an object in its path, such as the face of a building, it is usually rebuffed and forced to turn back on itself. This has the effect of setting up a whirling motion or eddying current at the corners, which eventually finds its way around or over the obstruction. The pressure of the wind is normally in the same direction as the path of the wind, which tends to push the wall of the building inwards, and will do so if sufficient resistance is not built into the structure. However, the effect of local eddies is often opposite in direction and force to that of the prevailing wind, producing a negative or suction force (see Fig. 9.3.1). In fact, the UK has more tornadoes than parts of America but they are much smaller, with less power and soon die out of energy.

Many factors must be taken into account before the magnitude and direction of wind pressures can be determined, including:

- height-to-width ratio of the building;
- length-to-width ratio of the building;
- plan shape of the building;
- approach topography;
- exposure of the building;
- proximity of surrounding structures.

Account must also be taken of any likely openings in the building, because the entry of wind will exert a positive pressure on any walls or ceilings encountered. These internal pressures must be added to or subtracted from the type of pressure anticipated acting on the external face at the same point.

All buildings are at some time subjected to wind pressures, but some are more vulnerable than others because of their shape, exposure or method of construction. One method of providing adequate resistance to wind pressures is to use materials of high density, so buildings that are clad with heavier traditional materials are less susceptible to wind damage than those using lightweight coverings. Factory buildings using lightweight claddings have therefore been taken to serve as an illustration of providing suitable means of resistance to wind pressures.

To overcome the problem of uplift or suction on roofs caused by the negative wind pressures, adequate fixing or anchorage of the lightweight coverings to the structural frame is recommended. Generally, sufficient resistance to uplift of the frame is inherent in the material used for the structural members; the problem is therefore to stop the covering being pulled away from the supporting member. This can be achieved by the quality or holding power of the fixings used, by the number of fixings employed or by a combination of both. Note that the whole roof considered as a single entity is at risk, and not merely individual sheets.

If the supporting member does not have sufficient self-dead load to overcome the suction forces, such as a timber plate bedded onto a brick wall, then it will be necessary to anchor the plate adequately to the wall. This can be carried

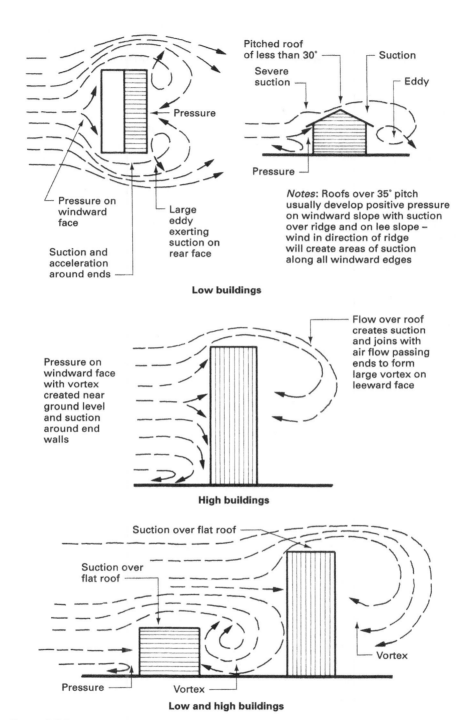

Pressure on windward face

Suction and acceleration around ends

Pressure

Large eddy exerting suction on rear face

Low buildings

Pitched roof of less than 30°

Severe suction

Suction

Eddy

Pressure

Notes: Roofs over 35° pitch usually develop positive pressure on windward slope with suction over ridge and on lee slope – wind in direction of ridge will create areas of suction along all windward edges

Pressure on windward face with vortex created near ground level and suction around end walls

Flow over roof creates suction and joins with air flow passing ends to form large vortex on leeward face

High buildings

Suction over flat roof

Suction over flat roof

Vortex

Pressure

Vortex

Low and high buildings

Figure 9.3.1   Typical effect of wind pressures around buildings

out by means of bolts or straps fixed to the plate and wall in such a manner that part of the dead load of the wall can be added to that of the supporting member. This is a requirement of the Building Regulations regarding the stability of structures.

Positive wind pressures tend to move or bend the wall forwards in the same direction as the wind; this tendency is usually overcome when using light structural framing by adding stiffeners called **wind bracing**. Wind braces are usually of a steel angle, circular hollow section or tension wire construction, fitted between the structural members where unacceptable pressures are anticipated. They take the form of cross-bracing, forming what is in fact a stiffening lattice within the frame (see Fig. 9.3.2). Although each area of the country has a predominant prevailing wind, buildings requiring wind bracing are usually treated the same at all likely vulnerable positions to counteract changes in wind direction and the effect of local eddies. Wind bracing support can also be supplied to reinforce the lightweight cladding, placed vertically between the horizontal sheeting rails at the midpoint of the span.

The immense destructive power of the wind, in the context of building works, cannot be overemphasised, and careful consideration is required from the design stage to actual construction on site. Readers should also appreciate that temporary works and site hutments are just as vulnerable to wind damage as the finished structure. Great care must be taken, therefore, when planning site layouts, plant positioning, erection of scaffolds and hoardings to ensure safe working conditions on building sites. If not suitably tied down, scaffold boards will lift in high winds and potentially fall from height.

Further details of the effects and assessment of wind loads on buildings can be obtained from the Building Research Establishment publication *Wind, floods and climate pack AP267*.

An additional reference for the determination of wind load values is BS EN 1991 Eurocode 1. *Actions on structures. General actions. Wind actions.* This publication provides a basis for the loading calculation. The calculation uses elements of the mean wind velocity that is an average, which is obtained from the wind climate, the variation of height above the topography, the nature of the terrain that the wind passes over and the exposure factor. The reader is advised to consult the document as the calculation is complex and beyond the scope of this text.

The Building Regulations also play an important role in regulating for excessive wind loadings. Figure 9.3.3 illustrates the zones of basic wind speed, a document that is used in Approved Document A of the Building Regulations. Part A provides the restrictions on mainly non-commercial properties that can be built in terms of height, location, exposure and basic wind speeds.

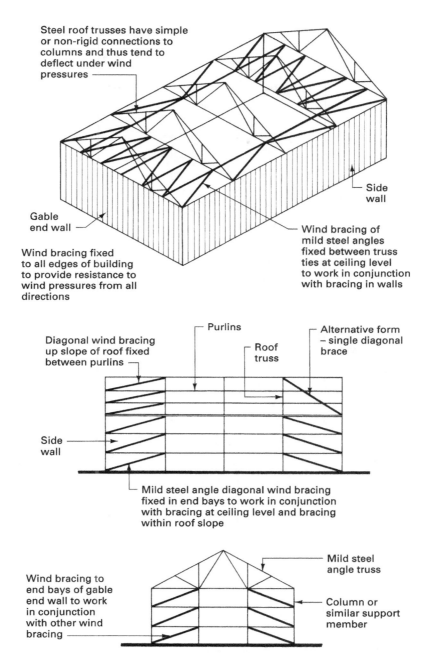

Steel roof trusses have simple or non-rigid connections to columns and thus tend to deflect under wind pressures

Side wall

Gable end wall

Wind bracing fixed to all edges of building to provide resistance to wind pressures from all directions

Wind bracing of mild steel angles fixed between truss ties at ceiling level to work in conjunction with bracing in walls

Purlins

Diagonal wind bracing up slope of roof fixed between purlins

Roof truss

Alternative form – single diagonal brace

Side wall

Mild steel angle diagonal wind bracing fixed in end bays to work in conjunction with bracing at ceiling level and bracing within roof slope

Wind bracing to end bays of gable end wall to work in conjunction with other wind bracing

Mild steel angle truss

Column or similar support member

Figure 9.3.2   Typical wind-bracing arrangements

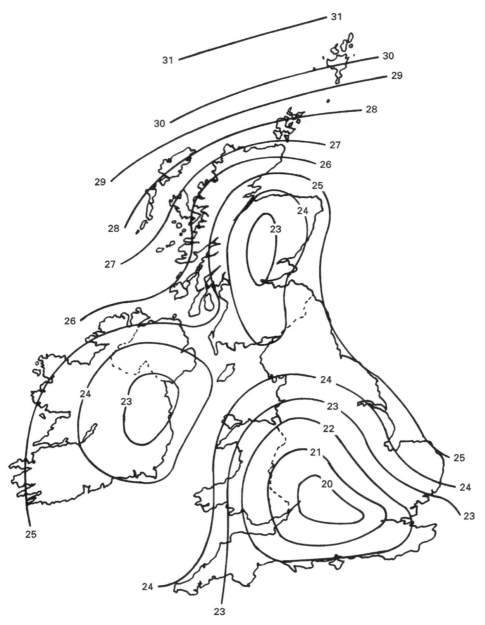

Figure 9.3.3   Zones of basic wind speed (m/s)

Driving rain   9.4

Wind-driven rain can cause material damage by dampness penetration into the structure, and loss of thermal function by penetration of the lightweight components by moisture, which reduces thermal efficiency. Rain can cause the building fabric to become sufficiently moist for the growth of fungi, mildew, mosses, moulds and other plant and insect life.

The quantity of rain falling on the vertical and sloping surfaces of domestic, industrial and commercial buildings is an important issue for exposure design considerations. The main concern is the method of disposal that will be used to remove the collected rainfall off the roof and into a suitable drainage medium. The source of rain penetration is not necessarily defective or damaged materials, as many building components such as tiles and window parts rely on sufficient overlapping to prevent moisture access. Total exclusion of moisture is optimum, but the nature of some building materials, such as tiles and porous bricks and stone, will inevitably permit some rain penetration in extreme conditions: hence the need for undertiling layers and purpose-made cavities in wall construction that are used to drain away any excess moisture that passes through the first layer.

The choice of external finishing may be influenced by architectural trends, but this should be balanced against the degree of exposure before specifying the external wall-facing products. The Building Research Establishment has published maps providing guidance on the driving rain index (DRI) for particular locations in the UK and Ireland, shown in Fig. 9.4.1. These can be used to establish in which areas a specification may need to be reinforced for the climatic conditions. The extremes of the west coast and areas of high contours are obviously the areas most exposed to driving rain. Values on the DRI are obtained by taking the mean annual wind speed (m/s), multiplying by the mean annual rainfall (mm) and dividing by 1,000 (m^2/s). The result is a series of contours linking areas with similar DRIs throughout the country. Correction factors may be applicable for topography, exposure, shelter, roughness of terrain, altitude and building height. Figure 9.4.2 shows annual mean driving rain roses for different parts of the country. The length of each radiating line indicates the mean DRI received from that direction. The direction of prevailing wind is therefore the longest of the lines. See also BS 5618: *Code of practice for thermal insulation of cavity walls (with masonry or concrete inner and outer leaves) by filling with urea-formaldehyde foam systems,* and *BRE Report 59: Directional driving rain indices for the UK – computation and mapping.*

An alternative approach is detailed in BS 8104: *Code of practice for assessing exposure of walls to wind driven rain.* This provides formulae, tables and maps to enable calculation of the quantity of wind-driven rain (l/m^2 per spell) at a given location on a wall. It is known as the **wall spell index**, where spell is the period for which wind-driven rain occurs on a vertical surface. The quantity of wind-driven rain can also be calculated on a yearly basis (l/m^2 per year), and this is known as the **wall annual index**. From these a **local spell index** and a **local annual index** can be established – the former useful for assessing the resistance of the structure to water penetration and the latter

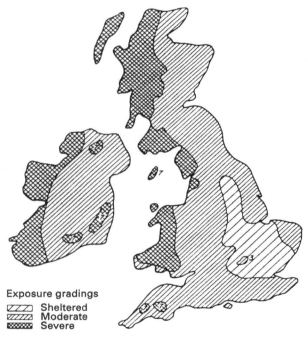

Exposure gradings
☐ Sheltered
☐ Moderate
☐ Severe

Sheltered exposure zone: areas where the DRI is 3 or less.
Moderate exposure zone: areas where the DRI is between 3 and 7.
Severe exposure zone: areas where the DRI is 7 or more.

Figure 9.4.1   Driving rain index (DRI)

appropriate when considering the quantity of moisture in a wall and its possible base for lichen, mosses and other growths that can have a deteriorating effect on the finishes to the envelope of a building.

The procedure for assessment can be followed using worked examples provided in BS 8104. Specialist maps for the UK are endorsed with data on sub-regions, spell and annual rose values, and geographical incremental data. A typical rose is shown in Fig. 9.4.3 with numbers corresponding to building orientation: for example, if a building faces west, the rose value in Fig. 9.4.3 is 7. Geographical increments will vary with location, and may add or reduce the value by a nominal amount.

From this information the **airfield spell** and **annual indices** can be established from tables. These are measures of driving rain (l/m²) occurring 10.000 m above ground in the middle of an airfield for the worst spell in three years and for an average annually, respectively. Factors for terrain roughness, topography, obstructions and wall characteristics compound the calculations to provide an assessment in litres/m². Further calculations can be used to determine the run-off in litres per metre width of wall, which is useful for assessing local flooding potential and the need for adjacent subsoil and surface drainage.

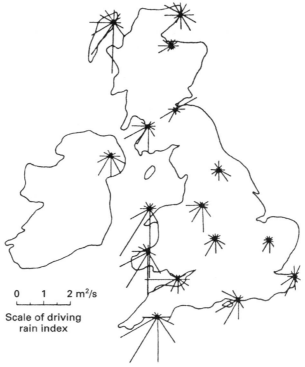

Figure 9.4.2 Driving rain rose diagram

Driving rain rose (BS 5618)
prevailing wind – south westerly

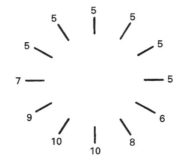

Spell and annual rose values (BS 8104)
e.g. building facing west = 7

Figure 9.4.3 **Rose diagrams**

BRE Report No. 59 and BS 8104 are complementary, whereas BS 5618 applies specifically to UF foam-filled cavity walls, where obviously there is no cavity space left to drain away any moisture. Therefore, when considering information, source identification is essential to avoid confusion.

The importance of categorising driving rain, whether by indices or by roses, is to determine suitable specification of construction details to resist the areas of orientation that will be subject to the most exposure. Details appropriate for sheltered exposure zones are unlikely to be acceptable in areas of severe exposure. Establishing standards for construction should be by consultation with the local building control authority, which will check this through a full plans Building Regulations application. Different areas of the country will have specific climatic conditions and possibly microclimates within those areas. For example, simple masonry cavity wall construction in a zone of severe exposure should maintain a 50 mm air space beyond any insulation included in the overall cavity width. This is essential to prevent dampness bridging the cavity.

Roof structures 9.5

Roof design and construction can be an extensive topic. Readers will need to develop an understanding by including specific aspects of roofing techniques at each element or level of study. In this section, the roofs considered are those suitable for a variety of large, clear-span commercial applications using structural engineered timber, steel framed and reinforced concrete. Roof construction techniques for smaller spans are described in Chapter 9.1 of this volume and Chapter 6 of the accompanying volume, *Construction Technology*.

LARGE-SPAN TIMBER ROOFS

A wide variety of timber roofs is available for both medium and large spans, and these can be classified under two headings:

- pitched roof trusses;
- glulam engineered timber.

PITCHED TRUSSES

Pitched roof trusses are two-dimensional triangulated designed frames spaced at 4.500 to 6.000 m centres with spans up to 30.000 m. The pitch should have a depth-to-span ratio of 1:5 or steeper and be chosen to suit roof coverings. The basic construction follows that of an ordinary domestic small-span roof truss, as shown in Fig. 6.1.7 of *Construction Technology*, except that the arrangement and number of struts and ties will vary according to the type of truss being used. Figure 9.5.1 illustrates the different types of truss design that are factory-manufactured using stress-graded timber and hydraulically pressed gang nail plate connections. This makes the joints very strong and, thanks to the detailed structural calculations involved in the truss manufacture, create a lightweight product. The flat-topped trusses can also be produced in this way, with pressed nail plates to connect the members.

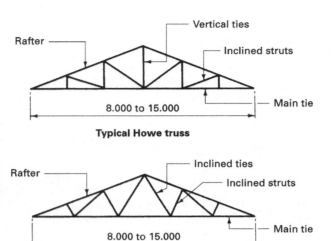

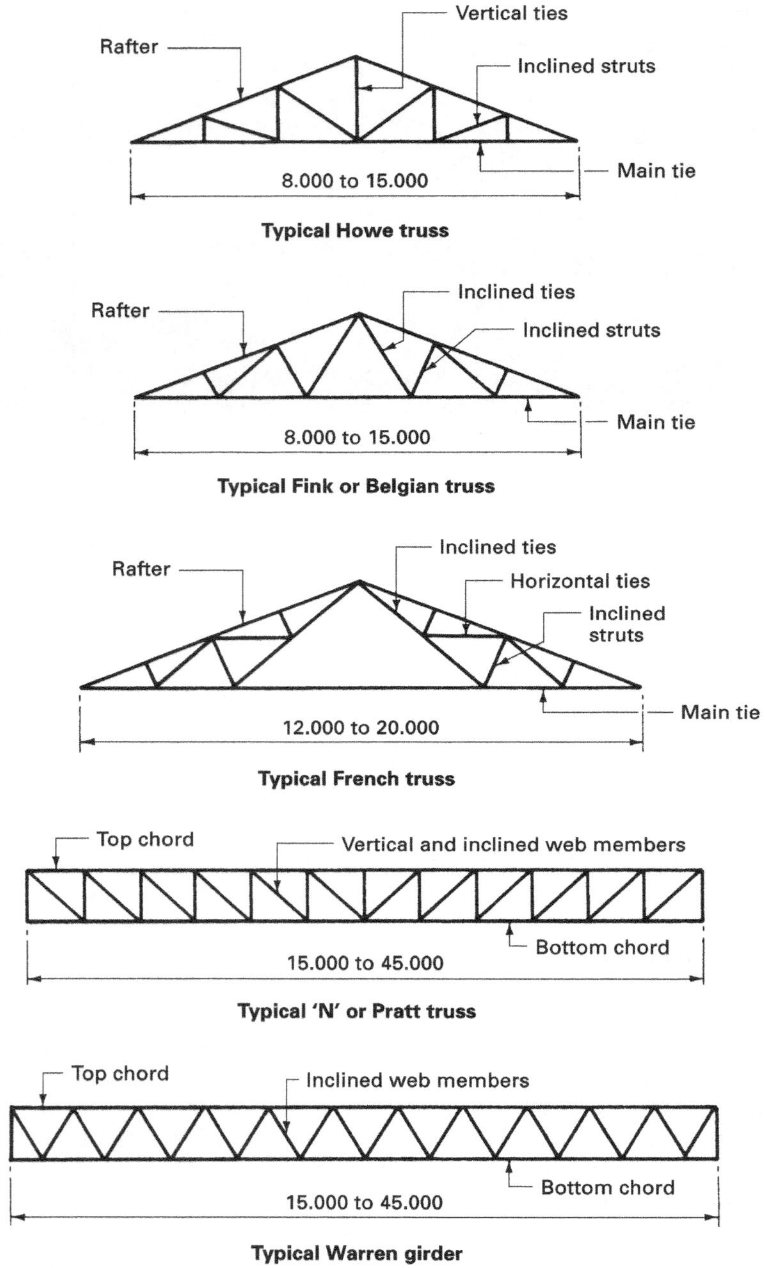

Vertical ties
Rafter
Inclined struts
Main tie
8.000 to 15.000
Typical Howe truss
Inclined ties
Rafter
Inclined struts
Main tie
8.000 to 15.000
Typical Fink or Belgian truss
Inclined ties
Rafter
Horizontal ties
Inclined struts
Main tie
12.000 to 20.000
Typical French truss
Top chord
Vertical and inclined web members
Bottom chord
15.000 to 45.000
Typical 'N' or Pratt truss
Top chord
Inclined web members
Bottom chord
15.000 to 45.000
Typical Warren girder

Figure 9.5.1 Typical large-span truss and girder types

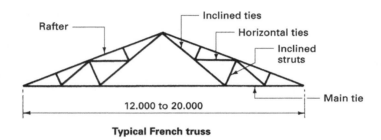

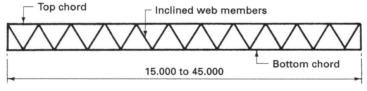

GLULAM BEAMS

Glulam is the process of constructing timber by gluing a series of sections together to form a larger timber section. The advantages of glulam are that it:

- makes a stronger beam;
- gives a higher strength-to-weight ratio;
- provide a more cost-effective and greener product than steel or concrete;
- enables much larger-depth beams to be constructed than a tree could provide;
- gives an aesthetic appearance to the structure;
- is made from a highly sustainable, reclaimable/recyclable material
- limits the amount of movement and distortion due to absorption of water by the timber.

These beams can be constructed as portal frames that contain an integral column and roof beam, or just as large-span roof beams that will need lateral restrain at the wall junction. Galvanised or stainless steel connections are required to cap the ends of the beams and provide a suitable fixing to a ridge, foundation, wall or column. Spans up to 60 m can be formed with large beams; the ridge connection is then vital in breaking down the structure for transport.

Figure 9.5.2 illustrates different connections for a curved glulam beam. Because the process involves clamping strips of timber into a former, virtually any shape can be produced. The roof purlins can be fixed using a normal cleating system and the roof covering completed over the structure. Figure 9.5.3 illustrates some of the variations that can be used within the portal frame design and the indicated spans that are feasible with this type of arrangement. The curves reduce the stresses throughout the portal and also provide an aesthetic appearance to the parts of the finished structure that remain exposed.

CHOICE OF TRUSS AND TIMBER

To decide on the most suitable and economic truss to be specified for any given situation, the following should be considered:

- availability of suitable timber in the sizes required;
- the loading conditions and the span carried;
- design and fabrication costs;
- transportation problems and costs;
- on-site assembly and erection problems and costs;
- roof-covering material availability and costs;
- architectural design considerations.

The final item may well outweigh some of the economic solutions for the preceding items.

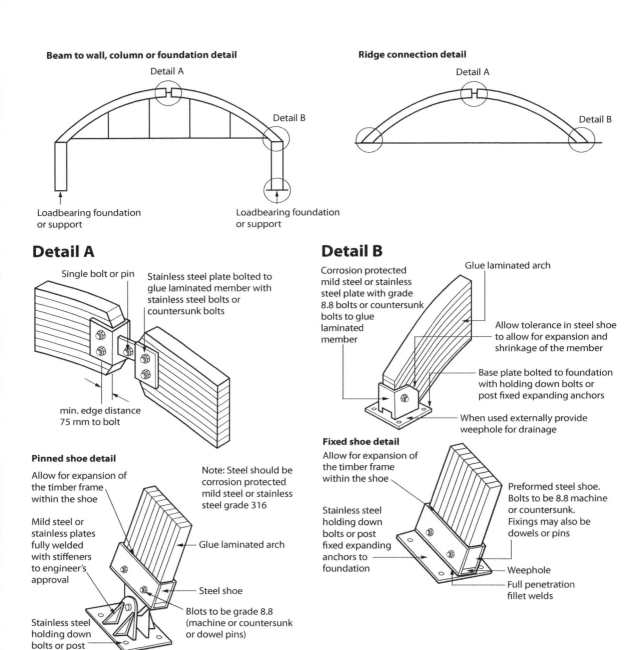

Beam to wall, column or foundation detail

Detail A

Detail B

Loadbearing foundation or support

Loadbearing foundation or support

Ridge connection detail

Detail A

Detail B

Detail A

Single bolt or pin

Stainless steel plate bolted to glue laminated member with stainless steel bolts or countersunk bolts

min. edge distance 75 mm to bolt

Pinned shoe detail

Allow for expansion of the timber frame within the shoe

Mild steel or stainless steel plates fully welded with stiffeners to engineer's approval

Stainless steel holding down bolts or post fixed expanding anchors to foundation

Note: Steel should be corrosion protected mild steel or stainless steel grade 316

Glue laminated arch

Steel shoe

Blots to be grade 8.8 (machine or countersunk or dowel pins)

Detail B

Corrosion protected mild steel or stainless steel plate with grade 8.8 bolts or countersunk bolts to glue laminated member

Glue laminated arch

Allow tolerance in steel shoe to allow for expansion and shrinkage of the member

Base plate bolted to foundation with holding down bolts or post fixed expanding anchors

When used externally provide weephole for drainage

Fixed shoe detail

Allow for expansion of the timber frame within the shoe

Stainless steel holding down bolts or post fixed expanding anchors to foundation

Preformed steel shoe. Bolts to be 8.8 machine or countersunk. Fixings may also be dowels or pins

Weephole

Full penetration fillet welds

Figure 9.5.2   Large glulam arch fixing details

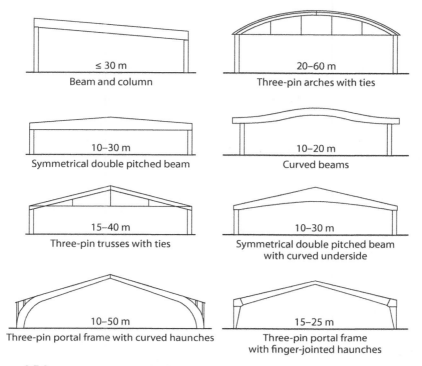

≤ 30 m
Beam and column

20–60 m
Three-pin arches with ties

10–30 m
Symmetrical double pitched beam

10–20 m
Curved beams

15–40 m
Three-pin trusses with ties

10–30 m
Symmetrical double pitched beam
with curved underside

10–50 m
Three-pin portal frame with curved haunches

15–25 m
Three-pin portal frame
with finger-jointed haunches

Figure 9.5.3 **The different glulam portal options**

STRENGTH GRADING

There are two types of strength grading employed in the UK: visual strength grading (VSG) and machine strength grading (MSG). The first relies on the experience and training of a competent inspector, and the latter uses a calibrated testing machine within a processing mill.

Suppliers and designers have to conform to several regulations and standards regarding the specification of structural timber. The main standards and regulations are:

- The Building Regulations and its approved documents, which are based on the Eurocode 5 *The design of timber structures*;

- BS EN 14081-2 *Timber structures, strength graded structural timber with rectangular cross section*;

- BS 4978 *Specification for visual strength grading of softwood* for visual grading undertaken in the UK;

- BS EN 1912 *Structural timber, strength classes, assignment of visual grades and species*;

- BS EN 338 *Structural timber strength classes*, which gives the characteristic values for timber.

Visual stress grading

Visual stress grading is carried out by the knot area ratio (KAR) method, in which the proportion of the cross-section occupied by the projected area of knots is assessed.

- Any pieces of timber where the KAR is less than one-fifth are graded **special structural** (SS).
- Pieces of timber where the KAR is between one-fifth and one-half are graded as **general structural** (GS) or SS, depending on whether a margin condition exists.

Any piece of timber with a KAR exceeding one-half is automatically rejected for structural work. All visually graded timber must be identified with:

- the name of the producer or manufacturer;
- whether the timber was dry- or wet-graded;
- the logo mark for the EU-notified body;
- the strength classification from BS EN 1912;
- whether the timber has had a preservative treatment.

Machine stress grading

Machine stress grading relies on the correlation between the strength of timber and its stiffness. These grading machines work on one of two principles: those that apply a fixed load and measure the deflection, and those that measure the load required to cause a fixed deflection. The gradings obtained are designated with the letter 'M' on the labelling, to indicate that it has been machine graded. Softwood that is processed in the UK or Europe will be classified to one of the 12 classifications for strength contained with BS EN 338. These are the classifications used within the guidance document produced by TRADA on load/span tables for structural timber components. The most common classes used in the UK are C16 and C24.

The most readily available species of structural softwoods are imported redwood, imported whitewood and imported commercial western hemlock. Other suitable structural softwoods are generally only available in small quantities. Structural softwoods are supplied in standard lengths, commencing at 1.800 m and increasing by 300 mm increments to lengths rarely exceeding 6.300 m.

When specifying stress-graded timber, the following points should be considered:

- species;
- section size;
- length;
- preparation requirements;
- stress grade;
- moisture content.

The reader should refer to the Trada publication Wood Information Sheet 4-7, which gives details on the current standards associated with the grading of timber for structural use.

JOINTING

Connections between structural timber members may be made by:

- nails;
- screws;
- glue and nails or screws;

- truss plates;
- bolts;
- bolts and timber connectors.

Nails and screws are usually used in conjunction with plywood gussets and, like the truss or gangnail plates, are usually confined to the small- to medium-span trusses. In all cases with large-span roofs, the structural integrity of the joint is vital to resist and transfer the loadings of the roof to the supporting walls or columns. In glulam construction we have seen that these joints are structurally designed to resist such loads.

The usual fixings, such as nails, screws and bolts, each have their own limitations. With nails, this is the ability to grip the timber and hold fast without pulling out; with screws, it is the pre-drilling required such that, in dense timbers, the screw is not sheared off while driving. Cut nails will generally have a greater holding power than round wire nails because of the higher friction set up by their rough, sharp edges and also the smaller disturbance of the timber grain. The joint efficiency of nails may be as low as 15 per cent because of the difficulty in driving sufficient numbers of them within a given area to obtain the required shear value. Screws have a greater holding power than nails, but are dearer in both labour and material costs.

Joints made with a rigid bar, such as a bolt, usually have low efficiency because of the low shear strength of timber parallel to the grain and the unequal distribution of bearing stress along the shank of the bolt. The weakest point in these connections is where the high stresses are set up around the bolt, and various methods have been devised to overcome this problem. The solution lies in the use of timber connectors that are designed to increase the bearing area of the bolts, as described below.

TOOTHED PLATE TIMBER CONNECTOR

Sometimes called **bulldog connectors**, toothed plate timber connectors are used to form an efficient joint without the need for special equipment. They are suitable for all types of connection, especially when small sections are being used. To form the connection, the timber members are held in position and drilled for the bolt to provide a bolt hole with 1.5 mm clearance. If the timber is not too dense, the toothed connectors can be embedded by tightening up the permanent bolts. In dense timbers, or where more than three connectors are used, the embedding pressure is provided by a high-tensile steel rod threaded at both ends. Once the connectors have been embedded, the rod is removed and replaced by the permanent bolt.

SPLIT RING TIMBER CONNECTORS

Split ring timber connectors are suitable for any type of structure, and timbers of all densities. They are very efficient and develop a high-strength joint. The split ring is a circular band of steel with a split tongue-and-groove vertical joint. A special boring and cutting tool is required to form the bolt hole and the grooves in the face of the timber into which the connector is inserted, making the ring independent of the bolt itself. The split in the ring ensures that a tight fit is achieved on the timber core, but at the same time it is sufficiently flexible to give a full bearing on the timber outside the ring when under heavy load. See Fig. 9.5.4 for typical timber connector details.

SHEAR PLATE TIMBER CONNECTOR

Shear plate timber connectors are counterparts of the split ring connectors. They are housed flush into the timber members and are used for demountable structures, as they can be unbolted and removed. A special tool is required to form the hole for the shear plate connector so it is held in place. Two units are used with one joint such that the plate connectors are flush with the outside of the timber to be connected.

LARGE-SPAN STEEL ROOFS

The roof types given for large-span timber roofs can also be designed and fabricated using standard structural steel sections. Span ranges and the spacings of the frames or lattice girders are similar to those given for timber roofs. Connections can be of traditional gusset plates to which the struts and ties are bolted or welded; alternatively, an all-welded construction is possible, especially if steel tubes are used to form the struts and ties. Large-span steel roofs can also take the form of space decks and space frames.

SPACE DECKS

A space deck is a structural roofing system designed to give large clear spans with wide column spacings. It is based on a simple, repeated unit consisting of an inverted pyramid frame, which can be joined to similar frames to give spans of up to 22.000 m for single spanning designs and up to 33.000 m for two-way spanning roofs. These basic units are joined together at the upper surface by bolting adjacent angle-framing together and by fixing threaded tie bars between the apex couplers (see Fig. 9.5.5). The angled framing then forms a deck that cladding materials can be laid over to weatherproof the envelope.

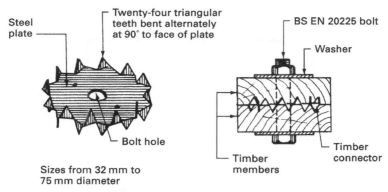

Double-sided round-toothed plate connectors

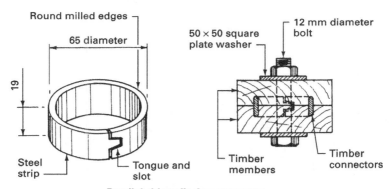

Parallel side split ring connector

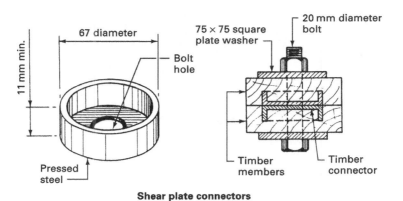

Shear plate connectors

Figure 9.5.4 Typical timber connectors. See also BS EN 912: *Timber fasteners. Specification for connections to timber*

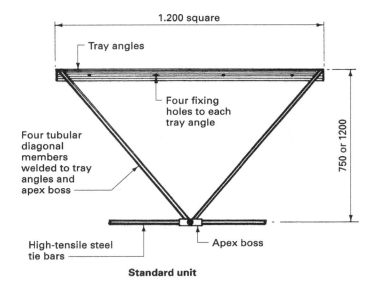

1.200 square

Tray angles

Four fixing
holes to each
tray angle

Four tubular
diagonal
members
welded to tray
angles and
apex boss

750 or 1200

High-tensile steel
tie bars

Apex boss

Standard unit

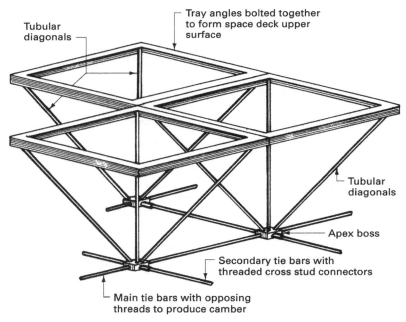

Tray angles bolted together
to form space deck upper
surface

Tubular
diagonals

Tubular
diagonals

Apex boss

Secondary tie bars with
threaded cross stud connectors

Main tie bars with opposing
threads to produce camber

Figure 9.5.5   Typical space deck standard units

Edge treatments include vertical fascia, mansard and cantilever. Rooflights can be fixed directly over the 1.200 m × 1.200 m modular upper framing (see Fig. 9.5.6). The roof can be laid to falls, or alternatively a camber to form the falls can be induced by tightening up the main tie bars placing the top into tension, which rises up forming a camber. The space deck roof can be supported by steel or concrete columns or fixed to padstones or ring beams situated at the top of loadbearing perimeter walls.

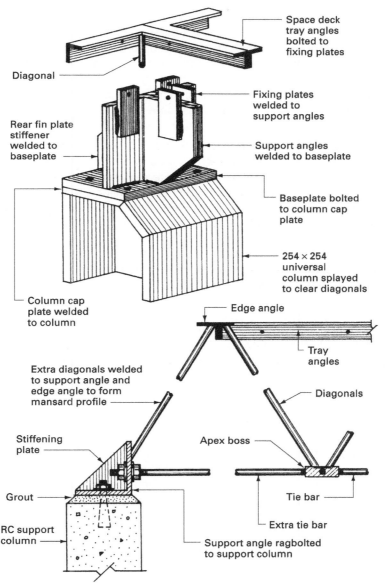

Figure 9.5.6   Typical space deck edge-fixing details

The most usual and economic roof covering for space decks is a high-density mineral-wool-insulated roofing board covered with three layers of built-up roofing felt, with a layer of reflective chippings. The chippings reflect some of the sun's heat energy off the roof surfaces. The main issue with three-layer built-up felt is the long-term lifespan of such a roof without any maintenance work. Improvements have been made in felt technologies that extend and guarantee the lifespan of a building's roof finish, and 30-year warranties are now feasible on some systems. Alternatively, composite double-skin decking may be used. As shown in Chapter 9.1 of this volume, a panel comprises a rigid urethane core with profiled coated-steel or aluminium sheet facings. The void created by the space deck structure can be used to house all forms of services, and the underside can be either left exposed or covered with an attached or suspended ceiling. The units are usually supplied with a basic protective paint coating, applied by a dipping process after the units have been degreased, shot-blasted and phosphate-treated to provide the necessary key.

The simplicity of the space deck unit format eliminates many of the handling and transportation problems encountered with other forms of large-span roof. This is because a complete roof can usually be transported on one lorry by stacking the pyramid units one inside the other, which is an economical and sustainable method of transporting a large roof structure. Site assembly and erection are usually carried out by a specialist subcontractor, who must have access to the whole floor area. Assembly is rapid, with a small labour force that assembles the units as beams in the inverted position, turns them over and connects the whole structure together by adding the secondary tie bars.

Two methods of assembly and erection can be used:

1. the deck is assembled on the completed floor immediately below the final position and lifted directly into its final position;
2. the deck is assembled outside the perimeter of the building and lifted in small sections to be connected together in the final position. This is generally more expensive than method 1.

Before assembly and erection the main contractor has to provide a clear, level and hard surface to the perimeter of the proposed building, as well as to the whole floor area to be covered by the roof. These surfaces must be capable of accepting the load from a 25-tonne mobile crane. The main contractor is also responsible for unloading, checking, storing and protecting the units during and after erection, and for providing all necessary temporary works and plant, such as scaffolds, ladders and hoists. The site procedures and main contractor responsibilities set out above in the context of space decks are generally common to all roofing contractors involving specialist subcontractors and materials.

SPACE FRAMES

Space frames are similar to space decks in their basic concept. However, they are technologically more advanced and are generally more flexible in their design and layout possibilities, because their main component is the connector (known as a node point) that joins together the chords and braces. Space frames are usually designed as a double-layer grid as opposed to the single-layer grid used mainly for geometrical shapes, such as a dome. A double-layer grid is relatively shallow when compared with other structural roof systems of similar loadings and span: the span-to-depth ratio for a space frame supported on all its edges would be about 1:20, whereas a space frame supported near the corners would require a ratio of about 1:15. A variety of systems is available to the architect and builder. The concept is illustrated in Figs 9.5.7 and 9.5.8. These show the early 'Nodus' system established by the British Steel Corporation (now Tata UK Ltd), which set the early standard in space frame technology.

Common methods now employ cast nodes that are machined to accept a screw-threaded tube at the correct angle. These are accurately machined so a framework can be assembled to form a roof structure. Figure 9.5.9 illustrates such a system. There are a number of different arrangements that can be created with this system and it provides excellent flexibility in designing three-dimensional structures with aesthetic features.

The claddings used in conjunction with a space-frame roof should not be unduly heavy, and normally any lightweight profiled decking would be suitable. As with the space decks described above, some of the main advantages of these systems are:

- they are constructed from simple standard prefabricated units;
- units can be mass-produced;
- the roof can be rapidly assembled and erected on site using semi-skilled labour;
- small sizes of components make storage and transportation easy.

Site works consist of assembling the grid at ground level, lifting the completed space frame and fixing it to its supports. The grid can be assembled on a series of blocks to counteract any ground irregularities, and during assembly the space frame will automatically generate the correct shape and camber. The correct procedure is to start assembling the space frame at the centre of the grid and work towards the edges, ensuring that there is sufficient ground clearance to enable the camber to be formed. Generally the space frame is assembled as a pure roof structure, but it is possible to install services and fix the cladding before lifting and fixing. Mobile cranes are usually employed to lift the completed roof structure, holding it in position while the columns are erected and fixed. Alternatively the grid can be constructed in an offset position around the columns that pass through the spaces in the grid; the completed roof structure is then lifted and moved sideways onto the support seatings on top of the columns.

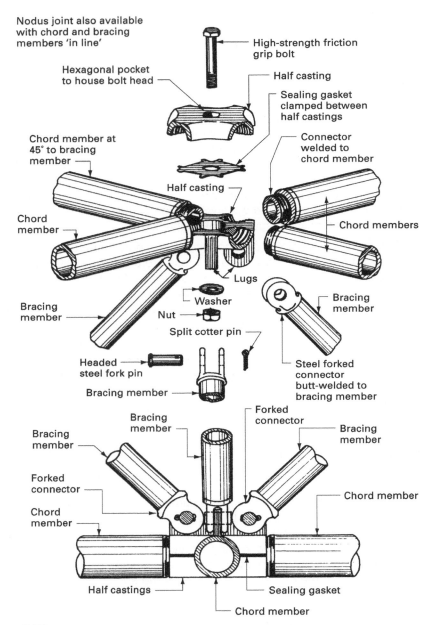

Nodus joint also available with chord and bracing members 'in line'

High-strength friction grip bolt

Hexagonal pocket to house bolt head

Half casting

Sealing gasket clamped between half castings

Chord member at 45° to bracing member

Connector welded to chord member

Half casting

Chord member

Chord members

Lugs

Washer

Nut

Bracing member

Bracing member

Split cotter pin

Headed steel fork pin

Steel forked connector butt-welded to bracing member

Bracing member

Forked connector

Bracing member

Bracing member

Bracing member

Bracing member

Forked connector

Chord member

Chord member

Chord member

Half castings

Sealing gasket

Chord member

Figure 9.5.7   Typical BSC (now Tata UK Ltd) 'Nodus' system joint details

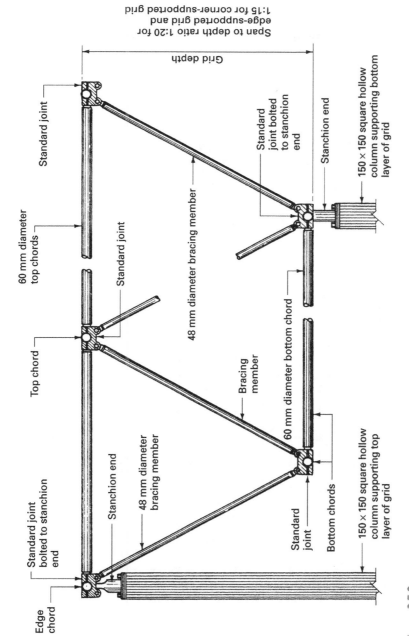

Span to depth ratio 1:20 for
edge-supported grid and
1:15 for corner-supported grid

Grid depth

Standard joint

Standard joint bolted to stanchion end

Stanchion end

150 × 150 square hollow column supporting bottom layer of grid

60 mm diameter top chords

Standard joint

48 mm diameter bracing member

Top chord

Bracing member

60 mm diameter bottom chord

Standard joint bolted to stanchion end

Stanchion end

48 mm diameter bracing member

Standard joint

Bottom chords

150 × 150 square hollow column supporting top layer of grid

Edge chord

Figure 9.5.8 Typical BSC (now Tata UK Ltd) 'Nodus' space frame details

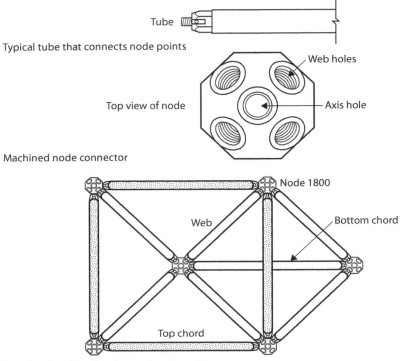

Tube

Typical tube that connects node points

Web holes

Top view of node

Axis hole

Machined node connector

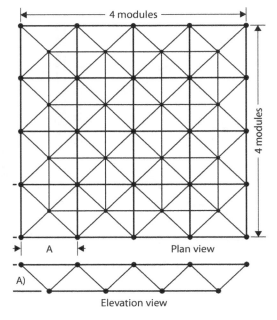

Node 1800

Web

Bottom chord

Top chord

Detail – plan view showing node orientation

4 modules

4 modules

A

Plan view

A)

Elevation view

Figure 9.5.9   The use of nodes and screw tubes to form a space frame

SHELL ROOFS

A shell roof can be defined as a structural curved skin covering a given plan shape and area. The main characteristics of a shell roof are:

- it is primarily a structural element, as the skin takes the loads;
- the basic strength of any particular shell is inherent in its completed shape;
- the quantity of material required to cover a given plan shape and area is generally less than for other forms of roofing due to curvature.

The basic materials that can be used in the formation of a shell roof are concrete, timber, steel and glass with aluminium frames. Concrete shell roofs consist of a thin, curved, reinforced membrane cast in situ over supporting timber formwork; timber shells are usually formed from carefully designed laminated timber; steel shells are generally formed using a single-layer grid. Concrete shell roofs, although popular, are often costly to construct because the formwork required is usually purpose-made from timber and is, in itself, a shell roof and has little chance of being reused to enable the cost of the formwork to be apportioned over several contracts.

A wide variety of shell roof shapes and types can be designed and constructed, but they can be classified under three headings:

- domes;
- vaults;
- saddle shapes and conoids.

DOMES

A dome in its simplest form consists of a half sphere, but it is also possible to construct domes based on the ellipse, parabola and hyperbola. Domes have been constructed by architects and builders over the centuries using individually shaped wedge blocks or traditional timber roof construction techniques. It is therefore the method of construction and the materials employed that have changed over the years, rather than the geometrical setting out.

Domes are either double-curvature shells, which can be rotational and are formed by a curved line rotating around a vertical axis, or translational domes, which are formed by a curved line moving over another curved line (see Fig. 9.5.10). Pendentive domes are formed by inscribing a polygon within the base circle and cutting vertical planes through the true hemispherical dome.

Any dome shell roof will tend to flatten because of the loadings, and this tendency must be resisted by the provision of stiffening ring beams or similar to all the cut edges. As a general guide, domes that rise in excess of one-sixth of their diameter will require a ring beam. Timber domes, like their steel counterparts, are usually constructed on a single-layer grid system and covered with a suitable thin-skin membrane.

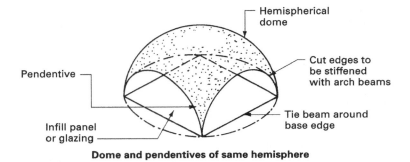

Dome and pendentives of same hemisphere

- Hemispherical dome
- Cut edges to be stiffened with arch beams
- Tie beam around base edge
- Pendentive
- Infill panel or glazing

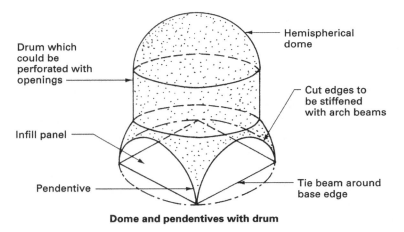

Dome and pendentives of different hemispheres

- Hemispherical dome
- Hemispherical dome
- Cut edges to be stiffened with arch beams
- Tie beam around base edge
- Infill panel or glazing
- Pendentive

Dome and pendentives with drum

- Hemispherical dome
- Drum which could be perforated with openings
- Cut edges to be stiffened with arch beams
- Infill panel
- Tie beam around base edge
- Pendentive

Figure 9.5.10   Typical dome roof shapes

VAULTS

Vaults are shells of single curvature and are commonly called **barrel vaults**. A barrel vault is basically a continuous arch or tunnel, and was first used by the Romans and later by the Normans in the UK. Geometrically a barrel vault is a cut half cylinder, which presents no particular setting-out problems. When two barrel vaults intersect, the lines of intersection are called **groins**. Barrel vaults, like domes, tend to flatten unless adequately restrained; in vaults, restraint will be required at the ends in the form of a diaphragm and along the edges (see Fig. 9.5.11).

From a design point of view, barrel vaults act as a beam, with the length being considered as the span. If this is longer than its width or chord distance, it is called a **long-span barrel vault**; conversely, if the span is shorter than the chord distance, it is termed a **short-span barrel vault**. Short-span barrel vaults, with their relatively large chord distances – and, consequently, large radii to their inner and outer curved surfaces – may require stiffening ribs to overcome the tendency to buckle. The extra stresses caused by the introduction of these stiffeners or ribs will necessitate the inclusion of extra reinforcement at the rib position; alternatively, the shell could be thickened locally about the rib for a distance of about one-fifth of the rib spacing (see Fig. 9.5.11).

In large barrel-vault shell roofs, allowances must be made for thermal expansion, and this usually takes the form of continuous expansion joints, as shown in Fig. 9.5.12, spaced at 30.000 m centres along the length. This will in fact create a series of individually supported abutting roofs weather-sealed together.

SADDLE SHAPES AND CONOIDS

These are similar to barrel vaults but are double-curvature shells as opposed to the single curvature of the barrel vault. There are two basic geometrical forms:

- a straight line is moved along a curved line at one end and a straight line at the other end, the resultant shape being cut to the required length;
- a straight line is moved along a curved line at one end and a different curved line at the other end.

Typical shapes are shown in Fig. 9.5.13.

Saddle shapes

Saddle-shaped shells are formed geometrically by moving a vertical parabola over another vertical parabola set at right angles to the moving parabola (see Fig. 9.5.14). The saddle shape created is termed a **hyperbolic paraboloid** because horizontal sections taken through the roof will give a hyperbolic outline and vertical sections will result in a parabolic outline. To obtain a more practical shape than the true saddle, the usual shape is that of a warped parallelogram or straight-line-limited hyperbolic paraboloid, which is formed by raising or lowering one or more corners of a square, as shown in Figs 9.5.14 and 9.5.15. By virtue of its shape, this form of shell roof has a greater resistance to buckling than dome shapes.

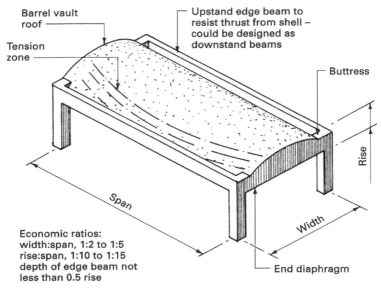

Barrel vault roof

Tension zone

Upstand edge beam to resist thrust from shell – could be designed as downstand beams

Buttress

Rise

Span

Width

Economic ratios:
width:span, 1:2 to 1:5
rise:span, 1:10 to 1:15
depth of edge beam not less than 0.5 rise

End diaphragm

Barrel vault principles

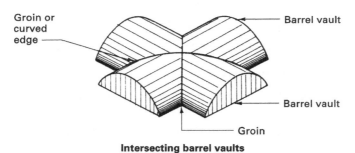

Groin or curved edge

Barrel vault

Barrel vault

Groin

Intersecting barrel vaults

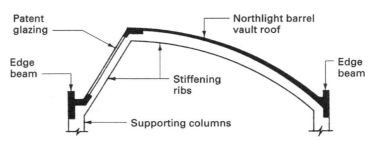

Patent glazing

Northlight barrel vault roof

Edge beam

Edge beam

Stiffening ribs

Supporting columns

Figure 9.5.11   Typical barrel vaults

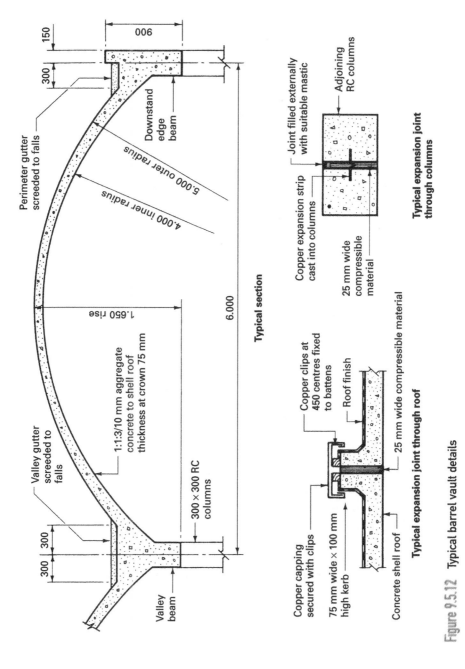

150

006

300

Perimeter gutter
screeded to falls

Downstand
edge beam

5.000 outer radius

4.000 inner radius

Typical section

1.650 rise

6.000

Valley gutter
screeded to
falls

1:1:3/10 mm aggregate
concrete to shell roof
thickness at crown 75 mm

300 × 300 RC
columns

300

300

Valley
beam

Joint filled externally
with suitable mastic

Adjoining
RC columns

Copper expansion strip
cast into columns

25 mm wide
compressible
material

**Typical expansion joint
through columns**

Copper clips at
450 centres fixed
to battens

Roof finish

25 mm wide compressible material

Copper capping
secured with clips

75 mm wide × 100 mm
high kerb

Concrete shell roof

Typical expansion joint through roof

Figure 9.5.12 Typical barrel vault details

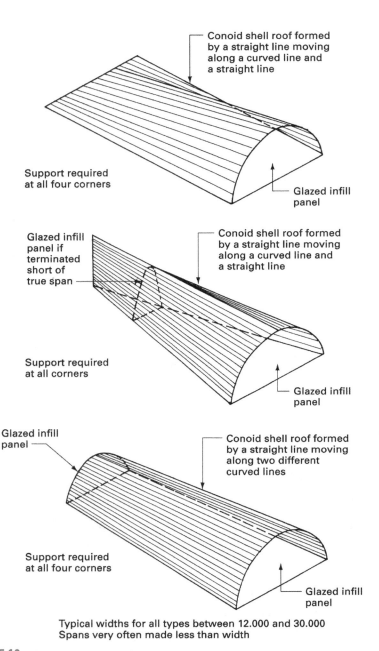

Conoid shell roof formed by a straight line moving along a curved line and a straight line

Support required at all four corners

Glazed infill panel

Glazed infill panel if terminated short of true span

Conoid shell roof formed by a straight line moving along a curved line and a straight line

Support required at all corners

Glazed infill panel

Glazed infill panel

Conoid shell roof formed by a straight line moving along two different curved lines

Support required at all four corners

Glazed infill panel

Typical widths for all types between 12.000 and 30.000
Spans very often made less than width

Figure 9.5.13 Typical conoid shell roof types

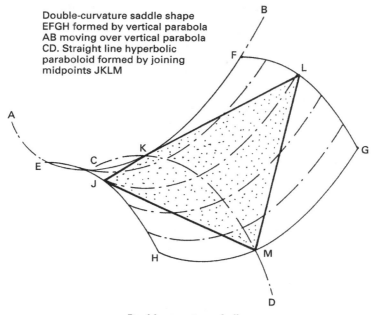

Double-curvature saddle shape EFGH formed by vertical parabola AB moving over vertical parabola CD. Straight line hyperbolic paraboloid formed by joining midpoints JKLM

Double-curvature shell

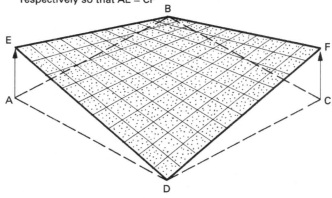

Straight line limited hyperbolic paraboloid formed by raising corners A and B of square ABCD to E and F respectively so that AE = CF

Hyperbolic paraboloid shell

Figure 9.5.14   Hyperbolic paraboloid roof principles

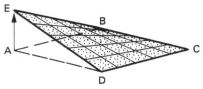

Straight line limited hyperbolic paraboloid formed by raising corner A of square ABCD to E

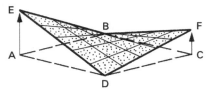

Straight line limited hyperbolic paraboloid formed by raising corners A and C of square ABCD to E and F respectively so that AE ≠ CF

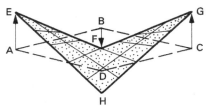

Straight line limited hyperbolic paraboloid formed by raising corners A and C and lowering corners B and D of square ABCD to EFG and H respectively so that AE = CG and BF = DH

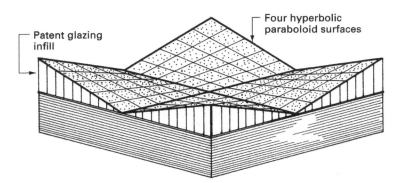

Hyperbolic paraboloids combined to form single roof

Figure 9.5.15   Typical hyperbolic paraboloid roof shapes

Hyperbolic paraboloid shells can be used singly or in conjunction with one another to cover any particular plan shape or size. If the rise – that is, the difference between the high and low points of the roof – is small, the result will be a hyperbolic paraboloid of low curvature acting structurally like a plate, which will have to be relatively thick to provide the necessary resistance to deflection. To obtain full advantage of the in-built strength of the shape, the rise to diagonal span ratio should not be less than 1:15; indeed, the higher the rise, the greater will be the strength and the shell can be thinner.

By adopting a suitable rise-to-span ratio it is possible to construct concrete shells with diagonal spans of up to 35.000 m with a shell thickness of only 50 mm. Timber hyperbolic paraboloid roofs can also be constructed using laminated edge beams with three layers of 20 mm-thick tongued-and-grooved boards. The top and bottom layers of boards are laid parallel to the edges but at right angles to one another, and the middle layer is laid diagonally. This is to overcome the problem of having to twist the boards across their width and at the same time bend them in their length.

CONSTRUCTION OF SHELL ROOFS

Concrete shell roofs are constructed on traditional formwork adequately supported to take the loads. When casting barrel vaults it is often convenient to have a movable form consisting of birdcage scaffolding supporting curved steel ribs to carry the curved plywood or steel forms. The form can then be moved in sections along the barrel vault enabling repetition of formwork use. Top formwork is not usually required unless the angle of pitch is greater than 45°. Reinforcement usually consists of steel fabric and bars of small diameter, the bottom layer of reinforcement being welded steel fabric, followed by small-diameter trajectory bars following the stress curves set out on the formwork, and finally a top layer of steel fabric. The whole reinforcement arrangement is wired together, and spacer blocks of precast concrete or plastic are fixed to maintain the required cover of concrete.

The concrete is usually specified as a mix with a characteristic strength of 25 or 30 N/mm^2. The concrete should preferably be placed in one operation in 1.000 m-wide strips, commencing at one end and running from edge beam to edge beam over the crown of the roof. A wet mix should be placed around the reinforcement, followed by a floated drier mix. Thermal insulation can be provided by laying high-density mineral-wool insulation boards over the completed shell before laying the roof covering.

FOLDED PLATE ROOFS

Folded plate roofing is another form of stressed skin roof, and is sometimes called **folded slab construction**. The basic design concept is to bend or fold a flat slab so that the roof will behave as a beam spanning in the direction of

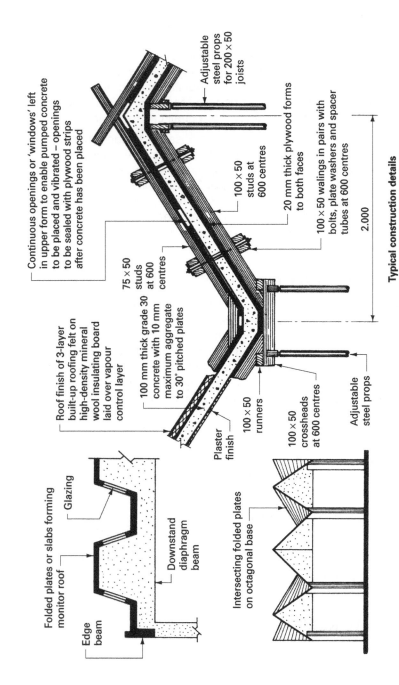

Continuous openings or 'windows' left in upper form to enable pumped concrete to be placed and vibrated – openings to be sealed with plywood strips after concrete has been placed

Adjustable steel props for 200 × 50 joists

100 × 50 studs at 600 centres

20 mm thick plywood forms to both faces

100 × 50 walings in pairs with bolts, plate washers and spacer tubes at 600 centres

2.000

Typical construction details

75 × 50 studs at 600 centres

Roof finish of 3-layer built-up roofing felt on high-density mineral wool insulating board laid over vapour control layer

100 mm thick grade 30 concrete with 10 mm maximum aggregate to 30° pitched plates

Plaster finish

100 × 50 runners

100 × 50 crossheads at 600 centres

Adjustable steel props

Folded plates or slabs forming monitor roof

Glazing

Downstand diaphragm beam

Edge beam

Intersecting folded plates on octagonal base

Figure 9.5.16 Typical folded plate roof details

the fold. To create an economic roof the overall depth of the roof should be related to span and width so that it is between 1/10 and 1/15 of the span or 1/10 of the width, whichever is the greater. The fold may take the form of a pitched roof, a monitor roof, or a multi-fold roof in single or multiple bays, with upstand or downstand diaphragms at the supports to collect and distribute the slab loadings (see Fig. 9.5.16). Formwork may be required to both top and bottom faces of the slabs. To enable concrete to be introduced and vibrated, openings or 'windows' can be left in the upper surface formwork, and these will be filled in with slip-in pieces after the concrete has been placed and vibrated.

TENSION ROOF STRUCTURES

Suspended or tensioned roof structures can be used to form permanent or temporary roofs. This is generally a system or network of cables, or in the temporary form they could be pneumatic tubes, which are used to support roof covering materials of the traditional form or continuous sheet membranes. With this form of roof the only direct stresses that are encountered are tensile stresses, and this – apart from aesthetic considerations – is their main advantage. Because of their shape and lightness, tension roof structures can sometimes present design problems in the context of negative wind pressures, and this is normally overcome by having a second system of curved cables at right angles to the main suspension cables. This will in effect prestress the main suspension cables. Figure 9.5.17 shows an outline principle in one direction only.

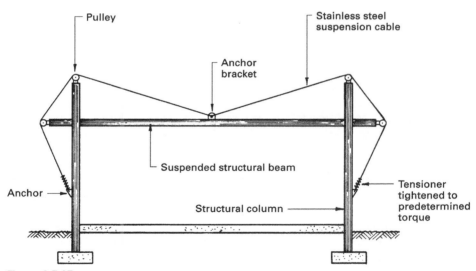

Figure 9.5.17   Tensioned roof structure

Partitions, doors and ceilings 9.6

Internal components and finishes that are normally associated with domestic dwellings such as brick walls, lightweight concrete block walls, plastering, dry linings, floor finishes and coverings, fixed partitions and the various trims are detailed in the associated volume, *Construction Technology*. Advanced studies of construction technology more appropriately concentrate on the components and finishes applied to buildings such as offices, commercial and institutional establishments. These include the partitions, sliding doors and suspended ceilings described in this chapter.

PARTITIONS

A partition is similar to a wall in that it is constructed to divide the internal space of a building. However, partitions are generally considered to be internal walls that are lightweight, non-loadbearing and demountable. They are used in buildings such as offices where it is desirable to have a system of internal division that can be altered to suit changes in usage, without excessive costs and with minimum interference to the building's services and ceilings. Note that any non-loadbearing partition could, if necessary, be taken down and moved; however, in most cases, moving partitions such as timber studwork with plasterboard linings will necessitate the replacement or repair of some of the materials and finishes involved, along with lengthy disruptions for cleaning and reinstatement of finishes. A true demountable partition can usually be taken down, moved and re-erected without any notable damage to the materials, components, finishes and surrounding parts of the building (for movable partitions, which usually take the form of a series of doors or door leaves hung to slide or slide and fold, see below). Many demountable partition systems are used in conjunction with a suspended ceiling; in this case, it is important that both systems are considered together in the context of their respective main functions, because one system may cancel out the benefits achieved by the other.

There are many patent demountable partition systems from which the architect, designer or builder can select a suitable system for any particular case. It would not be possible in a text of this nature to analyse or list them all, so only the general requirements of these types of partition are considered here. The composition of the types available range from glazed screens with timber, steel or aluminium alloy frames to panel construction with concealed or exposed jointing members. Low-level screens and toilet cubicles can also be considered in this context, and these can be attached to posts fixed at floor level or suspended from the structural floor above.

When selecting or specifying a demountable partition, you should take into account fixing and stability, sound insulation and fire resistance.

FIXING AND STABILITY

Points to be considered are bottom fixing, top fixing and joints between sections. The two common methods of bottom fixing are the horizontal base unit, which is often hollow to receive services, and the screw jack fixing: both methods are intended to be fixed above the finished floor level, which should have a tolerance not exceeding 5 mm. The top fixing can be of a similar nature, with a ceiling tolerance of the same magnitude, but problems can arise when used in conjunction with a suspended ceiling, because little or no pressure can be applied to the underside of the ceiling. A brace or panel could be fixed above the partition in the void over the ceiling to give the required stability; alternatively, the partition could pass through the ceiling, but this would reduce the flexibility of the two systems should a rearrangement be required at a later date. Where fire-rated compartmentation is required, a rated partition will have the required fire stopping placed above the partition within the suspended ceiling void. The joints between consecutive panels do not normally present any problems as they are an integral part of the design but, unless they are adequately sealed, the sound insulation and/or fire-resistance properties of the partition could be seriously impaired.

SOUND INSULATION

The degree of insulation required will depend largely on the usage of rooms created by the partitions. The level of sound insulation that can be obtained will depend on three factors: the density of the partition construction, the degree of demountability, and the continuity of the partition with the floor and ceiling. Generally, the less demountable a partition, the easier it will be to achieve an acceptable sound insulation level. Partitions fixed between the structural floor and the structural ceiling usually result in good sound insulation, whereas partitions erected between the floor and a suspended ceiling give poor sound insulation, mainly because of the flanking sound path over the head of the partition. In the latter case, two remedies are possible: either the partition could pass through the suspended ceiling, or a sound-insulating panel could be inserted in the ceiling void directly over the partition below.

FIRE RESISTANCE

The main restrictions on the choice of partition lie in the requirements of the Building Regulations Part B with regard to fire-resistant properties, spread of flame, fire stopping and providing a suitable means of escape in case of fire. The weakest points in any system will be the openings and the seals at the foot and head. Provisions at openings are a matter of detail and adequate construction, together with a suitable fire-resistant door or fire-resistant glazing. Fire stopping at the head of a partition used in conjunction with a suspended ceiling usually takes the form of a fire-break panel or barrier fixed in the ceiling void directly above the partition below. These may be produced from 25 mm galvanised wire mesh stitched to mineral-wool insulation in two layers of at least 50 mm thickness, or from two sheets of 12.5 mm plasterboard, as indicated in Fig. 9.6.1. Most proprietary demountable systems will give a half-hour or one-hour fire resistance with a class 1 or class 0 spread of flame classification.

Typical demountable partition details are shown in Figs 9.6.1 and 9.6.2.

SLIDING DOORS

Sliding doors may be used in all forms of building, from the small garage to large industrial structures. They may be incorporated into a design for any of the following reasons:

- as an alternative to a swing door, to conserve space or where it is not possible to install a swing door because of space restrictions;
- where heavy doors or doors in more than two leaves are to be used, and where conventional hinges would be inadequate;
- as a movable partition used in preference to a demountable partition because of the frequency with which it would be moved.

In most cases there is a choice of bottom or top sliding door gear, which in the case of very heavy doors may have to be mechanically operated. Top gear can be used only if there is a suitable soffit, beam or lintel over the proposed opening that is strong enough to take the load of the gear, track and doors. In the case of bottom gear, all the loads are transmitted to the floor. Top gear has the disadvantage of generally being noisier than its bottom gear counterpart, but bottom gear either has an upstanding track, which can be a hazard in terms of tripping, or it can have a sunk channel, which can become

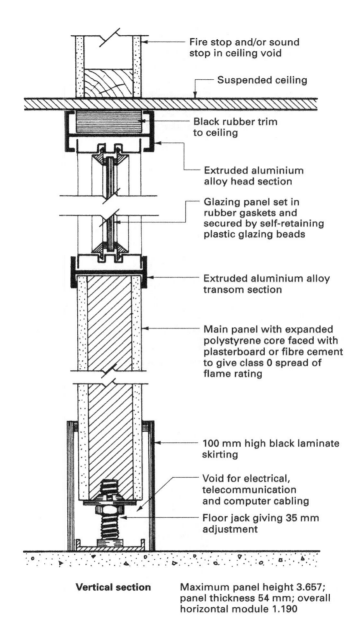

Fire stop and/or sound stop in ceiling void

Suspended ceiling

Black rubber trim to ceiling

Extruded aluminium alloy head section

Glazing panel set in rubber gaskets and secured by self-retaining plastic glazing beads

Extruded aluminium alloy transom section

Main panel with expanded polystyrene core faced with plasterboard or fibre cement to give class 0 spread of flame rating

100 mm high black laminate skirting

Void for electrical, telecommunication and computer cabling

Floor jack giving 35 mm adjustment

Vertical section   Maximum panel height 3.657; panel thickness 54 mm; overall horizontal module 1.190

Figure 9.6.1   Typical demountable partition details: 1 (courtesy: Tenon Contracts)

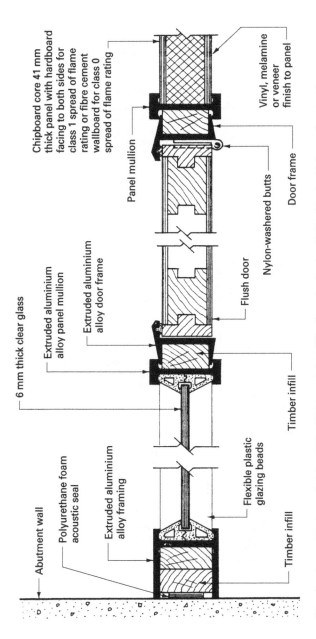

6 mm thick clear glass

Extruded aluminium alloy panel mullion

Extruded aluminium alloy door frame

Abutment wall

Polyurethane foam acoustic seal

Extruded aluminium alloy framing

Flexible plastic glazing beads

Timber infill

Chipboard core 41 mm thick panel with hardboard facing to both sides for class 1 spread of flame rating or fibre cement wallboard for class 0 spread of flame rating

Panel mullion

Vinyl, melamine or veneer finish to panel

Door frame

Nylon-washered butts

Flush door

Timber infill

Plan drawing   Maximum height 4.577   Width module 1.200   Sound reduction 27dB   Plastic skirting at floor level
Services can be accommodated within chipboard core and along continuous base ducts.

Figure 9.6.2   Typical demountable partition details: 2 (courtesy: Venesta International Components)

blocked with dirt and dust unless regularly maintained. In all cases it is essential that the track and guides are perfectly aligned, both vertically to one another and in the horizontal direction parallel to each other, to prevent stiff operation or the binding of the doors while being moved. When specifying sliding door gear, it is essential to choose the correct type to suit the particular door combination and weight range if the partition is to be operated efficiently.

Many types of patent sliding door gear and track are available to suit all needs, but these can usually be classified by the way in which the doors operate.

- **End-folding doors** Strictly speaking these are sliding and folding doors that are used for wider openings than the straight doors given above. They usually consist of a series of leaves operating off top gear with a bottom guide track, so that the folding leaves can be parked to one or both sides of the opening (see Fig. 9.6.3 for typical details).

- **Centre-folding doors** Like the previous type, these doors slide and fold to be parked at one or both ends of the opening, and have a top track with a bottom floor guide channel arranged so that the doors will pivot centrally over the channel. With this arrangement, the hinged leaf attached to the frame is only approximately a half-leaf (see typical details in Fig. 9.6.4).

Other forms of sliding doors include round-the-corner sliding doors, which use a curved track and folding gear arranged so that the leaves hinged together will slide around the corner and park alongside a side wall, making them suitable for situations where a clear opening is required. The number of leaves in any one hinged sliding set should not be fewer than three or more than five. Another application of the folding door technique is the folding screen, which is suitable for a limited number of lightweight leaves that can be folded back onto a side or return wall without the use of sliding gear, track or guide channels (see Fig. 9.6.5 for typical details). This relies on the three hinges that connect each of the screen leaves for the folding action. Clearly the weight of the screens must be carefully considered.

Acoustic folding partitions are also available. Produced in vinyl in a variety of colours, these act rather like a curtain and can be used to extend an office environment. They are fixed to a top track, which often requires the insertion of a calculated steel beam or a softwood beam, held in position by threaded drop rods from the soffit of the floor above. A bottom channel does not need to be inserted into the floor.

SUSPENDED CEILINGS

A suspended ceiling can be defined as a ceiling that is attached to a framework suspended from the main structure, thus forming a void between the ceiling and the underside of the main structure. Ceilings that are fixed, for example, to lattice girders and trusses while forming a void are strictly

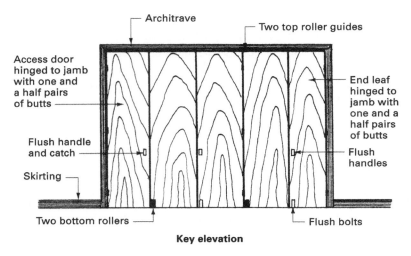

Architrave

Two top roller guides

Access door hinged to jamb with one and a half pairs of butts

End leaf hinged to jamb with one and a half pairs of butts

Flush handle and catch

Flush handles

Skirting

Two bottom rollers

Flush bolts

Key elevation

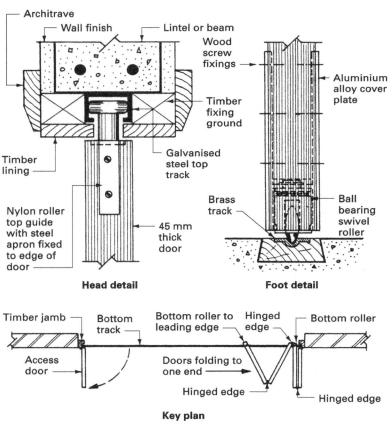

Architrave

Wall finish

Lintel or beam

Wood screw fixings

Aluminium alloy cover plate

Timber fixing ground

Galvanised steel top track

Timber lining

Nylon roller top guide with steel apron fixed to edge of door

45 mm thick door

Brass track

Ball bearing swivel roller

Head detail

Foot detail

Timber jamb

Bottom track

Bottom roller to leading edge

Hinged edge

Bottom roller

Access door

Doors folding to one end

Hinged edge

Hinged edge

Key plan

Figure 9.6.3   Typical end-folding door details (courtesy: Hillaldam Coburn)

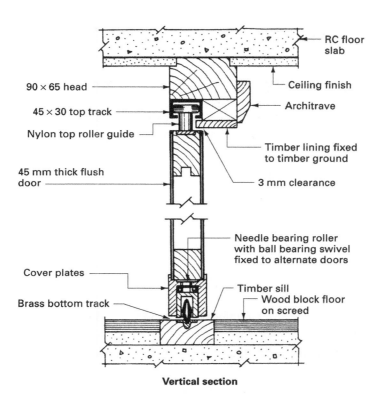

RC floor slab

90 × 65 head

45 × 30 top track

Nylon top roller guide

45 mm thick flush door

Cover plates

Brass bottom track

Ceiling finish

Architrave

Timber lining fixed to timber ground

3 mm clearance

Needle bearing roller with ball bearing swivel fixed to alternate doors

Timber sill

Wood block floor on screed

Vertical section

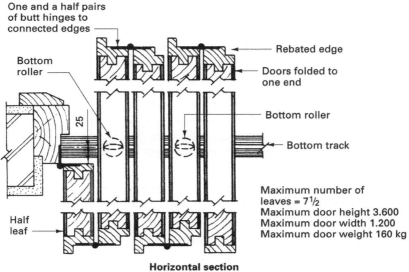

One and a half pairs of butt hinges to connected edges

Bottom roller

25

Half leaf

Rebated edge

Doors folded to one end

Bottom roller

Bottom track

Maximum number of leaves = $7\frac{1}{2}$
Maximum door height 3.600
Maximum door width 1.200
Maximum door weight 160 kg

Horizontal section

Figure 9.6.4   Typical centre-folding sliding door details (courtesy: Hillaldam Coburn)

speaking attached ceilings, although they may be formed using the same methods and materials as suspended ceilings. The same argument can be applied to ceilings that are fixed to a framework of battens attached to the underside of the main structure.

Some suspended ceilings contribute to the fire resistance of the structure above, and these are defined as fire-protecting suspended ceilings. Reference should be made to Building Regulations, Approved Document B, Appendix A, Table A3 for material limitations of these fire-protecting suspended ceilings in various applications.

Further reasons for including a suspended ceiling system in a building design are:

- to provide a finish to the underside of a structural floor or roof, generally for purposes of concealment;
- to create a void space suitable for housing, concealing and protecting services and light fittings;
- to add to the sound and/or thermal insulation properties of the floor or roof above and adjacent rooms;
- to provide a means of structural fire protection to steel beams supporting a concrete floor;
- to provide a means of acoustic control in terms of absorption and reverberation;
- to create a lower ceiling height to a particular room or space;
- to provide a fast and improved aesthetical installation to a refurbished building.

A suspended ceiling, for whatever reason it has been specified and installed, should:

- be easy to construct, repair, maintain and clean;
- comply with the requirements of the Building Regulations and, in particular, Approved Document B2 and associated Appendix A (these are concerned with flame spread over the surface, classification of performance and the limitations imposed on the use of certain materials with regard to fire, such as thermoplastic light diffusers);
- provide an adequate means of access for the maintenance of the suspension system and/or the maintenance of concealed services and light fittings;
- conform to a planning module that preferably should be based on the modular coordination recommendations set out in BS 6750: *Specification for modular coordination in building*, which recommends a first preference module of 300 mm.

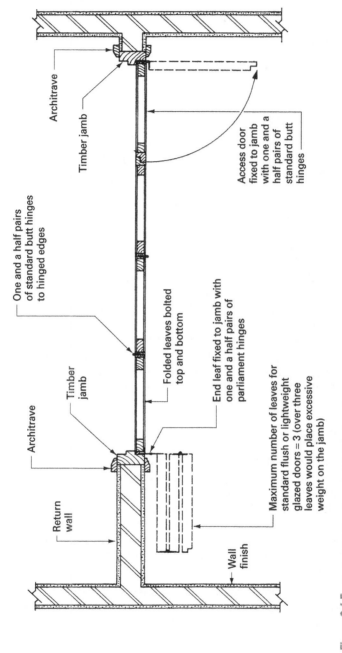

One and a half pairs
of standard butt hinges
to hinged edges

Architrave

Timber jamb

Access door
fixed to jamb
with one and a
half pairs of
standard butt
hinges

Architrave

Timber
jamb

Folded leaves bolted
top and bottom

End leaf fixed to jamb with
one and a half pairs of
parliament hinges

Return
wall

Maximum number of leaves for
standard flush or lightweight
glazed doors = 3 (over three
leaves would place excessive
weight on the jamb)

Wall
finish

Figure 9.6.5   Typical folding screen details: plan view

There are many ways in which to classify suspended ceilings. They can be classified by function (for example, sound absorbing or fire resisting) or by specification of the materials involved. Another simple, practical method of classification is to group the ceiling systems by their general method of construction as:

- jointless ceilings;
- jointed ceilings;
- open ceilings.

JOINTLESS CEILINGS

These are ceilings that, although suspended from the main structure, give the internal appearance of being a conventional ceiling. The final finish is usually of plaster applied in one or two coats to plasterboard or expanded metal lathing; alternatively, a jointless suspended ceiling could be formed by applying sprayed plaster or sprayed vermiculite-cement to an expanded metal background. Typical jointless ceiling details are shown in Fig. 9.6.6.

JOINTED CEILINGS

Jointed suspended ceilings are the most popular and common form because of their ease of assembly, installation and maintenance. They consist basically of a suspended metal framework to which the finish in a board or tile form is attached. The boards or tiles can be located on a series of inverted tee-bar supports, with the exposed supporting members forming part of the general appearance; alternatively the supporting members can be concealed by using various spring-clip devices or direct fixing. The common ceiling materials encountered are fibreboards, metal trays, fibre-cement materials and plastic tiles or trays (see Figs 9.6.7 and 9.6.8 for typical details).

OPEN CEILINGS

These suspended ceilings are largely decorative in function, but they can act as a luminous ceiling by installing light fittings in the ceiling void. The format of these ceilings can be an openwork grid of timber, with or without louvres or a series of closely spaced plates of polished steel or any other suitable material. The colour and texture of the sides and soffit of the ceiling void must be carefully designed to achieve an effective system. It is possible to line these surfaces with an acoustic absorption material to provide another function to the arrangement.

The fundamentals of applying decorative finishes in the form of paint and wallpaper have been considered in Chapter 8.10 of *Construction Technology*. Therefore, study at this advanced level progresses beyond basic principles, particularly in the context of painting timber and metals.

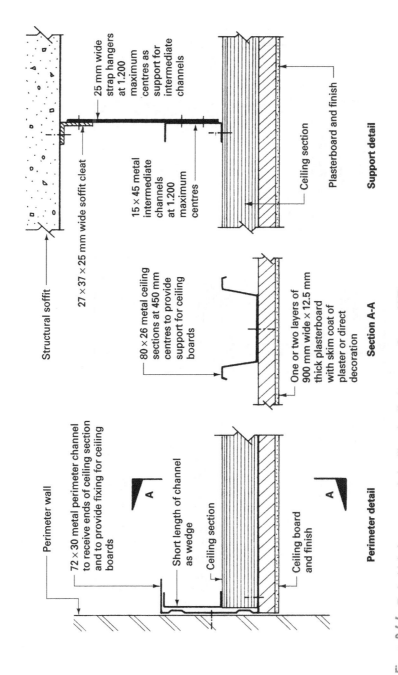

Structural soffit

25 mm wide strap hangers at 1.200 maximum centres as support for intermediate channels

27 × 37 × 25 mm wide soffit cleat

15 × 45 metal intermediate channels at 1.200 maximum centres

Ceiling section

Plasterboard and finish

Support detail

80 × 26 metal ceiling sections at 450 mm centres to provide support for ceiling boards

One or two layers of 900 mm wide × 12.5 mm thick plasterboard with skim coat of plaster or direct decoration

Section A–A

Perimeter wall

72 × 30 metal perimeter channel to receive ends of ceiling section and to provide fixing for ceiling boards

Short length of channel as wedge

Ceiling section

Ceiling board and finish

Perimeter detail

Figure 9.6.6   Typical jointless suspended ceiling details (courtesy: Gyproc M/F)

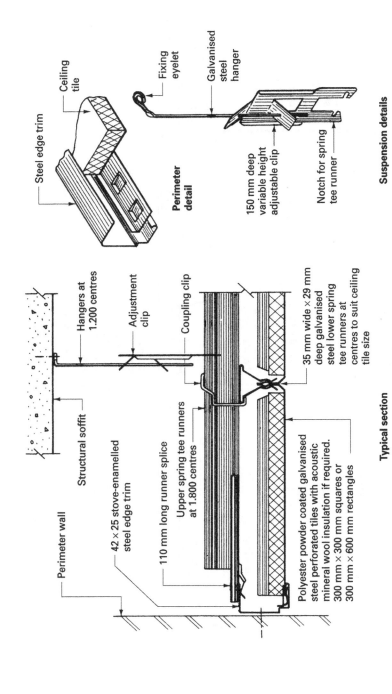

Ceiling tile

Steel edge trim

Fixing eyelet

Galvanised steel hanger

Perimeter detail

150 mm deep variable height adjustable clip

Notch for spring tee runner

Suspension details

Hangers at 1.200 centres

Adjustment clip

Coupling clip

Structural soffit

Perimeter wall

42 × 25 stove-enamelled steel edge trim

110 mm long runner splice

Upper spring tee runners at 1.800 centres

Polyester powder coated galvanised steel perforated tiles with acoustic mineral wool insulation if required. 300 mm × 300 mm squares or 300 mm × 600 mm rectangles

35 mm wide × 29 mm deep galvanised steel lower spring tee runners at centres to suit ceiling tile size

Typical section

Figure 9.6.7 Typical jointed suspended ceiling details (courtesy: Dampa (UK) Ltd)

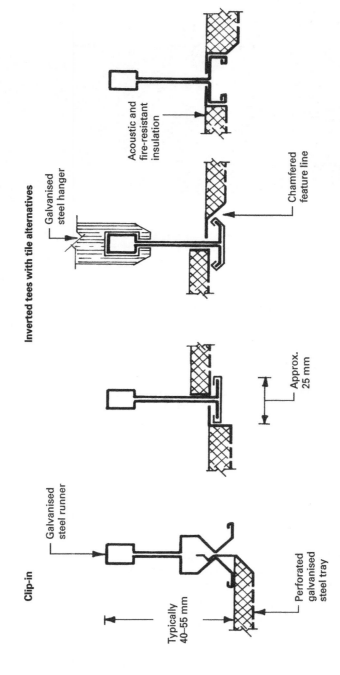

Clip-in

Galvanised
steel runner

Typically
40–55 mm

Perforated
galvanised
steel tray

Approx.
25 mm

Inverted tees with tile alternatives

Galvanised
steel hanger

Chamfered
feature line

Acoustic and
fire-resistant
insulation

Figure 9.6.8 Jointed suspended ceiling support profiles
Note: Dimensions vary between different manufacturers.

Painting and decorations 9.7

PAINTING TIMBER

Timber can be painted to prevent decay in the material by forming a barrier to the penetration of moisture, which gives rise to the conditions necessary for fungal attack to begin; alternatively, timber may be painted mainly to add colour. Timber that is exposed to the elements is more vulnerable to eventual decay than internal joinery items, but internal condensation can also give rise to damp conditions. Timber can also be protected with clear, water-repellent preservatives and varnishes to preserve the natural colour and texture of the timber.

The moisture content of the timber to which paint is to be applied should be as near as possible to that at which it will stabilise in its final condition. For internal joinery this would be in the region of 8–12 per cent, depending on the internal design temperature; for external timber, it would be in the region of 15–18 per cent, according to exposure conditions. If the timber is too dry when the paint is applied, any subsequent swelling of the base material due to moisture absorption will place unacceptable stresses on the paint film, leading to cracking of the paint barrier. Conversely, wet timber drying out after the paint application can result in blistering, opening of joints and the breakdown of the paint film.

New work should receive a four-coat application consisting of primer, undercoat and finishing coats. Traditional shellac knotting, used to reduce the risk of resin leaking and staining, should be applied before the primer, but in external conditions it may prove to be inadequate. Therefore timber selected for exposed or external conditions should be of a high quality, without or with only a few knots. Knotting and priming should preferably be carried out under the ideal conditions prevailing at the place of manufacture. Filling and stopping should take place after priming but before the application of the undercoat and finishing coats. It is a general misconception that a good key is necessary to achieve a satisfactory paint surface. It is not penetration that is required but a good molecular attraction of the paint binder to the timber that is needed to obtain a satisfactory result.

If the protective paint film is breached by moisture, decay can take place under the paint coverings; indeed, the coats of paint prevent the drying out of any moisture that has managed to penetrate the timber prior to painting. Any joinery items that in their final situation could be susceptible to moisture penetration should be treated with a paintable preservative. Preservation treatments include diffusion treatment using water-soluble borates, water-borne preservatives and organic solvents. The methods of application and chemical composition of these timber preservatives can be found in BS 8417: *Preservation of wood. Code of practice* and BS 8471: *Preservation of timber. Recommendations.*

PAINTING METALS

The application of paint to metals to provide a protective coat and for decorative purposes is normally confined to iron and steel, because most non-ferrous metals are left to oxidise and form their own natural protective coating. Corrosion of ferrous metals is a natural process that can cause disfiguration and possible failure of a metal component. Two aspects of painting these metals must be considered:

1. initial preparation and paint application;
2. maintenance of protective paint.

PREPARATION

The basic requirement is to prepare the metal surface adequately to receive the primer and subsequent coats of paint, because the mill scale, rust, oil, grease and dirt that are frequently found on metals are not suitable as a base for the applications of paint. Suitable preparation treatments are:

- **shot and grit blasting** – an effective method, which can be carried out on site, although it is usually considered to be a factory process;
- **phosphating and pickling** – a factory process involving immersion of the metal component in hot acid solutions to remove rust and scale;
- **degreasing** – a factory process involving washing or immersion using organic solvents, emulsions or hot alkali solutions followed by washing (often used as a preliminary treatment to phosphating);
- **mechanical** – a site or factory process using hand-held or powered tools, such as hammers, chisels, brushes and scrapers (to be effective, surface preparation must be thorough);

- **flame cleaning** – an oxyacetylene flame is used in conjunction with hand-held tools for removing existing coats of paint or loose scale and rust.

To obtain a satisfactory result, it is essential that, in conjunction with the above preparation processes, the correct primer is specified and used. For further reference, see BS EN ISO 12944-7: *Paints and varnishes. Corrosion protection of steel structures by protective paint systems. Execution and supervision of paintwork*.

MAINTENANCE

To protect steel in the long term, a rigid schedule of painting maintenance must be planned and carried out. This is particularly true of the many areas where access is difficult, such as pipes fixed close to a wall or adjacent members in a lattice truss. Access for future maintenance is an aspect of building that should be considered during the design and construction stages, not only in the context of metal components but also for any part of a building that will require future maintenance or inspection. To take full advantage of the structural properties of steelwork, and at the same time avoid the maintenance problems, weathering steels could be considered for exposed conditions.

WEATHERING STEELS

Weathering steels do not rust or corrode like normal steels, but interact with the atmosphere to produce a layer of sealing oxides. The colour of this protective layer will vary from a lightish brown to a dark purple-grey depending on the degree of exposure, amount and type of pollution in the atmosphere and orientation. The best-known weathering steel is called 'Cor-Ten', which is derived from the fact that it is CORrosion resistant with a high TENsile strength. Cor-Ten is not a stainless steel but a low-alloy steel with a lower proportion of non-ferrous metals than stainless steel, which makes it more expensive than mild steel but cheaper than traditional stainless steels.

Weathering steels can be used as a substitute for other steels except in wet situations, marine works or in areas of high pollution (unless protected with paint or similar protective applications, which defeats the main objective of using this material). Jointing can be by welding or friction-grip bolts. Note that the protective coating will not form on weathering steels in internal situations because the formation of this coat is a natural process of the wet and dry cycles encountered with ordinary weather conditions.

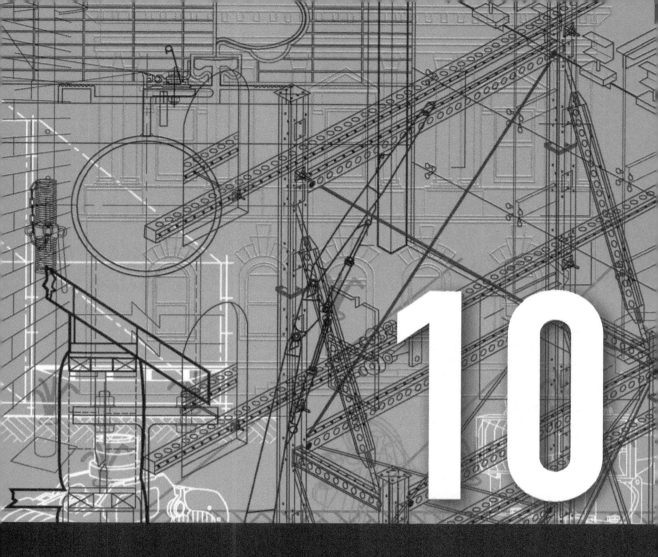

10

Stairs for commercial and industrial applications

Concrete stairs 10.1

Any form of stairs is primarily a means of providing circulation of the occupants and communication between the various levels within a building. Apart from this primary function, stairs may also be classified as a means of escape in case of fire; if this is the case, the designer is severely limited by necessary regulations as to the choice of materials, position and sizing of the complete stairway. Stairs that do not fulfil this means of escape function are usually called **accommodation** or **service stairs**, and as such are not restricted by the limitations given above for escape stairs.

Escape stairs have been covered in Chapter 5.3 of this volume under fire, and it is necessary only to reiterate the main points.

1. They must be constructed from non-combustible materials.

2. The stairway needs protecting by a fire-resisting enclosure.

3. The stairway must be separated from each main floor area by a set or sets of self-closing, fire-resisting doors.

4. Limitations as to riser heights, tread lengths and handrail requirements must be based on Part K of the Building Regulation Approved Documents.

5. In common with all forms of stairs, all riser heights must be equal throughout the rise of the stairs.

It should be appreciated that points 2 and 3 listed above can prove to be inconvenient to people using the stairway for general circulation within the building, such as having to pass through self-closing doors, but in the context of providing a safe escape route this is an essential requirement.

The use of in-situ concrete in the manufacture of stairways and stairs has all but been superseded by the use of precast concrete in their manufacture. The use of in-situ concrete in stair manufacture is time-consuming, requires specialist shuttering joiners, needs curing time to achieve initial strength, and is not an efficient or economical method of producing this structural element.

PRECAST CONCRETE STAIRS

Most commercial solutions for access into and egress out of a building now utilise the precast concrete approach. This has several advantages:

- factory-produced quality control of the finished product;
- faster construction period as units can be lifted and positioned in one day, enabling access to further storeys of the building;
- reduction in labour costs as units can be positioned and fixed by semi-skilled operatives;
- seamless integration into a hollow-rib poured floor with reinforcement already factory-installed;
- pockets can be formed for handrails and fittings, which can be factory-installed;
- saving in site space, because formwork storage and fabrication space are no longer necessary;
- stairway enclosing shaft can be used as a space for hoisting or lifting materials during the major construction period.

In common with the use of all precast concrete components, the stairs must be designed to be a repetitive element and exist in sufficient quantities to be an economic proposition.

The straight-flight stairs spanning between landings can have a simple bearing or, by leaving projecting reinforcement to be grouted into preformed slots in the landings, can be given a degree of structural continuity. With the advent of modern methods of construction, the use of precast concrete hollow floors, planks and hollow-rib steel-decking floors, the bearing of the top of the precast unit must be carefully considered. The bearing location may require additional steel work to support the top of the precast concrete unit while a poured slab system integrates the stair reinforcement within its construction. Figure 10.1.1 illustrates some typical examples using precast concrete hollow and solid floor units, a steel hollow-rib decking system; it also shows the connection of the stairs reinforcement to ensure continuity. The steel hollow-rib floor requires the insertion of a supporting steel beam to provide bearing for the unit and safe access while the floor is reinforced then poured up to the precast concrete unit.

Cranked slab precast concrete stairs are usually formed as an open well stairway. The bearing for the precast landings to the in-situ floor or to the structural frame is usually in the form of a simple bearing, as described above for straight-flight precast concrete stairs. The infill between the two

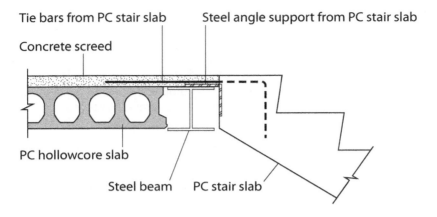

Tie bars from PC stair slab

Steel angle support from PC stair slab

Concrete screed

PC hollowcore slab

Steel beam   PC stair slab

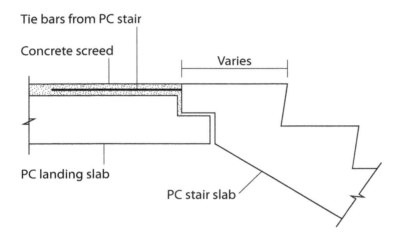

Tie bars from PC stair

Concrete screed

Varies

PC landing slab

PC stair slab

Figure 10.1.1   Integration of precast concrete stairs with PC floor

adjacent flights, in an open well plan arrangement at floor and intermediate landing levels, can be of in-situ concrete, with structural continuity provided by leaving reinforcement projecting from the inside edge of the landings (see Fig. 10.1.2). There are of course modern variations on Fig. 10.1.2 as precast concrete elements can be used for the supporting landings and beams along with structural steel beams and columns. Figures 10.1.2 and 10.1.3 have been retained within this edition as many existing in-situ concrete structures still exist.

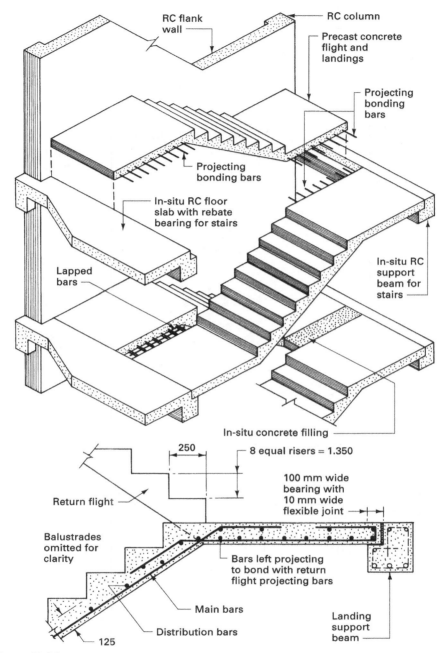

RC flank wall

RC column

Precast concrete flight and landings

Projecting bonding bars

Projecting bonding bars

In-situ RC floor slab with rebate bearing for stairs

Lapped bars

In-situ RC support beam for stairs

In-situ concrete filling

250

8 equal risers = 1.350

100 mm wide bearing with 10 mm wide flexible joint

Return flight

Balustrades omitted for clarity

Bars left projecting to bond with return flight projecting bars

Main bars

Distribution bars

Landing support beam

125

Figure 10.1.2   Precast concrete cranked slab stairs

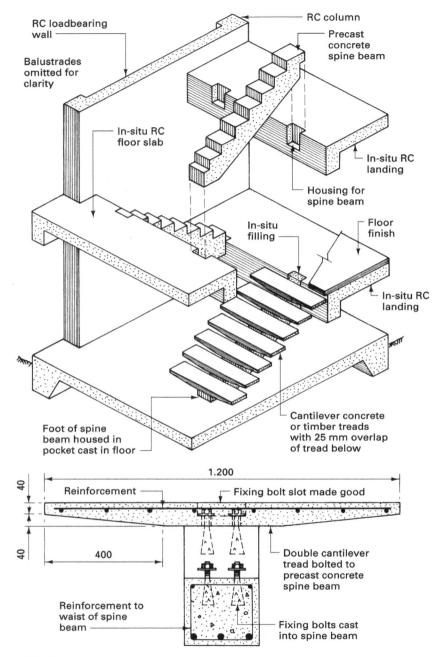

RC loadbearing wall

Balustrades omitted for clarity

RC column

Precast concrete spine beam

In-situ RC floor slab

In-situ RC landing

Housing for spine beam

In-situ filling

Floor finish

In-situ RC landing

Foot of spine beam housed in pocket cast in floor

Cantilever concrete or timber treads with 25 mm overlap of tread below

1.200

40

Reinforcement

Fixing bolt slot made good

40

400

Double cantilever tread bolted to precast concrete spine beam

Reinforcement to waist of spine beam

Fixing bolts cast into spine beam

Figure 10.1.3 Precast concrete open-riser stairs

Remember that, when precast concrete stair flights are hoisted into position, different stresses may be induced from those that will be encountered in the fixed position. To overcome this problem, the designer can either reinforce the units for both conditions or, as is more usual, provide definite lifting points in the form of projecting lugs or by using any holes cast in to receive the balustrading.

Precast open-riser stairs are a form of stairs that can be both economic and attractive, consisting of a central spine beam in the form of a cut string supporting double cantilever treads of timber or precast concrete. The foot of the lowest spine beam is located and grouted into a preformed pocket cast in the floor, whereas the support at landing and floor levels is a simple bearing located in a housing cast into the slab edge (see Fig. 10.1.3). Anchor bolts or cement in sockets are cast into the spine beam to provide the fixing for the cantilever treads. The bolt heads are recessed below the upper tread surface and grouted over with a matching cement mortar for precast concrete treads, or concealed with matching timber pellets when hardwood treads are used. The supports for the balustrading and handrail are located in the holes formed at the ends of the treads and secured with a nut and washer on the underside of the tread. The balustrade and handrail details have been omitted from all the stair details so far considered for reasons of clarity. Modern open-tread precast treads are often used in conjunction with glass balustrades to give an aesthetic appearance of light into an access area.

Spiral stairs in precast concrete work are based on the stone stairs found in many historic buildings, such as Norman castles and cathedrals, consisting essentially of steps that have a 'keyhole' plan shape rotating round a central core. Precast concrete spiral stairs are usually open-riser stairs with a reinforced concrete core, or a concrete-filled steel-tube core. Holes are formed at the extreme ends of the treads, to receive the handrail supports in such a manner that the standard passes through a tread and is fixed to the underside of the tread immediately below. A hollow spacer or distance piece is usually incorporated between the two consecutive treads (see typical details in Fig. 10.1.4). In common with all forms of this type of stairs, precast concrete spiral stairs are limited as to minimum diameter and total rise when being considered as escape stairs, and therefore they are usually installed as accommodation stairs. The beauty of the system is that each stair tread unit is identical to the next, which allows for economies in the casting process. However, an open-tread precast spiral stair cannot be considered fire-resistant as it has no fire-rated soffit or enclosure for compartmentation.

The finishes that can be applied to a concrete floor can also be applied in the same manner to an in-situ or precast concrete stairway. Care must be taken, however, with the design and detail, because the thickness of finish given to stairs is generally less than the thickness of a similar finish to floors. For reasons of safety, and for compliance with the Building Regulations Approved Documents, it is necessary to have equal-height risers throughout the

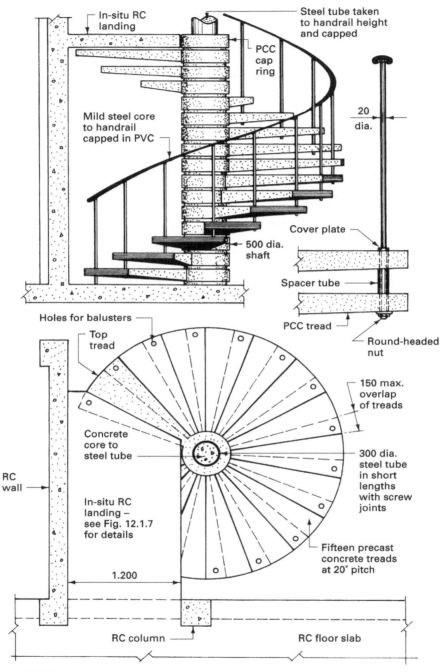

In-situ RC landing

Steel tube taken to handrail height and capped

PCC cap ring

Mild steel core to handrail capped in PVC

20 dia.

Cover plate

500 dia. shaft

Spacer tube

PCC tread

Round-headed nut

Holes for balusters

Top tread

150 max. overlap of treads

Concrete core to steel tube

300 dia. steel tube in short lengths with screw joints

RC wall

In-situ RC landing – see Fig. 12.1.7 for details

1.200

Fifteen precast concrete treads at 20° pitch

RC column

RC floor slab

Figure 10.1.4   **Precast concrete spiral stairs**

Remember that, when precast concrete stair flights are hoisted into position, different stresses may be induced from those that will be encountered in the fixed position. To overcome this problem, the designer can either reinforce the units for both conditions or, as is more usual, provide definite lifting points in the form of projecting lugs or by using any holes cast in to receive the balustrading.

Precast open-riser stairs are a form of stairs that can be both economic and attractive, consisting of a central spine beam in the form of a cut string supporting double cantilever treads of timber or precast concrete. The foot of the lowest spine beam is located and grouted into a preformed pocket cast in the floor, whereas the support at landing and floor levels is a simple bearing located in a housing cast into the slab edge (see Fig. 10.1.3). Anchor bolts or cement in sockets are cast into the spine beam to provide the fixing for the cantilever treads. The bolt heads are recessed below the upper tread surface and grouted over with a matching cement mortar for precast concrete treads, or concealed with matching timber pellets when hardwood treads are used. The supports for the balustrading and handrail are located in the holes formed at the ends of the treads and secured with a nut and washer on the underside of the tread. The balustrade and handrail details have been omitted from all the stair details so far considered for reasons of clarity. Modern open-tread precast treads are often used in conjunction with glass balustrades to give an aesthetic appearance of light into an access area.

Spiral stairs in precast concrete work are based on the stone stairs found in many historic buildings, such as Norman castles and cathedrals, consisting essentially of steps that have a 'keyhole' plan shape rotating round a central core. Precast concrete spiral stairs are usually open-riser stairs with a reinforced concrete core, or a concrete-filled steel-tube core. Holes are formed at the extreme ends of the treads, to receive the handrail supports in such a manner that the standard passes through a tread and is fixed to the underside of the tread immediately below. A hollow spacer or distance piece is usually incorporated between the two consecutive treads (see typical details in Fig. 10.1.4). In common with all forms of this type of stairs, precast concrete spiral stairs are limited as to minimum diameter and total rise when being considered as escape stairs, and therefore they are usually installed as accommodation stairs. The beauty of the system is that each stair tread unit is identical to the next, which allows for economies in the casting process. However, an open-tread precast spiral stair cannot be considered fire-resistant as it has no fire-rated soffit or enclosure for compartmentation.

The finishes that can be applied to a concrete floor can also be applied in the same manner to an in-situ or precast concrete stairway. Care must be taken, however, with the design and detail, because the thickness of finish given to stairs is generally less than the thickness of a similar finish to floors. For reasons of safety, and for compliance with the Building Regulations Approved Documents, it is necessary to have equal-height risers throughout the

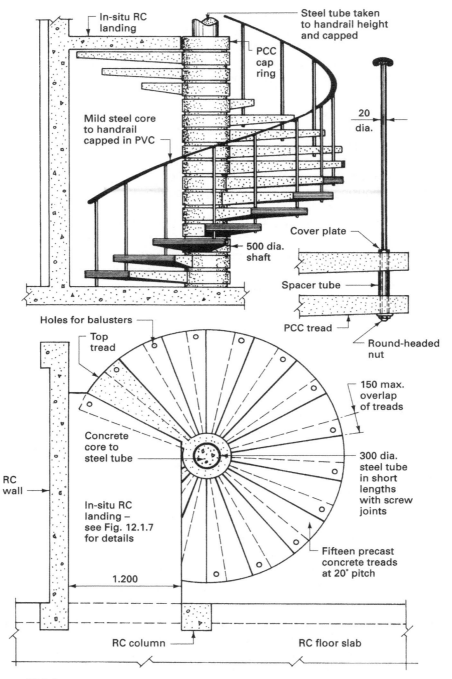

Figure 10.1.4 **Precast concrete spiral stairs**

stairway rise. Therefore, when casting the stairs, it may be necessary to have the top and bottom risers of different heights from the remainder of the stairs, to allow for the floor finishes that are yet to be applied.

Precast concrete stairs in a commercial application have a light screeding applied of a latex levelling compound if the surface of the risers and treads is not smooth. A vinyl non-slip floor finish is applied over this after rubbing down any imperfections. Stair nosings are highlighted, usually by using an aluminum stair nosing that has a raised grip pattern within it, which is drilled and screwed down to the top of the tread. These are produced in a variety of cross-sections that have inserts already factory-fitted to them or removable inserts that can be renewed or manufactured in a high-visibility material.

Metal stairs   10.2

Metal stairs can be constructed to be used as escape stairs or accommodation stairs, both internally and externally. This application tends to be for industrial or institutional building projects where a hard-wearing heavy-duty staircase suits better than a traditional timber, precast concrete or glass staircase. Most metal stairs are manufactured from mild steel, in straight flights with intermediate half-space landings. The materials used for the treads and landings tend to be steel plates with an impression or raised anti-slip pattern within their surface for safety in use or, on industrial applications, open-mesh grids that allow any liquids to fall through. Spiral stairs in steel are also produced, but their use as escape stairs is limited by size and the number of people likely to use the stairway in the event of a fire. Aluminium alloy stairs are also manufactured, which are lighter in construction, and are used almost exclusively as internal accommodation stairs.

As building layouts vary, stairs are often purpose-made to suit the particular situation. Concrete, being a flexible material at the casting stage, generally presents little or no problems in this respect, whereas purpose-made metal stairs can be more expensive and take longer to fabricate in the workshop. Consideration has to be given to the size of the component that will need to be transported and craned into position, along with the ease of bolting it into place. Metal spiral stairs have the distinct advantage that they do not need temporary support as they are normally built around a central support. All steel stairs have the common disadvantage of requiring regular maintenance in the form of painting as a protection against corrosion, especially in external applications, although a metal stair can be galvanised to protect it from rust and corrosion.

Most metal stairs are supplied in a form that requires some site fabrication. This is usually carried out by the supplier's site erection staff, the main contractor having been supplied with the necessary setting-out information, to enable preparatory work to be completed before the stairs are ready to be installed. This information would cover:

- foundation pads
- holding-down bolts
- secondary support steelwork at landings
- any special cast-in fixings
- any pockets to be left in the structural members or floor slabs.

STEEL ESCAPE STAIRS

A typical general arrangement for steel escape stairs is illustrated in Fig. 5.3.10. This shows a structural steel support frame and a stairway composed of steel plate strings, with preformed treads giving an open riser format. The treads for this type of stair are bolted to the strings and can be of a variety of types, ranging from perforated steel open-mesh grids to patterned steel treads with bent nosings incorporated into the treads as part of the manufacturing process. Handrail balustrades or standards can be of steel square or tubular sections, bolted or welded to the upper surface of a channel string or to the side of a channel or steel plate string. Fig. 10.2.1 shows typical steel escape stair components, and should be read in conjunction with the stair arrangement shown in Fig. 5.3.10.

STEEL SPIRAL STAIRS

Steel spiral stairs may be allowed as an internal or external means of escape stairs if they are to accommodate no more than 50 people. The minimum overall diameter is 1.6 m and the design must be in accordance with BS 5395-2: *Stairs, ladders and walkways. Code of practice for the design of helical and spiral stairs*, to satisfy Building Regulation classification as an escape stair. Spiral stairs give a very compact arrangement, and can be the solution in situations where the plan area is limited by special requirements. In common with all steel external escape stairs, the tread and landing plates should have a non-slip surface and be self-draining to allow safe use, with the stairway circulation width completely clear of any opening doors.

Two basic forms are encountered: those with treads that project from the central pole or tube on a cantilever basis, and those that have riser legs for support. The usual plan format is to have 12 or 16 treads to complete one turn around the central core, terminating at floor level with a quarter-circle landing or square landing. The standards, like those used for precast concrete spiral stairs, pass through one tread and are secured on the underside of the tread immediately below, giving strength and stability to both handrail and steps. Handrails are continuous and usually convex in cross-section, of polished metal, painted metal or plastic covered. Typical details are shown in Fig. 10.2.2.

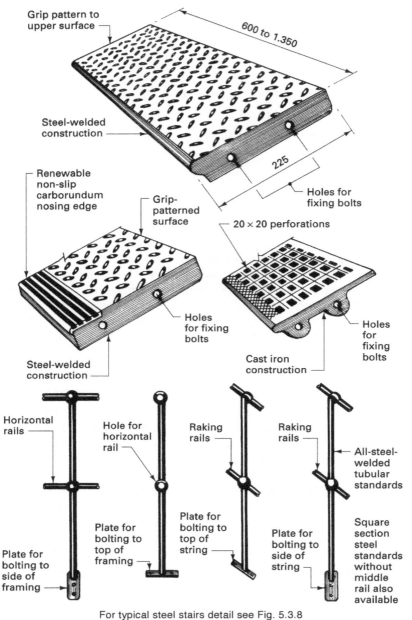

Grip pattern to upper surface

600 to 1.350

Steel-welded construction

225

Holes for fixing bolts

Renewable non-slip carborundum nosing edge

Grip-patterned surface

20 × 20 perforations

Steel-welded construction

Holes for fixing bolts

Cast iron construction

Holes for fixing bolts

Horizontal rails

Hole for horizontal rail

Raking rails

Raking rails

All-steel-welded tubular standards

Plate for bolting to top of framing

Plate for bolting to top of string

Plate for bolting to side of string

Square section steel standards without middle rail also available

Plate for bolting to side of framing

For typical steel stairs detail see Fig. 5.3.8

Figure 10.2.1 Typical steel stairway components

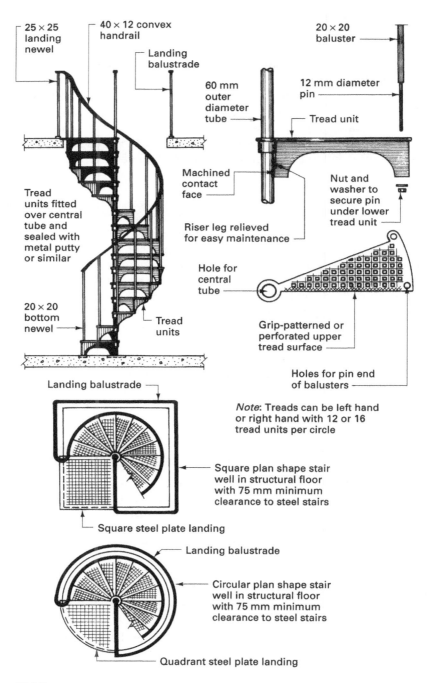

25 × 25 landing newel

40 × 12 convex handrail

Landing balustrade

20 × 20 baluster

60 mm outer diameter tube

12 mm diameter pin

Tread unit

Machined contact face

Nut and washer to secure pin under lower tread unit

Tread units fitted over central tube and sealed with metal putty or similar

Riser leg relieved for easy maintenance

Hole for central tube

20 × 20 bottom newel

Tread units

Grip-patterned or perforated upper tread surface

Holes for pin end of balusters

Landing balustrade

Note: Treads can be left hand or right hand with 12 or 16 tread units per circle

Square plan shape stair well in structural floor with 75 mm minimum clearance to steel stairs

Square steel plate landing

Landing balustrade

Circular plan shape stair well in structural floor with 75 mm minimum clearance to steel stairs

Quadrant steel plate landing

Figure 10.2.2   Typical steel spiral stairs

STRING-BEAM STEEL STAIRS

String-beam steel stairs are used mainly to form accommodation stairs that need to be light and elegant in appearance. This is achieved by using small sections and an open riser format. The strings can be of mild steel tube, steel channel, steel box or small universal beam sections fixed by brackets to the upper floor surfaces or landing edges to act as inclined beams. The treads (which can be of hardwood timber, precast concrete, glass or steel) are supported by plate, angle or tube brackets welded to the top of the string beam. Balustrading can be fixed through the ends of the treads, or supported by brackets attached to the outer face of the string beam. Typical details are shown in Fig. 10.2.3.

PRESSED STEEL STAIRS

Pressed steel stairs are accommodation stairs, made from light, pressed metal, such as mild steel. Each step is usually pressed as one unit, with the tread component recessed to receive a filling of concrete, granolithic, terrazzo, timber or any other suitable material. This provides a hard-wearing tread that can easily be replaced and maintained in the future. The strings are often in two pieces, consisting of a back plate to which the steps are fixed and a cover plate to form a box section string. The cover plate is site-welded using a continuous MIG (metal inert gas) process. The completed strings are secured by brackets or built into the floors or landings, and provide the support for the balustrade. Stairs of this nature are generally purpose-made to the required layout, and site assembled and fixed by a specialist subcontractor, leaving only the tread finishes and decoration as builder's work. Typical details are shown in Fig. 10.2.4.

ALUMINIUM ALLOY STAIRS

Aluminium alloy stairs are usually purpose-made to suit individual layout requirements, with half- or quarter-space landings from aluminium alloy extrusions. They are suitable for accommodation stairs in public buildings, shops, offices and flats. The treads have a non-slip nosing with a general tread covering of any suitable floor-finish material. Format can be open or closed riser, the latter having greater strength. The two-part box strings support the balustrading and are connected to one another by small-diameter tie rods that, in turn, support the tread units. The flights are either secured by screwing to purpose-made baseplates or brackets fixed to floors and landings, or located in preformed pockets and grouted in. When the stairs are assembled, they are very light and can usually be lifted and positioned by two operatives, without the need for lifting gear. No decoration or maintenance is required except for routine cleaning. Typical details are shown in Fig. 10.2.5.

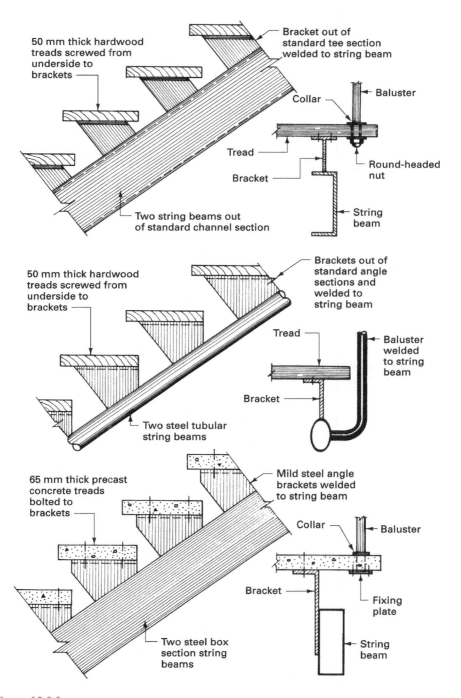

Figure 10.2.3 Typical string-beam steel stairs details

50 mm thick hardwood treads screwed from underside to brackets

Bracket out of standard tee section welded to string beam

Collar

Baluster

Tread

Bracket

Round-headed nut

String beam

Two string beams out of standard channel section

50 mm thick hardwood treads screwed from underside to brackets

Brackets out of standard angle sections and welded to string beam

Tread

Baluster welded to string beam

Bracket

Two steel tubular string beams

65 mm thick precast concrete treads bolted to brackets

Mild steel angle brackets welded to string beam

Collar

Baluster

Bracket

Fixing plate

String beam

Two steel box section string beams

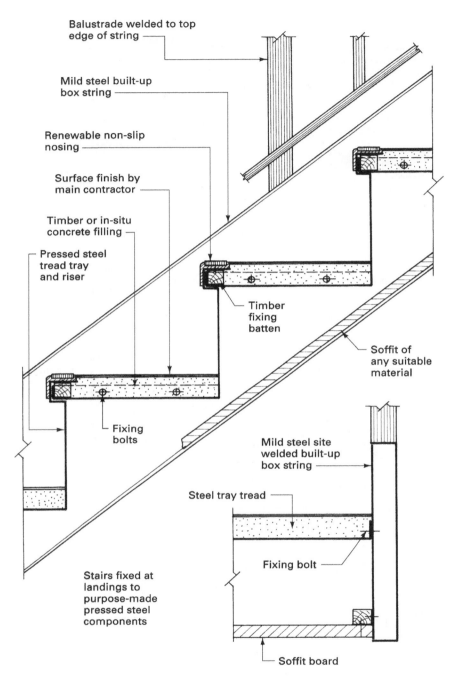

Balustrade welded to top edge of string

Mild steel built-up box string

Renewable non-slip nosing

Surface finish by main contractor

Timber or in-situ concrete filling

Pressed steel tread tray and riser

Timber fixing batten

Fixing bolts

Soffit of any suitable material

Mild steel site welded built-up box string

Steel tray tread

Fixing bolt

Stairs fixed at landings to purpose-made pressed steel components

Soffit board

Figure 10.2.4 'Prestair' internal pressed steel stairs

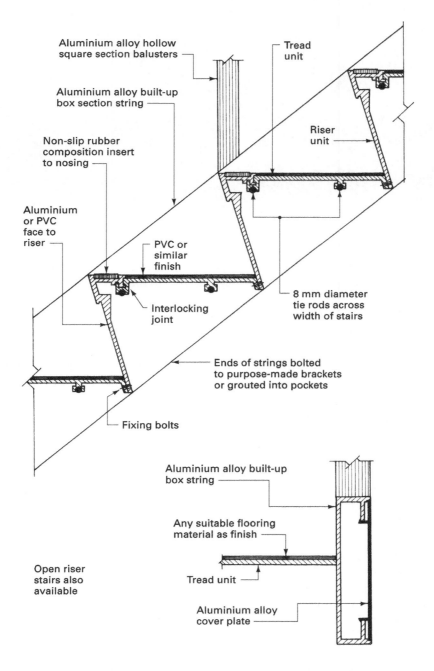

Aluminium alloy hollow
square section balusters

Tread
unit

Aluminium alloy built-up
box section string

Riser
unit

Non-slip rubber
composition insert
to nosing

Aluminium
or PVC
face to
riser

PVC or
similar
finish

Interlocking
joint

8 mm diameter
tie rods across
width of stairs

Ends of strings bolted
to purpose-made brackets
or grouted into pockets

Fixing bolts

Aluminium alloy built-up
box string

Any suitable flooring
material as finish

Open riser
stairs also
available

Tread unit

Aluminium alloy
cover plate

Figure 10.2.5   'Gradus' aluminium alloy stairs

Glass staircases 10.3

The commercial application of glass in a retail environment has been capitalised on by the Apple Stores throughout the world, which feature unique staircases manufactured entirely out of glass, held together with specially designed stainless steel fittings. Spiral, helical and straight flights can be enabled using glass components. Apple holds the US patent for these unique engineering and architectural marvels.

The renowned architect Eva Jiricna incorporates glass within staircase designs to maximum effect to produce beautiful glass stairs which can be considered artwork.

The key to the use of glass is the bonding of several layers together to form a glass laminate, which is polished after the bonding process.

Specifically, the glass treads consist of the following layers:

- an 8 mm extra clear Depp Glass Diamond Plate layer;
- a DuPont SentryGlas® Plus ionoplast interlayer;
- a 15 mm extra clear Diamond Plate layer;
- another DuPont SentryGlas® Plus ionoplast interlayer;
- another 8 mm extra clear Diamond Plate layer.

The tread finishes with a thickness totalling 55 mm. The treads have a diamond pattern pressed into the finished glass, which acts as a non-slip finish. The strength lies within the bonding that is used to join the layers of glass together. This product, manufactured by DuPont, is specifically used for high-strength applications as it is 100 times stronger than standard laminate adhesives, and has five times the tear strength.

The depth of the tread will depend on the application, the amount of foot traffic that it is expected to carry and the width of the staircase that might be required for Building Regulation safety purposes. Figure 10.3.1 illustrates a

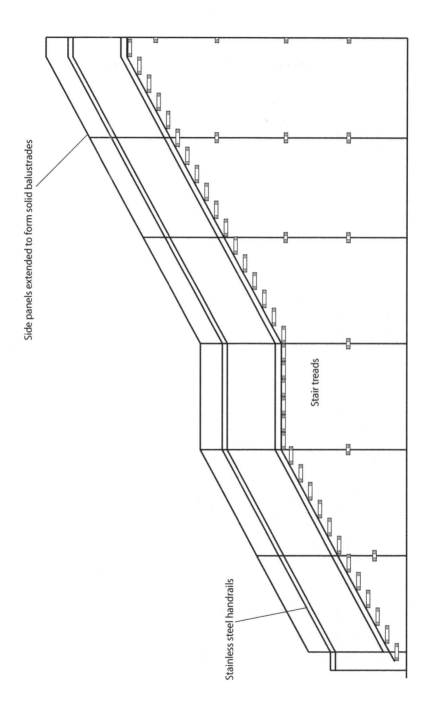

Side panels extended to form solid balustrades

Stair treads

Stainless steel handrails

Figure 10.3.1 Transparent staircase

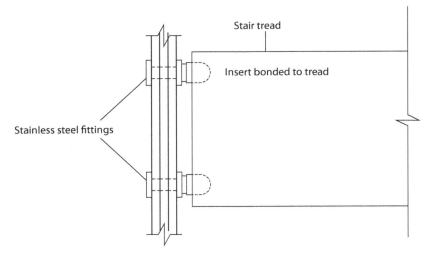

Stair tread

Insert bonded to tread

Stainless steel fittings

Figure 10.3.1   *Continued*

side view of the staircase and a tread detail of the fixing. The stair is supported by the glass side panels, which extend past the stair treads to form a balustrade. The side panels are fixed in turn to each other and to short, glass fin panels at 90° to the side panels, which supply rigidity to the structure in the form of 'T's.

The stair treads have a fixing inserted into them during the lamination process to securely bond the stainless steel fitting to the stair treads, on all three sides of the fitting. A housing fitting is then fixed to the insert using two stainless steel countersunk bolts. This housing enables a circular fitting that is fixed through the side wall panels to locate into the stair tread housing. This is finally secured using stainless steel grub screws so it cannot pull out or upwards. A complete glass structure must be designed structurally by an engineer to ensure that it is safe, maintenance can be undertaken easily, and the finished staircase conforms to all relevant regulations and standards, covering areas such as fire, structural stability and impact tests.

There are of course many variations to the use of glass as an access material. Here are just some of the combinations that can be used in the form of a 'composite' construction:

■ incorporating it with structural steel enables several options, such as a central steel string supporting glass stair treads;

■ cantilevered glass stair treads can be built into a structural wall;

■ glass stair treads can be used with precast concrete strings;

■ glass stair treads can be used with glulam timber.

GLASS BALUSTRADES

The use of glass commercially in retail and commercial complexes must be carefully considered with regard to the impact of human traffic against such surfaces. Laminated and toughed safety glass must be used in the correct locations in accordance with standards and regulations on its classification, use and method of fixing. The use of glass commercially as a safety barrier has to be carefully considered under BS 6262-4 Glazing for Buildings *Code of practice for safety related to human impact*, which covers many of the recommendations contained within the Building Regulations Approved Document Part N. The code of practice in BS 6180:2011 *Barriers in and about buildings* further extends the use of glass infill panels and the methods of restraint that can be employed. BS EN 12600:2002 *Glass in building. Pendulum test. Impact test method and classification for flat glass* partially replaces BS 6206 and provides a testing method using a pendulum to impact from a known distance onto a sample of glass. There are three classifications under this standard:

- Class 3 – material that conforms to the requirements of clause 4 when tested by the method given in clause 5 at a drop height of 190 mm;
- Class 2 – material that conforms to the requirements of clause 4 when tested by the method given in clause 5 at drop heights of 190 mm and 450 mm;
- Class 1 – material that conforms to the requirements of clause 4 when tested by the method given in clause 5 at drop heights of 190 mm, 450 mm and 1,200 mm.

Clause 4 is the test requirements and clause 5 is the test method.

Mechanical access

10.4

This section examines the use of lifts, escalators and travelators to move people in commercial buildings. It gives an insight into the technology around the installation of such mechanical access methods, rather than a detailed investigation into how they work.

Large, multi-storey buildings sometimes have to hand over a large section of planned floor space to the vertical movements of people who use the building. The design space requirements for mechanical access for high-rise commercial buildings will depend on:

- the number of occupants who use the building;
- the height of the building;
- the capacity to be lifted;
- the predicted frequency of use;
- the floor plan and layouts;
- the speed of travel required at peak times.

Similarly, at airports a large number of people have to move quickly and efficiently across several terminals. This can be achieved with a horizontal escalator or travelator, which is in effect a moving walkway, the speed of which can be carefully controlled.

With any mechanical movement of occupants it needs to be smooth without any sensation of acceleration or deceleration and must make occupants feel safe and secure in its use, access and egress. To this end many safety features are built into such mechanical machinery to include, emergency stop buttons and emergency telephone services. There are primarily three principal methods of movement around a multi-storey building: lifts, escalators and travelators. The latter is for horizontal movement across large distances, and is not a primary method of vertical movement. However, travelators can be inclined at an angle to lift to one storey.

Modern lift systems developed following Otis's invention of a safe braking system in the event of failure in the 1850s. This device made lifts safer should problems occur with the lift cable, and its integration led to faster and more efficient lift systems.

Lifts for vertical movement can be classified into two categories:

1. those that utilise a cable with counterbalance weights, known as traction lifts;

2. those that use a hydraulic-action ram to push the lift up and then release the pressure to bring the lift down.

The latter can only be used for a number of storeys of lift due to the length of the hydraulic ram required and its capacity.

TRACTION-LIFT SYSTEMS

The main traction-lift system uses cables to suspend the lift car on one end and a counterbalance weight on the other.

Figure 10.4.1 illustrates the basic principle of the traction-lift system. A car carrying a limited number of passengers is fixed by a suspension cable, which is looped over a traction wheel. This wheel is connected via a gearbox to an electric motor. When the call button is pressed, the lift will descend or rise to the floor required. The doors open automatically and the user can then select the floor they require.

Such lifts need a lift motor room above the lift car for the equipment to operate. The whole structure is enclosed in a lift shaft, which can now be constructed using precast concrete units. A large counterbalance offsets the weight of the occupants, so the lift motor has to do little work as most of this is done by gravitational forces.

Technological developments in electric motor design mean that a gearless direct drive can be placed in the lift shaft itself; the counterbalance weights are replaced by directly fixing the suspension cables to the structure. This is helpful for building designers, as they no longer have to find space for the counterbalance weights and lift motor room, and is especially beneficial in refurbishment projects, where wheelchair access requires more space within the lift car.

The lift itself has become an aesthetic feature in many buildings. For example, Lloyds of London places the lift system on the outside of the building, and uses lift cars manufactured in toughened glass, to create a unique and iconic addition to the building.

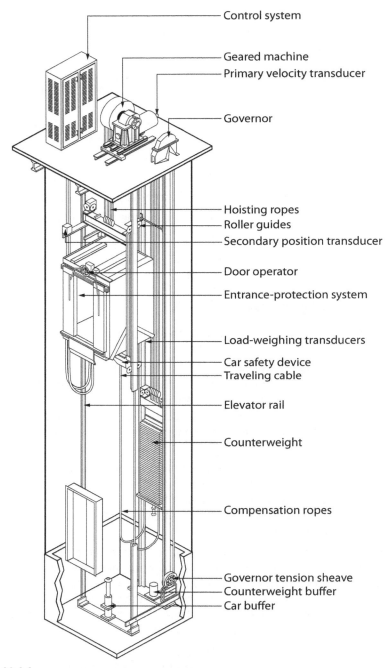

- Control system
- Geared machine
- Primary velocity transducer
- Governor
- Hoisting ropes
- Roller guides
- Secondary position transducer
- Door operator
- Entrance-protection system
- Load-weighing transducers
- Car safety device
- Traveling cable
- Elevator rail
- Counterweight
- Compensation ropes
- Governor tension sheave
- Counterweight buffer
- Car buffer

Figure 10.4.1   Basic cable-operated traction-lift components

HYDRAULIC-ACTION RAM SYSTEMS

The alternative lift system utilises a hydraulic ram for its lifting mechanism.

Figure 10.4.2 illustrates such a system set-up. A pit must be provided for the ram when it is in its closed position, so that the ram components are collapsed into the cylinder when on the ground floor. Adjacent to the hydraulic cylinder is the pump room and control panel that feeds the hydraulic oil to the cylinder that raises and lowers the lift car.

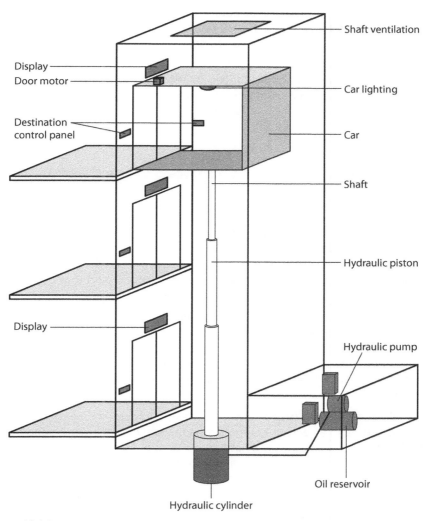

Figure 10.4.2   **Basic hydraulic-ram-operated lift components**

As this system does not utilise a counterbalance weight and gravitational forces, it has higher energy costs. Another disadvantage with this system is that it can only reach a limited number of storeys, due to the capacity of the hydraulic cylinder and the length of the hydraulic ram.

However, advantages are that the building's structure does not have to be designed for supporting counterbalance weights, and the force of the lift is carried by the ground rather than the supporting frame. This may reduce the installation costs, although savings need to be offset against reductions in the strength of the supporting structure.

The installation of a lift requires the construction of the lift shaft. In high-rise buildings this is usually a concrete structure, either cast in situ or precast and fabricated into units that are assembled on site. The concrete structure will be able to support the side rails required to keep the lift perfectly vertical during use; any non-solid structure would deflect and move with the loading.

A lift shaft can have a secondary function by offering fire protection for services. Services are often distributed by adding some additional space next to the lift shaft enclosure and forming a vertical riser. Fire collars are usually installed around the services to separate the fire zones into fire compartments.

ESCALATORS

An escalator is essentially a moving set of stairs attached to a chain-driven belt that rotates continuously and lifts the user from one level to another. Escalators are associated with a number of different locations, including the Tube system in London and department stores. The use of an escalator is limited to about five storeys; after this point, it usually becomes more economical to install a lift system. Escalators are often limited to the lower ground floors of commercial buildings or offices.

The beauty of an escalator is that it is fabricated as a whole unit within the manufacturer's facility. All the contractor has to do is provide the structural supports at the top landing and the service access pits required by some manufacturers at the bottom. Escalators can be installed quickly and efficiently into a building and arrive with all protection installed to glass and steel panels.

In an escalator, an electric motor powers the steps through a drive-gear system that moves the handrail and the steps simultaneously, at the same speed, to ensure its safe operation. The escalator is built around a steel truss, which forms the basis of an inclined beam supported at two points. Figure 10.4.3 shows the components of a typical escalator. The major components are:

- truss frame – a designed, structural component that takes the weight of the equipment and live loads and transfers it to the support points on the structure. This element is hidden by an enclosure when in use;

- handrail drive – the mechanism that drives the handrail at the same speed as the stairs;

- comb plate – a device that removes any foreign objects from the stair treads;

- balustrades – a side panel that encases the passenger and prevents any falls. These can be manufactured from a variety of materials, from transparent glass through to solid panels;

- handrails – constructed of black synthetic rubber;

- skirting – the internal skirting against the steps that protects users from the stair mechanisms, which often has a sloping top to stop anything dropped getting trapped;

- landing plates – the area at the top and bottom that is set aside for access to and egress from the escalator. The landing plates are often hinged and key-locked as they provide access to the servicing of the moving machinery.

The escalator must be serviced at regular intervals in order to:

- lubricate all the moving parts;

- remove dust and debris carried through the combs and deposited into the landing pits;

- service the gear mechanism and motor.

The relevant British Standard here is BS EN 115-1:2008 *Safety of escalators and moving walks, construction and installation*. This includes such standards as handrail heights, pitch, fire protection, and inspection and testing.

TRAVELATORS

While escalators are used for vertical rise, travelators (or moving walkways) are a safe and easy way to speed up short journeys on one level. A travelator is essentially an escalator that has no steps, installed either horizontally or at an inclined, shallow angle. The development of travelators has been established in retail shopping, where users can push their trolley onto the moving walkway and exit onto a mezzanine floor above.

Within airports or other public transport infrastructures, travelators are used for the speedy transfer of passengers from gate to gate; set at a faster-than-walking pace, they help passengers reach their destination more quickly. Advances in this technology now include module-based elements and above-ground drive mechanisms (with a ramp projection of 150 mm above the flat floor area) that reduce the structural costs associated with their installation.

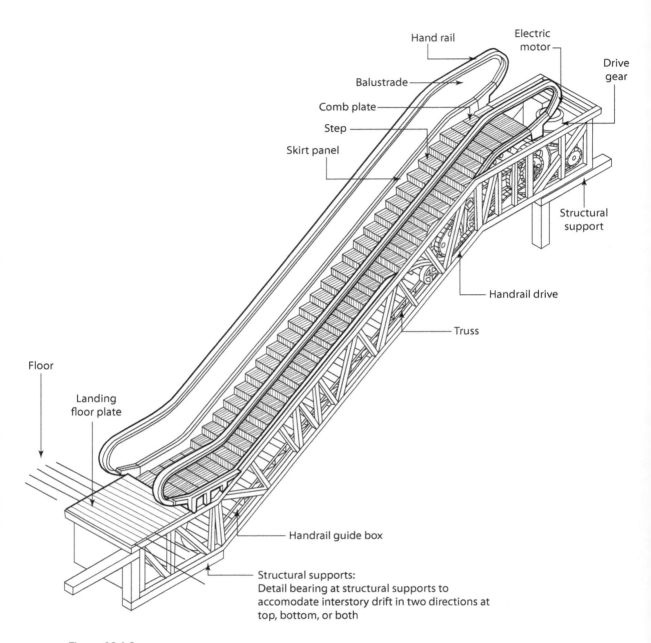

Hand rail

Electric motor

Drive gear

Balustrade

Comb plate

Step

Skirt panel

Structural support

Floor

Handrail drive

Landing floor plate

Truss

Handrail guide box

Structural supports:
Detail bearing at structural supports to accomodate interstory drift in two directions at top, bottom, or both

Figure 10.4.3   **Basic elements of an escalator unit**

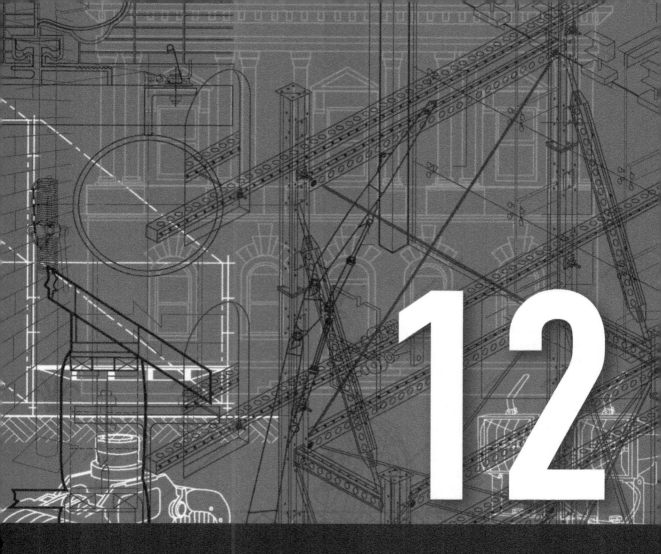

12

External works

Roads, pavings and slabs 12.1

Building contractors are not normally engaged to construct major roads or motorways, as this is the province of the civil engineer. However, they can be involved in the laying of small estate roads, service roads and driveways, and it is within this context that the student of construction technology would consider roadworks.

Before considering road construction techniques and types, it is worth considering some of the problems the designer encounters when planning road layouts. The width of a road can be determined by the anticipated traffic flow, volume and speed. Lane width is calculated from an allowance of maximum vehicle width of 2.5 m plus 0.5 m minimum clearance between vehicles, to provide a minimum lane width of 3 m. The major layout problems occur at road junctions and at the termination of cul-de-sac roads. At right-angle junctions, it is important that vehicles approaching the junction from any direction have a clear view of approaching vehicles intending to join the main traffic flow. Most planning authorities have layout restrictions at such junctions in the form of triangulated sight lines, which give a distance and area in which an observer can see an object when both are at specific heights above the carriageway. Within this triangulated area, street furniture or any other obstruction is not allowed (see Fig. 12.1.1). Angled junctions, by virtue of their distinctive layout, do not present the same problems. At any junction, a suitable radius should be planned so that vehicles filtering into the main road should not have to apply a full turning lock. The actual radius required will be governed by the anticipated vehicle types that will use the road.

Terminations at the end of cul-de-sacs must be planned to allow vehicles to turn round. For service roads this is usually based on the length and turning circle specifications of refuse collection vehicles, which are probably the largest vehicles to use the road. Typical examples are shown in Fig. 12.1.2.

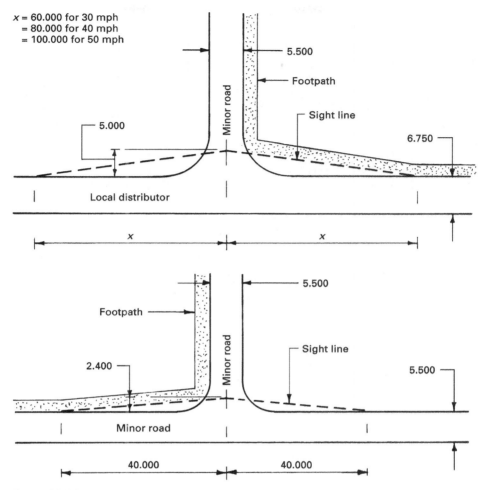

Figure 12.1.1 Typical road junctions

The construction of roads can be considered under two headings:

1. preparation or earthworks – the excavation, improvement or importation of engineering soils or fill materials;

2. pavement construction – the construction of the permanent highway and pedestrian structures.

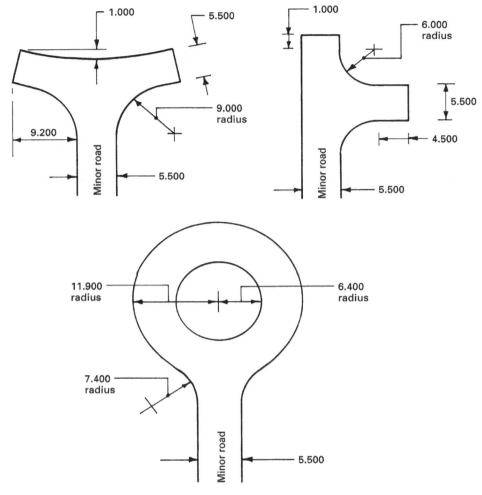

Figure 12.1.2 Typical road turn-around and terminations

Before undertaking any roadwork, a thorough soil investigation should be carried out to determine the nature of the subgrade – the soil immediately below the topsoil that will ultimately carry the traffic loads from the pavement above. Soil investigations should preferably be carried out during the winter, when subgrade conditions will be at their worst. Trial holes should be taken down to at least 1 m below the proposed formation level. To ensure that a good pavement design can be formulated, these investigations need to provide information on the elasticity, plasticity, cohesion and internal friction properties of the subgrade.

EARTHWORKS

Earthworks includes removing topsoil, scraping and grading the exposed surface to the required formation level, preparing the subgrade to receive the pavement, and forming any embankments and/or cuttings. The strength of the subgrade will generally decrease as the moisture content increases. An excess of water in the subgrade can also cause damage by freezing, causing frost heave in fine sands, chalk and silty soils. Conversely, thawing may cause a reduction in subgrade strength, giving rise to failure of the pavement above. Tree roots, particularly those of fast-growing deciduous trees, such as poplar and willow, can also cause damage in heavy clay soils, by extracting vast quantities of water from the subgrade down to a depth of 3 m.

The pavement covering will give final protection to the subgrade from excess moisture, but during the construction period the subgrade should be protected by a waterproof surfacing, such as a sprayed bituminous binder with a sand cover applied at a rate of 1 litre per square metre. If the subgrade is not to be immediately covered with the sub-base of the pavement, it should be protected by an impermeable membrane, such as 500-gauge plastic sheeting with 300 mm side and end laps.

PAVEMENT CONSTRUCTION

'Pavement' is both a general term for any paved surface, and the term applied to the whole construction of a road, including pedestrian areas. Road pavements can be classified as flexible pavements that, for the purpose of design, are assumed to have no tensile strength and consist of a series of layers of materials to distribute the wheel loads to the subgrade. The alternative form is the rigid pavement of which, for the purpose of design, the tensile strength is taken into account, and which consists of a concrete slab resting on a granular base.

FLEXIBLE PAVEMENTS

The sub-base for a flexible pavement is laid directly onto the formation level. It should consist of a well-compacted, granular material, such as a quarry overburden or crushed rocks. The actual thickness of sub-base required is determined by the cumulative number of standard axles (msa) to be carried (msa = millions of standard axles, where a standard axle is taken as 8,200 kg) and the CBR (California bearing ratio) of the subgrade. The CBR is an empirical measure in which the thickness of the sub-base is related to the strength of the subgrade and to the amount of traffic the road is expected to carry. A fully flexible pavement design life should be at least 20 years. For pavement construction in the UK, the main reference is Volume 7 of the *Design Manual for Roads and Bridges: Pavement Design and Maintenance*, administered by the Highways Agency and available from their website at www.dft.gov.uk/ha/standards.

This manual provides graphical presentations on the principle shown in Fig. 12.1.3 for determining road-construction specification relative to design traffic load (msa) and overall bituminous layer thickness.

The subgrade is covered with a sub-base, a base course and a wearing course; the last two components are collectively known as the **surfacing**. The sub-base can consist of any material that remains stable in water, such as crushed stone, dry lean concrete or blast furnace slag. You can also use compacted dry-bound macadam in a 75 to 125 mm-thick layer with a 25 mm thick overlay of firmer material, or a compacted wet mix macadam in 75 to 150 mm thick layers. The material you choose should not be affected by frost and should be well compacted in layers, giving a compacted thickness of between 100 and 150 mm for each layer.

The base course of the surfacing can consist of rolled asphalt, dense tarmacadam, dense bitumen macadam or open-textured macadam, and should be applied to a minimum thickness of 60 mm. Base courses are laid to the required finished road section providing any necessary gradients or crossfalls, ready to receive the thinner wearing course, which should be laid within three days of completing the base course. The wearing course is usually laid by machine and provides the water protection for the base layers. It should also have non-skid properties, reasonable resistance to glare, good

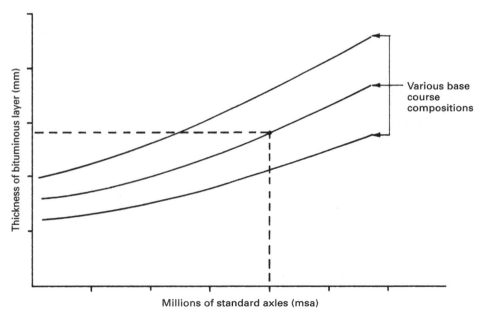

Figure 12.1.3 Design thickness for pavements

riding properties and a good life expectancy. Materials that give these properties include hot-rolled asphalt, bitumen macadam, dense tar surfacing and cold asphalt (see Fig. 12.1.4 for typical details). Existing flexible road surfaces can be renovated quickly and cheaply by the application of a hot tar or cut-back bitumen binder, with a rolled layer of gravel, crushed stone or slag chippings applied immediately after the binder and before it sets.

Pavements that contain a nominal amount of cement to bind materials in the sub-base and have bituminous materials for surfacing vary sufficiently from a truly flexible specification. These are often known appropriately as **semi-flexible** or **flexible composite**.

The above treatments are termed bound surfaces, but flexible roads or pavements with unbound surfaces can also be constructed. These are suitable for light vehicular traffic where violent braking and/or acceleration is not anticipated, such as driveways to domestic properties. Unbound pavements consist of a 100 to 200 mm-thick base of clinker or hardcore laid directly onto the formation level of the subgrade. This is covered with a well-rolled layer of screeded gravel to pass a 40 mm ring, with sufficient sand to fill the small voids, to form an overall consolidated thickness of 25 mm. This will give a relatively cheap flexible pavement. However, to to be really efficient, it should have adequate falls to prevent the ponding of water, and should be treated each spring with an effective weedkiller.

RIGID PAVEMENTS

A rigid pavement is a form of road using a concrete slab laid over a base layer. The subgrade is prepared as described above for flexible pavements, and should be adequately protected against water. The base layer is laid over the subgrade, and must form a working surface from which to case the concrete slab and enable work to proceed during wet and frosty weather, without damage to the subgrade. Generally, granular materials such as crushed concrete, crushed stone, crushed slag and suitably graded gravels are used to form the base layer. The thickness will depend on the nature and type of subgrade: weak subgrades usually require a minimum thickness of 150 mm, while normal subgrades require a minimum thickness of only 80 mm.

The thickness of concrete slabs used in rigid pavement construction will depend on the condition of the subgrade and intensity of traffic, and on whether the slab is to be reinforced. With a normal subgrade using a base layer 80 mm thick, the slab thickness would vary from 125 mm for a reinforced slab carrying light traffic to 200 mm for an unreinforced slab carrying a medium to heavy traffic intensity. The usual strength specification is 28 MN/m^2 at 28 days, with not more than 1 per cent test cube failure rate. The mix design should be based on a mean strength of between 40 and 50 MN/m^2, depending on the degree of quality control possible on or off site. To minimise the damage that frost and de-icing salts can cause, the water/cement ratio

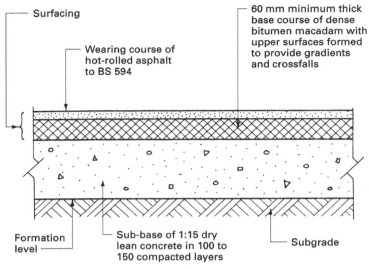

Surfacing

Wearing course of
hot-rolled asphalt
to BS 594

60 mm minimum thick
base course of dense
bitumen macadam with
upper surfaces formed
to provide gradients
and crossfalls

Formation
level

Sub-base of 1:15 dry
lean concrete in 100 to
150 compacted layers

Subgrade

Typical semi-flexible pavement

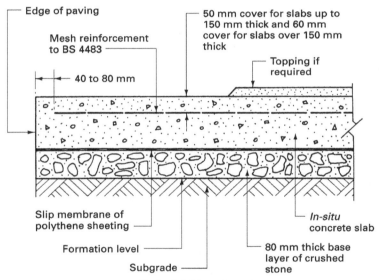

Edge of paving

Mesh reinforcement
to BS 4483

40 to 80 mm

50 mm cover for slabs up to
150 mm thick and 60 mm
cover for slabs over 150 mm
thick

Topping if
required

Slip membrane of
polythene sheeting

Formation level

Subgrade

In-situ
concrete slab

80 mm thick base
layer of crushed
stone

Typical rigid pavement

References:
BS 594-1 and 2: *Hot rolled asphalt for roads and other paved areas. Specifications*
BS 4483: *Steel fabric for the reinforcement of concrete*

Figure 12.1.4   Typical pavement details

should not exceed 0.5 by weight, and air entrainment to at least the top 50 mm of the concrete should be specified. The air-entraining agent used should produce 3–6 per cent of minute air bubbles in the hardened concrete, preventing saturation of the slab by capillary action.

Before the concrete is laid, the base layer should be covered with a slip membrane of polythene sheet, which will also prevent grout loss from the concrete slab. Concrete slabs are usually laid between pressed steel road forms, which are positioned and fixed to the ground with steel stakes. These side forms are designed to provide the guide for hand tamping or to provide for a concrete train consisting of spreaders and compacting units. Curved or flexible road forms have no top or bottom flange and are secured to the ground with an increased number of steel stakes (see Fig. 12.1.5 for examples of the typical road form).

Reinforcement can be included in rigid pavement constructions to prevent the formation of cracks and to reduce the number of expansion and contraction joints required. Reinforcement is generally in the form of a welded-steel fabric complying with the recommendations of BS 4483: *Steel fabric for the reinforcement of concrete*. If bar reinforcement is used instead, it should consist of deformed bars at spacings not exceeding 150 mm. The cover of concrete over the reinforcement will depend on the thickness of concrete: for slabs under 150 mm thick, the minimum cover should be 50 mm; for slabs over 150 mm thick, the minimum cover should be 60 mm.

Joints used in rigid pavements may be either transverse or longitudinal, and are included in the design to:

- limit the size of slab;
- limit the stresses due to subgrade restraint;
- make provision for slab movements, such as expansion, contraction and warping.

The spacing of joints will be governed by slab thickness, presence of reinforcement, traffic intensity and the temperature at which the concrete is placed.

Five types of joint are used in rigid road and pavement construction.

1. **Expansion joints** These are transverse joints at 36 to 72 m centres in reinforced slabs, and at 27 to 54 m centres in unreinforced slabs.

2. **Contraction joints** These are transverse joints that are placed between expansion joints at 12 to 24 m centres in reinforced slabs, and at 4.5 to 7.5 m centres in unreinforced slabs, to limit the size of slab bay or panel. Note that every third joint should be an expansion joint.

3. **Longitudinal joints** These are similar to contraction joints and are required where slab width exceeds 4.5 m.

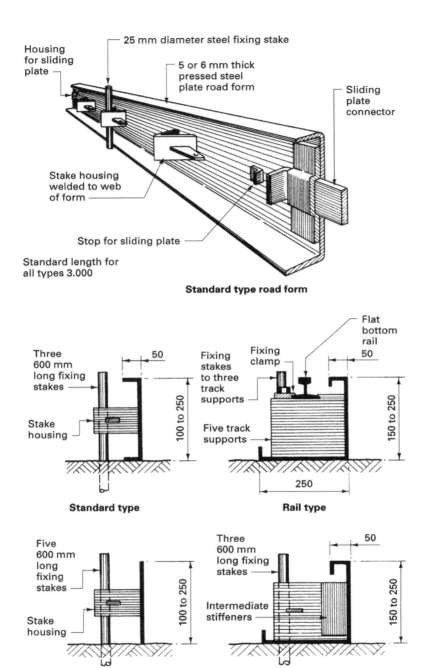

Housing for sliding plate

25 mm diameter steel fixing stake

5 or 6 mm thick pressed steel plate road form

Sliding plate connector

Stake housing welded to web of form

Stop for sliding plate

Standard length for all types 3.000

Standard type road form

Three 600 mm long fixing stakes

Stake housing

50

100 to 250

Standard type

Fixing stakes to three track supports

Fixing clamp

Flat bottom rail

50

Five track supports

150 to 250

250

Rail type

Five 600 mm long fixing stakes

Stake housing

100 to 250

Flexible type

Three 600 mm long fixing stakes

Intermediate stiffeners

50

150 to 250

Track type

Figure 12.1.5   Typical steel road form details

4. **Construction joints** The day's work should normally be terminated at an expansion or contraction joint but, if this is impossible, a construction joint can be included. These joints are similar to contraction joints but with the two portions tied together with reinforcement. Construction joints should not be placed within 3 m of another joint, and should be avoided wherever possible.

5. **Warping joints** These are transverse joints that are sometimes required in unreinforced slabs to relieve the stresses caused by vertical temperature gradients within the slab, if they are higher than the contractional stresses. The detail is similar to contraction joints, but has a special arrangement of reinforcement.

Typical joint details are shown in Fig. 12.1.6.

Road joints can require fillers and/or sealers: the former need to be compressible, whereas the latter should protect the joint against the entry of water and grit. Suitable materials for fillers are soft, knot-free timber, impregnated fibreboard, chipboard, cork and cellular rubber. The most common sealing compounds are resinous compounds, rubber-bituminous compounds and straight-run bitumen compounds containing fillers. The sealed surface groove used in contraction joints to predetermine the position of a crack can be formed while casting the slab, or can be sawn into the hardened concrete using water-cooled circular saws. Although slightly dearer than the formed joint, sawn joints require less labour and generally give a better finish.

The curing of newly laid rigid roads and pavings is important to maintain the concrete strength and avoid the formation of surface cracks. Curing precautions should start as soon as practicable after laying, preferably within 15 minutes of completion, by covering the newly laid surface with a suitable material to protect it from the rapid drying effects of the sun and wind. This form of covering will also prevent unsightly pitting of the surface due to rain. Light covering materials, such as waterproof paper and plastic film, can be laid directly onto the concrete surface, ensuring that they are adequately secured at the edges. Plastic film can give rise to a smooth surface if the concrete is wet; this can be avoided by placing raised bearers over the surface to support the covering. Heavier coverings, such as tarpaulin sheets, will need to be supported on frames of timber or light metalwork so that the covering is completely clear of the concrete surface. Coverings should remain in place for about seven days in warm weather, and for longer periods in cold weather.

The design of rigid pavements follows a similar procedure to that described for flexible pavements. Graphical extracts from Volume 7 of the *Design Manual for Roads and Bridges* resemble the illustration in Fig. 12.1.2, but with concrete thickness in the vertical axis.

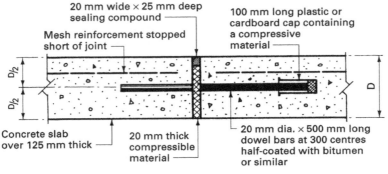

20 mm wide × 25 mm deep
sealing compound

100 mm long plastic or
cardboard cap containing
a compressive
material

Mesh reinforcement stopped
short of joint

Concrete slab
over 125 mm thick

20 mm thick
compressible
material

20 mm dia. × 500 mm long
dowel bars at 300 centres
half-coated with bitumen
or similar

Expansion joint

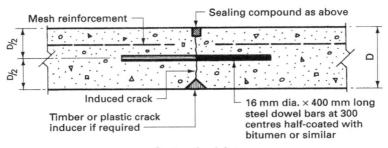

Mesh reinforcement

Sealing compound as above

Induced crack

Timber or plastic crack
inducer if required

16 mm dia. × 400 mm long
steel dowel bars at 300
centres half-coated with
bitumen or similar

Contraction joint

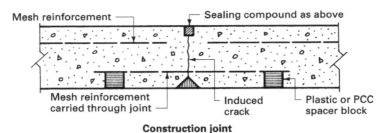

Mesh reinforcement

Sealing compound as above

Mesh reinforcement
carried through joint

Induced
crack

Plastic or PCC
spacer block

Construction joint

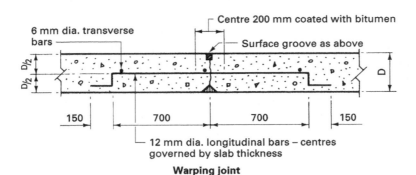

6 mm dia. transverse
bars

Centre 200 mm coated with bitumen

Surface groove as above

150   700   700   150

12 mm dia. longitudinal bars – centres
governed by slab thickness

Warping joint

Figure 12.1.6   Typical rigid road joint details

Rigid composite or **semi-rigid** road construction can be specified to combine the benefits of a smooth surface with a solid concrete sub-base. The concrete is reinforced throughout and surfaced with a bituminous wearing course of hot-rolled or porous asphalt.

DRAINAGE

Road drainage consists of directing the surface water to suitable collection points and conveying the collected water to a suitable outfall. The surface water is encouraged to flow off the paved area by crossfalls, which must be designed with sufficient gradient to cope with the volume of water likely to be encountered during a heavy storm, to prevent vehicles skidding or aquaplaning. A minimum crossfall of 1:40 is generally specified for urban roads and motorways, whereas crossfall specifications of between 1:40 and 1:60 are common for service roads. The run-off water is directed towards the edges of the road, where it is conveyed by gutters or drainage channels at a fall of about 1:200 in the longitudinal direction to discharge into road gullies, and from there into the surface water drains.

Road gullies are available in clayware and precast concrete, with or without a trapped outlet. If final discharge is to a combined sewer, the trapped outlet is required, and in some areas the local authority will insist on a trapped outlet gully for all situations. Spacing of road gullies depends on the anticipated storm conditions and crossfalls, but common spacings are 25 to 30 m. The gratings are usually made from cast iron, slotted and hinged to allow easy flow into the collection chamber of the gully and to allow for access for suction cleaning (see Fig. 12.1.7). Roads that are not bounded by kerbs can be drained by having subsoil filter drains beneath the verge, or can be drained directly into a swale or stream running alongside the road – see sustainable drainage systems later.

The sizing and layout design of a road drainage system is based on anticipated rainfall. For example, given a rainfall (R) intensity of 40 mm/h, the quantity (Q) of water in litres/second running off a road, car-park or similar surface can be shown as:

$$Q = \frac{APR}{3,600}$$

where A is the surface area to be drained, and P is the surface permeability (asphalt and concrete = 0.85 to 0.95).

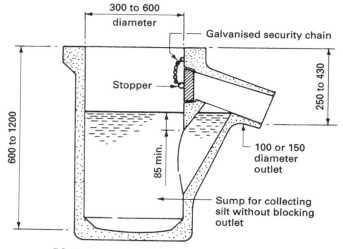

BS 5911 Unreinforced concrete street gully

300 to 600 diameter

Galvanised security chain

Stopper

250 to 430

600 to 1200

85 min.

100 or 150 diameter outlet

Sump for collecting silt without blocking outlet

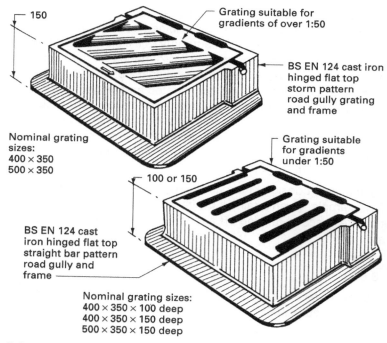

150

Grating suitable for gradients of over 1:50

BS EN 124 cast iron hinged flat top storm pattern road gully grating and frame

Nominal grating sizes:
400 × 350
500 × 350

Grating suitable for gradients under 1:50

100 or 150

BS EN 124 cast iron hinged flat top straight bar pattern road gully and frame

Nominal grating sizes:
400 × 350 × 100 deep
400 × 350 × 150 deep
500 × 350 × 150 deep

References:
BS 5911-6: *Concrete pipes and ancillary concrete products.
Specification for road gullies and gully cover slabs*
BS EN 124: *Gully tops and manhole tops for vehicular and pedestrian
areas. Design requirements, type testing, marking, quality control*

Figure 12.1.7 Typical road gully and grating details

If a road has an area of 1,000 m² and a permeability of 0.9, then:

$$Q = \frac{1,000 \times 0.9 \times 40}{3,600} = 10 \text{ litres/s} \quad \text{or} \quad 0.010 \text{ m}^3/\text{s}$$

$$Q = V \quad A$$

where V is the velocity of water flowing in drain (min = 0.75 m/s for self-cleansing) and A is the area of water flowing in pipe (m²) – allow half-full bore. That is:

$$0.010 = 0.75 \quad A$$

therefore

$$A = \frac{0.010}{0.75} = 0.0133 \text{ m}^2$$

Area (A) for half-bore = $\frac{1}{2}\pi r^2$; therefore r (radius) = 0.092 m and diameter of drain = 0.184 m. The nearest commercial drain is 225 mm or possibly a non-standard pipe of 200 mm.

For a more detailed perspective of drainage design, the reader is recommended to resource Chapter 8 of *Building Services, Technology and Design*, by Roger Greeno, published jointly by CIOB Publishing and Pearson Education (1997).

SUSTAINABLE DRAINAGE SYSTEMS (SUDS)

In the past, surface water was drained from paved areas using sub-surface channels and pipes. This system prevented flooding locally by conveying the water away as quickly as possible. However recently, particularly in highly populated urban areas, there has been an increase in paved areas, leading to an increase in the amount of storm water present within the drainage system. This has led to changes in the natural water-flow patterns, often leading to flooding problems further downstream within the catchment area.

Recent changes in planning policy now require all developers to minimise flood risk within the catchment – to deal with the surface water locally, where it is created – by mimicking the original drainage patterns of the site, present before the development. Managing run-off volumes and flow rates can minimise the impact of urbanisation on flooding.

Two important terms used in sustainable drainage are infiltration and attenuation. Infiltration is the process of run-off being absorbed into the soil, while attenuation means controlling run-off entering the public sewer by slowing it down and temporarily storing it, thus reducing local flooding.

Other benefits to this sustainable approach in external works and pavement design include:

■ encouraging natural groundwater recharge;

■ protecting or improving water quality;

- providing a habitat for wildlife in urban watercourses;
- providing a sympathetic environmental setting to enhance the needs of the local community.

Sustainable drainage systems (SUDS) are a key technique in achieving these benefits. The system itself can be made up of one or more structures built to manage surface water run-off, and, used in conjunction with good management of the site, include the following accepted methods of 'natural' control:

- filter strips – gently sloping vegetated surface areas of ground that permit infiltration by slowing and filtering the flow of storm water (see Fig. 12.1.8);
- swales – long, shallow, open channels with vegetated surfaces that allow water to drain evenly off impermeable areas; they also retain and permit infiltration to the subsoil below (see Fig. 12.1.8);
- filter drains – act like swales, but comprise a perforated pipe buried in a trench, which is backfilled with a permeable granular material; they too store and direct the run-off as well as allowing infiltration, and to prevent the silting up of the pipe the trench is lined with a fine permeable geo-textile liner (see Fig. 12.1.8);
- permeable surfaces – commonly permeable asphalts, gravelled areas or paving blocks, which allow surface water to be absorbed into voids within the pavement construction; this run-off can be attenuated or directed through collection pipes into storage areas constructed below the ground, where it can then be infiltrated into the surrounding subsoil;
- infiltration devices – soakaways, infiltration trenches and infiltration basins or wetlands, which drain water directly into the ground, and can easily be integrated into landscaped areas; these are often installed on large-scale projects with high volumes of run-off (see Fig. 12.1.9).

Some of these methods are illustrated in Figs 12.1.8 and 12.1.9.

FOOTPATHS AND PEDESTRIAN AREAS

Footpaths and pedestrian areas can be constructed from a wide variety of materials or, for a large area, can consist of a mixture of materials to form attractive layouts. Widths of footpaths will be determined by local authority planning requirements, but a width of 1.2 m is usually considered to be the minimum in all cases. Roads are usually separated from the adjacent footpath by a kerb of precast concrete or natural stone. The kerb is set at a higher level than the road so that it marks the boundary of both road and footpath, and acts as a means of controlling the movement of surface water, by directing it along the gutter into the road gullies (see Fig. 12.1.10).

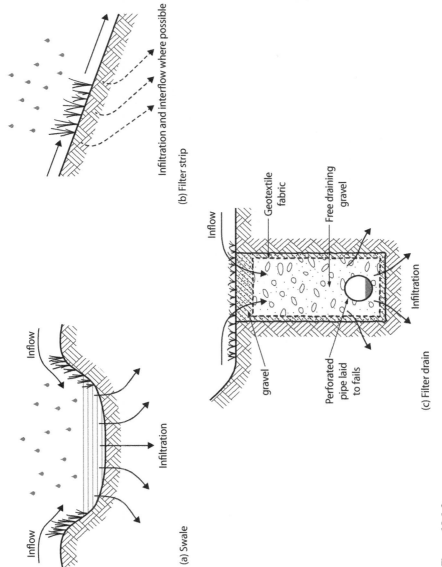

Inflow

Inflow

Infiltration

(a) Swale

Infiltration and interflow where possible

(b) Filter strip

Inflow

Geotextile fabric

Free draining gravel

gravel

Perforated pipe laid to fails

Infiltration

(c) Filter drain

Figure 12.1.8   Sustainable drainage systems: a) swale, b) filter strip and c) filter drain

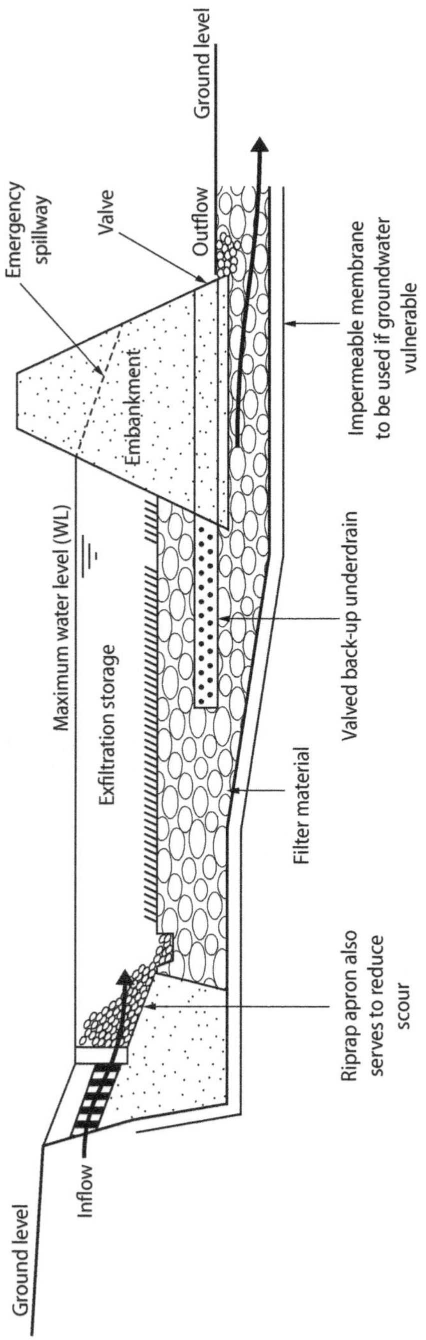

Ground level
Inflow
Riprap apron also serves to reduce scour
Maximum water level (WL)
Exfiltration storage
Filter material
Valved back-up underdrain
Embankment
Emergency spillway
Valve
Outflow
Ground level
Impermeable membrane to be used if groundwater vulnerable

Figure 12.1.9 Typical infiltration basin to attenuate storm water outflows for a small housing development

Flexible footpaths, like roads, are those in which, for design purposes, no tensile strength is taken into account. They are usually constructed in at least two layers: an upper wearing course laid over a base course of tarmacadam. Wearing courses of tarmacadam and cold asphalt are used, the latter being more expensive but having better durability properties (see Fig. 12.1.10 for typical details). Gravel paths similar in construction to unbound road surfaces are an alternative to the layered tarmacadam footpaths. Loose cobble areas can make an attractive edging to a footpath as an alternative to the traditional grass verge. The 30 to 125 mm diameter cobbles are laid directly onto a hardcore or similar bed and are handpacked to the required depth.

Unit pavings are a common form of footpath construction, consisting of a 50 to 75 mm-thick base of well-compacted hardcore, laid to a minimum crossfall of 1:60 if the subgrade has not already been formed to falls. The unit pavings can be of precast concrete flags or slabs laid on a 25 mm-thick bed of sand, a mortar bed of 1:5 cement:sand, or a weak dry cement:sharp sand mortar. Alternatively, each unit can be laid on five mortar dots of 1:5 cement:sand mix, one dot in each corner and one in the centre of each flag or slab. Mortar dot fixing is favoured by many designers because it facilitates easy levelling, and the slabs are easy to lift and re-lay if the need arises. Paving flags can have a dry butt joint – a 12 to 20 mm-wide soil joint to encourage plant growth – or they can be grouted together with a 1:3 cement:sand grout mix. Component specification is to BS 7263-1: *Precast concrete flags, kerbs, channels, edgings and quadrants. Precast, unreinforced concrete paving flags and complementary fittings. Requirements and test methods*. See also BS EN 1340: *Concrete kerb units. Requirements and test methods*.

Brick pavings of hard, well-burnt clay bricks or concrete pavers can be used to create attractive patterned and coloured areas. They can be laid on their bed face or laid on edge, and can be set in a bed of sharp sand or dry weak mortar, with dry or filled joints. Take care to select bricks that have adequate resistance to wear, frost and sulphate attack. Bricks with a rough texture will also give a reasonably good non-slip surface.

Small granite setts of square or rectangular plan format make a very hard-wearing unit paving. Setts should be laid in a 25 mm-thick sand bed, with a 10 mm wide joint, to a broken bond pattern. The laid setts should be rammed well, and the joints should be filled with chippings and grouted with a cement:sand grout.

Firepath pots are suitable for forming a surface required for occasional vehicle traffic, such as a firefighting appliance. Firepaths consist of 100 mm-deep precast concrete hexagonal or round pots with a 175 mm-diameter hole in the middle, which may be filled with topsoil for growing grass or with any suitable, loose filling material. These have the additional bonus of being permeable and allowing free drainage to the underlying subgrade. The pots are laid directly onto a 150 mm-thick base of sand-blinded compacted hardcore.

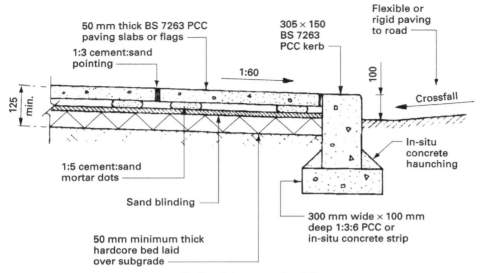

50 mm thick BS 7263 PCC paving slabs or flags

1:3 cement:sand pointing

1:60

305 × 150 BS 7263 PCC kerb

Flexible or rigid paving to road

100

125 min.

Crossfall

In-situ concrete haunching

1:5 cement:sand mortar dots

Sand blinding

300 mm wide × 100 mm deep 1:3:6 PCC or in-situ concrete strip

50 mm minimum thick hardcore bed laid over subgrade

Paving slabs on mortar dots

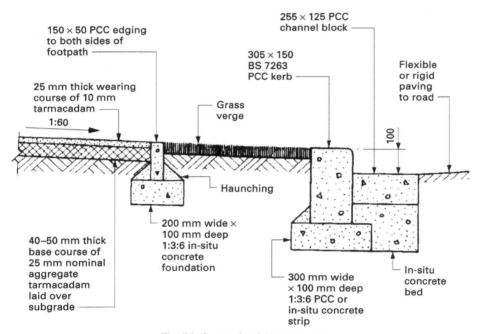

150 × 50 PCC edging to both sides of footpath

255 × 125 PCC channel block

305 × 150 BS 7263 PCC kerb

Flexible or rigid paving to road

25 mm thick wearing course of 10 mm tarmacadam

Grass verge

1:60

100

Haunching

200 mm wide × 100 mm deep 1:3:6 in-situ concrete foundation

40–50 mm thick base course of 25 mm nominal aggregate tarmacadam laid over subgrade

300 mm wide × 100 mm deep 1:3:6 PCC or in-situ concrete strip

In-situ concrete bed

Flexible footpath with grass verge

Figure 12.1.10   Typical footpath details

Cobbled pavings and footpaths can be laid to form a loose cobble surface as previously described, or the oval cobbles can be hand set into a 50 mm-thick bed of 1:2:4 concrete, with a small maximum aggregate size laid over a 50 mm-thick compacted-sand base layer.

Rigid pavements consisting of a 75 mm-thick unreinforced slab of in-situ concrete laid over a 75 mm-thick base of compacted hardcore can also be used to form footpaths. Like their road counterparts, these pavings should have expansion and contraction joints incorporated into the construction. Formation of these joints would be similar to those shown in Fig. 12.1.6; for expansion joints, the maximum centres would be 27 m; for contraction joints, the maximum centres would be 3 m.

ACCOMMODATION OF SERVICES

The services that may have to be accommodated under a paved area could include:

- public or private sewers;
- electrical supply cables;
- gas mains;
- water mains;
- telecommunications cables;
- television relay cables;
- district heating mains.

In planning the layout of these service, coordination between the various undertakings and bodies concerned is essential to formulate a logical, economical plan and installation programme.

Sewers are not generally grouped with other services. Because of their lower flexibility, they are given priority of position. They can be laid under the footpath or verge, or under the carriageway. Most are laid under the footpath or verge so that repairs will cause the minimum of disturbance, and because the reinstatement of a footpath is usually cheaper and easier than that of the carriageway. The specification for ducts, covers and access positions for any particular service will be determined by the undertaking or board concerned.

Services that can be grouped together are often laid in a common trench, starting with the laying of the lowest service, backfilling until the next service depth is reached, and then repeating the procedure until all the required services have been laid. The selected granular backfilling materials should be placed in 200 to 250 mm well-compacted layers. All services should be kept at least 1.5 m clear of tree trunks, and any small tree roots should be cut, square-trimmed and tarred. Typical common service trench details are shown in Fig. 12.1.11.

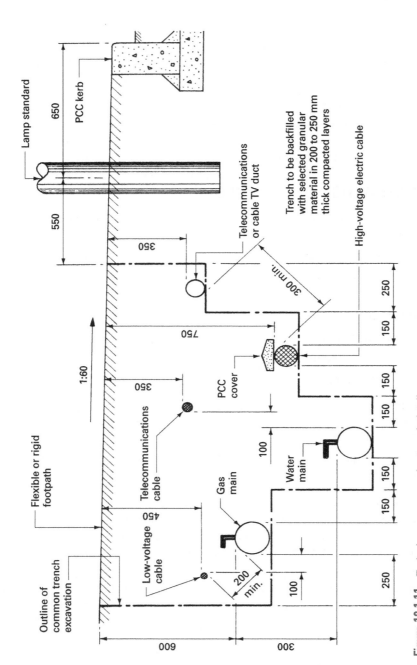

Lamp standard

650

PCC kerb

550

350

1:60

Flexible or rigid footpath

450

Low-voltage cable

200 min.

100

Outline of common trench excavation

600

300

Telecommunications cable

350

Gas main

Telecommunications or cable TV duct

750

PCC cover

300 min.

100

Water main

Trench to be backfilled with selected granular material in 200 to 250 mm thick compacted layers

High-voltage electric cable

250

150

150   150

150

150

250

Figure 12.1.11   Typical common service trench details

Bibliography

Barry, R. (1996) *The Construction of Buildings*. Oxford: Blackwell Scientific Publications

Boughton, B. (1979) *Reinforced Concrete Detailer's Manual*. Crosby Lockwood Staples. Blackwell Scientific Publications

Fisher Cassie, W. and Napper, J. H. (1960) *Structure in Building*. The Architectural Press

Holmes, R. (1995) *Introduction to Civil Engineering Construction*. 3rd revised edition. College of Estate Management.

Handisyde, C. C. (1967) *Building Materials*. The Architectural Press.

Leech, L. V. (1988) *Structural Steelwork for Students*. Oxford: Butterworth-Heinemann Ltd

Llewelyn Davies, R. and Petty, D. J. *Building Elements*. The Architectural Press. 1956-60

McKay, W. B. (1956-60) *Building Construction*, Vols 1 to 4. 4th edition. Longman

Smith, G. N. (1987) *Elements of Soil Mechanics for Civil and Mining Engineers*. 5th edition. Wiley-Blackwell

West, A. S. (1972) *Piling Practice*. Newnes-Butterworth

Whitaker, T. (1970) *The Design of Piled Foundations*. Pergamon Press

GENERAL

A Guide to Scaffolding Construction and Use. Scaffolding (GB) Ltd.

BSP. Pocket Book. The British Steel Piling Co. Ltd.

Building Regulations, Approved Documents. ODPM.

Construction Safety. Construction Industry Publications Ltd.

Data sheets of the British Precast Concrete Federation.

Design Manual for Roads and Bridges, Vol. 7. The Stationery Office Ltd.

DoE Construction Issues: 1 to 17. The Stationery Office Ltd.

Drained Joints in Precast Concrete Cladding. The National Buildings Agency.
Glass. Pilkington UK Ltd.
Handbook on Structural Steelwork. The British Constructional Steelwork Association Ltd.
Lifting Operations and Lifting Equipment Regulations 1998. The Stationery Office Ltd.
Lighting for Building Sites. Electricity Association Services Ltd.
Loss Prevention Standards. Loss Prevention Certification Board.
Mitchell's Building Series. Pearson Education.
The Confined Spaces Regulations 1997. The Stationery Office Ltd.
The Construction (Health, Safety and Welfare) Regulations 1996. The Stationery Office Ltd.
The Fire Precautions (Workplace) Regulations 1997. The Stationery Office Ltd.
The Work at Height Regulations 2005. The Stationery Office Ltd.

OTHER READING

Relevant advisory leaflets. The Stationery Office Ltd.
Relevant AJ Handbooks. The Architectural Press.
Relevant BRE Digests. Construction Research Communications Ltd.
Relevant British Standards, Codes of Practice and EuroNorms. British Standards Institution.
Relevant manufacturers' catalogues and technical guides contained in the Barbour Index and Building Products Index libraries.

Index